INVESTIGATING THE UNIVERSE

ASTROPHYSICS AND SPACE SCIENCE LIBRARY

A SERIES OF BOOKS ON THE RECENT DEVELOPMENTS
OF SPACE SCIENCE AND OF GENERAL GEOPHYSICS AND ASTROPHYSICS
PUBLISHED IN CONNECTION WITH THE JOURNAL
SPACE SCIENCE REVIEWS

VOLUME 91

INVESTIGATING THE UNIVERSE

Papers presented to Zdeněk Kopal
on the occasion of his retirement, September 1981

WITH A FOREWORD BY M. K. V. BAPPU,
PRESIDENT OF THE INTERNATIONAL ASTRONOMICAL UNION

Edited by

F. D. KAHN

Department of Astronomy, The University of Manchester, England

D. REIDEL PUBLISHING COMPANY

DORDRECHT : HOLLAND / BOSTON : U.S.A.
LONDON : ENGLAND

Library of Congress Cataloging in Publication Data

CIP

Main entry under title:

Investigating the universe.

(Astrophysics and space science library ; v. 91)
Includes index.
1. Astrophysics-Addresses, essays, lectures. 2. Astronomy-Addresses, essays, lectures. 3. Kopal, Zdeněk, 1914– I. Kopal, Zdeněk, 1914– . II. Kahn, F. D. (Franz Daniel). III. Series.

QB461.5.I56 523.01 81-12013
AACR2

ISBN-13: 978-94-009-8536-0 e-ISBN-13: 978-94-009-8534-6
DOI: 10.1007/978-94-009-8534-6

Published by D. Reidel Publishing Company,
P.O. Box 17, 3300 AA Dordrecht, Holland.

Sold and distributed in the U.S.A. and Canada
by Kluwer Boston Inc.,
190 Old Derby Street, Hingham, MA 02043, U.S.A.

In all other countries, sold and distributed
by Kluwer Academic Publishers Group,
P.O. Box 322, 3300 AH Dordrecht, Holland.

D. Reidel Publishing Company is a member of the Kluwer Group.

TABLE OF CONTENTS

Section V: Modelling of Galaxies

ACKNOWLEDGEMENTS

It is evident that a volume like this cannot be produced without much painstaking work by the publishers and the typists involved. The Editor is especially grateful to the staff of Reidel's for their assistance and courtecy, and to Barbara Barlow, Ellen Carling and Joanne Suthers for their constant willingness to help.

CONTRIBUTORS

in alphabetical order

M. K. V. Bappu	Indian Institute of Astrophysics, Bangalore
A. H. Batten	Dominion Astrophysical Observatory, Victoria B.C.
E. Budding	Department of Astronomy, University of Manchester
J. Cantó	Institute of Astronomy, National University of Mexico
M. A. Dopita	Mount Stromlo and Siding Spring Observatories, Canberra
J. E. Dyson	Department of Astronomy, University of Manchester
C. Goudas	Department of Mathematics, University of Patras
J. Hazlehurst	Hamburg Observatory
R. A. James	Department of Astronomy, University of Manchester
F. D. Kahn	Department of Astronomy, University of Manchester
M. Kitamura	Tokyo Astronomical Observatory
J. Meaburn	Department of Astronomy, University of Manchester
D. A. Mendis	University of California at San Diego
M. G. Smith	Royal Observatory, Edinburgh
P. Stewart	Department of Mathematics, University of Manchester

PROFESSOR ZDENĚK KOPAL

FOREWORD

Professor Zdeněk Kopal is sixty-seven this year even though his scientific activity, enthusiasm and springy step hardly betray the advancement in years. He came to Manchester as Professor of Astronomy thirty years ago after a very fruitful association of fourteen years with the Harvard Observatory. Much impressed with the young man, Harlow Shapley, who with characteristic insight had recognised in Kopal the qualities that have since made him an outstanding leader in eclipsing binary research, had invited him over as a Research Associate. In the subsequent decade Kopal set about the task of introducing analytical rigour in the solution of orbital elements that hitherto had depended exclusively on the semigraphical procedures introduced by Russell and exploited fully by Shapley. These first efforts stimulated publication of the first of his many books on eclipsing variables; the *Introduction to the Study of Eclipsing Variables* summarized these iterative methods and remains a classic in this field. Soon after the appearance of this volume in print, Kopal gave a course on this subject for the graduate students at Harvard. I was one of those who had the opportunity to attend it and learn much on the need of care and precision in the practice of photoelectric photometry and the importance of exploiting such data to the fullest extent with methods of increasing resolving power. The RCA 1P21 photomultiplier tube had just begun to revolutionize photoelectric photometry and make it accessible to the several small instruments scattered all over the world. It was a tide in astronomical practice that could be fully swung over to maximize the availability of the basic astrophysical parameters, and much of the happy position we are in today in this regard stems from the lively role Kopal has played as a member and subsequently as President of IAU Commission 42.

Since then at Manchester, Kopal has had a very flourishing school whose contributions have continuously taken advantage of the increasing new facilities for quick and elaborate computation. The transition from the time to the frequency domain was timely from the standpoint of available means of automatic computation. The large machines can solve for the elements of a system in a fantastically short interval. Kopal's recent efforts have been to achieve similar goals with simpler computation facilities.

Many years ago, that pioneer of eclipsing binary analysis, Henry Norris Russell, delighted in having his undergraduate class solve the elements of a system with the aid of a slide rule. We have come a full circle today from this simple graphical procedure to something that can be done in the same time with great elegance and analytical rigour. This

F. D. Kahn (ed.), Investigating the Universe, ix–x.

transformation is undoubtedly the result of the Kopal era in the study of eclipsing binaries.

This Festschrift Volume is an expression of admiration, appreciation and gratitude from the many who have been Professor Kopal's students and collaborators. May the years ahead hold for him much happiness and continued creative scientific activity.

M. K. V. Bappu,
President,
International Astronomical Union

DYNAMICS OF THE GALACTIC FOUNTAIN

F. D. Kahn
Department of Astronomy, University of Manchester.

The visitor to Zdenek's office will have noticed a cupboard fixed to the outside wall. It is filled with theses written by former students of the Department. Even though the building is immensely strong and the cupboard well made I have often worried that the weight of knowledge stored inside it would one day bring it down.

Much the same problem faced us when planning this collection of papers. Had we invited contributions from everyone who qualified then everyone would have accepted. The resulting volume would have been much too large even for Reidel's, our generous publisher. What is worse, Zdenek would have been unable to lift it up and read it. We therefore had to make a somewhat arbitrary selection of authors, and now apologise to those whom we could not invite. But we are sure that they join us in wishing Zdenek well. Of course we also thank all the contributors who have written such interesting and topical papers.

In presenting this volume to you, Zdenek, we ask you to regard it as a token of the great success achieved by the Department that you founded in the autumn of 1951. May you have a long, happy and active retirement.

Under ordinary circumstances I should have broken off here and got on with the astronomy. But circumstances are far from ordinary. In the spring of 1951 I married Carla, having got to know her a few months earlier. There followed a union uniquely blessed by love and happiness. We had many children, and even a grandchild, and through our different activities, notably the astronomical ones, we acquired a worldwide circle of very good friends. All seemed set to go on in this way when one morning in January this year Carla suffered a severe stroke, and then died on the evening of the next day.

It is a commonplace that life can be cruel. But fortunately fate only very rarely administers so devastating a blow. This paper deals with the first scientific work that I have attempted since Carla's death. I dedicate it to her memory, may she rest in peace.

F. D. Kahn (ed.), Investigating the Universe, 1–28.

I. THE HOT INTER-CLOUD MEDIUM IN THE GALAXY

In this paper I shall discuss the properties of the hot inter-cloud medium (ICM); to do so something needs to be said about the input of energy, mainly from supernova explosions, about radiative processes which cause substantial cooling, and about the dynamical consequences of having gas in the Galaxy which is so hot that its thermal velocity far exceeds the escape velocity from the galactic disk, but not from the Galaxy as a whole. There is an intricate inter-relation between these various aspects, so that I shall cautiously approach the problem by a series of successive approximations.

The most striking result of having so much gas that cannot be confined is that there develops a fountain which carries some of the interstellar matter far from the galactic plane. This effect has been extensively discussed, in the last two years, in a series of papers by Bregman. His model is two-dimensional and allows for the fountain to spray both upwards from the plane and outward, away from the axis. But eventually the gas loses its thermal energy, and therefore, to maintain pressure, it condenses into cool clouds which fall back inwards and towards the galactic plane. Bregman's model shows a very encouraging correspondence between theory and observation, and may well settle some difficult astrophysical problems, such as the origin of the absorption lines seen in distant quasars.

But in this paper I shall take a different point of view, and explore the consequences. If the gas is to accelerate in a radial direction, as well as perpendicular to the plane, then it must have a cooling time that is sufficiently long. The most likely origin of the hot gas is in supernova remnants. Most of a supernova remnant will have lost its thermal energy by the time that its internal pressure has dropped to interstellar values. At that stage the remaining hot gas will itself have a cooling time whose length is of the order of ten times the age of the remnant. During this time the gas can rise only a relatively short distance from the plane. If there exist pressure changes in the radial direction on a length scale comparable with the distance to the galactic centre, then the pressure gradient will still be overwhelmingly in the upward direction. A one-dimensional description will therefore be quite adequate. But of course it does not at first sight give anything like as good an account of the observed properties of the halo gas, notably of the high velocity clouds. This point will be taken up again later.

The first requirement for a model is a suitable description of the gravitational field in and above the galactic disk. As is well known the overall field of the Galaxy is such that the speed V of a star in a circular orbit is independent of the orbital radius, and equals some 250 km s^{-1}. The mass distribution that gives rise to the corresponding gravitational field is widely believed to be due to a spherical halo of Population II objects. Thus the mass M(r) within a radial distance r is given by

$$M(r) = V^2 r/G \equiv \mu r \tag{1}$$

where $\mu = 9.4 \times 10^{21}$ gm cm^{-1}.

When considering the equilibrium of the ICM in our model we are concerned mainly with the component g_z of the gravitational field perpendicular to the galactic plane. The contribution made by the spherical halo population to this component is denoted by $g_z^{(h)}$; another contribution $g_z^{(d)}$ comes from the mass in the disk of the Galaxy, or Population I, and I shall deal with that shortly.

At a height z above the galactic plane the gravitational potential is

$$\mathcal{V}(z, R_*) = -\tfrac{1}{2} G\mu \log(1 + z^2/R_*^2) \tag{2}$$

with respect to the point, at radial distance R_*, in the plane directly underneath. Provided z is small in comparison with the galactocentric distance R_*,

$$g_z^{(h)} = \frac{G\mu z}{R_*^2 + z^2} \doteq \frac{G\mu z}{R_*^2} = \frac{V^2 z}{R_*^2} \tag{3}$$

At the Sun's distance from the centre R_* is some 10 kpc and so

$$g_z^{(h)} = 6.9 \times 10^{-31} z. \tag{4}$$

To estimate the contribution made to g_z by matter in the galactic disk, consider the typical dispersion $\bar{\sigma}$ of z-components of velocity of extreme Population I objects; typically $\bar{\sigma}$ is 10 km s^{-1}. These objects fill a layer that rises to a height D $\sim$ 100pc above the plane. Given a uniform distribution of Population I mass near the plane it follows that, for z sufficiently small,

$$g_z^{(d)} = \frac{\bar{\sigma}^2}{D^2} z \equiv \gamma z, \tag{5}$$

and, with the values adopted here,

$$\gamma = 1.1 \times 10^{-29} \quad s^{-2}. \tag{6}$$

Close to the galactic plane the disk population contributes overwhelmingly to g_z. The necessary mass density is

$$\rho = \gamma/4\pi G = 1.3 \times 10^{-23} \text{ gm cm}^{-3} \tag{7}$$

Models of the Galaxy in common use ascribe some 20 per cent of the mass to the disk population. In the spirit of this series of rough estimates it follows that the total mass of the disk population out to galactocentric distance 10 kpc is thus about 5.6×10^{43} gm. Assume now that the disk material is also distributed in such a way that the mass out to radius r is proportional to r. Then the local value of the half surface density at the Sun's distance from the centre is

$$\Sigma_d(R_*) = 5.0 \times 10^{-3} \text{ gm cm}^{-2}. \tag{8}$$

A comparison with relation (7) indicates that the half-thickness of this mass distribution is

$$Z = \Sigma_d(R_*)/\rho_d = 3.8 \times 10^{20} \text{ cm} = 127 \text{ pc}. \tag{9}$$

The gravitational potential at height z, with respect to a point in the galactic plane directly below, is therefore

$$\vartheta \doteqdot -\tfrac{1}{2}\gamma z^2 \tag{10}$$

if $z < Z$, and

$$\vartheta = -\{\gamma Z(z - Z) + \tfrac{1}{2}\gamma Z^2 + (V^2/2R_*^2)z^2\}, \tag{11}$$

if $z > Z$, in reasonable approximation.

The hot ICM has a temperature T of around 10^6K. It is fully ionized so that $\bar{m}$ the mean particle mass, is 10^{-24}gm, and the isothermal sound speed is

$$c_s = \sqrt{kT/\bar{m}} \doteqdot 120 \text{ km s}^{-1} \tag{12}$$

If the gas were to remain uncooled it would be expected to rise to a height z such that

$$kT/\bar{m} = c_s^2 \sim \vartheta(z) \tag{13}$$

The time needed to establish hydrostatic equilibrium would be of the order of

$$t_{eq} = z/c_s \tag{14}$$

But the hot gas will radiate following the excitation of various ion species by electron impact. A useful approximation is that the rate of loss of energy by radiation, per ion, is

$$L = \lambda n T^{-1/2} \tag{15}$$

at temperature T, where n is the density of atoms plus ions, and

$$\lambda \doteqdot 1.33 \times 10^{-19} \tag{16}$$

in CGS units. The formula is applicable in the range of temperatures between 2×10^5K and 10^8K.

The cooling law can be most usefully formulated in terms of the adiabatic parameter

$$K \equiv P/\rho^{5/3}, \tag{17}$$

which is related to the specific entropy S by

$$S = (3k/2m) \log K . \tag{18}$$

The density is given by

$$\rho = n m_a , \tag{19}$$

where m_a is the average mass per atom, and then the rate of loss of energy per unit mass can be expressed by

$$T \frac{dS}{dt} = -\frac{L}{m_a} = -\frac{\lambda n}{m_a} T^{-1/2} = -\frac{\lambda \rho}{m_a^2} T^{-1/2} . \tag{20}$$

Here d/dt denotes rate of change following the motion of the fluid. Then using relation (18), as well as the equation of state

$$P = (k/\bar{m}) \rho T, \tag{21}$$

it can be shown that

$$\frac{d}{dt} K^{3/2} = -q , \tag{22}$$

where $$q \equiv \frac{\lambda k^{1/2}}{m_a^2 \bar{m}^{1/2}} = 4 \times 10^{32} (cm^6 gm^{-1} s^{-4}).$$

Relation (22) implicitly assumes that there are no other significant sources of pressure support, such as a magnetic field.

The cooling time for an element of the hot gas is thus

$$t_c = \kappa^{3/2}/q = P^{3/2}/q\rho^{5/2}. \qquad (23)$$

So there is no possibility for the hot ICM to establish hydrostatic equilibrium if

$$t_{eq} \equiv \frac{z}{c_s} > t_c = \frac{P^{3/2}}{q\rho^{5/2}} = \frac{c_s^3}{q\rho} \qquad (24)$$

or

$$c_s^4 < q\rho z, \qquad (25)$$

where z is given by equation (13) and ρ is the density of the hot medium in the galactic disk. It may be preferable to express this inequality in terms of the interstellar pressure P so as to read

$$c_s^6 < qPz. \qquad (26)$$

With typical values $P = 10^{-12}$ dyne cm^{-2}, $c_s = 10^7$ cm s^{-1}, this means that cooling occurs too fast for hydrostatic equilibrium to be possible in any halo whose height z is larger than or of the order of

$$c_s^6/qP = 4 \times 10^{21} \text{cm} = 1.3 \text{ kpc}. \qquad (27)$$

On the other hand equations (12) and (13) lead to an estimated scale height of some 1.1×10^{22} cm = 3.7 kpc for a static halo, if the sound speed is $c_s \sim 100$ km s^{-1}. Thus the inter-cloud medium is too hot to be confined by the gravitational field to the layer occupied by Population I. The gas rises away from the plane towards an equilibrium structure with considerably greater thickness, but loses its thermal energy long before the new configuration is reached. The inter-cloud medium can never attain equilibrium in a static structure. The correct description of the halo must be dynamic.

II. INPUT OF HEAT INTO THE INTER-CLOUD MEDIUM

Supernova explosions are the most likely source of energy for the inter-cloud medium. The growth of a supernova remnant is well described by the famous Sedov solution. In this model it is assumed that the energy input occurs in a small volume and affects initially only a small mass. But as the disturbance spreads it entrains progressively more interstellar matter. A basic requirement of the Sedov solution is that the interstellar gas be uniformly distributed before the shock wave overtakes it. Here is a great difficulty. The very existence of the inter-cloud medium shows that large regions of space contain gas whose density is much below the interstellar average. In fact the bulk of the interstellar mass resides in cool clouds that occupy only a small fraction of the available volume. It is certainly not possible to argue that the shock from an expanding supernova remnant will by-pass these dense clouds and heat only the inter-cloud medium. If that were so the ICM would be expelled from the galactic disk in a very short time, and would immediately need to be replaced. Therefore the cool clouds must also be heated, since they are the only possible source of mass for the galactic fountain.

An immediate consequence is that there must be a means whereby the thermal energy left behind by a shock wave is conducted or convected into the cool clouds, and then causes them to heat up, expand and become part of the ICM. Various schemes have been proposed for this transport of heat. There are also modified solutions of the Sedov type which mimic the way in which matter in the clouds partially evaporates and joins the expanding bubble of hot gas behind the shock. But at present there is only limited understanding of the conduction process: the interstellar magnetic field will clearly be important since thermal conduction across the lines of force is always very small. Even the rate at which conduction occurs along the lines of force may be severely inhibited by various plasma instabilities which can arise if the field strength is low enough. So, knowing full well that the model is inadequate, I shall nevertheless assume that the unshocked interstellar gas is uniform, at density ρ_o, and that it is cool.

Such a procedure yields the most conservative estimate for the rate of input of energy into the ICM. The existence of regions of high density in the interstellar gas must mean that some parts of the medium are protected from the input of heat. The available energy is therefore shared among only a fraction of the available gas, with consequences that are the same as if the medium had been uniform, but with a density lower than ρ_o, the actual smoothed-out value. The ICM contains that part of the gas from various supernova remnants which takes the longest time to cool, that is the gas which attains the largest value of $K (\equiv P/\rho^{5/3})$ behind the blast wave. The mass of gas which can be raised above any given K increases as ρ_o is decreased. By overestimating the effective ρ_o one underestimates the contribution of a supernova explosion in heating the inter-cloud medium.

The Sedov solution applies only during the so-called Phase II of a supernova remnant, which occurs when enough time has elapsed that the mass entrained much exceeds the mass ejected by the original explosion, but not so much time has passed that cooling has become important. At time t the shock radius is then given by

$$r = at^{2/5} \tag{28}$$

where

$$a^5 \doteq 2E/\rho_o \quad , \tag{29}$$

and E is the energy released by the explosion. The shock speed

$$\dot{r} = \tfrac{2}{5}\, a\, t^{-3/5} \tag{30}$$

is so much larger than the sound speed in the undisturbed medium ahead that the shock can be taken to be strong. The post-shock pressure and density are, respectively,

$$P_s = \tfrac{3}{4}\rho_0 \dot{r}^2 \qquad \text{and} \qquad \rho_s = 4\rho_0 \;, \tag{31}$$

and the adiabatic parameter just behind the shock is

$$K_s = \frac{P_s}{\rho_s^{5/3}} = \frac{3a^2}{25 \times 2^{10/3}\, \rho_0^{2/3}}\, t^{-6/5} . \tag{32}$$

In the absence of thermal conduction an element of gas will retain the value of K it attained behind the shock until radiative losses become important for it. At time t a mass

$$M_s = (4\pi/3)\, \rho_0 a^3 t^{-6/5} \tag{33}$$

has passed through the shock. In the subsequent evolution of the gas in the remnant, the mass M within which the adiabatic parameter exceeds some given value K_* satisfies the relation

$$K_* M = \frac{\pi}{25 \times 2^{4/3}}\, \rho_0^{1/3} a^5 = \frac{\pi}{25 \times 2^{1/3}}\, \frac{E}{\rho_0^{2/3}} \doteq \frac{0.1\,E}{\rho_0^{2/3}} . \tag{34}$$

Eventually radiative cooling sets in, beginning in the outer parts of the remnant. At that time the Sedov solution becomes inapplicable, but relation (34) is still valid for any fluid element that passed through the shock during Phase II, and has not yet started to lose energy by radiation at a significant rate.

When an uncooled mass M has come to equilibrium under external pressure P, a range dK of the adiabatic parameter corresponds to a mass

$$dM = \frac{0.1\,E\, dK}{\rho_0^{2/3}\, K^2} \;, \tag{35}$$

and occupies an element of volume

$$dV = \frac{dM}{\rho} = \frac{0.1\,E\, dK}{\rho_0^{2/3}\, P^{3/5}\, K^{7/5}} . \tag{36}$$

The volume occupied by mass $M = 0.1\, E/\rho_0^{2/3} K_*$ is

$$V = \tfrac{5}{2}\left(\frac{0.1\,E}{\rho_0^{2/3}}\right) K_*^{-2/5} P^{-3/5} = \tfrac{5}{2} K_*^{3/5}\, M P^{-3/5} . \tag{37}$$

The smoothed-out density of the uncooled gas is $\rho \equiv M/V$, and the resulting pressure density relation is

$$P = \left(\frac{5}{2}\right)^{5/3} K_* \rho^{5/3} . \tag{38}$$

In the inter-cloud medium P is 10^{-12} dyne cm^{-2} and the density ρ is close to 10^{-26} gm cm^{-3}. Agreement with these values requires that

$$K_* = 4.7 \times 10^{30} . \tag{39}$$

The mass of hot gas left over from the supernova remnant, with adiabatic parameter larger than K_*, is

$$M = \frac{0.1\, E}{\rho_0^{2/3} K_*} = 6.4 \times 10^{35} \text{ gm} = 320\, M_\odot , \tag{40}$$

where E has been taken to be 3×10^{51} erg.

There are two points to check. The rate of loss of mass from the ICM into the galactic fountain is of the order of 10^{-19} gm cm^{-2} s^{-1} on each side of the galactic disk, and therefore 6×10^{26} gm s^{-1} for the disk out as far as the Sun's distance from the centre. One supernova explosion every thirty years, or 10^9 s, can clearly supply this mass. The time needed to cool the gas with $K < K_*$ is

$$t_c = K_*^{3/2}/q = 2.5 \times 10^{13} \text{ s} \doteqdot 850\,000 \text{ years.}$$

This is far greater than the lifetime of any recognisable supernova remnant, and certainly far greater than the duration of the Sedov phase. But it is a reasonable estimate for the length of time required to distribute the heated gas within the general inter-cloud medium. The calculation only holds, of course, if the different parts of the remnant gas do not mix with one another up to this stage. But at some time the mixing must occur. To simplify the further work, and because it makes little difference in the subsequent calculation, it will now be taken that the gas is well mixed once it becomes part of the ICM. Provisionally it will be assumed that a unit mass of gas is injected with an amount of energy $\mathcal{E}_0 = 5P/2\rho = 2.5 \times 10^{14}$ erg gm^{-1}. The adiabatic parameter for this mixed-up gas is

$$K = P/\rho^{5/3} = \left(\tfrac{2}{5}\mathcal{E}_0\right)^{5/3} P^{-2/3} = 2.15 \times 10^{31} \tag{41}$$

and the cooling time

$$t_c = K^{3/2}/q = 2.5 \times 10^{14} \text{ s} = 8.3 \times 10^6 \text{ years} \tag{42}$$

about ten times longer than the time needed for the gas from a supernova remnant to come to pressure equilibrium with the ICM. But to understand how a change in $\mathcal{E}_0$ affects the nature of the flow, I shall also consider models for which this parameter takes the values 1.0, 1.5 and 2.0 x 10^{14} erg gm^{-1}

III. A MODEL FOR THE OUTFLOW FROM THE DISK

Three processes dominate the production of the galactic fountain: the injection of hot gas into the ICM, the loss of energy by radiation and the downward pull of gravity on the upward flowing gas. The characteristic cooling time for the gas is small compared with the time needed to reach maximum height in the fountain, at least according to the approach adopted in this paper (and in contrast with systems studied by Bregman) It seems reasonable then to separate the various effects, so that gravity is unimportant close to the galactic plane, but the balance between injection and cooling dominates the flow there, and in particular determines the existence of a sonic point in the flow. Above the sonic point the gas rapidly loses thermal energy. In the region where gravity becomes important the gas is essentially devoid of heat, and the flow becomes pressureless. By thus splitting the regimes, the discussion of the fountain is made much simpler.

In a plane-parallel model let the injection of hot gas occur in such a way that the rate at which mass is added, per unit volume, is

$$(\Phi_0/D)\, e^{-z/D}$$

at height z above the plane, z being positive. The energy per unit mass of the injected gas is taken to be $\mathcal{E}_0$, independent of the position where injection takes place. The pressure in the galactic plane is Π, and the flow velocity there is zero. The conservation laws lead to the relation

$$\rho u = \int_0^z (\Phi_0/D)\, e^{-z/D}\, dz = \Phi_0(1 - e^{-z/D}) \tag{43}$$

for the mass flow, and

$$P + \rho u^2 = \Pi \tag{44}$$

for the momentum flow. The energy flux is

$$W = \Phi\left(\tfrac{1}{2}u^2 + \tfrac{5}{2}\frac{P}{\rho}\right) = \tfrac{5}{2}\Pi u - 2\Phi u^2 \tag{45}$$

The rate of gain of energy, per unit volume, from injections by supernovae is

$$\Phi \frac{\mathcal{E}_0}{D}\, e^{-z/D} = \mathcal{E}_0 \frac{d\Phi}{dz}, \tag{46}$$

and the rate of loss of energy, per unit volume, due to radiative cooling

$$\lambda n^2 T^{-1/2} = \frac{\lambda}{m_a^2}\left(\frac{k}{\bar{m}}\right)^{1/2} \rho^{5/2} P^{-1/2} \equiv q \rho^{5/2} P^{-1/2}, \tag{47}$$

where q = 4 x 10^{32}, as before. The equation of energy balance is thus

$$\frac{dW}{dz} = \left(\tfrac{5}{2}\Pi - 4\Phi u\right)\frac{du}{dz} - 2u^2 \frac{d\Phi}{dz}$$

$$= \mathcal{E}_0 \frac{d\Phi}{dz} - q \frac{\rho^{5/2}}{P^{1/2}}$$

or

$$\left(\tfrac{5}{2}\Pi - 4\Phi u\right)\frac{du}{dz} = \left(\mathcal{E}_0 + 2u^2\right)\frac{d\Phi}{dz} - q \frac{\rho^{5/2}}{P^{1/2}} \tag{48}$$

At the sonic point $\Phi = \Phi_{sp}$ and $u = u_{sp} = 5\Pi/8\Phi_{sp}$ so that

$$\left(\mathcal{E}_0 + 2u_{sp}^2\right)\frac{d\Phi}{dz} = q \frac{\rho^{5/2}}{P^{1/2}} = \frac{q\,\Phi_{sp}^{5/2}}{u_{sp}^{5/2}\left(\Pi - \Phi u_{sp}\right)^{1/2}} \tag{49}$$

In the range of parameters that are treated here, the cooling is not very important in the gas until it comes near the sonic point. For a zero-order description it is therefore helpful to neglect the cooling term. If v is the gas velocity at height z in the uncooled flow,

$$2v^2 - \frac{5\Pi}{2\Phi} v + \mathcal{E}_0 = 0, \tag{50}$$

so that

$$\Phi = \frac{5\Pi v}{2(2v^2 + \mathcal{E}_0)} = \frac{5\Pi v}{4(v^2 + v_c^2)} \tag{51}$$

where $2v_c^2 \equiv \mathcal{E}_0$

An uncooled flow accelerates at all heights, and, to meet the boundary condition at large z, it must reach a sonic point, where $v = v_c$, as Φ tends to Φ_0. Note that Φ_0 equals $5\Pi/8v_c$ and that

$$\frac{d\Phi}{\Phi\, dz} = \frac{\Phi_0 - \Phi}{\Phi D} = \frac{(v - v_c)^2}{2D\, v\, v_c} \tag{52}$$

Both Φ and v are known in terms of z, and it is convenient to express dz in terms of dv as follows. From relation (51)

$$\frac{d\Phi}{\Phi} = \frac{dv}{v} - \frac{2v\,dv}{v^2 + v_c^2} = \frac{(v_c^2 - v^2)\,dv}{v(v^2 + v_c^2)}, \tag{53}$$

and then combining (52) and (53) leads to

$$dz = \frac{2D\, v_c (v + v_c)\, dv}{(v^2 + v_c^2)(v_c - v)} . \tag{54}$$

This result can be used to calculate, correct to first order, the integrated loss of energy by cooling in a unit column up to height z, or equivalently up to the level where the velocity parameter is v. The energy loss is found to be

$$-\Delta W = \int \frac{q\, \Phi^{5/2}}{v^{5/2} (\Pi - \Phi v)^{1/2}}\, dz$$

$$= \frac{5^{5/2}}{8} q D \Pi^2 v_c \int \frac{(v_c + v)\, dv}{(v_c^2 + v^2)^3 (4 v_c^2 - v^2)^{1/2} (v_c - v)} \tag{55}$$

after some manipulation. The term $(v_c - v)$ is the important factor in the denominator. The aim is now to find the level where the sonic point occurs: it is expected to be at a height where v is not much less than v_c. The calculation is simplified without too much loss of accuracy by setting v equal to v_c for all factors save, of course, $v_c - v$ in the integrand. Then

$$-\Delta W = \frac{5^{5/2} q D \Pi^2}{2^5 \times 3^{1/2} v_c^5} \log \frac{1}{1 - v/v_c} ; \tag{56}$$

correct to zero order the mass flux at this level is Φ_o. The accumulated cooling therefore has reduced the stagnation enthalpy per unit mass by an amount

$$-\Delta \mathcal{E} = \frac{-\Delta W}{\Phi_o} = \frac{5^{3/2} q D \Pi}{4 \times 3^{1/2} v_c^4} \log \frac{1}{1 - v/v_c} . \tag{57}$$

The energy loss brings about a change which moves the sonic point down to a finite level. Its height can be found, to first order, by means of equation (49). Substituting zero order terms into this equation gives, after some manipulation,

$$\frac{2}{D} (v_c - v)^2 = \frac{5^{1/2} q \Phi_o}{3^{1/2} v_c^3} = \frac{5^{3/2} q \Pi}{8 \times 3^{1/2} v_c^4} . \tag{58}$$

Now set

$$\frac{16 \times 3^{1/2} v_c^6}{5^{3/2} q \Pi D} \equiv N^2 \equiv \frac{2 \times 3^{1/2} \mathcal{E}_o^3}{5^{3/2} q \Pi D} , \tag{59}$$

and then

$$1 - v/v_c = 1/N , \tag{60}$$

and from relation (57)

$$-\Delta \mathcal{E} = \frac{4 v_c^2}{N^2} \log N = \frac{2\mathcal{E}_o}{N^2} \log N . \qquad (61)$$

The interpretation of these results is that, when cooling is allowed for, a flow in which material is injected with stagnation enthalpy $\mathcal{E}_o$ will reach its sonic point when the stagnation enthalpy in the gas has dropped to $\mathcal{E}_o - \Delta\mathcal{E}$. The mass flux at any level is given by

$$\frac{\Phi}{\Phi_o} = \frac{2 v v_c}{v^2 + v_c^2}$$

so that at the sonic level

$$1 - \frac{\Phi}{\Phi_o} = \frac{(v_c - v)^2}{v^2 + v_c^2} \doteqdot \frac{1}{2}\left(1 - \frac{v}{v_c}\right)^2 = \frac{1}{2N^2} . \qquad (62)$$

The table below shows values of N, Δ and Φ_o / Φ_{sp}, where Φ_{sp} is the mass flux at the sonic level, together with the velocity u_{sp} there, and the terminal velocity u_∞ of the flow (gravity being ignored). The tabulation is for four different values of $\mathcal{E}_o$, the stagnation enthalpy at injection. In the calculation the value of Π is taken to be 10^{-12} dyne cm^{-2}, and D set equal to 100pc.

$\mathcal{E}_o$ (erg gm^{-1})	N	$\Delta\mathcal{E}$	$\mathcal{E}_{sp} \equiv \mathcal{E}_o - \Delta\mathcal{E}$	Φ_o/Φ_{sp}	$u_{sp} = \sqrt{\mathcal{E}_{sp}/2}$	$u_\infty \doteqdot \frac{8}{5} u_{sp}\left(1 - \frac{1}{2N^2}\right)$
1.0×10^{14}	1.61	3.7×10^{13}	6.3×10^{13}	1.19	5.6×10^6	7.3×10^6
1.5×10^{14}	2.95	3.7×10^{13}	1.13×10^{14}	1.06	7.5×10^6	1.13×10^7
2.0×10^{14}	4.54	2.9×10^{13}	1.71×10^{14}	1.03	9.2×10^6	1.42×10^7
2.5×10^{14}	6.35	2.3×10^{13}	2.27×10^{14}	1.01	1.07×10^7	1.69×10^7

The height z_{sp} of the sonic point is given in terms of N by

$$1 - \frac{\Phi}{\Phi_o} = e^{-z_{sp}/D} = \frac{1}{2N^2} , \qquad (63)$$

so that $z_{sp} = D(2 \log N + \log 2)$ and equals 164pc, 285pc, 372pc and 439pc in the four tabulated cases.

The gas goes on cooling above the sonic point, and when it has lost all its thermal energy its velocity tends to u_∞. To estimate the thickness of the layer within which this final cooling takes place it is easiest to evaluate du/dz at the sonic point. The thickness of the cooling layer is then approximately

$$Z_{cool} = \frac{u_{sp} + u_\infty}{2(du/dz)_{sp}} \tag{64}$$

For this calculation write equation (48) in the form

$$\left(\tfrac{5}{2}\Pi - 4\Phi u\right)\frac{du}{dz} = \left(\varepsilon_0 + 2u^2\right)\frac{d\Phi}{dz} - \frac{q\,\Phi^{5/2}}{u^{5/2}(\Pi - \Phi u)^{1/2}}, \tag{65}$$

remembering that the coefficient of du/dz, as well as the right-hand side, vanish at $z = z_{sp}$. Further, using dashes to denote d/dz, write

$$\frac{u'}{u} = \frac{\xi}{D}, \tag{66}$$

and recall that

$$\frac{\Phi'}{\Phi} = \frac{1}{D(2N^2-1)}, \qquad \frac{\Phi''}{\Phi} = -\frac{1}{D^2(2N^2-1)}, \tag{67}$$

all at $z = z_{sp}$. Then differentiate equation (65) with respect to z at $z = z_{sp}$; after some manipulation one then obtains the equation

$$4\xi^2 + \xi\left(\frac{34}{3} + \frac{5\varepsilon_0}{3u^2}\right)\left(\frac{1}{2N^2-1}\right)$$
$$-\left(2 + \frac{\varepsilon_0}{u^2}\right)\left\{\frac{1}{2N^2-1} + \frac{10}{3(2N^2-1)^2}\right\} = 0. \tag{68}$$

In the four cases given in the table the values of Z_{cool} are, respectively, 66pc, 200pc, 330pc and 450pc. From these results, together with the estimates for Z_{sp}, it follows that the wind has entirely lost pressure by the time it reaches a height of 230pc, 485pc, 702pc and 889 pc, respectively. The gravitational field has only a small effect on the flow up to this level.

But the cool wind will eventually be pulled back by gravity. The potential of the field in which the gas moves is given by

$$\mathcal{V} = 4\times 10^{-9} z + 3.5\times 10^{-31} z^2, \tag{69}$$

to good enough approximation, and the maximum height z_{max} that would be attained by a cold flow which starts with velocity u_∞ from height z=0 is given by

$$u_\infty^2 = 8\times 10^{-9} z_{max} + 7\times 10^{-31} z_{max}^2. \tag{70}$$

In the present representative four cases the values of z_{max} are 1570pc, 2980pc, 4065pc and 5093pc. The maximum height reached in the various cases is some six times larger than the height at which the gas has cooled off entirely. The system of approximations that has been used here is therefore entirely appropriate.

Finally the mass loss rate that is consistent with a value 10^{-12} dynes cm^{-2} for Π is given by Π/u_∞, and in the four cases equals 1.4×10^{-19}, 0.88×10^{-19}, 0.70×10^{-19} and 0.59×10^{-19} gm cm^{-2} s^{-1}.

A pressureless flow rising from the galactic plane with speed u_∞ has speed

$$u = \left(u_\infty^2 - 8 \times 10^{-9} z - 7 \times 10^{-31} z^2\right)^{1/2} \tag{71}$$

at height z, and takes a time

$$t_u = \int_0^{z_{max}} \frac{dz}{u} = 1.20 \times 10^{15} \int \frac{d\zeta}{(A - \zeta - \zeta^2)^{1/2}} \tag{72}$$

to reach height z_{max}; here $A \equiv 1.1 \times 10^{-14} u_\infty^2$ and the integration is taken from $\zeta = 0$ up to the value of ζ for which the denominator vanishes. Thus

$$t_u = 1.20 \times 10^{15} \cos^{-1} \frac{1}{2(A + \frac{1}{4})^{1/2}} ; \tag{73}$$

in the four examples this time equals 1.19×10^{15}, 1.41×10^{15}, 1.50×10^{15} and 1.56×10^{15} s. A part of the fountain that extends to its maximum height thus contains a column density of 1.67×10^{-4}, 1.24×10^{-4}, 1.05×10^{-4} and 0.92×10^{-4} gm cm^{-2}. Typically the upward stream contains some 10^{-4} gm cm^{-2} or 5×10^{19} atoms cm^{-2}. But the wind zone will not, in general, reach its full height. Somehow the gas has to come back down. In a one-dimensional model the material descends in a sheet of cool, compressed gas. When such a sheet is on the way towards the galactic plane, the upflow terminates at a shock just below the sheet.

IV. THE DESCENT OF A COOL SHEET TOWARDS THE GALACTIC PLANE

In this description of the downward motion of the cooled gas, let ζ measure the distance below height z_{max}, let w be the downward velocity of the sheet and Σ the mass per unit area of the sheet. As before Φ and u are, respectively, the mass flux and the upward velocity in the fountain, and $\vartheta^{(m)}$ is the gravitational potential with respect to height z_{max}. The gas in the sheet is taken to be so cool that its thermal speed is much less than w, the speed of mass motion. This assumption will be justified later.

The equation of motion for the flow in the fountain is

$$u \frac{du}{d\zeta} = \frac{d\vartheta^{(m)}}{d\zeta} , \tag{74}$$

and the equation for conservation of mass in the sheet is

$$\varpi \frac{d\Sigma}{d\zeta} = \Phi \frac{u + \varpi}{u} \; ; \tag{75}$$

the right-hand side here is simply the rate at which the sheet picks up mass from the fountain. Finally the equation of conservation of momentum is

$$\varpi \frac{d}{d\zeta} \Sigma \varpi = \Sigma \frac{d\mathcal{V}^{(m)}}{d\zeta} - \Phi (u + \varpi), \tag{76}$$

with contributions on the right-hand side from gravity and from the momentum carried by the material absorbed from the stream. To write the equations with u as independent variable, note that

$$u \, du = g \, d\zeta \equiv d\mathcal{V}^{(m)}, \tag{77}$$

and that g is a function of ζ or equivalently, of u. With this change, equation (75) becomes

$$\varpi \frac{d\Sigma}{du} = \frac{\Phi}{g} (u + \varpi), \tag{78}$$

and equation (76)

$$\varpi \frac{d}{du} \Sigma \varpi = \Sigma u - \frac{\Phi u}{g} (u + \varpi). \tag{79}$$

After elimination of the terms that contain Φ, and some manipulation,

$$\varpi \frac{d}{du} \Sigma (u + \varpi) = \Sigma (u + \varpi), \tag{80}$$

and then it follows from equations (78) and (80) that

$$\frac{d}{d\Sigma} \Sigma (u + \varpi) = \frac{g \Sigma}{\Phi} . \tag{81}$$

In a region where the gravitational field is due essentially to the mass in the galactic disk, g is constant. Equation (81) can then be easily integrated, giving

$$u + \varpi = \frac{g \Sigma}{2 \Phi}, \tag{82}$$

and then from equation (78)

$$\varpi \frac{d}{du} (u + \varpi) = \frac{1}{2} (u + \varpi) . \tag{83}$$

the solution for which w = 0 when u = 0 is

$$w = \tfrac{1}{2} u, \tag{84}$$

and the speed of the descending sheet, at any point, is one half the local upward velocity of the gas in the fountain. From relation (82) the surface density in the sheet is

$$\Sigma = 3\Phi u/g \, . \tag{85}$$

But u/g is the time that would be taken by the gas in the fountain to travel from the present level of the sheet to height z_{max}. Typically Σ can be up to three times the column density in the wind, and so up to 3×10^{-4} gm cm^{-2}. This value is about the same as the smoothed-out surface density of the cold interstellar matter on one side of the galactic plane. But 3×10^{-4} gm cm^{-2} is the maximum possible surface density in the sheet. On average the mass contained in sheets in the halo would be about half the mass of the interstellar gas in the disk of the Galaxy.

A model having a uniform value of g will serve for cases where u_∞ is small, so that the fountain does not rise too high, and the contribution of the disk to the gravitational field remains dominant. This condition applies if z_{max} is much less than 10^{22} cm, or u_∞ much less than 10^7 cm s^{-1}

Such a description of the motion of the sheet can be adapted to a case where the gravitational field has the more general form

$$g = g_0 + g_1 z, \tag{86}$$

with $g_0 = 4 \times 10^{-9}$ cm s^{-2}, $g_1 = 7 \times 10^{-31}$ s^{-2} in the present application. At the fountain top

$$g = g_* = g_0 + g_1 z_{max} \tag{87}$$

Elsewhere g can be expressed as a function of u. The gravitational potential is now

$$\mathcal{V}^{(m)} = g_* \zeta - \tfrac{1}{2} g_1 \zeta^2, \tag{88}$$

and so

$$u^2 = 2g_* \zeta - g_1 \zeta^2 \tag{89}$$

and

$$g = g_* \left(1 - \frac{g_1 u^2}{g_*^2}\right)^{1/2} . \tag{90}$$

Close to the galactic plane $g = g_0$ and $u = u_\infty$ so that the formula becomes

$$g = g_* \left(1 - \gamma u^2\right)^{1/2}, \tag{91}$$

with

$$\gamma = \frac{1}{u_\infty^2}\left(1 - \frac{g_0^2}{g_*^2}\right) . \tag{92}$$

The factor 1/g in equation (78) is therefore to be expanded

$$\frac{1}{g} = \frac{1}{g_*}\left\{1 + \frac{1}{2}\left(1 - \frac{g_0^2}{g_*^2}\right)\frac{u^2}{u_\infty^2} + \ldots\ldots\right\}$$

$$\equiv \frac{1}{g_*}\left(1 + \lambda u^2\right), \qquad \text{say.} \tag{93}$$

Next expand

$$\varpi = \frac{u}{2}\left(1 + \omega_2 u^2 + \ldots\ldots\right) \tag{94}$$

and then from equation (80)

$$\log \Sigma (u + \varpi) = 2\int \frac{du}{u}\left(1 - \omega_2 u^2 + \ldots\ldots\right)$$

or

$$\Sigma (u + \varpi) = C u^2 e^{-\omega_2 u^2}$$

$$= C u^2\left(1 - \omega_2 u^2 + \ldots\ldots\right) \tag{95}$$

From these last two equations it follows that

$$\frac{3u}{2}\Sigma\left(1 + \frac{\omega_2 u^2}{3} + \ldots\ldots\right) = C u^2\left(1 - \omega_2 u^2 + \ldots\ldots\right)$$

or

$$\Sigma = \frac{2C}{3} u\left(1 - \frac{4}{3}\omega_2 u^2 + \ldots\ldots\right) \tag{96}$$

Returning then to equation (78)

$$\frac{u}{2}\left(1 + \omega_2 u^2 + \ldots\ldots\right)\frac{2C}{3}\left(1 - 4\omega_2 u^2 + \ldots\ldots\right)$$

$$= \frac{\Phi}{g_*}\left(1 + \lambda u^2 + \ldots\right)\frac{3u}{2}\left(1 + \frac{\omega_2 u^2}{3} + \ldots\right) \tag{97}$$

giving $C = 9\Phi/2g_*$, as before,

and

$$\omega_2 = -\frac{3\lambda}{10} = -\frac{3}{20u_\infty^2}\left(1 - \frac{g_0^2}{g_*^2}\right) \tag{98}$$

So

$$\omega = \frac{u}{2}\left\{1 - \frac{3u^2}{20u_\infty^2}\left(1 - \frac{g_0^2}{g_*^2}\right)\right\} \tag{99}$$

and

$$\Sigma = \frac{3\Phi u}{g_*}\left\{1 + \frac{u^2}{5u_\infty^2}\left(1 - \frac{g_0^2}{g_*^2}\right)\right\} \tag{100}$$

There is an obvious similarity here with the corresponding results for a uniform gravitational field.

The downward acceleration of the cool sheet is

$$f = g\frac{\omega d\omega}{u du} = \frac{1}{4}g_*\left\{1 - \frac{11u^2}{10u_\infty^2}\left(1 - \frac{g_0^2}{g_*^2}\right)\right\}. \tag{101}$$

Viewed from the rest frame of the Galaxy the gravitational acceleration is

$$g = g_*\left\{1 - \frac{1}{2}\frac{u^2}{u_\infty^2}\left(1 - \frac{g_0^2}{g_*^2}\right)\right\} \tag{102}$$

at the position of the sheet. Viewed from the rest frame of the sheet, the effective gravitational acceleration is

$$g_{eff} = g - f = \frac{3g_*}{4}\left\{1 - \frac{3u^2}{10u_\infty^2}\left(1 - \frac{g_0^2}{g_*^2}\right)\right\}. \tag{103}$$

Thus the relation between g_{eff} and g becomes

$$g_{eff} = \frac{3g}{4}\left\{1 + \frac{u^2}{5u_\infty^2}\left(1 - \frac{g_0^2}{g_*^2}\right)\right\}, \tag{104}$$

and the effective gravity on the matter in the sheet never differs much from three-quarters of the local gravitational acceleration, as seen from the rest frame of the Galaxy.

V. THE COOLING OF THE SHEET

The gas from the fountain joins the descending sheet after passing through a shock. With respect to this shock the speed of inflow of the gas is

$$V_0 = u + \omega \doteq \frac{3}{2}u, \tag{105}$$

the rate of mass flow is

$$\varphi = \Phi(1 + \omega/u) \doteq \frac{3}{2}\Phi \tag{106}$$

the momentum flow is

$$\varpi = \varphi V_0 = (\Phi/u)(u+\omega)^2 \doteqdot \frac{9}{4}\Phi u, \tag{107}$$

and the gas speed immediately post-shock is

$$v_s = \frac{1}{4}V_0 = \frac{3}{8}u . \tag{108}$$

The sound speed just behind the shock is

$$c_s = \frac{\sqrt{5}}{4}V_0 = \frac{3\sqrt{5}}{8}u, \tag{109}$$

and the time scale for gravitational effects to become important behind the shock is

$$t_g = \frac{c_s}{g_{eff}} \doteqdot \frac{\sqrt{5}\,u}{2g} . \tag{110}$$

Again u/g is of the order of the time that would be needed for the fountain to rise from the present level of the sheet to height z_{max}, or twice the time that the sheet has taken to descend to its present position. The cooling time for the newly shocked gas is

$$t_c = K_s^{3/2}/q , \tag{111}$$

where

$$K_s \equiv \frac{P}{\rho^{5/3}} \qquad \frac{3\,\varpi^{8/3}}{2^{16/3}\,\varphi^{10/3}} , \tag{112}$$

and so

$$t_c = \frac{3^{3/2}\varpi^4}{2^8\varphi q} = \frac{3^{9/2}u^4}{2^{11}\Phi^4 q} . \tag{113}$$

The cooling time t_c is much shorter than the dynamical time t_g if

$$u^3 \ll 16.3\,\Phi q/g . \tag{114}$$

To illustrate the cooling process, take a sheet at height 1000pc above the galactic plane; the local value of g is 6.0×10^{-9} cm s^{-2}. In the four cases that have been considered in this paper the mass flux Φ was 1.4×10^{-19}, 0.88×10^{-19}, 0.70×10^{-19} and 0.59×10^{-19} gm cm^{-2} s^{-1}. Even with the smallest of these values for Φ the inequality (114) reduces to

$$u^3 \ll 6.4 \times 10^{22}\ \mathrm{cm^3\,s^{-3}} ; \tag{115}$$

it easily holds in all four cases. The heated gas that joins the sheet will cool enough in so short a time that gravity produces no appreciable effect while it is still hot. There is no need to include a gravitational term in describing the structure of the cooling layer. But once the gas has cooled off its dynamical time scale becomes rather shorter and the cooling time loses its relevance. Gravitational acceleration is more important in the structure of the cooled layer, and the thermal speed within it small compared with its speed of mass motion.

The mass of gas, per unit area, in the cooling layer is

$$\phi t_c = \frac{3}{2}\Phi t_c = \frac{3^{11/2}u^4}{2^{12}q} \doteq 0.10\,\frac{u^4}{q} \,. \qquad (116)$$

The stream velocity u, at 1 kpc height, is 66, 99, 131 and 160 km s^{-1}, respectively. The corresponding mass per unit area in the cooling layer is 4.7×10^{-7}, 2.4×10^{-6}, 7.4×10^{-6} and 1.6×10^{-5} gm cm^{-2}. The post-shock temperature is given by

$$\frac{kT_s}{\bar{m}} = \frac{3}{16}\left(\frac{3}{2}u\right)^2 = \frac{27}{64}u^2, \qquad (117)$$

so that T_s equals, respectively, 1.31×10^5, 2.95×10^5, 5.17×10^5 and 7.71×10^5K. In the first case the gas is rather too cool for us to be able to use our simplified law of radiative energy loss. It is also not hot enough to show the high excitation absorption lines seen in the halo.

To describe the temperature distribution in the cooling layer, note that an element of gas with adiabatic parameter K takes time $K^{3/2}/q$ to cool off, and therefore the mass per unit area of warm gas, having adiabatic parameter less than K, is

$$\Sigma(K) = K^{3/2}\varphi/q \,. \qquad (118)$$

The relation between K and the temperature in the cooling layer follows from

$$\varpi = P + \rho v^2, \qquad (119)$$

or

$$\frac{P}{\rho} = \frac{kT}{\bar{m}} = \frac{\varpi}{\rho} - v^2 = \frac{\varpi v}{\varphi} - v^2 \equiv V_o^2(\eta - \eta^2), \qquad (120)$$

where v is the flow velocity relative to the shock and $\eta = v/V_o$, so that

$$K^{3/2} = \left(\frac{P}{\rho}\right)^{3/2}\rho^{-1} = \frac{V_o^4}{\varphi}\eta^{5/2}(1-\eta)^{3/2}. \qquad (121)$$

Then, from relation (118) and (120)

$$\Sigma(K) = \frac{K^{3/2}\varphi}{q} = \frac{V_o^4}{q}\left(\frac{kT}{\bar{m}V_o^2}\right)^{5/2}\frac{1}{1-\eta} = \frac{2}{3}\left(\frac{kT}{\bar{m}}\right)^{5/2}\frac{1}{uq(1-\eta)}. \qquad (122)$$

The value of η runs from 1/4 to zero in the cooling layer, so that $1/1-\eta$ lies between 1.33 and 1. With reasonable accuracy this factor can be set equal to 1.17 and then $2/3(1-\eta)$ is roughly 0.78. The table below shows how much mass per unit area the sheet contains in various ranges of temperature. The surface densities are in gm cm^{-2}.

$\mathcal{E}_o$	0 to 2	to 4	to 6	to 8 x 10^5K
1.0×10^{14}	4.7×10^{-7}			
1.5×10^{14}	8.2×10^{-7}	1.6×10^{-6}		
2.0×10^{14}	6.2×10^{-7}	2.9×10^{-6}	3.9×20^{-6}	
2.5×10^{14}	5.1×10^{-7}	2.4×10^{-6}	5.0×10^{-6}	3.5×10^{-6}

An estimate of the thickness of the cooling layer is given by

$$\Delta z_c \sim v_s t_c = \frac{3^{11/2} u^5}{2^{14} \Phi q} \doteq 0.026 \frac{u^5}{\Phi q} \tag{118}$$

In the present four cases the values appropriate to height 1000pc are, respectively, 2pc, 23pc, 119pc and 385pc. The composite layer can reasonably be treated as thin, except perhaps in the fourth case. The mass of gas per unit area calculated for the range 2 to 4 x 10^5K seems about right in comparison with values that have been observed in the halo between these temperatures.

VI. STABILITY AND DEPARTURE FROM THE PLANE PARALLEL FLOW

If the flow remains plane parallel then the motion of the sheet as a whole can readily be shown to be stable. The treatment given here applies only to cases where g is uniform. It seems a reasonable inference that the one-dimensional flow remains stable even when the gravitational field has a more general form.

Once again ζ denotes distance below height z_{max} and t the time interval since the sheet set out from height z_{max}. In the unperturbed flow the disk has a downward velocity $w = gt/4$, the upward velocity of the gas in the fountain, at the level of the sheet, is $u = gt/2$, and the mass per unit area of the disk is $\Sigma = 3\Phi t/2$.

In a perturbation let Σ_1 be the first-order increment in surface density, and ζ_1, u_1 and w_1 the first-order increments in ζ, u and w. Then

$$\Sigma_1 = \rho\zeta_1 = \frac{\Phi\zeta_1}{u} = \frac{2\Phi}{gt}\zeta_1 , \tag{119}$$

$$u_1 = \zeta_1\frac{du}{d\zeta} = \frac{g\zeta_1}{u} = \frac{2\zeta_1}{t} \tag{120}$$

and

$$w_1 = \frac{d\zeta_1}{dt} \tag{121}$$

The momentum equation for the disk is, in first order,

$$\Sigma_1 \frac{d\omega}{dt} + \omega \frac{d\Sigma_1}{dt} + \Sigma \frac{d\omega_1}{dt} + \omega_1 \frac{d\Sigma}{dt}$$

$$= \Sigma_1 g - u_1 \frac{d\Sigma}{dt} - u \frac{d\Sigma_1}{dt} \qquad (122)$$

After substitution from relations (119) to (121) there results the equation

$$t \frac{d^2\zeta_1}{dt^2} + 2 \frac{d\zeta_1}{dt} = 0, \qquad (123)$$

with the solution

$$\zeta_1 = A/t + B. \qquad (124)$$

There are obviously no growing solutions for ζ_1. The physical reason is as follows. If the disk were to move ahead of its zero-order location, it would become more massive. But the increased force due to gravity is more than compensated for by the additional momentum flux from the fountain, which occurs when the downward speed of the sheet is increased.

The next problem involves the stability of the sheet, allowing for departures from a plane-parallel flow. The entire stratification will obviously be rather unstable, with the well-cooled gas lying over the much hotter layers where the radiative losses occur. It is hard to estimate just what the final temperature will be in the cooled sheet, for this depends on the radiation field to which the material is exposed, and little is known about that. But high velocity clouds are principally detected by the 21 cm line of H, and the cooled sheets should probably be identified with such clouds. Therefore the gas may be expected to cool below 10^4K, and may easily become as cold as 100 K, the typical temperature of atomic hydrogen in the galactic disk. A layer of gas at 100 K, in the effective gravitational field 3g/4, typically 4.5×10^{-9} cm s^{-2}, will have a scale height $H = 4kT_c/3m_a g$, where $m_a = 2 \times 10^{-24}$gm is the average particle mass in the gas. With the present parameters $H = 1.56 \times 10^{18}$cm or 0.5pc, and is in every case considered very much smaller than the thickness of the cooling layer below.

A Rayleigh-Taylor instability cannot be avoided and would be disastrous for the continued existence of the sheet in such a case. One would expect instead that the sheet breaks up into lumps whose linear dimensions are of order H. If this were to happen the whole structure of the fountain would change in character. The upward flowing gas would not be stopped at a descending sheet: instead it would shock on encountering a set of cooled clouds which are on the way down, and would flow round them. In this fashion we come back to a fountain more like that treated by Bregman, but rather more limited in extent.

But this treatment makes no allowance for the effect of the magnetic field. In the galactic disk the lines of force, on average, run along circles parallel to the plane and concentric with the Galaxy. During the one-dimensional motion in the fountain the magnetic field strength remains proportional to the density; let $B = \rho b$. An estimate for B can be taken from a paper on pulsar twinkling by Andrew Hall, where he postulates $B = 10^{-7}$ gauss and an electron density $n_e = 10^{-2}$ cm^{-3} in the ICM. The corresponding value of ρ is 2×10^{-26} gm cm^{-3} and so **b** equals 5×10^{18} gauss cm^3 gm^{-1}. The magnetic field has no noticeable effect on the motion in the fountain if

$$\frac{B^2}{4\pi\rho} = \frac{b^2\rho}{4\pi} \ll u^2 \tag{125}$$

or

$$\frac{u^3}{\Phi} \gg \frac{b^2}{4\pi} = 2 \times 10^{36} \text{ cm}^5 \text{ s}^{-1} \text{ gm}^{-1}. \tag{126}$$

The limit thus set on u depends only weakly on Φ, which is of the order of 10^{-19} gm cm^{-2} s^{-1}. So magnetic fields do not seriously affect the motion in the fountain until u falls below 6 km s^{-1}, and can safely be neglected in the dynamics of the upward flow.

But if the temperature in the sheet falls low enough, the gas pressure eventually becomes negligible by comparison with the magnetic pressure. It is easy to calculate the structure of a plane-parallel cool layer supported by the magnetic field alone. The magnetic pressure is

$$\frac{B^2}{8\pi} = \frac{b^2\rho^2}{8\pi} ; \tag{127}$$

the equation of hydrostatic support, with an effective gravitational field 3g/4, becomes

$$\frac{d}{dz}\frac{b^2\rho^2}{8\pi} = -\frac{3g}{4}\rho . \tag{128}$$

Hence

$$dz = -\frac{b^2}{3\pi g}\, d\rho , \tag{129}$$

$$\rho = \frac{3\pi g}{b^2}(z_0 - z) , \tag{130}$$

where z_0 is the height of the top level in the cold layer, and the increment in the surface density of the sheet is expressed in terms of an increment in height by

$$d\Sigma = \rho dz = \frac{3\pi g}{b^2}(z_0 - z)\, dz. \tag{131}$$

The total surface density of the sheet is

$$\Sigma = \frac{3\pi g}{2b^2}(\Delta z)^2 \; ; \tag{132}$$

here Δz is the thickness of the cold sheet. With the surface density given by

$$\Sigma \doteq \frac{3\Phi u}{g}$$

it is then found that

$$\Delta z = \left(\frac{2\Phi u}{\pi}\right)^{1/2} \frac{b}{g}. \tag{133}$$

But the thickness of the radiating layer is

$$\Delta Z_c \sim 0.026 \frac{u^5}{\Phi q}, \tag{134}$$

and therefore the ratio of thicknesses is

$$\frac{\Delta z}{\Delta Z_c} = 30.7 \frac{\Phi^{3/2} bq}{g u^{9/2}}. \tag{135}$$

In the four representative cases the ratio equals 111, 8.84, 1.78, 0.56. For a given value of the momentum flux Π the ratio of thicknesses varies like $u_\infty^{-3/2} u^{-9/2}$, or roughly as u_∞^{-6}. If u_∞ is small enough the radiating layer will be much thinner than the layer of cold gas above it. The Rayleigh-Taylor instability will still upset the equilibrium of the plane parallel stratification. A disturbance with a wavelength λ that is small compared with Δz or ΔZ_c can still grow to large amplitude, but will only cause a corrugation of the bottom surface of the sheet of cold gas. A disturbance whose wavelength is comparable with or larger than Δz, and therefore certainly larger than ΔZ_c, will be prevented from growing to large amplitude because in doing so it would distort shock, and the system would thus be stabilised. So the one-dimensional flow pattern will be disturbed, but not destroyed, provided that u_∞, or equivalently $\mathcal{L}_0$, is small enough. The critical value of $\mathcal{L}_0$ seems to be about 2.5×10^{14} erg, the largest value considered in this paper, as it happens. This conclusion is only weakly dependent on the assumed values of the pressure Π, or of the parameter $b(\equiv B/\rho)$ in the intercloud medium.

The perturbation velocity picked up by the gas during the growth of the Rayleigh-Taylor instability is given by

$$(\Delta v)^2 \sim 2 \times \tfrac{3}{4} g \, \Delta Z_c \doteqdot 0.04 \, \frac{g u^5}{\Phi q} \; . \qquad (136)$$

An estimate for the velocity dispersion thus introduced into the gas is

$$\sigma \sim \frac{\Delta v}{2} \sim 0.1 \left(\frac{g u^5}{\Phi q} \right)^{1/2} , \qquad (137)$$

and in the four cases we are treating this equals 1.16, 4.03, 9.10 and 16.3 km s^{-1}. The values at the upper end of the range correspond to an $\mathcal{E}_o$ of 2 or 2.5 x 10^{14} erg gm^{-1}, and are of the same order as the velocity dispersion within individual high velocity clouds.

Finally two very obvious points. A strictly one-dimensional model can account for the existence of high-velocity clouds at high galactic latitudes, but not in directions close to the galactic plane, particularly so in the direction of the anti-centre. Further it is obviously idle to expect that a plane-parallel model will serve for extended regions of the Galaxy. The very existence of spiral structure shows that the rate of star formation varies considerably from place to place in the galactic disk, and there must be similar variations in the rate at which supernova explosions subsequently inject energy into the ICM. It is probably still reasonable to expect that the length scale on which these variations occur, typically 3 kpc or more, will be large in comparison with the height above the plane at which the upgoing stream passes through its sonic point. The wind is therefore still launched in a near-vertical direction. But the maximum height it reaches will depend on the position in the Galaxy with respect to the spiral pattern, and the descending sheet will not generally be set parallel to the galactic plane. The shock ahead of the sheet will be oblique with respect to the upward flow, and the shocked gas will retain a velocity component along the shock front. To take an extreme case, for a sheet inclined at an angle $\pi/4$ to the horizontal, descending at 80 km s^{-1} into a rising wind moving at 160 km s^{-1}, the transverse velocity will be 120 $\sqrt{2}$ km s^{-1}, and the component of this velocity parallel to the plane will be 120 km s^{-1}. At a pinch the model can thus even account for high velocity clouds in the anti-centre.

VII. CONCLUSIONS

There are obvious differences between the present version of the galactic fountain, and that described by Bregman. Here the motion is more orderly, and more nearly confined to a direction perpendicular to the galactic plane. The cooled gas coming down towards the plane remains in a sheet and is not expected to fragment into separate clouds. The difference can be led back to the assumed value of $\mathcal{E}_o$, the original injection energy per unit mass into the ICM.

When $\mathcal{E}_o$ is large, the temperature in the ICM is high, and the gas streaming away upwards from the plane stays hot long enough to experience the effect of the horizontal pressure gradient that arises from the variation in the rate of injection of energy with position in the Galaxy. The gas acquires a velocity component in a radial direction and is propelled out to a larger distance from the axis of the Galaxy. This is the basic difference between Bregman's fountain and the model described in this paper.

The orderly descent of the cold gas in the sheet depends on the ratio of the thickness of the cold layer to that of the radiating layer beneath. The thickness of the former is determined by the magnetic pressure; the ratio depends very sensitively on the speeds involved, and much less sensitively on the magnetic flux linked with the sheet. It can be confidently stated that the cooled sheet stays intact when the injection energy $\mathcal{E}_0$ is rather less than 2.5×10^{14} erg gm^{-1}, for any reasonable assumption about the magnetic linkage. Conversely it can be stated with equal confidence that the descending gas breaks up into clouds when $\mathcal{E}_0$ is rather larger, as Bregman postulates.

A low value of $\mathcal{E}_0$ seems appropriate to the Sun's vicinity in the Galaxy, where the star formation rate and the supernova rate are modest. Conditions are different, though, elsewhere in the Galaxy, and in other galaxies, where there is more supernova activity. The fountain there is more like Bregman's, and the resulting halo is more extensive and lumpy.

A final remark has to be made about chemical abundances in the ICM. If the metals are less abundant, then the rate of cooling on excited ions is correspondingly reduced. The injection energy $\mathcal{E}_0$ increases, and cooling processes in the upstreaming gas become slower. All these effects favour the growth of much larger haloes.

It has been found that there are severe problems in accounting for the absorption spectrum of QSO's on the basis of the intervening galaxy hypothesis, if it is assumed that the halo seen in our part of the Galaxy is typical for all galaxies. But it is now clear that haloes around younger galaxies can be very much more extensive. This should help remove the difficulty.

COMMENTS AND REFERENCES

SECTION I.

The recent interest in galactic fountains has been very much stimulated by the theoretical work of J. N. Bregman (1980) Astrophys. J., 236, p.577. His researches are further described in various later papers, including a preprint (1981) on Galactic Absorption Coronae. Earlier ideas on galactic fountains occur, for example, in the paper by P. R. Shapiro and G. B. Field (1976) Astrophys. J., 205, p.762.

The model for the gravitational field of the Galaxy, which is used in this paper, resembles that employed by J. A. Sellwood (1980) Astron. Astrophys., 89, p.296.

The data on the cooling of hot interstellar gas come from the calculations by J. C. Raymond, D. P. Cox and B. W. Smith (1976), Astrophys. J., 204, p.290. The simple all-purpose formula was suggested by F. D. Kahn (1976), Astron. Astrophys., 50, p.145.

SECTION II.

The similarity model for the development of a supernova explosion is based on the solution given by L. I. Sedov (1959) in his book *Similarity and Dimensional Methods in Mechanics*, Cleaver Hume Press, London, Chapter 4. A variant on the Sedov solution has been described by J. P. Chieze and B. Lazareff (1981), Astron. Astrophys., 95, p.194. The difference is that here allowance is made for the lumpy nature of the interstellar medium, and for the need to conduct heat into the interstellar clouds in order to evaporate them, and to add some of the material to the hot supernova remnant.

Properties of supernova remnants are reviewed by F. D. Kahn (1975) in the reports of the 14th International Conference on Cosmic Rays at Munich, Volume 11, p.3566, and by R. A. Chevalier (1977), Annual Reviews Astron. Astrophys., 15, p.175.

The transfer of heat from the hot ICM to cool clouds is discussed in the paper by C. F. McKee and J. P. Ostriker (1977), Astrophys. J., 218, p.148; the physics of heat conduction within supernova remnants is considered by F. D. Kahn (1980), Highlights of Astronomy, 5, p.365.

SECTION V.

The evidence for the existence of highly ionized species in the galactic halo comes from B. D. Savage and K. S. de Boer (1979), Astrophys J. (Letters), 230, L77.

SECTION VI.

For an account of the properties of high-velocity clouds see, for example, A. N. M. Hulsbosch (1975), Astron. Astrophys., 40, p.1. The model for the magnetic field linked with the ICM is based on that used by A. N. Hall (1980), Monthly Notices, 190, p.353, in his work on plasma instabilities and pulsar twinkling.

SECTION VII.

Recent results on QSO absorption lines are reviewed by M. G. Smith in his contribution to this volume.

OPTICAL OBSERVATIONS OF INTERSTELLAR SHOCKWAVES

M. A. Dopita
Mount Stromlo and Siding Spring Observatories
Research School of Physical Sciences
Australian National University, Canberra.

1. INTRODUCTION

When first asked to contribute to this volume to mark Professor Kopal's retirement after thirty years as Head of the Department of Astronomy at Manchester, my mind cast back to my first meeting with him, a scant ten years ago. Like most things connected with this man, this could never be classified as commonplace. Having applied to undertake a post-graduate course at Manchester, I received a letter in London asking me to attend at a certain date and time at the tomb of Isaac Newton in Westminster Abbey. This I did, to find several others milling about, eyeing the stone work, inscriptions and each other. At five minutes past the appointed hour in swept Prof. Kopal up the main aisle, resplendent in fur coat and accompanied by two or three of his staff, and proceeded to interview us hopeful candidates in a very business-like manner in the front pew. This lasted until the winter's light faded and a verger appeared before us to end the interview by enquiring "whether the gentlemen would be staying for Evensong". Thus began a very warm professional association!

In preparing this review, I rapidly became aware of the enormous task that I had undertaken. This is in main due to the explosion of high quality observational data in the last ten years as instrumental techniques became at last sufficient for the demands put upon them. Although shocks are observed optically in novae, supernovae, accretion disks and columns, ring nebulae about WR stars and in HII regions and planetary nebulae, I have been forced as much by space restrictions as by personal interest to consider only three classes of object, supernova remnants, Herbig-Haro objects and active galactic nuclei. Even so, what riches! Much of what I have perforce treated superficially is covered in the excellent recent reviews of McKee and Hollenbach (1980) of Chevalier (1977) and of McCray and Snow (1979)

F. D. Kahn (ed.), Investigating the Universe, 29–59.

2. YOUNG SUPERNOVA REMNANTS

The evolution of supernova remnants (SNR) has for a long time been divided into four phases. These are, briefly;

1. A free expansion, terminated when the mass of swept up interstellar medium is approximately equal to the mass of the ejecta. The recent review by Chevalier (1977) describes the various types of instability accompanying the formation of the reverse shock-wave which may break the shell into fragments in this phase.

2. The adiabatic blast wave phase in which radiative energy losses are still unimportant. This phase is described by similarity solutions (Taylor 1950, Sedov 1959), and these solutions (after a furious theoretical debate sparked by Isenberg 1977) now appear to have been proven to be stable against radial perturbations.

3. The radiative phase (Cox 1972b) which begins when the shock is slow enough so that the cooling time scale becomes shorter than the expansion time scale. Most optically visible SNR are presumed to be in this phase.

4. The momentum conserving ('snowplough') phase described by Oort (1946), terminated when the expansion velocity becomes of order of the random turbulent or thermal motions in the interstellar medium (ISM).

The last ten years have seen an explosion of data on SNR's, and much of these are incompatible with this simple picture. However, a wealth of new information has been gathered on nucleosynthesis in massive stars, the mechanism and energy released in supernova explosions, the structure of the interstellar medium and chemical abundance gradients in our own and external galaxies. In this discussion, young and evolved SNR are distinguished by whether they are still in phase 1, or in later phases (although this distinction is still somewhat fuzzy).

2.1 Young SNR of Type I

Supernovae of Type I are universally accepted to be the explosion of a Population II star of low mass. These are the most spectacular optical events, and can probably be used as standard candles with a peak photographic magnitude $M_{pg} = -19.8 + 5 \log (H_0/50 \text{ km s}^{-1} \text{ Mpc})$ (Tammann 1978). In a systematic search for optical counterparts of Galactic Supernova remnants, van den Bergh (1978) has identified only twenty-nine. Amongst these, only two, Tycho's and Kepler's can be certainly identified as remnants of Type I SNe on the basis of historical observation, and a third, SN1006 is probably Type I on the basis of its very great optical luminosity and distance from the galactic plane (Clark and Stephenson 1977).

Morphologically and spectrally, Tycho and the remnant of SN1006,

G327.6 + 14.5, have very similar structure. They both consist of a very sharp, faint, apparently thin arc of a shell. Their spectra are remarkable in that only the Balmer lines of hydrogen are seen, and the decrement is very large (Kirshner and Chevalier 1978; Schweizer and Lasker 1978). The radial velocities and velocity dispersions are very low, less than 100 km s^{-1}, despite apparent transverse velocities of several thousand km s^{-1}. (van den Bergh 1973, 76, Kamper and van den Bergh 1978) In the case of Tycho, the measured value of 5600 km s^{-1}, at an assumed distance of 6 kpc is 2/5 the average expansion rate, just as predicted by the Sedov theory. The X-ray temperatures are in the range 4 x 10^7 ^oK (Davison et al. 1976) to >8.4 x 10^7 ^oK (Pravdo and Smith 1979). These figures imply a blast wave velocity of only about 2000 km s^{-1}. However, if the Tycho SNR is only 2.3 kpc away and is nearing the end of Phase 1 these velocity estimates may be reconciled; Chevalier, Kirshner and Raymond (1980).

In this, and an earlier paper (Chevalier and Raymond 1978), a convincing theory capable of explaining the peculiarities of these SNR was proposed. According to this, a collisionless shock propagates into a neutral medium at very high velocity. Initially hydrogen atoms, essentially at rest in this medium, are collisionally excited by the hot electron gas giving rise to the Balmer spectrum. After a limited number of excitations they are ionised and join the hot stream. Forbidden lines are not strong because the collision strengths are very small for the high post-shock electron temperatures. At lower shock velocities charge exchange between the shocked ions and the unshocked neutral gas becomes important, and so, whilst maintaining an essentially pure Balmer emission spectrum, the lines become very much broadened by the Doppler effect.

Regions of faint, high velocity Balmer emission have been recognised for some years in evolved SNR such as the Cygnus Loop (Kirshner and Taylor 1976; Doroshenko and Lozinskaya 1977) and in IC443 (Lozinskaya 1979) where the diffuse Hα emission is found over at least the LSR range of velocities -400 to +240 km s^{-1}. Raymond, Davis et al. (1980) detected a well defined filament 4-6 arc min outside the bright optical regions of the Cygnus Loop. This also showed a pure Balmer spectrum with widths of order 200 km s^{-1}, in striking confirmation of the theory.

The evidence that the shocks are propagating into a neutral medium is therefore quite strong. Chevalier, Kirshner and Raymond (1980) have shown that material behind the shock is unable to pre-ionise the gas ahead, since the escaping ionising photon density is too low. If the SN itself produced a pulse of UV photons sufficient to ionise the medium about it, then recombination could not have taken place in the time available, and non-radiative shocks would be invisible. Thus, the claim by Kahn (1974) that SN produce a negligible ionising flux appears to be vindicated.

2.2 Young SNR of Type II

There is increasing evidence that the remnants of Type II SNe can furnish important evidence on both the physics of the explosion, and the chemical makeup of the massive precursor star. The most detailed models of SN II light curves are by Falk and Arnett (1977) and show that the explosion follows roughly three phases; first, a pulse of radiation possibly strong in the UV following the escape of the shock from the photosphere; second, a fairly extended period of diffusive energy release giving an exponential luminosity decline and fairly low photospheric temperatures, $12000 \gtrsim T \gtrsim 5000^{o}K$. The final phase is a rapid decline in luminosity associated with recombination and envelope transparency, with some observational evidence for a longer living core.

The fact that recombination can take place is most important, since it ensures that material deep inside the star may fragment in a thermally unstable fashion to cool clouds which stand a better chance of being observed optically than at X-ray energies.

Perhaps the best studied SNR of this type is Cas A, which exploded about 1653-1671 A.D. (Kamper and van den Bergh 1976) but which was not observed at the time presumably due to the high extinction and the lower absolute luminosity of Type II SNe. The optical remnant is dominated by two different kinds of filaments distinguishable both in their dynamics and spectra. These are unusually termed the 'quasi-stationary flocculi' (QSF) and the 'fast moving knots' (FMK). The QSF have spectra with strong hydrogen Balmer lines, [NII] $\lambda 6583 \sim 2\text{-}3 \times H\alpha$. The [OI] lines are fairly strong but the [OIII] are weaker than $H\beta$. Lines of [NI], [SII], [FeII], [HeI] and [CaII] appear in similar strength to $H\beta$ (Kirshner and Chevalier 1977). The QSF have a low velocity dispersion (~ 200 km s^{-1}) and Kamper and van den Bergh (1976) find an overall expansion rate, from proper motions of about 150 km s^{-1}. This velocity is probably indicative of the velocity of a shock propagating into pre-existing dense ($10^{2}\text{-}10^{4}$ cm^{-3}) cloudlets (Bychov 1973). However, to give a quantitative description of the spectrum, shock models (Dopita 1977, Raymond 1976, 1979) would require nitrogen to be overabundant with respect to hydrogen by a large factor. Such nitrogen enriched cloudlets would have had to originate from CNO-processed material blown off at low velocity from the pre-supernova star and indicate a possible catastrophic pre-supernova mass loss (Chevalier and Kirshner 1978).

The flocculi seen in Puppis A seem to be of a very similar nature, though at a later stage in their evolution. These too have very strong [NII] with respect to $H\alpha$ (Dopita, Mathewson and Ford 1977) and in some filaments [NII]$\lambda 6584$ achieves a strength greater than 10 times that of $H\alpha$ (Danziger et al. 1981). The flocculi form an expanding cloud with expansion velocity roughly proportional to distance from the explosion centre, with a maximum velocity of ~ 300 km s^{-1} (Elliott 1978). Since some filaments are found beyond the outer boundary of the X-ray or radio radio remnants, it would seem that the nitrogen rich cloudlets have been accelerated, along the lines suggested by McKee, Cowie and Ostriker (1978).

It is possible that both the Pup A and Cas A nitrogen-rich flocculi are remnants of a ring nebula about a WN star. Parker (1978) found evidence for a nitrogen enrichment in NGC 6888, one such nebula, with a N/O ratio (derived from his observations) of 1/2 by number. Some He enrichment was also found. These observations are important since they indicate that the shocked stellar wind can cool, probably as a result of thermal conduction.

The FMK of Cas A are quite different from the QSF. Their spectra are very peculiar, dominated by the [OIII], [OII], [OI], and [SII] lines with no sign of hydrogen helium or nitrogen emission. In one FMK, lines of [ArIII], [ArIV] and [ArV] are visible (Kirshner and Chevalier 1977). The space velocities are enormous, ranging from 440 to 8460 km s^{-1} (van den Bergh and Dodd 1976). These knots fade, and new ones appear over relatively short time scales. The number continuously visible since 1951 has declined exponentially with an e-folding time scale of 25 years (Kamper and van den Bergh 1976).

It is clear that the FMK have their origin deep within the chemically processed layers of a massive star. Peimbert (1971) sets the oxygen to hydrogen abundance ratio at least 40 times greater than Orion, whilst Chevalier and Kirshner (1978) suggest it may be greater than 500. Models of pre-supernova stars (Weaver et al. 1978) show that the layers within the carbon burning shell and outside the silicon burning shell are dominated by oxygen provided $M_* \gtrsim 25\ M_\odot$. Argon and sulphur enhancements occur in the inner regions close to the core that will form a neutron star.

A second galactic example of a young oxygen-rich SNR has been recently found (Goss et al. 1979), the filled radio shell SNR G292.0 + 1.8 (MSH 11-54). Murdin and Clark (1979) find oxygen filaments with a velocity dispersion of ~2030 km s^{-1}. A crude diagnostic indicates $A_V \sim 2.0$ and $\log T_e \sim 4.5$ for these filaments. A high resolution X-ray map has recently been obtained for this SNR (I. Tuohy, private communication), and shows an elongated bipolar loop structure with a bright equatorial ring. The oxygen flocculi are scattered on the eastern side of this equatorial region, and these observations are unequivocal evidence for the importance of equatorial symmetry in the explosion. Presumably, rotation moderates the collapse, either by allowing more rapid collapse in the polar direction, or by amplifying the magnetic field pressures in the equatorial plane.

Examples of oxygen rich SNR's are also seen in the LMC, and these have much the same structure as those in the Galaxy. The SNR N132D in the bar of the LMC has been shown to contain a ring of [OIII] emitting material ~6pc in diameter expanding as an annulus at 2250 km s^{-1} (Lasker 1980, 1978; Danziger and Dennefeld 1976 a,b). A larger, outer shell some 32pc in diameter is also seen, but has a much lower velocity dispersion. It is possible that the remnant is a result of rapidly consecutive explosions of the components of a massive binary star.

A much smaller ~1.6pc diameter annulus, strong in [OIII], was recently found to be associated with the radio and X-ray remnant 0540-69.3 (Mathewson et al. 1980). This has a velocity dispersion of ~3000 km s^{-1}. Lines of [OIII], [OII], [SII] and interestingly enough [NII] are observed in the spectrum. The [NII] emission forms a faint 4 arc sec (0.8pc) annulus within the [OIII] ring and may be coming from a region analagous to the QSF of Cass A.

3. EVOLVED SUPERNOVA REMNANTS

In their later stages of evolution, the optical spectra of SNR's are remarkably similar, showing [SII], [OI], [OII] and [NII] lines often as strong or stronger than Hα, [OIII] and [NeIII] lines of similar strength to Hβ and frequently, emission in lines of [CaII], [MgI], [FeII], [NI], HeI and HeII. This spectrum is recognised as being due to shockwaves of modest (50-200 km s^{-1}) velocity (Cox 1972 a,b, Raymond 1976, 1979, Dopita 1976, 1977, Shull and McKee 1979).

The dimensions are such that the remnant is dominated by swept-up interstellar matter, so by study of the spectrum and the velocity field, we may infer physical conditions and abundances in the interstellar medium (ISM). With a model of the ISM we can infer the kinetic energy delivered by the SN, and statistical arguments can be used to compute the supernova rate in nearby galaxies. Since the most rapid advances are being made in the field of SNR's in external galaxies, the problem of identification of these will first be reviewed.

3.1 Identification of SNR's in External Galaxies

The technique of optical identification of SNR's in external galaxies was pioneered by Mathewson and Clarke (1972, 1973 a,b,c). A non-thermal radio source is first identified with an optical nebulosity. Plates of this nebulosity in the emissions of Hα + [NII], [SII] and in the continuum are then intercompared. Objects equally bright in all three are distant galaxies, those which appear much brighter at Hα are HII regions and those which are absent in the continuum, but about equally bright at Hα and [SII] are probably SNR's. The exact spectral discriminants by which SNR's can be distinguished from photo-ionised HII regions has been investigated by D'Odorico and collaborators (D'Odorico and Sabbadin 1976, Daltabuit et al. 1976, D'Odorico 1978); nearly all SNR were found to have a Hα/[SII] (λ6717 + 31) ratio less than 2.5.

An alternative to using non-thermal radio emission to locate the position of the SNR by X-ray observations of a spatially extended soft source. Recently Mathewson, Dopita and Tuohy have detected a further nine new optical SNR's in the Magellanic Clouds guided by Long and Helfand's Einstein Observatory (HEAO-B) survey. However, large diameter remnants are difficult to detect owing to the rapid decline of both the X-ray flux and the non-thermal radio flux with diameter.

In galaxies further away from the Magellanic Clouds, identifications rely very heavily on spectrophotometry or determinations of internal velocity dispersion in the optical. Seven SNR were thus classified in M31 by Kumar (1976). More recently, D'Odorico et al. (1978) identified three in M33 of which two are definately detected as radio sources (Israel and van der Kruit 1974, Goss et al. 1980). This list was enlarged for these and other galaxies by D'Odorico et al. (1980) and, for M31, by Blair et al. (1981). In M31, five radio sources have been identified as SNR using the VLA at 20 cm (Dickel et al. 1981) and two as X-ray sources (van Speybroeck et al. 1979, Blair et al. 1981).

3.2 Measurement of Abundances and Abundance Gradients

The general methods whereby the gaseous phase abundances can be determined from a radiating shock were outlined by Dopita (1977). His grids still remain the only general study on the effects of abundance, but other models (Raymond 1976, 1979, Shull and McKee 1979) are more sophisticated and deal with the effects of self-ionisation in a more consistent manner. However, all three independent models give agreement to about 0.3 in the logarithmic abundances of He, O, N, Ne and S.

In essence, the technique consists of obtaining the shock conditions from the density-sensitive [SII] $\lambda 6731/\lambda 6717$ ratio or the [OII] $\lambda 3726/\lambda 3729$ ratio and from the predominantly temperature-sensitive ratios [OIII]$\lambda 5007$ to [OII]$\lambda 3726 + 29$ and [OIII]$\lambda 5007$ to [OI]$\lambda 6300$ which have been given as diagnostic plots. In fact, the abundances are only very weakly dependent on these parameters provided low ionisation sp species such as HI, HeI, [OII], [OI], [NII], [NI] and [SII] are used. This is because the structure of the shock in the region where these species emit is invariant on the shock conditions to first order.

These techniques were first applied to the LMC and SMC SNR's (Dopita 1976; Dopita, Mathewson and Ford 1977). These systems were found to be ordered according to chemical enrichment, SMC, LMC and the Galaxy, but, due to incomplete spectral coverage, there was a wide scatter. However, there was no evidence for an abundance gradient in the Magellanic Clouds, a result also found by examination of HII region spectra. (Peimbert and Torres-Peimbert 1974; Dufour 1975; Dufour and Harlow 1977 and Pagel et al. 1978).

The discovery of many SNR in M31 and M33 has made practicable the measurement of the absolute abundance gradient in the arms of spiral galaxies, as has been done for many years using HII regions (see review by Pagel 1981). So far this has only been accomplished in M33 (Dopita, D'Odorico and Benvenuti 1980) with the result that a pronounced gradient was obtained. No galactic enrichment model seemed to be able to fit the observations, however. The absolute abundances derived are very similar to those found in nearby HII regions, but a disagreement arises in the case of nitrogen. The HII region observations (Pagel, private communication) agree very well with the SNR's for oxygen and nitrogen in the outer parts of the Galaxy, but in the inner regions nitrogen seems to be enhanced as a secondary element according to the SNR work, but not according to HII region measurements. Data on HII regions in other galaxies (Smith 1975, Edmunds and Pagel 1978 and Alloin et al. 1979) paint a rather confusing picture about the N and O relative abundances. Irregular and Scd galaxies tend to have lower N/O than the Galaxy even when the oxygen abundance is the same, and as in the case of M33, N/O variations across the face of the galaxy are quite small. If correct, this indicates that nitrogen is essentially primary but with a yield relative to oxygen that varies greatly galaxy to galaxy (Edmunds and Pagel 1978, Pagel 1980).

Although N/O gradients have been measured using SNR's only in the case of M33, much more can be said about N/S gradients, since these have been determined both in M31 (Blair, Kirshner and Chevalier 1981) and in our Galaxy (Dopita, D'Odorico and Benvenuti 1981). This ratio furnishes essentially the same information as the N/O ratio, since both O and S are theoretically expected to be primary elements and, furthermore, the O/S ratio is found to be constant (Dopita, D'Odorico and Benvenuti 1980.

Figure 1 shows all the available data on evolved SNR's in our own and external galaxies (Dopita, D'Odorico and Benvenuti 1981). Neither the [SII]λ6731 Å nor the [NII]λ6548 + 6583 Å lines should be influenced by collisional de-excitation effects. From the remarks made at the beginning of this section, it should be evident that these lines are sensitive to abundance. Furthermore, owing to the strength of lines of other elements, principally [OII], the possibility of saturation sets in only at the high abundance extreme, and the ratios with Hα otherwise scale as the ionic abundance. Thus Fig. 1 demonstrates that the nitrogen abundance changes more rapidly than the sulphur abundance as we go from galaxy to galaxy. Furthermore, where an abundance gradient is measurable, the relative abundances of nitrogen and sulphur appear to vary in the same manner. Thus the SNR evidence is for a secondary nucleosynthetic component of nitrogen, a conclusion which is almost independent of the detailed shock-wave modelling.

3.3 Physical Conditions in Evolved Remnants

The evidence that SN blast waves run into an inhomogeneous interstellar medium is very strong. The wildly discordant blast wave velocities derived using X-ray data and data in other spectral regions was early proof that the Sedov (1959) model of a self-similar blast

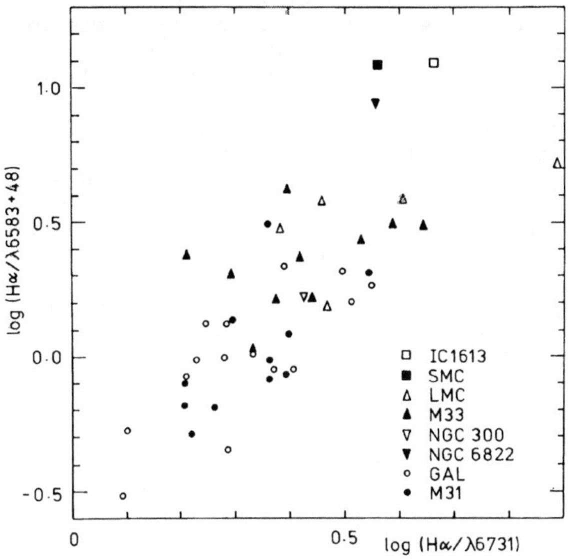

Fig. 1 Abundance sensitive line ratios in SNR's. Note that all galaxies fall on a single sequence with a (possibly real) scatter of 0.2 dex. Abundance gradients are evident in M31, M33 and the galaxy. The more rapid variation in the Hα/[NII] ratio is possibly indicative of a secondary nucleosynthetic component of enrichment.

wave running into a uniform ISM had problems. In IC443, for example, the soft X-ray observations can be fitted by a thermal spectrum with $T_e \approx 10^7$ °K which suggests a blast wave velocity close to 800 km s^{-1} (Malina et al. 1976, Parkes et al. 1977) in a uniform medium. However, the mean optical expansion velocity was found to be ~65 km s^{-1} in the NE part of the remnant, but some faint filaments had radial velocities as great as ~150 km s^{-1} (Lozinskaya 1969, 1975). Optical spectrophotometry gives velocities in the range 65-90 km s^{-1} when the models of Raymond (1979) are used (Fesen and Kirshner 1981). If only the very faintest Hα filaments are used in a Kompaneets (1960) (see also Gulliford 1974) solution for a point explosion in a medium with exponential density distribution, then the expansion velocities are higher, 150-280 km s^{-1} in the NE and 280-400 km s^{-1} in the SN. (Lozinskaya, 1979) A similar dichotomy is seen in the Cygnus Loop. The X-ray data (Gorenstein et al. 1971, Rappaport et al. 1974, Burginyon et al. 1975) give a blast wave velocity of about 400 km s^{-1}. In contrast with the extremely localised optical nebulosity, the X-ray image forms much more of a complete shell (Kayat 1978). Spectrophotometry both in the visible (Cox 1972 a,b, Miller 1974, Raymond 1976) and in the far UV (Benvenuti et al. 1979, 80, 81, Raymond,

Black et al. 1980, 81) are consistent with shock velocities in the range 80-130 km s^{-1}. Again, regions of faint hydrogen emission show higher velocity dispersions, of order 200 km s^{-1} (Kirshner and Taylor 1976, Doroshenko and Lozinskaya 1977, Raymond, Davis et al. 1980).

McKee and his collaborators (McKee and Cowie 1975, McKee and Ostriker 1977, McKee et al. 1978) have proposed that the solution to these problems lies in a cloudy ISM. As the blast wave velocity is proportional to the square root of the driving pressure divided by the local density, the blast wave evolves in the lowest density (intercloud) component. Cloudlets are therefore surrounded by a hot jacket of shocked gas which has two effects; first, electron conduction heats up their surfaces and 'evaporates' cloudlet material into the intercloud region and second, a slow shock is driven into the cloud by the effects of surface pressure. In front of the cloud a 'main' shock is driven by both the thermal and dynamical pressure, which, at the stagnation point can reach about 3.5 times the intercloud pressure (McKee and Cowie 1975, Sgro 1975). Behind and to the sides of the cloud, this pressure has only its thermal component, so 'secondary' shocks are driven at lower velocities. If the cooling time is longer than the time taken for the main shock to cross the cloudlet, this never becomes radiative and the cloudlet is completely evaporated into the intercloud medium. This will apply for small or low density cloudlets.

Thermal evaporation of or from cloudlets profoundly modifies the evolution of the blast wave by increasing the post-shock density and preventing the rapid deceleration predicted by Sedov (1959) theory. In Sedov theory, if V_B is the blast wave velocity and r_B its radius, $V_B \propto r_B^{-3/2}$ in a uniform medium. In 'evaporative' theory, the density $\rho \propto r_B^{-5/3}$ and $V_B \propto r_B^{-2/3}$. X-ray observations of Galactic SNR's certainly suggest that there is a density variation along the lines suggested by 'evaporative' theory (McKee and Ostriker 1977).

Perhaps the strongest evidence in support of the 'evaporative' model comes from the extragalactic SNR. Since we can see SNR of many different diameters at a constant and known distance, the variation of properties with diameter can be successfully determined. Such a study was made by Dopita (1979), for the LMC remnants which combined high resolution ($V \approx 25$ km s^{-1}) spectrophotometry of the [OII] doublet with absolute surface photometry at Hα.

The velocity data were used to derive a mean shock velocity, and this was combined with the [OII] and, in some cases the [SII] density sensitive ratios to estimate the pre-shock density. The Hα photometry provided an independent estimate of a product of these two variables, so that the results could be checked for consistency.

No systematic trend of pre-shock density with diameter was found, but these ranged from 2.7 - 230 cm^{-3} with a mean of about 12 cm^{-3}, much higher than would be estimated from HI column densities. This fact alone suggests the optical emission is arising from clouds or dense cloudlets.

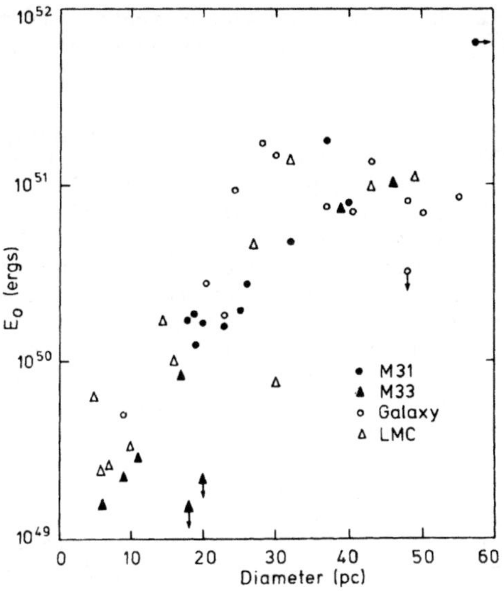

Fig. 2 The explosive energy E_o in SNR's derived from Sedov theory. Although this theory is clearly inapplicable, probably due to the cloudy nature of the Interstellar Medium, the close similarity of the sequences in these four galaxies is strong evidence in favour of similar structure in the interstellar medium of those galaxies. After Blair, Kirshner and Chevalier 1981.

The mean shock velocity, V_S, was found to vary but slowly with radius, r_s. For SNR's with radii larger than 5 pc, the relationship $V_s \propto r_s^{-1/2}$ gave a good fit. If we use the Sedov (1959) solution modified to deal with the case of a strong explosion propagating in a medium whose density $\rho(r)$ varies about the explosion centre in the form $\rho(r) = Ar^{-\omega}$ (Chevalier 1976, Newman 1977, Isenberg 1977), then the age, t, radius r_s, blast wave velocity, V_s, and explosion kinetic energy, E_o are related through:

$$t = \left(\frac{2}{5-\omega}\right) \frac{r_s}{V_s}$$

$$E_o = \left(\frac{Ar_s}{t^2}\right)^{(5-\omega)}$$

$$V_s = \left(\frac{2}{5-\omega}\right)\left(\frac{\alpha E_o}{A}\right) r_s^{-(3-\omega)/2}$$

These imply that, in order to have $V_s \propto r_s^{-1/2}$, then $\omega = 2$, which is in contradiction to the pre-shock density measurements. If these equations are used to derive E_o, then this is a strongly increasing function of radius.

Blair, Kirshner and Chevalier (1981) have found an almost identical relationship for M31, using the post-shock pressure as given by the [SII] ratio to derive the driving pressure of the blast wave and hence the explosion energy. They also used Dopita, D'Odorico and Benvenuti's (1980) observations in M33 and galactic data from Daltabuit, D'Odorico and Sabbadin (1976) to show that these galaxies too follow essentially the same relation (Fig. 2). Blair et al. prefer a model in which the [SII] emitting regions are dominated by magnetic pressure, either due to direct compression of the interstellar field (McKee and Hollenbach 1980), or by field amplification (Reynolds and Chevalier 1981). However, although these are consistent with the pressure data, the velocity would decline as $r_s^{-3/2}$ for both these models.

Dopita (1979) believes that the only way a consistent interpretation can be made of the data is to invoke the full evaporative solution (McKee and Ostriker 1977). The evaporation of smaller cloudlets by the passage of the blast wave increases the density of the intercloud region. This affects the velocity evolution of the blast wave in the same way as a pre-shock density which decreases with diameter would. Only when the pressure driving the cloudlet shocks has decreased sufficiently following the passage of the blast wave do cloudlet shocks become radiative. Higher velocity shocks do appear to exist closer to the blast wave, as indicated by their relatively strong [OIII] emission (Lasker 1977). The radiative region and the blast wave should draw together in larger diameter SNR. Since the pressure falls very rapidly inside the remnant, smaller diameter optical SNR appear to be driven by relatively small pressures, and Sedov energies derived are low. The optical velocity dispersion is determined more by the point at which the cloudlet shocks become radiative, and so is only weakly dependent on diameter.

In this model, the physical conditions within the cloudlet and the position of the cloudlet within the SNR together determine the shock pressure and the velocity dispersion and so the accurate determination of explosion energy becomes almost impossible. However, for the LMC remnants this is estimated to lie in the range $51.3 \gtrsim \log E_o \gtrsim 50.3$ with a probable value near $\log E_o \approx 50.8$ ergs. The properties of the blast wave can be much better investigated in X-rays or using very high excitation optically emitting species such as [FeXIV] λ5303 Å, since these probe the intercloud medium more directly.

Long and Helfand (1979) find that the X-ray data support a decrease in density of the emitting material with diameter. The pre-shock density computed with the Sedov relations is found to fall off roughly as $r_s^{-5/2}$. Densities derived range from 35 to 0.18 cm^{-3}, and in all cases are lower than the optical values, as expected in a cloudlet model. Both the optical and X-ray ω are somewhat higher than the evaporative value (5/3), so perhaps some modification of evaporative theory is needed, such as a change in evaporative efficiency with post-shock temperature.

Shklovsky (1967) was first to suggest that the hot X-ray emitting component of SNR could be studied optically in the coronal lines of iron. The [FeXIV] line at $\lambda 5303$ Å, for example, is emitted with peak efficiency by a thermal plasma of 2×10^6 °K (e.g., Raymond 1976). It was not until 1973 that this line was successfully observed in the Cygnus loop (Woodgate et al. 1974). Since then it has possibly been seen in Vela (Woodgate et al. 1975, Murdin et al. 1978) and has certainly been detected in N49 in the LMC (Danziger and Dennefeld 1976, Murdin et al. 1978), Puppis A (Clark et al. 1979, Lucke et al. 1979) and RCW 103 (Tuohy et al. 1981).

In three cases where spatial data are available, Cygnus, N49 and RCW 103, the [FeXIV] emission appears to come from a quite smooth shell of larger diameter than that defined by the bright optical shell. In these cases, the overall diameter seems to be closely similar to that of the X-ray shell, but the [FeXIV] emission is rather more concentrated into a hollow shell, presumably because [FeXIV] is emitted in a limited range of temperatures (Woodgate et al. 1977, Dopita and Mathewson 1979, Tuohy et al. 1981). The spatial separation of the main optical filaments from the coronal emission line region is in good agreement with the predictions of a cloudlet shock model.

Puppis A, on the other hand, is very peculiar. The nitrogen anomaly of this remnant has already been mentioned in section 2.2 above. Exceedingly strong [FeXIV] emission was found to be associated with one filament in particular (Clark et al. 1979, Lucke et al. 1979). In early 1980, Dopita, Mathewson and Tuohy secured direct images of this filament in [FeXIV] and [OIII]. These are shown in Figure 3. Each picture element is ~0.8 arc sec square, so subtends a distance at Puppis A of 2.5×10^{16} cm. Although the [FeXIV] features are reasonably in accord with the high resolution X-ray image being prepared by Winkler and his co-workers, including the fairly sharp outer boundary of the remnant, there is very little correlation with the [OIII] image. Furthermore, the [OIII] filaments are in general much finer and have a knotty appearnce compared with the coronal line emitting regions.

A developed thermal instability is necessary to explain the abundance peculiarities of the filaments, but it would seem that in addition, magnetic fields must be invoked to explain the results of Figure 3. It is possible that the filaments were formed in the region of the contact discontinuity where magnetic fields are high and the Rayleigh-Taylor instability serves to amplify the field (Gull 1973 a,b, 1975, Chevalier 1977). This would produce magnetic sheets or filaments which would collapse and cool until supported by the magnetic field

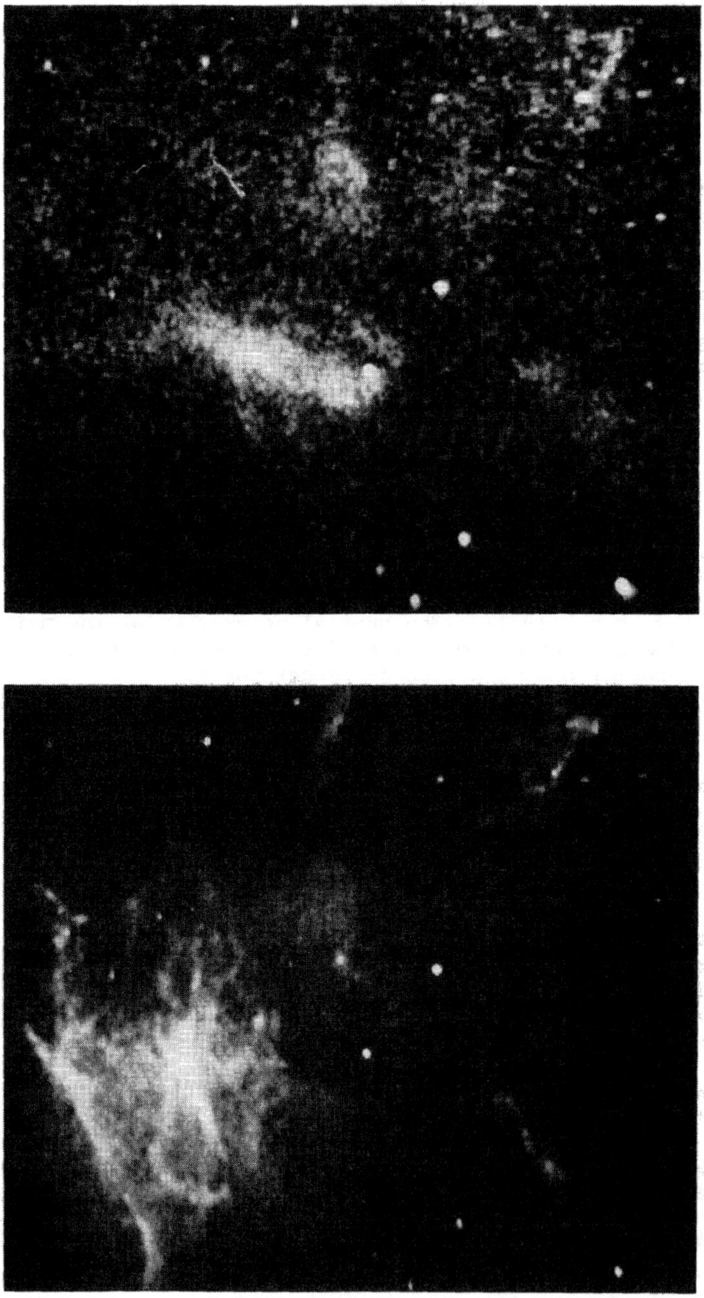

Fig. 3 IPCS direct frames the X-ray bright region of Puppis A in the [OIII]λ5007 Å line (left) and the [FeXIV]λ5303 Å line (right). These lines measure gas at $T_e \sim 3 \times 10^4$ °K respectively and the complete lack of spatial correlation coupled with the more diffuse nature of the [FeXIV] filaments suggests a developed thermal instability in this SNR.

pressure. These could continue to accrete matter by operation of the thermal instability, electron conductivity being damped by the magnetic field.

4. THE HERBIG-HARO OBJECTS

For many years following their discovery by Herbig in 1948 (Herbig 1951) and Haro (1950) the nature of the semi-stellar nebulous objects now known as Herbig-Haro (HH) objects was enigmatic. It was early understood that by the study of these objects we could gain valuable insight into the mechanism of star formation, for they occur in the vicinity of dense clouds of dust and molecules, of reflection and emission nebulosities and in known T-Tauri associations.

Early spectrophotometric studies (Böhm 1956, Osterbrock 1958) recognised their spectral peculiarities, best described by Herbig (1969);

"The H emission lines are strong and [OI] and [SII] are unusually intense. The [NII] lines are also strong, and in those Objects that are not too heavily reddened [OII] (occurs) as well... A number of weaker emission lines (are seen) that are not ordinarily found in appreciable strength in gaseous nebulae: H and K of CaII, the infrared [CaII] lines, MgI λ4571 and lines of [FeII] and [FeIII]."

It is only over the last ten years or so that a growing interest in these objects, coupled with the advent of modern detector technology, has permitted a rapid and systematic accumulation of data.

Do HH objects emit intrinsically or are they just reflection nebulae? The latter viewpoint was advanced principally by the Stroms and co-workers (Strom, Strom and Grasdalen 1974; Strom, Grasdalen and Strom 1974; Strom, Strom and Kinman 1974; Vrba, Strom and Strom 1975; Strom, Strom and Grasdalen 1975). Together these papers paint a picture of enormously obscured protostars, emitting as IR objects around which the veil of dust is occasionally rent allowing a beam of radiation to escape and be scattered thus permitting us "by an uncharacteristic kindness on nature's part...to observe a very early phase of stellar evolution". Whilst this model can explain the high linear polarisation observed in some objects, rapid optical variability (Herbig 1969, Bohm, Perry and Schwartz 1973) and even high velocity dispersions by virtue of multiple scattering (Friedjung 1975), it is of little use in explaining the spectra.

The idea that HH objects may be due to circumstellar shocks was first put forward by Schwartz (1975), in a study of T-Tauri and its environs. He noted that Burnham's nebula, a few arc seconds from the star, had a spectrum very similar to known HH objects and supernova remnants. Later papers have developed on this idea (Böhm, Siegmund and

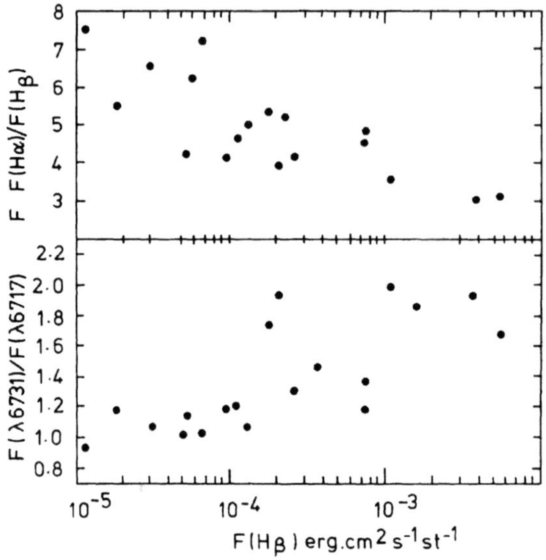

Fig. 4 Correlation of Balmer Decrement and density sensitive [SII] ratio with absolute surface brightness in Hβ for HH objects. High density objects show greater scatter because of greater variation in shock conditions and geometry, whereas a greater scatter in Balmer decrement occurs for low surface brightness objects because these are cooling cloudlets at different temperatures.

Schwartz 1976, Schwartz 1976, Böhm 1977). The virtue of the shock hypothesis is that theoretical shock models can be produced and compared with observation. This was accomplished by Raymond (1976, 79) and by Dopita (1978). Some difficulties are encountered associated with the fact that shock velocities are somewhat low and so shocks are moving into an unionised medium. In this case the low collision cross-section of the neutral atoms and the very rapid cooling due to excitation of Ly-α by electron impact ensures that the thermalisation time scale for the neutrals becomes comparable to, or longer than the cooling time scale of the electron gas. Thus, a distribution function for the neutrals cannot be defined. Both Raymond and Dopita devised ways of overcoming the problem, and in the case of HH1, excellent agreement of theory and observation was found by both authors with shock velocities of order 89-100 km s^{-1} and pre-shock densities in the range 250-350 cm^{-3}.

Dopita (1978) also discovered a correlation, on the basis of extensive spectrophotometry, which gives very strong support to the shock model. He found that the square of the local gas density as indicated by the SII λ6731/λ6717 Å ratio varies very nearly as the reddening corrected surface flux at Hβ. (Fig. 4). This type of relation can only occur if the filling factor of HH nebulae is approximately constant (found to be of order 1%), a natural result of shock waves since these generate a thin compressed radiating layer. However, this result is explicable in the case of other mechanisms or models only if very special geometry or other ad-hoc assumptions are introduced.

Other correlations were found however, which are not so readily explicable; namely an inverse correlation between intrinsic Hα/Hβ ratio and Hβ flux and also, though with larger scatter, between this same ratio and an excitation parameter. The generation of Hα/Hβ ratios larger than about four is difficult in a steady-flow shock. Furthermore, detailed shock models fail to produce [NI] and [OI] lines of such a great intensity as observed in objects of large Hα/Hβ ratios. This problem has been exacerbated by the discovery of even more extreme examples of low excitation HH objects (Böhm, Brugel and Mannery 1980).

In a study of southern HH objects (Schwartz 1977 a,b, Schwartz and Dopita 1980) found two further correlations which at first sight are even more remarkable (Fig. 5).

Using the density given by the red [SII] pressure and deriving from this a shock pressure using the expressions given by McKee and Cowie (1975), they found an inverse correlation between shock pressure

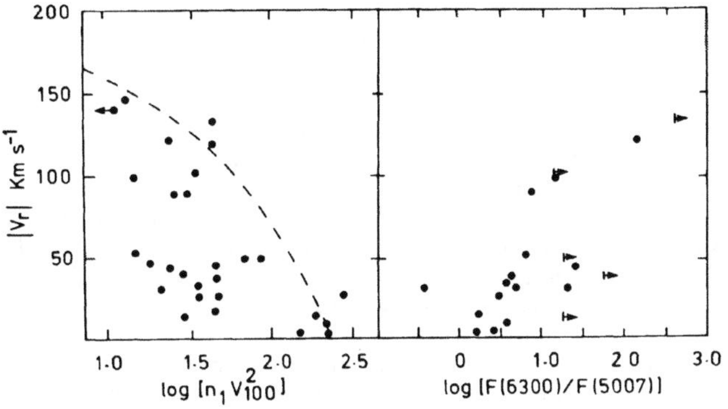

Fig. 5 (Left) The correlation for HH objects between modulus of radial velocity and shock pressure measured in units of cm^{-3} $(10^2$ km $s^{-1})^2$. The dotted line is the upper envelope given for a stellar wind of 200 km s^{-1} (see text). (Right) The correlation of an excitation parameter with modulus of radial velocity. Note low excitation correlates with high radial velocity. After Schwartz and Dopita (1980).

and modulus of the radial velocity. No HH shock driven at high pressure has high radial velocity whereas a substantial fraction of low pressure objects have high velocities. Furthermore, a correlation was found between the modulus of radial velocity and excitation as measured by the parameter log [F(λ6300)/F(λ5007)]. Low-excitation objects have high radial velocities. Indeed, for one very low excitation object, HH29, a fully space motion corrected velocity has been determined to be about 150 km s^{-1} using the proper-motion measurement of Cudworth and Herbig (1979). As both Schwartz and Dopita (1980) and Böhm, Brugel and Mannery (1980) point out, in many cases the radial velocity exceeds the maximum permitted by simple shock models by a generous margin and thus it is impossible that the preshock material is stationary.

Schwartz and Dopita (1980) have developed a model by Schwartz (1978) in which all of these apparently contradictory observational correlations in HH objects may be reconciled. In this, a dense shell around a protostar is broken up into cloudlets as a strong stellar wind pushes out. At early phases cloudlets are slow moving, and the stellar wind ram pressure produces a bow shock, the surface pressure of which drives a strong shock into the cloudlet. This would be the case in, for example, HH1 or HH2. At this phase, a much fainter high radial velocity or high velocity dispersion component of high excitation may be visible as the intercloud medium is shocked and accelerated.

Once the strong shock has propagated through the dense cloudlet, producing a dense flattened pancake of ionised material, it is steadily accelerated outwards by the continued action of the stellar wind, cooling and recombining as it proceeds. Surface pressure and excitation decrease both as the distance from the protostar and the space velocity of the cloud increase, whereas the filling factor of the cloud may also increase. High Balmer Decrements will occur in cooling plasmas in the 10-20000°K range, and it is significant that the high decrements are observed in those low density objects for which evolution (expansion) times become shorter than post-shock recombination times (Dopita 1978). These cooling plasmas would also have very strongly enhanced [NI] and [OI] lines (Dopita, Mason and Robb 1976) in qualitative agreement with what is observed. However, detailed models are required to describe such time-dependent shocks and these are being investigated (Dopita and Schwartz 1981).

Since the radial velocity does not measure the full space velocity of the cloudlets, a true space velocity/pressure relation is defined approximately by the upper envelope of the points in Figure 4. A plausible theoretical relation is shown on the figure as a dotted line, derived for a stellar wind of 200 km s^{-1} and a density in the vicinity of of the shocked cloudlets (assumed constant) of $60/\beta$ cm^{-3}, where β (McKee and Cowie 1975) is the ratio of stagnation pressure in the bow shock to ram pressure in the stellar wind, possibly of order three.

A problem with this model is that the mass-loss rates derived

Fig. 6 Prof. Bart Bok's "Valentine's Night" photograph of HH46 and HH47 (Mercury 8, No. 4, 1978). A T-Tau star is buried in the bright nucleus of the cometary like HH46 and seems to be surrounded by a ring of dark matter. HH47 is approaching at ~ 140 km s^{-1}, and a very faint counter-jet is visible on the right hand edge of the dark cloud. I thank Prof. Bok for supplying the print and permission to reproduce it.

from these wind parameters are rather high, about 10^{-5} $M_\odot$ yr^{-1}. In the few cases where optical candidates exist for the exciting stars these turn out to be of T-Tauri type (Schwartz 1975, Dopita 1978, Cohen and Schwartz 1979), and these stars, which lie close to the deuterium-burning main sequence (Mazzitelli and Moretti 1980), are rarely more luminous than 10 $L_\odot$. Mass loss rates as great as 10^{-5} $M_\odot$ yr^{-1} imply mechanical luminosities of ~100 $L_\odot$. This requirement is relaxed if outflows are non-isotropic. (Schwartz and Dopita 1980). Canto and Rodriguez (1980) have shown that in an undisturbed medium with a density gradient, a stellar wind can be focussed to the region of lowest density. According to their model, the HH object is formed by the cylindrical shock at the convergence of the stream. Kahn (1980) has examined the stability of the shocked stellar wind flow in this model against the Kelvin-Helmholtz instability and shows that the flow will break up only if there is appreciable cooling in the shocked layer. With the Canto and Rodriguez model, the mass loss rate needed to excite HH1 is lowered to 2×10^{-7} $M_\odot$ yr^{-1} and the mechanical luminosity to 0.3 $L_\odot$; a value entirely consistent with those found in 'normal' T-Tauri stars (Kuhi 1964).

Millimetric radio observations by Lada et al. (1974) in the molecular lines of carbon monoxide, carbon monosulphide, formaldehyde and ammonia and the more recent study by Loren et al. (1979) in the formaldehyde and carbon monoxide lines have shown that cloud densities are very variable in the vicinity of HH objects. Maximum densities are usually found closer to the infrared sources associated with HH objects (Strom, Strom and Grasdalen 1974; Strom, Grasdalen and Strom 1974, Elias 1978) rather than at the position of the HH objects themselves. Densities of the cool material range up to nearly 10^5 cm^{-3} and are in general two orders of magnitude higher than those characterising the pre-shock density in the HH object. Such a large difference is probably more indicative of the break-up of an accretion structure around the embedded pre-main sequence star rather than being due to accidental large-scale gradients.

A promising new technique for probing shock conditions in the denser regions close to the stars has been opened up by Elias (1980). He finds vibrationally excited H_2 emission in the 2-2.5 μm band, which can be so strong as to account for most of the broad-band K emission. The $V = 1 \rightarrow 0$ S(0), S(1) S(2), Q(1), Q(3) and Q(5) lines are seen and the $V = 2 \rightarrow 1$ S(1) line is also observed. The surface brightness and line ratios are consistent with shocks of ~15 km s^{-1} moving into material with a density of ~10^4 mol. cm^{-3}. This gives a driving pressure similar to that derived from the optical spectrum.

Data are available in the far UV for only one object, HH1 (Ortolani and D'Odorico 1980). Although an emission spectrum was seen, including OIV λ1406 Å, CIV λ1549 Å and CIII]λ1909 Å lines, the relative intensities did not agree well with shock model calculations. In contradiction with SNR's, the CIV line was much stronger than the models, and this may indicate that the UV emission arises principally from a region of lower absorption outside the optical knot, the shocked intercloud region, for example. However, until a UV satellite is launched which can penetrate deeper than the IUE, little or no progress in this field can be expected.

Given then that the balance of evidence from the line emission is strongly in favour of a shock interpretation, what can be gleaned from the continuum, which is usually rather strong in HH objects? Although some objects such as HH24 show an abnormally strong continuum they also show unusually strong polarisation (Strom, Grasdalen and Strom 1974) and Hα is usually stronger relative to the forbidden lines in these objects. Gyul'budagyan (1975) argued that HH objects should be divided into two classes, those intrinsic emission sources which produce 'normal' HH spectra and extrinsic sources such as HH 24 which are genuine reflection nebulosities. Convincing evidence to support this point of view was presented by Schmidt and Miller (1979). They found that in HH24, the continuum is highly linearly polarised and is due to a reflection nebula whereas the polarisation found in the emission lines is much lower and is consistent with in situ emission. Thus the reflection nebula versus shock excitation point of views are very satisfactorily reconciled. Recent observations by Brugel, Böhm and Mannery (1981) on the blue continua of HH objects disturb this conclusion and suggest that the debate is not yet closed. They find strongly rising continua into the blue, which can be approximately fitted by power law spectra, $F_\lambda \propto \lambda^{-n}$, with $2 \leq n \leq 3$. For all except HH-24, the ratio of continuum flux to Hβ seemed to be constant. Both results cause problems; or if the continuum is due to scattered light, it is much too blue to be scattered T-Tauri starlight. If characterised by a black body temperature, then this must be at least as high as 40000°K. Furthermore, it is difficult to understand how it can remain in constant ratio with Hβ. If local however, what atomic process could give rise to such a blue continuum? In this author's opinion, the best process is likely to be the two-photon continuum of hydrogen. This follows a power law almost exactly in the λ3000-λ6000 Å range with exponent n, as defined above, equal to 3.28. This operates against the bound-free Paschen continuum of hydrogen which has an exponent of roughly -1.1 over the same wavelengths, so a mixture of continua dominant in two-photon emission can have power laws in the same range as those observed. In order to give a Balmer jump of the size seen in HH1, the two photon continuum would have to be enhanced over its recombination-cascade value by a factor of about fourteen. Some collisional enhancement of the two photon continuum had already been noted in the Cygnus Loop SNR (Benvenuti et al. 1981), and the very large enhancements seen in HH objects are probably the signature of shocks moving into a neutral or partially ionised medium since collisional excitation of Ly-α is the dominant post-shock coolant in this case. The UV spectrum (Ortolani and D'Odorico 1980) should peak at about λ1450 Å, but the quality of the data is not sufficient to confirm this.

Further polarisation measurements are required. The anomalous continuum strength and polarisation of HH-24 has already been noted, however, in other objects in which the ratio of Hβ the continuum is constant, the line and continuum polarisations should be similar (and, in general, small). Schmidt and Vrba (1975) have measured low polarisations in HH1 and HH2, which suggest that, in at least these two objects, local emission is the source of the continuum.

Has a causal connection between T-Tauri and HH objects been definitely established? Certainly in the few cases where the exciting stars have been optically identified these turn out to be of T-Tauri type (Herbig 1974, Schwartz 1975, Dopita 1978, Cohen and Schwartz 1979). Also the energy distributions of those IR sources apparently associated with HH objects are comparable with T-Tauri stars (Strom, Grasdalen and Strom 1974). However, discrepant examples may exist. Canto et al. (1981) claim an association of R Monocerotis, NGC 2261 (its reflection nebula) and HH39. Imhoff and Mendoza (1974) estimated a luminosity of 660 $L_{\odot}$ and spectral type A5I for this object. However, it has brightened considerably in the last thirty-five years and Joy (1945) classified it as a Ge T-Tauri variable. Thus, this object may be a T-Tauri star undergoing an FU-Orionis type eruption. Only three FU Orionis variables have been observed extensively; V1057 Cygni, V1515 Cygni and FU Orionis itself (Herbig 1977). However, one of these (V1057 Cygni) was seen to be a T-Tauri star before its outburst.

Larson (1980) has shown that such outbursts can only be caused by transient heating of a thin outer layer of the star. He shows rather convincingly that the secular instability of a (consecutive) rotating star to bar-like deformations could convert enough rotational energy to heating of the outer layers. This will cause matter ejection and the loss of angular momentum to the star. Thus the HH-objects may be excited by episodic rather than continuous mass-loss as suggested by Dopita (1978). Herbig (1977) has argued from the observed number of FU Orionis stars that the phenomenon is recurrent, and must occur at least every 10^4 years in most T-Tauri stars. Since the dynamical lifetimes of most HH objects is of order 10^2-10^3 years, then we might expect one to ten percent of T-Tauri stars to be associated with HH-objects. More complete statistics of both classes of object may resolve this question.

5. SHOCK HEATED GAS IN GALACTIC NUCLEI

Amongst the narrow-line radio galaxies, Seyferts and some nearby elliptical galaxies with emission line nuclei are to be found several with spectral characteristics strongly reminiscent of radiative shock waves. However, very few cases have been found where shock excitation can be unambiguously identified as the emission mechanism. In general, the effects of a strong nuclear source of UV photons has to be taken into account, and many of the observed spectra of emission-line radio galaxies (Osterbrock and Miller 1975, Costero and Osterbrock 1977) are better modelled by the photoionisation of a power law spectrum (MacAlpine 1971) even when there is no clear evidence for a non-thermal continuum in the visible. The principal differences between shock and photoionised models is that the latter have large [OIII]λ5007/λ4363 ratios ratios, the [OIII] lines are frequently stronger and the [OII] lines weaker, respectively than the shock models, and lines of very high excitation such as [NeV], [CaV] and [FeVII] are often present.

The elliptical galaxy NGC1052 is one object in which we can say that shock excited gas has been positively identified. Minkowski and Osterbrock (1959) were first to suggest this might be the case, and Koski and Osterbrock (1976) found that the [OIII] temperature sensitive ratio (λ4363/λ5007 Å) implies an electron temperature greater than 3×10^4 °K, a common characteristic of radiative shocks. Fosbury et al. (1978) produced a detailed shock model and found that a shock velocity of 130 km s^{-1} moving into a medium of density 3 cm^{-3} gave a good account of the observations. Approximately 5×10^5 $M_\odot$ of ionised gas is present. Although a compact (0.14 pc diameter) radio and infrared core object is seen, the presence of a large mass of HI in the galaxy made it difficult to decide whether accretion or expulsion is powering the shocks. The accretion model has been made more attractive by the radio observations of Bottinelli and Gouguenheim (1980). They found a very massive (4.6×10^9 $M_\odot$) HI halo associated with the nearby galaxy NGC1042 which, with NGC1035 and NGC1047, forms a loose cluster. This HI seems to form a bridge between the galaxies, and a broader HI component is associated with NGC1052 itself. The most likely model is therefore that gas accretes from 1042 to 1052, and undergoes an accretion shock about the compact nucleus. The optical emission can be maintained by an accretion rate of 240 $M_\odot$ yr^{-1}. The pressure support for this accretion shock comes from the relativistic gas expelled by the nuclear source and the magnetic field pressure, measured to lie in the range $2\text{-}4 \times 10^{-10}$ dyne cm^{-2} (Fosbury et al. 1978). Presumably the relativistic gas is in turn generated by the settling of some of the accreted gas into an accretion disk about the central massive compact object.

What seems to set NGC1052 apart from many other active galaxies is that very little, if any, ionising radiation escapes from this nuclear object. Recent IUE observations by Penston et al. (1980) show only the upturn at 2000 Å expected from the A-F stellar population, with no evidence for a source of ionising flux. Weak emission lines of CII]λ2326 Å and CIII]λ1909 Å were seen, as predicted by the shock models.

Other possible examples of the NGC1052 type are NGC4278, NGC4105 and NGC5846. Amongst the emission line radio galaxies, PKS2322-12 has a very similar spectrum to NGC1052 (Costero and Osterbrock 1977) and is probably shock excited. Danziger et al. (1979) have observed PKS0521-36, an elliptical with an optically variable continuum sometimes classified as a BL Lac object. This too, despite the strong continuum, has a superficial similarity to NGC1052 in its emission spectrum and a similar quantity, 10^5 $M_\odot$, of ionised material.

Another example of a galaxy with a strong non-thermal continuum in the optical, yet with an emission line spectrum typical of shock excitation is 3C 371 (Miller 1975). This galaxy, like PKS 0521-36 exhibits rapid light variations as well as variable polarisation.

These galaxies seem to suggest that the nuclear activity in NGC1052 is a very mild example of the same phenomenum which powers the BL Lac objects.

The possibility of a continuum of activity from objects principally shock excited up to photon ionised galaxies of much higher emission line luminosities such as narrow line Seyfert galaxies has been suggested by Heckman (1980). He finds that many normal galaxies, particularly those of early Hubble type, are characterised by low luminosity nuclear emission in which [OII] λ3727 Å is at least as strong as [OIII] λ5007 Å and in which [OI] λ6300 Å is at least a third as strong as [OIII] λ5007 Å. These low ionisation nuclear emission-line regions (LINERS) have spectra which can be quite well matched by models involving a shock wave that moves at ~100 km s^{-1} into a gas of near solar composition with densities 10-100 cm^{-3}. Other models either fail to describe the spectrum or can only do so with the introduction of ad-hoc assumptions.

Galaxies with broad line nuclei and QSO's which are best modelled as photoionised by a power law (MacAlpine 1971, 1972, 1974) may well derive these ionising photons by very high velocity cloud-cloud collisions close in to the galactic nucleus as suggested by Daltabuit and Cox (1972). The only serious attempt to model these shocks (V_s ~1000 km s^{-1}) is by Daltabuit, MacAlpine and Cox (1978). Whilst this was in part successful, important discrepancies exist between the computed line and continuum spectra and the observations, which may be corrected by different geometries, velocities, densities, lifting of modelling approximations and treatment of non-equilibrium shocks. This said, it is clear that the prospect for an unambiguous shock model with so many variables is not good. However, a model such as this, which offers the potential to at once explain the photoionising source and the emission line spectrum in terms of a single physical origin is clearly worth further investigation.

A third class of objects which show filamentary emission structure close to the nucleus are galaxies apparently accreting gas from a cooling intra-cluster X-ray emitting reservoir. The best studied examples are NGC1275, a Seyfert galaxy in the Perseus cluster and M87, the giant elliptical at the core of the Virgo cluster. The existence of a massive hot halo in both these galaxies is incontrovertible. M87 is at the centre of an extended X-ray source (Gorenstein et al. 1977, Malina, Lamton and Bowyer 1976, Lawrence 1978) which seems to have a core radius of 12-15 arc min centred on M87 in soft X-rays, but which extends over about 2 degrees about the cluster centroid in harder wavebands. Emission in the 6.7 keV FeXXV line has been detected both in the Perseus and Virgo clusters (Serlemitsos et al. 1977) and the

0.65 KeV OVIII line in M87 (Canizares et al. 1979). The source of this hot gas is presumably ram-pressure stripping of cluster members. Near the massive galaxy, the density of this hot gas is higher and the temperature lower. Mathews (1978) and Mathews and Bregman (1978) find that in the massive potential well of M87, the hot gas is likely to cool in a thermally unstable fashion (Field 1965) and accrete towards the galactic nucleus at rates in the vicinity of 30 $M_\odot$ yr^{-1}. Such gas would be expected to become visible as sheets or filaments, which, since it is cooling and recombining, would emit with a spectrum characteristic of radiative shocks. Arp (1967) discovered nuclear emission in M87, and Ford and Butcher (1979) have made a very complete study of the distribution and spectrum of this emission, and find strong evidence to support the above model. The ionised mass of $\sim 10^4$ $M_\odot$ required is not inconsistent with the accretion rate estimated by Mathews and Bregman. Since optical filaments tend to occur along one side of the nuclear radio-emitting lobes and these are displaced away from the nucleus to the side opposite the filaments, it is tempting to assume that these two phenomena are connected; that the ram pressure of accretion pushes the radio lobes back, and that the optical filaments form at the resultant accretion shock in the same general manner as NGC1052. Here too, the accretion flow is the probable power source of the compact nuclear source and the optical and radio synchrotron jets.

Very similar observational results have been obtained for the NGC1275 low velocity system (Kent and Sargent 1979). This too shows a complex of [SII] strong, velocity broadened filaments. Cooling moderated accretion has been considered by Cowie and Binney (1977) and Fabian and Nulsen (1977). Subsequently, the effects of rotation on this accretion flow were considered (Cowie, Fabian and Nulsen 1980). In this, flows are thermally unstable only for a restricted range of initial parameters. Much of the velocity spread of 700 km s^{-1} (Rubin et al. 1978) is taken to be the result of Keplerian motion, and an accretion flow rate of 200 $M_\odot$ yr^{-1} is estimated.

So far we have seen evidence for shock heated gas accreted from a close encounter between galaxies as in the case of NGC1052 or from a cooling pool of hot gas as in M87 or NGC1275. The dark-lane ellipticals, of which the best-studied example is Cen A (NGC5128), appear to represent the most spectacular type of accretion - the cannibalism of one galaxy by another. The most likely model is one in which an elliptical galaxy has captured a smaller gas and dust rich second galaxy (Baade and Minkowski 1954). This capture is likely to have taken place of order 5 x 10^8 years ago (Tubbs 1980) and the dust is either distributed in a distorted disk, or an elongated tubular structure (Rodgers and Harding 1980).

Since these models allow large amounts of gas to be transported towards the nucleus, a natural source of power is available to drive the optical (Blanco et al. 1975, Dufour and van den Bergh 1978) and radio (Rees 1977) jets via an accretion disk. The non-spherically symmetric gravitational field of the prolate elliptical galaxy could also cause precession of the disk and jets (Ekers et al. 1978).

Shock excited gas has been definitely identified in the nucleus and in a region, knot X, some 30 arc sec south east of the nucleus (Graham 1979, Phillips 1981), and this is associated with a non-circular velocity field.

Peculiar emission line ratios possibly associated with shock heated gas are also seen in the outer NW optical jet, where young stars and HII regions are also observed (Osmer 1978). Within the inner Hα emitting region of this jet is a faint X-ray jet which extends all the way to the bright X-ray nucleus (Schreier et al. 1979). Estimates of the cooling time of this X-ray jet range as low as 10^6 years. This raises the interesting possibility that the NW jet is matter which has cooled in a thermally unstable fashion and is now moving ballistically from the nucleus. Radial velocity measurements indicate that the NW jet is moving at several hundred km s^{-1} towards us, and is therefore on the nearside of the galaxy. No evidence for an optical counter-jet has been found, but the presence of the shock heated knot X in the approximate position of the expected counter-jet may simply indicate that the counter jet is at present suffering a more severe interaction with the dark lane material, and so becomes radiative at a much earlier phase.

I would like to thank Ian Tuohy and Mark Phillips for their careful reading of this manuscript and their valuable suggestions for its improvement.

REFERENCES

Arp, H.C.: 1967, Astrophys. J. Letters, 1, L1.
Baade, W., and Minkowski, R.: 1954, Astrophys. J. 119, 215.
Benvenuti, P., D'Odorico, S., and Dopita, M.A.: 1979, Nature, 277, 99.
, 1980, Astrophys. J. 238, 601.
Benvenuti, P., Dennefeld, M., D'Odorico, S., Dopita, M.A. and Greve, A.: 1981, Astron. Astrophys. (in press).
Bergh, S. van den.: 1973, Publ. Astron. Soc. Pacific, 85, 335.
———: 1976, Astrophys. J. Letters, 208, L17.
———: 1978, Astrophys. J. Suppl. Ser. 38, 119.
Bergh, S. van den and Dodd, W.W.: 1970, Astrophys. J. 162, 485.
Blair, W.P., Kirshner, R.P., and Chevalier, R.A.: 1981, Astrophys. J. (in press).
Blanco, V.M., Graham, J.A., Lasker, B.M., and Osmer, P.S.: 1975, Astrophys. J. Letters, 198, L63.
Böhm, K.H.: 1956, Astrophys. J. 123, 379.
———: 1978, Astron. Astrophys. 64, 115.
Böhm, K.H., Brugel, E.W., and Mannery, E.: 1980, Astrophys. J. Letters, 235, L37.

Böhm, K.H., Perry, J.F., and Schwartz, R.D.: 1973, Astrophys. J. 179, 149.
Böhm, K.H., Siegmund, W., and Schwartz, R.D.: 1976, Astrophys. J. 203, 399.
Bottinelli, L., and Gouguenheim, L.: 1980, Astron. Astrophys. 88, 108.
Brugel, E.W., Bohm, K.H., and Mannery, E.: 1981, Astrophys. J. (in press)
Burginyon, G.A., Hill, P.W., and Seward, F.D.: 1975, Astrophys. J. 200, 163.
Bychkov, K.V.: 1973, Astron. Zh. 50, 907, (Sov. Astron A.J. 17, 577).
Canizares, C.R., Clark, G.W., Markert, T.H., Berg, C., Smedira, M., Bardas, D., Schnopper, H., and Kalata, K.: 1979, Astrophys. J. Letters, 234, L33.
Cantó, J., and Rodriguez, L.F.: 1980, Astrophys. J. 239, 982.
Cantó, J., Rodriguez, L.F., Barral, J.F., and Carral, P.: 1981, Astrophys. J. (Feb. 15)
Chevalier, R.A.: 1976, Astrophys. J. 207, 872.
Chevalier, R.A.: 1977, Ann. Rev. Astron. Astrophys. 15, 175.
Chevalier, R.A., and Kirshner, R.P.: 1978, Astrophys. J. 219, 931.
Chevalier, R.A., Kirshner, R.P., and Raymond, J.C.: 1980, Astrophys. J. 235, 186.
Chevalier, R.A., and Raymond, J.C.: 1978, Astrophys. J. Letters, 225, L27.
Clark, D.H., Murdin, P., Zarnecki, J.C., and Culhane, J.L.: 1979, Monthly Notices Roy. Astron. Soc., 188, 11P
Clark, D.H., and Stephenson, F.R.: 1977, 'The Historical Supernovae', Pergamon Press, ISBN 0-08-020914-9.
Cohen, M., and Schwartz, R.D.: 1979, Astrophys. J. Letters, 233, L77.
Costero, R., and Osterbrock, D.E.: 1977, Astrophys. J. 216, 408.
Cowie, L.L., and Binney, J.: 1977, Astrophys. J. 215, 723.
Cowie, L.L., Fabian, A.C., and Nulsen, P.E.J.: 1980, Monthly Notices Roy. Astron. Soc., 191, 399.
Cox, D.P.: 1972a, Astrophys. J. 178, 143.
Cox, D.P.: 1972b, Astrophys. J. 178, 159.
Cudworth, K.M., and Herbig, G.: 1979, Astron. J. 84, 548.
Daltabuit, E., D'Odorico, S., and Sabbadin, F.: 1976, Astron. Astrophys. 52, 93.
Daltabuit, E., MacAlpine, G.M., and Cox, D.P.: 1978, Astrophys. J. 219, 372.
Danziger, I.J., and Dennefeld, M.: 1976a, Publ. Astron. Soc. Pacific, 88, 44.
——— : 1976b, Astrophys. J. 207, 394.
Danziger, I.J., Dopita, M.A., Elliott, K.H., Goss, W.M., Milne, D.K., and Wilson, I.R.: 1981, Astron. Astrophys. (in press)
Danziger, I.J., Fosbury, R.A.E., Goss, W.M., and Ekers, R.D.: 1979, Monthly Notices Roy. Astron. Soc., 188, 415.
Davison, P.J.N., Culhane, J.L., Mitchell, R.J.: 1976, Astrophys. J. Letters, 206, L37.
Dickel, H.R., Felli, M., D'Odorico, S., and Dopita, M.A.: 1981 (in preparation).
D'Odorico, S.: 1978, Mem. Soc. Astron. Ital. 49, 485.

D'Odorico, S., Benvenuti, P., and Sabbadin, F.: 1978, Astron. Astrophys. 63, 63.
D'Odorico, S., Dopita, M.A., and Benvenuti, P.: 1980, Astron. Astrophys. Suppl. 40, 67.
D'Odorico, S., and Sabbadin, F.: 1976, Astron. Astrophys. 50, 315.
Dopita, M.A.: 1976, Astrophys. J. 209, 395.
————: 1977, Astrophys. J. Suppl. 33, 437.
————: 1978, Astrophys. J. Suppl. 37, 117.
————: 1979, Astrophys. J. Suppl. 40, 455.
Dopita, M.A., D'Odorico, S., and Benvenuti, P.: 1980, Astrophys. J. 236, 628.
————: 1981, in preparation).
Dopita, M.A., Mason, D.J., and Robb, W.D.: 1976, Astrophys. J. 207, 102.
Dopita, M.A., Mathewson, D.S., and Ford, V.L.: 1977, Astrophys. J. 214, 179.
Dopita, M.A., Mathewson, D.S., and Tuohy, I.: 1981, (in preparation).
Dopita, M.A., and Schwartz, R.D.: 1981, (in preparation).
Doroshenko, V.T., and Lozinskaya, T.A.: 1977, Pisma Astron. Zh. 3.
Dufour, R.J.: 1975, Astrophys. J. 195, 315.
Dufour, R.J., and Bergh, S. van den.: 1978, Astrophys. J. Letters, 226, L73.
Dufour, R.J., and Harlow, W.V.: 1977, Astrophys. J. 216, 706.
Edmunds, M.G., and Pagel, B.E.J.: 1978, Monthly Notices Roy. Astron. Soc. 185, 77.
Ekers, R.D., Fanti, R., Lari, C., and Parma, P.: 1978, Nature, 276, 588.
Elias, J.H.: 1980, Astrophys. J. (in press).
Elliott, K.H.: 1978, Mem. Soc. Astron. Ital. 49, 477.
Fabian, A.C., and Nulsen, P.E.J.: 1977, Monthly Notices Roy. Astron. Soc. 180, 479.
Falk, S.W., and Arnett, W.D.: 1977, Astrophys. J. Suppl. 33, 515.
Fesen, R.A., and Kirshner, R.P.: 1981, Astrophys. J. (in press).
Field, G.B.: 1965, Astrophys. J. 142, 531.
Ford, H.C., and Butcher, H.: 1979, Astrophys. J. Suppl. 41, 147.
Friedjung, M.: 1975, Observatory, 95, 51.
Gorenstein, P., Fabricant, D., Topka, K., Tucker, W., and Harnden, Jr., F.R.: 1977, Astrophys. J. Letters, 216, L95.
Gorenstein, P., Harris, B., Guosky, H., Giacconi, R., Novick, R., and van den Bout, P.: 1971, Science, 172, 329.
Goss, W.M., Ekers, R.D., Danziger, I.J., and Israel, F.P.: 1980, Astron. Astrophys.
Goss, W.M., Shaver, P.A., Zealey, W.J., Murdin, P., and Clark, D.H.: 1979, Monthly Notices Roy. Astron. Soc. 188, 357.
Graham, J.A.: 1979, Astrophys. J. 232, 60.
Gull, S.F.: 1973a, Monthly Notices Roy. Astron. Soc. 161, 47.
————: 1973b, Monthly Notices Roy. Astron. Soc. 162, 135.
————: 1975, Monthly Notices Roy. Astron. Soc. 171, 263.
Gulliford, P.: 1974, Astrophys. Space Sci. 31, 241.
Gyul'budagyan, A.L.: 1975, Astrofixika, 11, 511 (Eng. trans. Astrophysics 11, 341).
Haro, G.: 1950, Astron. J. 55, 72.

Heckman, T.M.: 1980, Astron. Astrophys. 87, 152.
Herbig, G.H.: 1951, Astrophys. J. 113, 697.
————: 1969 in 'Non-Stable Phenomena in Variable Stars' ed L. Detre (Dordrecht: Reidel) p.75.
————: 1974, Lick Obs. Bull. No. 658.
————: 1975, Astrophys. J. 2-0, 1.
————: 1977, Astrophys. J. 217, 693.
Imhoff, C.L., and Mendoza, V.E.E.: 1974, Rev. Mex. Astron. Astrof. 1, 25.
Isenberg, P.A.: 1977, Astrophys. J. 217, 597.
Israel, F.P., and van der Kruit, P.C.: 1974, Astron. Astrophys. 32, 363.
Joy, A.H.: 1945, Astrophys. J. 102, 168.
Kahn, F.D.: 1974, I.A.U. Symp. #60, 'Galactic Radio Astronomy' eds. F.J. Kerr and S.C.S. Simonson, D. Reidel; p.329.
————: 1980, Astron. Astrophys. 83, 303.
Kamper, K., and Bergh, S. van den.: 1976, Astrophys. J. Suppl. 32, 351.
————: 1978, Astrophys. J. 224, 851.
Kayat, M.A.: 1978, Mem. Soc. Astron. Ital. 49, 619.
Kent, S.M., and Sargent, W.L.W.: 1979, Astrophys. J. 230, 667.
Kirshner, R.P., and Chevalier, R.A.: 1977, Astrophys. J. 218, 142.
Kirshner, R.P., and Chevalier, R.A.: 1978, Astron. Astrophys. 67, 267.
Kirshner, R.P., and Taylor, K.: 1976, Astrophys. J. Letters, 208, L83.
Kompaneets, A.S.: 1960, Soviet Phys. Dokl. 5, 46.
Kuhi, L.V.: 1964, Astrophys. J. 140, 1409.
Kumar, C.K.: 1976, Publ. Astron. Soc. Pacific, 88, 323.
Lada, C.J., Gottlieb, M.M., Litvak, M.M., and Lilley, A.E.: 1974, Astrophys. J. 194, 609.
Larson, R.D.: 1980, Monthly Notices Roy. Astron Soc. 190, 321.
Lasker, B.M.: 1977, Publ. Astron. Soc. Pacific, 89, 474.
Lasker, B.M.: 1978, Astrophys. J. 223, 109.
————: 1980, Astrophys. J. 237, 765.
Lawrence, A.: 1978, Monthly Notices Roy. Astron. Soc. 185, 423.
Long, K.S., D'Odorico, S., Charles, P.A., and Dopita, M.A.: 1980, Astrophys. J. (in press).
Long, K.S., and Helfand, D.J.: 1979, Astrophys. J. Letters, 234, L77.
Loren, R.B., Evans, N.J.II and Knapp, G.R.: 1979, Astrophys. J. 234, 932.
Lozinskaya, T.A.: 1969, Astron. Zh. 46, 245.
————: 1975, Astron. Zh. 52, 39.
————: 1979, Australian J. Phys. 32, 113.
Lucke, R.L., Zarnecki, J.C., Woodgate, B.E., Culhane, J.E., and Socker, D.G.: 1979, Astrophys. J. 228, 763.
MacAlpine, G.M.: 1971, Ph.D. Thesis - University of Wisconsin.
————: 1972, Astrophys. J. 175, 11.
————: 1974, Astrophys. J. 204, 694.
McCray, R., and Snow, T.P.: 1979, Ann. Rev. Astron. Astrophys. 17, 213
McKee, C.F., and Cowie, L.L.: 1975, Astrophys. J. 195, 715.
McKee, C.F., Cowie, L.L., and Ostriker, J.P.: 1978, Astrophys. J. Letters, 219, L23.
McKee, C.F., and Hollenbach, D.: 1980, Ann. Rev. Astron. Astrophys. 18, 219.

McKee, C.F., and Ostriker, J.P.: 1977, Astrophys. J. 218. 148.
Malina, R., Lampton, M., and Bowyer, S.: 1976, Astrophys. J. 207, 894.
Malina, R., Lampton, M., and Bowyer, S.: 1976, Astrophys. J. 209, 678.
Mathews, W.G.: 1978, Astrophys. J. 219, 413.
Mathews, W.G., and Bregman, S.N.: 1978, Astrophys. J. 219, 413.
Mathewson, D.S., and Clarke, J.N.: 1972, Astrophys. J. Letters, 178, L105.
——————: 1973a, Astrophys. J. 179, 89.
——————: 1973b, Astrophys. J. 180, 725.
——————: 1973c, Astrophys. J. 182, 697.
Mathewson, D.S., Dopita, M.A., Tuohy, I.R., and Ford, V.L.: 1980, Astrophys. J. (in press).
Mazzitelli, I., and Moretti, M.: 1980, Astrophys. J. 235, 955.
Miller, J.S.: 1975, Astrophys. J. Letters, 200, L55.
Minkowski, R., and Osterbrock, D.: 1959, Astrophys. J. 206, 247.
Murdin, P., and Clark, D.H.: 1979, Monthly Notices Roy. Astron. Soc. 189, 501.
Murdin, P., Clark, D.H., and Culhane, J.L.: 1978, Monthly Notices Roy. Astron. Soc. 183, 79p.
Newman, W.I.: 1977, Astrophys. Space Sci. 47, 99.
Oort, J.H.: 1946, Monthly Notices Roy. Astron. Soc. 106, 159.
Ortolani, S., and D'Odorico, S.: 1980, Astron. Astrophys. 83, L8
Osmer, P.S.: 1978, Astrophys. J. Letters, 226, L79.
Osterbrock, D.E.: 1958, Publ. Astron. Soc. Pacific, 70, 399.
Osterbrock, D.E., and Miller, J.S.: 1975, Astrophys. J. 197, 535.
Pagel, B.E.J.: 1980, Paper presented at the NATO Conference on Normal Galaxies Cambridge, August 1980.
Pagel, B.E.J., and Edmunds, M.G.: 1981, Ann. Rev. Astron. Astrophys. 19 (in press).
Pagel, B.E.J., Edmunds, M.G., Fosbury, R.A.E., and Webster, B.L.: 1978, Monthly Notices Roy. Astron. Soc. 184, 569.
Parker, R.A.R.: 1978, Astrophys. J. 224, 873.
Parkes, G.E., Charles, P.A., Culhane, J.L., and Ives, J.C.: 1977, Monthly Notices Roy. Astron. Soc. 179, 55.
Peimbert, M.: 1971, Astrophys. J. 170, 261.
Peimbert, M., and Torres-Peimbert, S.: 1974, Astrophys. J. 193, 327.
Penston, M.V., Fosbury, R.A.E., Snijders, M.A.J., and Boksenburg, A.: 1980, Proc. 2nd IUE Conf. Tubingen, Germany, 26-28 March, ESA SP-157 p.293.
Phillips, M.M. 1981, Monthly Notices Roy. Astron. Soc. (in press).
Pravdo, S.H., and Smith, B.W.: 1979, Astrophys. J. Letters, 234, L195.
Rappaport, S., Doxsey, R., and Solinger, A.: 1974, Astrophys. J. 194, 329.
Raymond, J.C.: 1976, Thesis, Univ. of Wisconsin, Madison.
: 1979, Astrophys. J. Suppl. 39, 1.
Raymond, J.C., Black, J.H., Dupree, A.K., Hartmann, L., and Wolff, R.S.: 1980, Astrophys. J. 238, 881.
Raymond, J.C., Black, J.H., Dupree, A.K., Hartmann, L., and Wolff, R.S.: 1981, Astrophys. J. (in press).
Raymond, J.C., Davis, M., Gull, T.R., and Parker, R.A.R.: 1980, Astrophys. J. Letters, 238, L21.

Rees, M.J.: 1977, Physica Scripta. 17, 193.
Rodgers, A.W., and Harding, P.: 1980, Astrophys. J. Letters, 236, L17.
Rubin, V.C., Ford, W.K., Peterson, C.J., and Lynds, C.R.: 1978, Astrophys. J. Suppl. 37, 235.
Schmidt, G., and Miller, J.S.: 1979, Astrophys. J. Letters, 234, L191.
Schmidt, G.D., and Vrba, F.J.: 1975, Astrophys. J. Letters, 201, L33.
Schreier, E.J., Feigelson, E., Delvaille, J., Giacconi, R., Grindlay, J., and Schwartz, D.A.: 1979, Astrophys. J. Letters, 234, L39.
Schwartz, R.D.: 1975, Astrophys. J. 195, 631.
———: 1976, Publ. Astron. Soc. Pacific 88, 159.
———: 1977a, Astrophys. J. Letters, 212, L25.
———: 1977b, Astrophys. J. Suppl. 35, 161.
———: 1978, Astrophys. J. 223, 884.
Schwartz, R.D., and Dopita, M.A.: 1980, Astrophys. J. 236, 543.
Schweizer, F., and Lasker, B.M.: 1978, Astrophys. J. 220, 167.
Sedov, L.: 1959, 'Similarity and Dimensional Methods in Mechanics' New York, Academic Press.
Serlemitsos, P.J., Smith, B.W., Boldt, E.A., Holt, S.S., and Swank, J.H.: 1977, Astrophys. J. Letters, 211, L63.
Sgro, A.G.: 1975, Astrophys. J. 197, 621.
Shklovsky, I.E.: 1967, Soviet Astron. -A.J. 10, 594.
Shull, M., and McKee, C.F.: 1979, Astrophys. J. 227, 131.
Strom, K.M., Strom, S.E., and Grasdalen, G.L.: 1974, Astrophys. J. 187, 83.
Strom, K.M., Strom, S.E., and Kinman, T.D.: 1974, Astrophys. J. Letters, 191, L93.
Strom, S.E., Grasdalen, G.L., and Strom, K.M.: 1974, Astrophys. J. 191, 111.
Strom, S.E., Strom, K.M., and Grasdalen, G.L.: 1975, Ann. Rev. Astron. Astrophys. 13, 187.
Tammann, G.A.: 1978, Mem. Soc. Astron. Ital. 49, 315.
Taylor, G.I.: 1950, Proc. Roy. Soc. London, A201, 159.
Tubbs, A.D.: 1980, Astrophys. J. 241, 969.
Tuohy, I., Dopita, M.A., Mathewson, D.S., and Garmire, G.: 1981 (in preparation).
van Speybroeck, L., Epstein, A., Forman, W., Giacconi, R., Jones, C., Liller, W., and Smarr, L.: 1979, Astrophys. J. Letters, 234, L45.
Vrba, F.J., Strom, S.E., and Strom, K.M.: 1975, Publ. Astron. Soc. Pacific, 87, 337.
Woodgate, B.E., Angel, J.R.P., and Kirshner, R.P.: 1975, Astrophys. J. 200, 715.
Woodgate, B.E., Kirshner, R.P., and Balon, R.J.: 1977, Astrophys. J. Letters, 218, L129.
Woodgate, B.E., Stockman, H.S., Angel, J.R.P., and Kirshner, R.P.: 1974, Astrophys. J. Letters, 188, L79.

THE YOUNG PHENOMENA WITHIN THE LARGE MAGELLANIC CLOUD

John Meaburn
Department of Astronomy, University of Manchester,
Manchester, England.

ABSTRACT

The most pertinent observations of the very youngest phenomena occurring within the Large Magellanic Cloud are presented. A picture emerges where giant (20-220pc diam.) shells are formed within $\lesssim 10^6$ yr by successive supernova explosions, stellar winds and radiation pressure. These chisel their supergiant counterparts out of the >3kpc sheets of HI within periods of 10^7 yr, by a process of stochastic self-propagating star formation. The massive 30 Doradus nebula and R136 become the nucleus of the young spiral structure formed in this way. Interstellar velocities over a range of 400km s^{-1} within this central complex could be driven predominantly by the stellar wind and radiation pressure of R136 alone, supplemented by all other sources.

1. INTRODUCTION

The Large and Small Magellanic Clouds (LMC and SMC, respectively) have formed at a distance of $\simeq$ 55kpc from the Galaxy within a common envelope of neutral interstellar matter which is connected to the galactic plane by a neutral filament 'the HI stream' (Mathewson et al, 1974). Curiously, the very young ($<10^7$yr old) phenomena within the LMC form a distinct spiral structure with the massive 30 Doradus nebular complex as its nucleus, which is, however, displaced from the rotation centre of that galaxy, whereas the very old ($>10^9$ yr old) stars form a barred structure.

As many new observations have recently been made of all of the phenomena within the young spiral arms and 30 Doradus, an attempt here will be made to present, and analyse the significance of, the most pertinent of these. Some emphasis will be placed on the work of the author and his associates in this field.

F. D. Kahn (ed.), Investigating the Universe, 61–94.

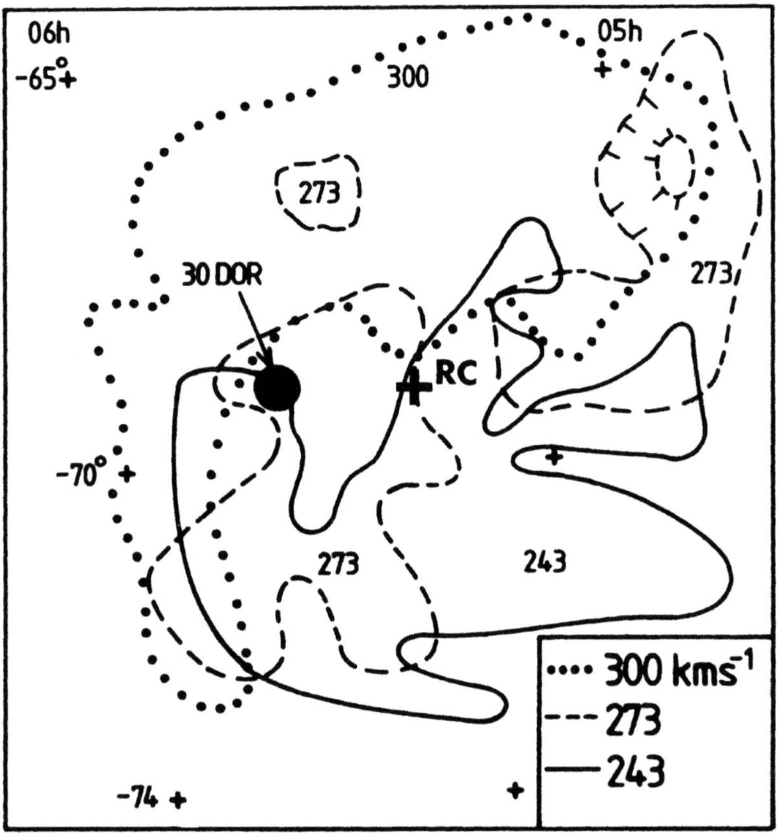

Figure 1 The > 3kpc Sheets of HI. The rotation centre RC and different velocities are shown.

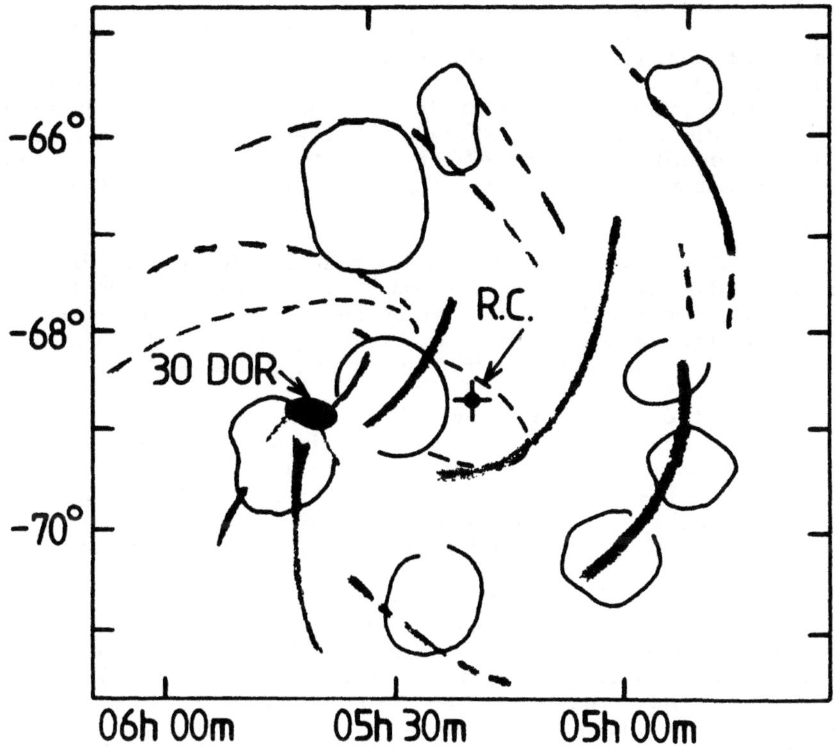

Figure 2 The Young Spiral Arms (shaded) are shown compared to the Supergiant Shells in Figure 3.

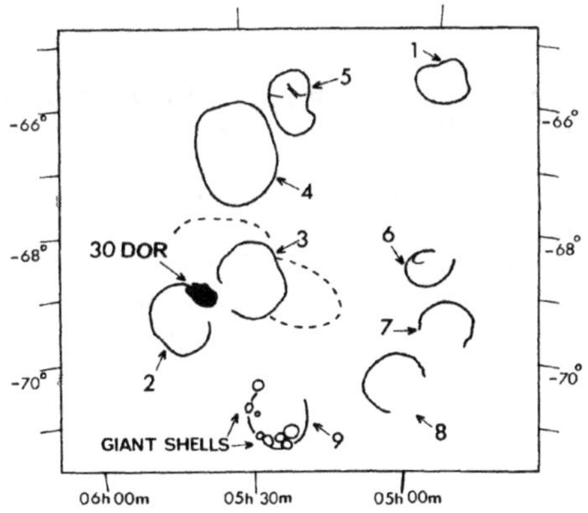

Figure 3 The 9 Supergiant Shells of the LMC.

2. YOUNG SPIRAL ARMS

One of the most curious aspects of the distribution of neutral hydrogen HI within the LMC is summarised in Figure 1. Here the three large-scale (>3kpc across) sheets of HI with distinctly separate heliocentric radial velocities, V_{HEL}, are shown (McGee and Milton, 1966). These overlap in the vicinity of the massive (see Section 4) 30 Doradus nebular complex, well displaced from the rotation centre RC at α = 5h 20m, δ = -69° of the LMC.

Within these sheets of HI many of the young (<10^7yr old) objects that can be used as tracers of spiral arms have formed, for instance, OB associations, dense HII regions, supernova remnants, blue supergiant stars and the giant ($\lesssim$200pc diam.) and supergiant ($\simeq$1000kpc diam.) interstellar shells (of complex origin - see Sections 3 and 4).

In Figure 2 the young spiral arms identified most clearly by Schmidt-Kaler (1979) from the distribution of many of these young phenomena are shown to be emanating again from the region of 30 Doradus, leaving the distinct impression that this is acting as the nucleus of the young LMC. The supergiant shells to be discussed in Section 3 are also shown in Figure 2 to be intimately associated with this large-scale structure.

The deep Hα + [NII] Schmidt plate in Davies, Elliott and Meaburn (1976) helps to substantiate the reality of these spiral features. This was obtained with a 5 hour exposure with the SRC 1.2-m Schmidt through a 38cm square mosaic interference filter (Meaburn, 1976 and 1978) of 100 Å bandwidth. Here the HII regions, supernova remnants and giant shells emitting Hα and [NII] within these arms are particularly apparent. Similarly, the complementary ultra-violet map of Carruthers and Page (1978) from 1250 Å to 1600 Å reveals most clearly the presence of the OB association and blue supergiant stars along these young arms.

3. SUPERGIANT SHELLS

The combination (Garner and Meaburn, 1979) of unsharp masking and high contrast copying (Malin, 1977, 1978) of four 5-hour Hα + [NII] and [SII] plates of the LMC with the 1.2-metre SRC Schmidt has revealed (Meaburn, 1980) at least nine supergiant interstellar shells within the LMC whose parameters are given in Table 1. These have diameters from 600 to 1400pc and are shown numbered in Figure 3 and against the young spiral arms in Figure 2. More may exist, particularly in the vicinity of 30 Doradus where further faint filaments are contained within the dashed line in Figure 3. A high contrast, masked print of the 5-hour Hα +[NII] photograph of SGS LMC 2 and 3 is shown in Figure 4 and of SGS LMC 4 and 5 in Figure 5. In Figure 6 the optical filaments delineating SGS LMC 4 are compared (Goudis and Meaburn, 1978) to contours of the 21cm emission line from HI (McGee and Milton, 1966). It can be seen here how the HII filaments are on the inside edge of a much larger shell of neutral hydrogen whose HI mass is given in Table 1, whereas in Figure 7 the

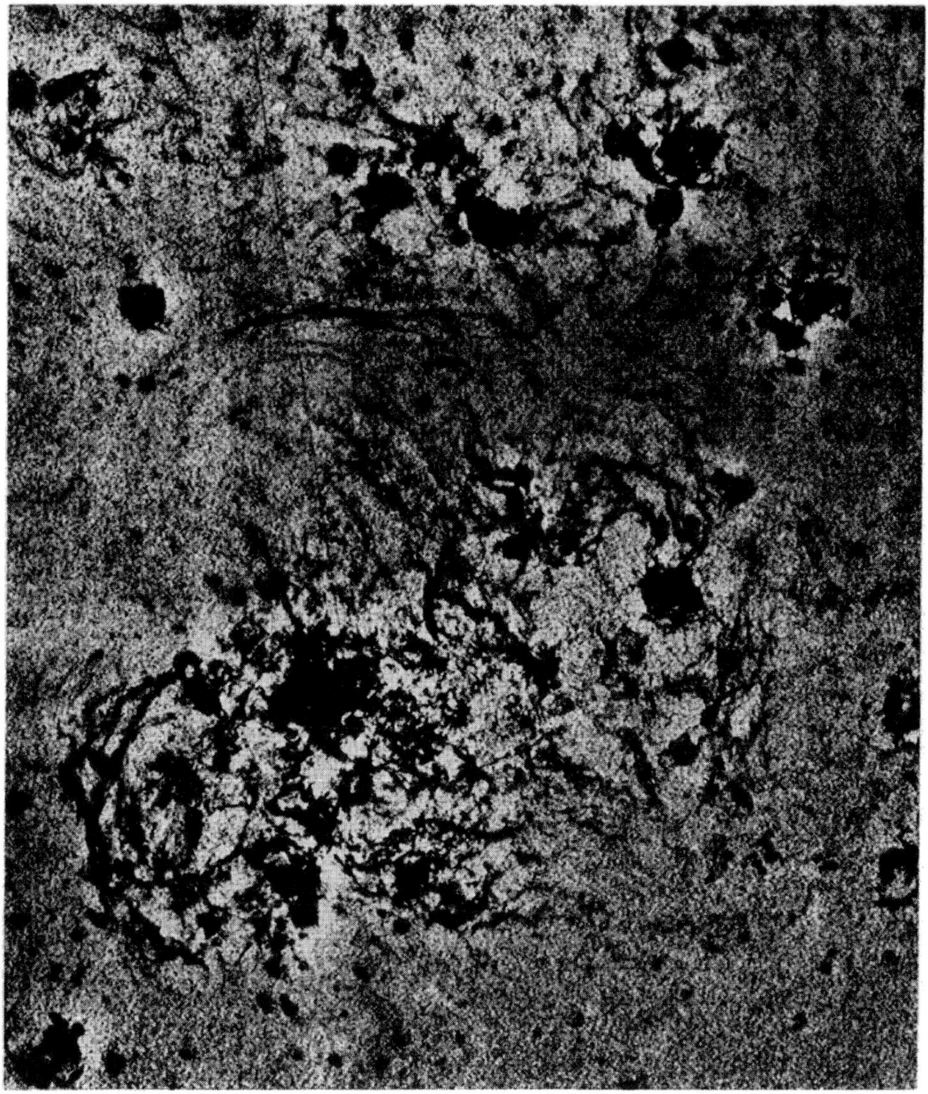

Figure 4 Supergiant shells LMC 2 and 3 (see Figure 3).

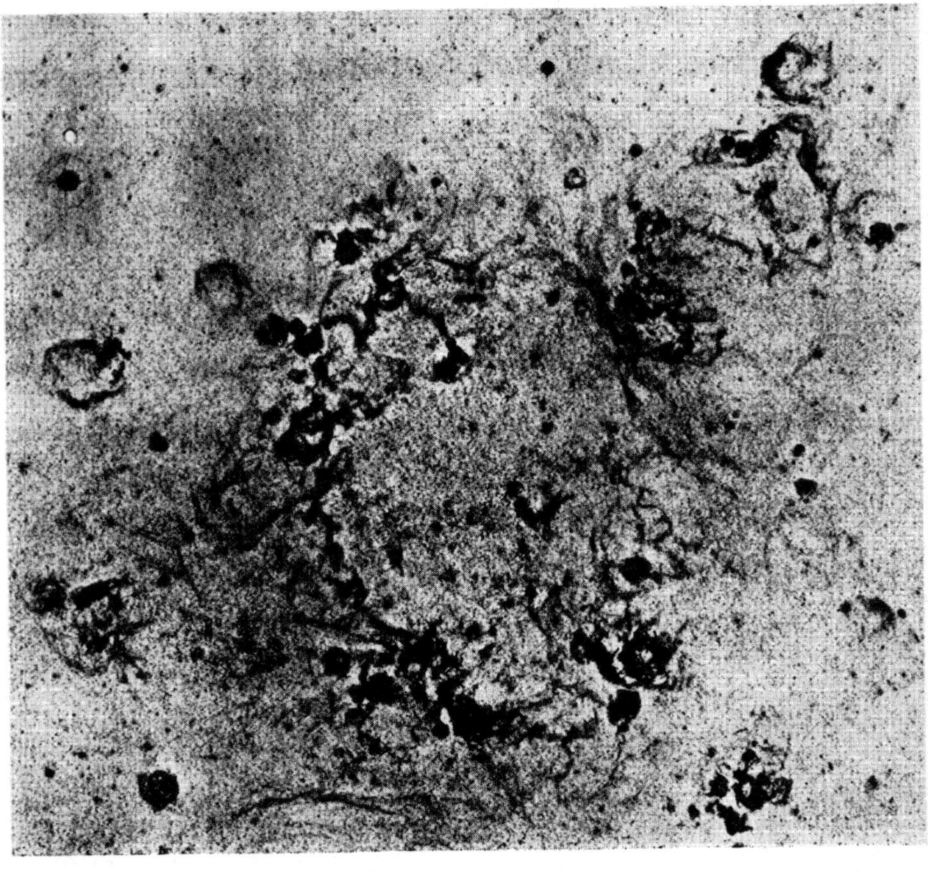

Figure 5 Supergiant Shells LMC 4 and 5 (see Figure 3).

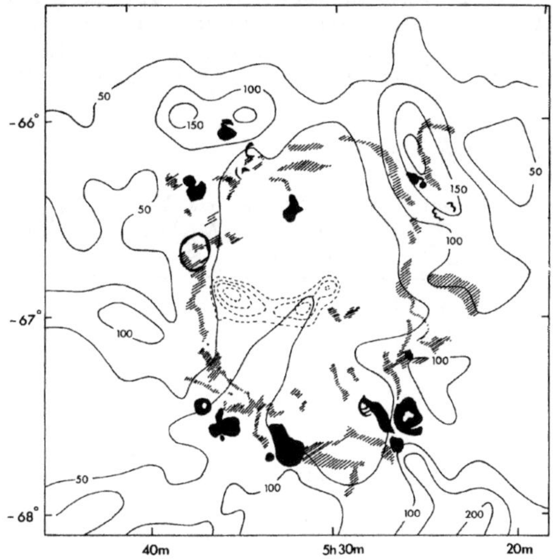

Figure 6 SGS LMC 4 compared to HI central UV contours.

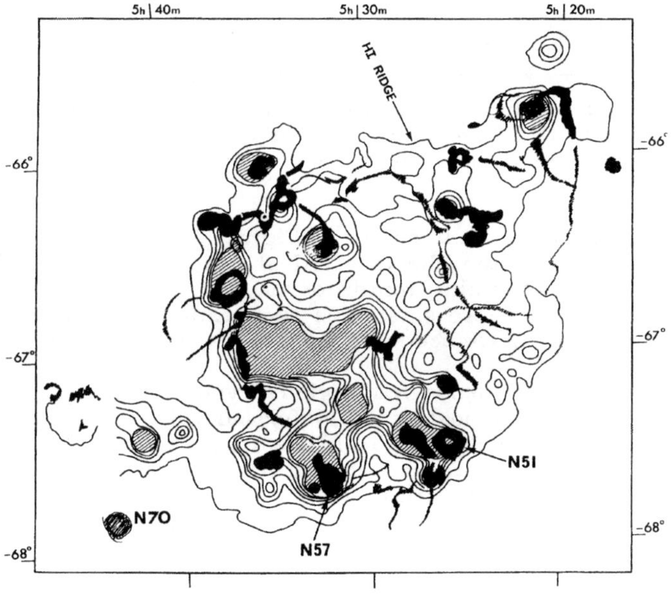

Figure 7 SGS LMC 4 and 5 compared to UV contours.

optical filaments of SGS LMC 4 and 5 are sketched against the contours of the UV emission, from 1250 to 1600 Å, of Page and Carruthers (1978). This is being emitted by the relatively young, hot stars listed in Table 1 (from Lucke and Hodge, 1970 and Lucke, 1974) within the central volumes of SGS LMC 4 and 5 and the most recent stars within their inner circumferences.

Table 1 Supergiant Shells of the LMC

Name	Figure	Dimensions pc	Blue Stars in Centre	HI Mass ($M_\odot$) in Shell	N_{HI} cm^{-3}	Centre (1950)
SGS LMC 1	2 and 3	700	49	-	3	5h00m -65°40'
SGS LMC 2	2, 3 and 4	900	>824	2.5×10^7	-	5h42m -69°30'
SGS LMC 3	2, 3 and 4	1000		-	-	5h30m -69°20'
SGS LMC 4	2, 3 and 5	1400 x 1000	>416	1.7×10^7	0.8	5h31m -66°50'
SGS LMC 5	2, 3 and 5	800	25	-	-	5h22m -66°10'
SGS LMC 6	2 and 3	604	-	-	-	4h58m -68°45'
SGS LMC 7	2 and 3	800	59	-	-	4h54m -69°35'
SGS LMC 8	2 and 3	900	71	-	-	5h02m -70°30'
SGS LMC 9	2 and 3	890	-	-	-	5h25m -71°05'

It is interesting that measurements of V_{HEL} of two of the large HII filaments in SGS LMC 2 (see Figure 4) demonstrate that each exists in the separate >3kpc HI sheets which overlap in the vicinity (shown in Figure 1).

Generally the supergiant shells in Figure 3 have an outer shell of neutral material (within the sheets of HI shown in Figure 1) which surrounds associations containing hundreds of OB and Wolf Rayet stars. The inside surfaces of these shells are then radiatively ionized by these hot stars to produce the ionized filaments apparent in Figures 4 and 5.

Furthermore, many dense HII regions, young (10^3 to 10^5yr old) supernova remnants, the most recently formed OB associations and giant (20-260pc diameter - see Section 3) filamentary shells are also found within these perimeters (see Figures 5 and 6) which must then be the regions of the most recent star formation.

3. GIANT SHELLS

From eight photographs through the author's Hα + [NII] and [SII] interference filters, combined with the SRC 1.2-metre Schmidt, 85 giant (20-260pc diameter) filamentary shells have been identified (Meaburn,

1980) within the LMC. The histograms in Figures 8a and b (Meaburn, 1980) illustrate that these giant shells (GS) are distinctly separate phenomena from their supergiant counterparts.

At least 16 of the giant shells in Figure 8b are strong non-thermal (α between -0.25 and -0.59 where $S \alpha \nu^{\alpha}$) radio continuum sources (Milne et al, 1980). In this case supernova explosions must be dominant in their formation (single or successive explosions - see Sections 5(i) and 5(ii)). The remainder must have a more complex origin for their radiatively ionized filamentary structures surround OB associations and very complex motions occur within them.

Giant shells of all types are sketched in Figures 3, 6, 7 and 12 and can be seen in Figures 4, 5, 9, 10 and 11, whereas, examples which contain significant OB associations are shown in Figures 9, 10, 11 and 12 and their parameters are listed in Table 2. In Figure 9 three $\simeq$60pc diameter filamentary shells can be seen (Meaburn and Walsh, 1981) projecting from a dense neutral ridge where very active (and most recent?) star formation is occurring (Elliott et al, 1978). The principal shell, N44, surrounds an OB association containing $\simeq$40 blue stars (Lucke, 1974) and evidence of a fine neutral absorbing ridge around its outer perimeter is also apparent. N70, shown in Figure 10, is $\simeq$120pc diameter and surrounds 8 blue supergiant stars. However, its radio emission possibly has a significant non-thermal component indicating that supernovae have played some part in its formation. N59 in Figure 11 (see Meaburn, Terrett and Blades, 1981), N51 in Figure 12 (see Meaburn and Terrett, 1980) and N57 in Figures 5, 6 and 7 (see Meaburn and Blades, 1980) are all strongly thermal ($\alpha > -0.2$) radio continuum sources whose filaments are radiatively ionized by the supergiant OB (and Wolf-Rayet?) blue stars listed in Table 2.

The relationship of N51 and N44 (and other giant shells) to the HI distribution is illustrated in Figure 12. These giant shells are shown to exist on either side of a dense HI ridge just resolved by the 14.5' 21cm beamwidth.

The dynamics of N57 (Meaburn and Blades, 1980), N70 (Blades et al, 1980), N51 (Meaburn and Terrett, 1980 and Lasker, 1980) and N59 (Meaburn, Terrett and Blades, 1981) have all been investigated by observations of the profiles of their optical emission lines and interstellar absorption lines in the central stars. A summary of these motions is included in Table 2 and examples of the variations of V_{HEL} of separate velocity components in the emission lines are shown in Figure 13 for the centre of N57 and in Figure 14 for north-south lines across N59.

Simple radially expanding shells produced either by stellar wind bubbles, or single supernova explosions in a reasonably uniform medium are expected to have doubled lines over the centres, systematically changing to single lines on the mean V_{HEL} of the objects at their edges. As is apparent in Table 2 and Figures 13 and 14 (and the references cited here) much more complicated motions are present. For example,

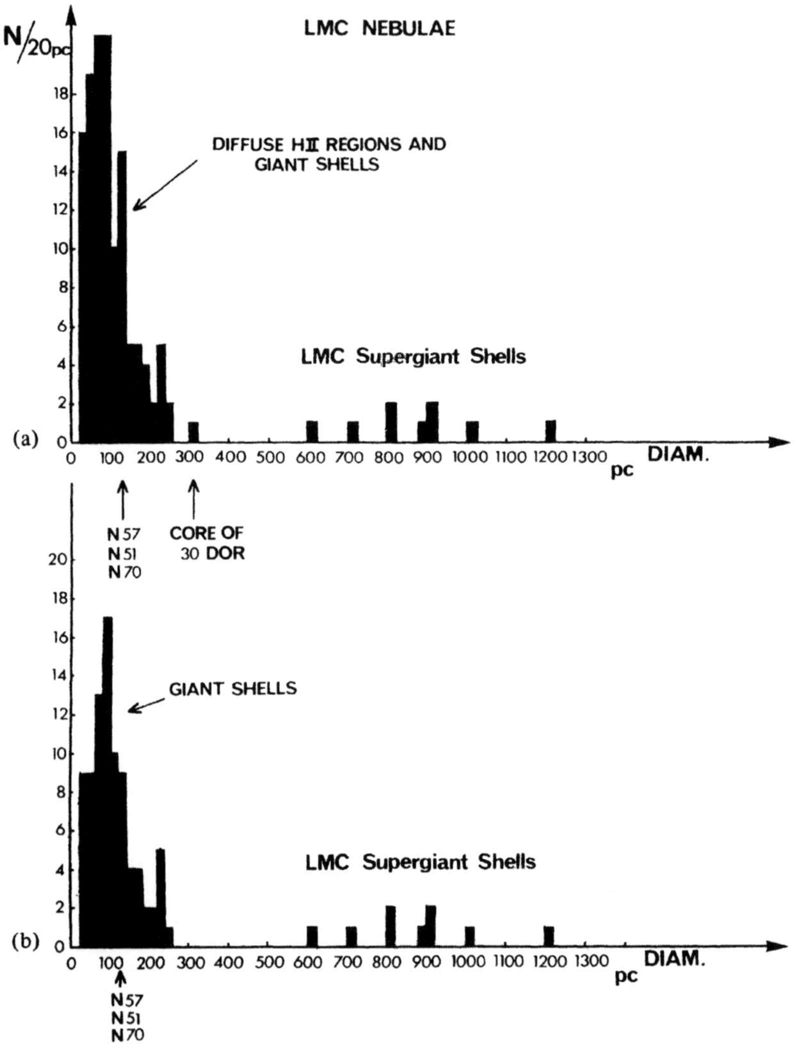

Figure 8 Statistics of nebulae in LMC.

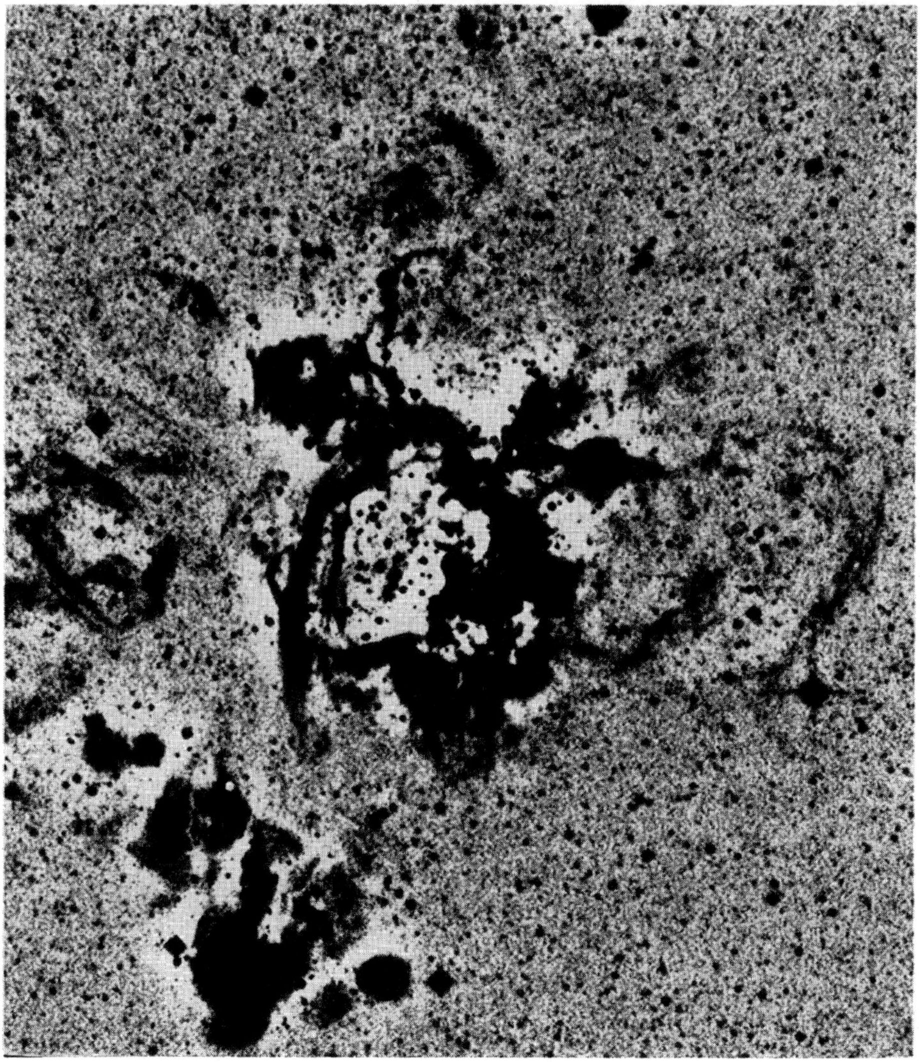

Figure 9 A deep print of N44. Two other 60pc diameter giant shells project from a dense neutral ridge.

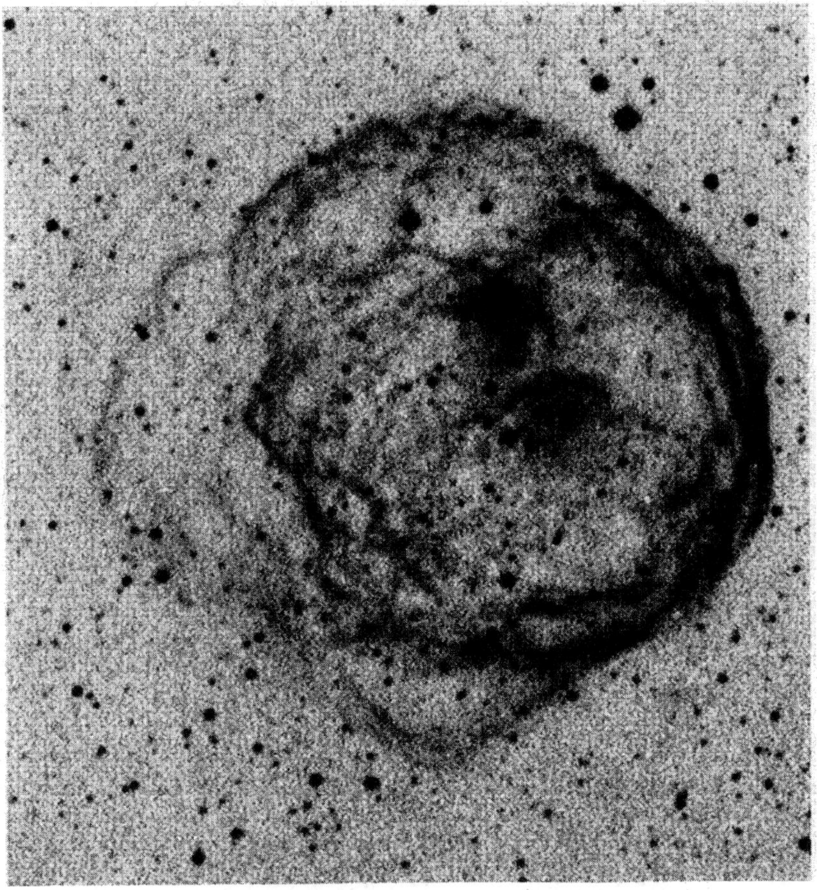

Figure 10 Giant Shell N70.

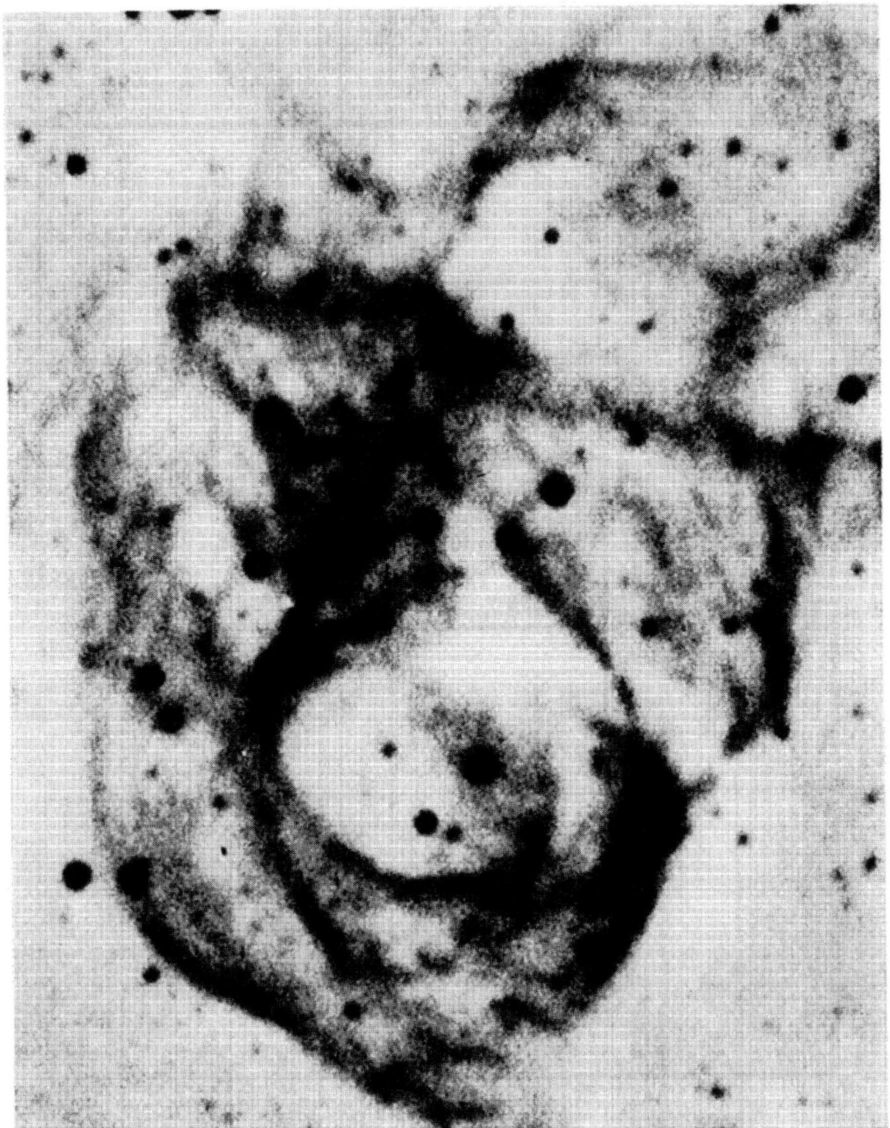

Figure 11 Giant Shell N59.

Table 2 Parameters of Some Giant Shells.

Shell	Radio Source		Dimen-sions	Motions V_{HEL} (km s^{-1})					Central Blue Stars	
Henize (1956) and Davies, Elliott, Meaburn (1970)	$S \alpha \nu^{\alpha}$ name	α	pc	Emission Lines Centre	Edge	Absorp-tion lines Centre	N_{HI} cm^{-3}	>3kpc HI Sheets	Associ-ation	Num-ber
N57 (DEM 229) Figs 5,6,7 & 13	MC57	+0.1	160 x 105	277 294 310 339 342	Faint edge 300 310 320 bright edge 305	273	0.8	246 276 301	LH 76	13
N51 (DEM 741) Figs 5,6,7 & 12	MC46	-0,13	176 x 113	265 325	305	-	0.8	276 302	LH 54	12
N59 (DEM 241) Figs 11 & 14	MC64	0.0	40	255 299 334	Faint edge 288 322 bright edge 310	-	0.8	273 301		9
N70 (DEM 302) Figs 5,6,7 & 10		-0.3	120	310 328	272 to 316 Four components	258 wing to 320	0.8	276 300	LH114	8
N44 (DEM 199) Figs 9 & 12		0.0?	50 (+ 2 of 60)	-	-	-	-	248 272 298	LH 47	40

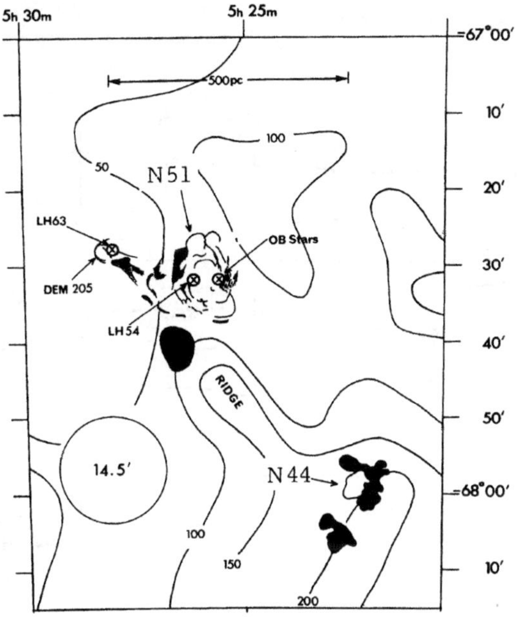

Figure 12 Giant Shells N51 and N44 compared to the HI distribution (14.5' HI beamwidth).

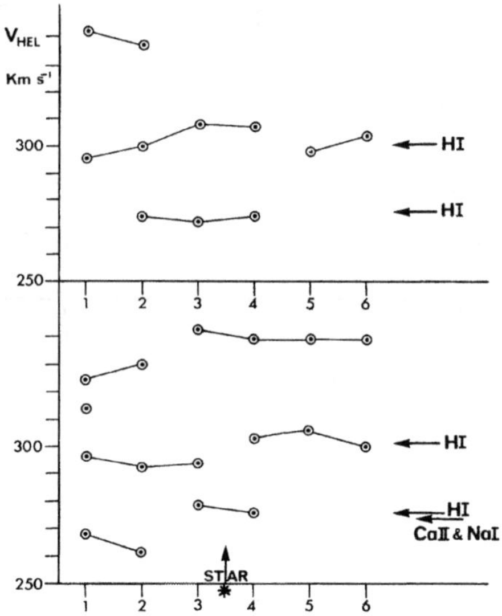

Figure 13 Trends in V_{HEL} of components in [OII] profiles over centre of N57.

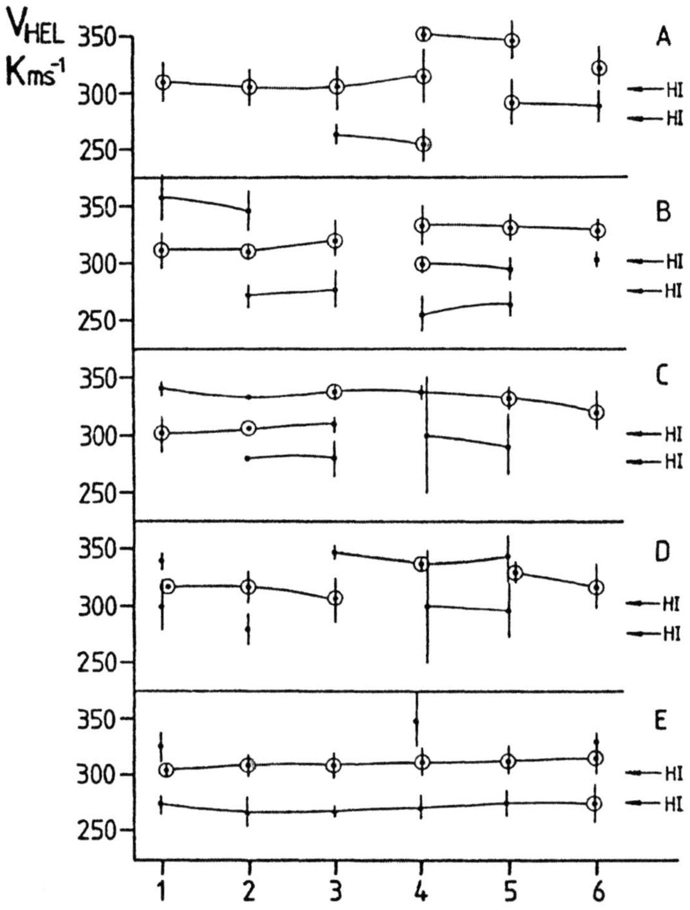

Figure 14 Trends in V_{HEL} of components in [OII] profiles in north-south lines over N59 (see Figure 11).

for N57 five components each with values of V_{HEL} over a range of 277-342 km s^{-1} are found over the nebular centre, and on one edge three components within a range of V_{HEL} from 300-320 km s^{-1} are still present.

Similar results are found for N59, but here there is a wing out to -100 km s^{-1} with respect to the mean V_{HEL} over the central star which implies (wind-driven?) outflowing material (perhaps described by Equation 10 in Section 5(i)). For N51 and N70 some systematic trends could indicate expanding shells, but the profiles of N70 are still split on one edge (each filament in Figure 10 appears to be in a separate >3kpc HI sheet shown in Figure 1).

In Table 2 the radio spectral indices α are taken from Milne et al (1980) and McGee et al (1978).

4. 30 DORADUS - NUCLEUS OF YOUNG LMC

The massive complex of HI, HII, stars, dust and molecules associated with the 30 Doradus nebula, though displaced from the rotation centre, appears in Figures 1, 2 and 3 to be the nucleus of the young (<10^7yr old) spiral arms of the LMC (Schmidt-Kaler and Isserstedt, 1976; Schmidt-Kaler and Feitzinger, 1976 and Schmidt-Kaler, 1976).

A summary of the approximate parameters of the various constituents of this region is given in Table 3 and a general view of the nebula in the light of Hα and [NII] is shown in the 1.2m Schmidt photograph in Figure 15, whereas the filamentary structure of the ionized material of the core is illustrated in Figure 16. These photographs were obtained in the light of [OIII] at the f/8 focus of the Anglo Australian Telescope (Cantó et al, 1980). It is notable that the dense central filaments in Figure 16 are continuous with extremely large low density filaments shown in Figure 4, which delineate supergiant shells LMC 2 and 3 (see Table 1).

Within the central OB association of 30 Doradus is the unique stellar-like object R136 whose bolometric magnitude is brighter than -14^m (Feitzinger et al, 1980). This is shown in the masked print of an optical continuum photograph taken at the f/15 focus of the Anglo Australian Telescope and shown in Figure 17. Here its dominance of the surrounding OB association is clearly apparent.

Israel and Koorneef (1979), and Koorneef and Mathis (1981), from their UV observations emphasize that a large, and possibly dominant, proportion of the ionizing photons (listed in Table 3) for 30 Doradus, originate within this object. If so its total mass >1000 $M_{\odot}$. Walraven and Walraven (1975) and Feitzinger et al (1980) have shown that this concentration of early type stellar mass is within a volume of <0.2pc³.

Furthermore, Cassinelli et al (1981) have suggested that R136 may, in fact, be a single supermassive star of $3x10^3 M_{\odot}$ with a surface temperature of 60 000K, but which can only exist for <10^5yrs. Also, they

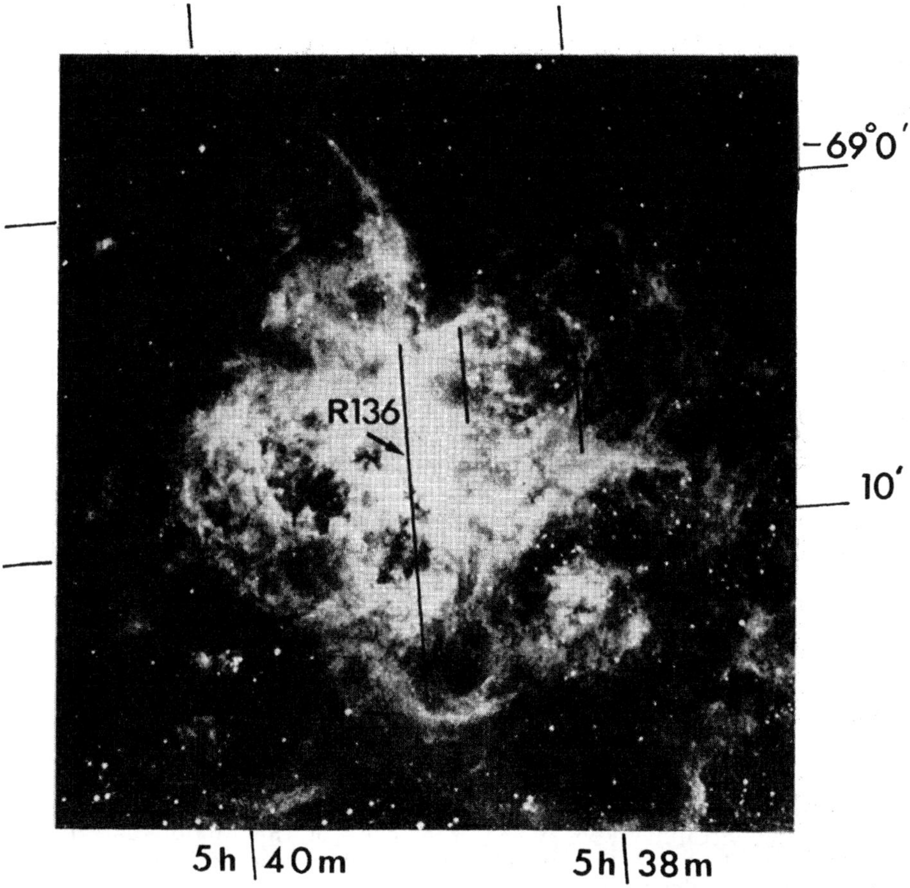

Figure 15 Hα photograph of 30 Doradus. The velocities in Figure 19 were obtained along the longest north-south line.

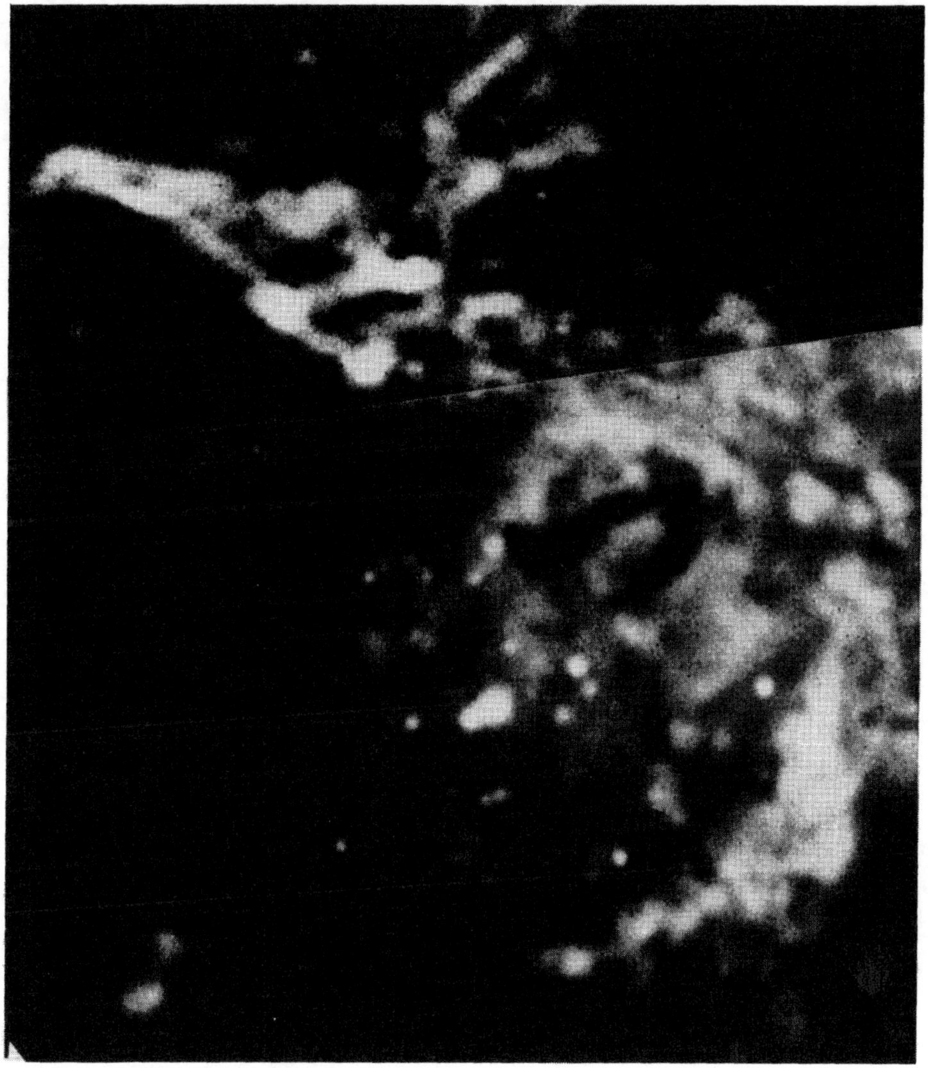

Figure 16 A masked [OIII] photograph of the core of 30 Doradus obtained at the f/8 focus of the AAT.

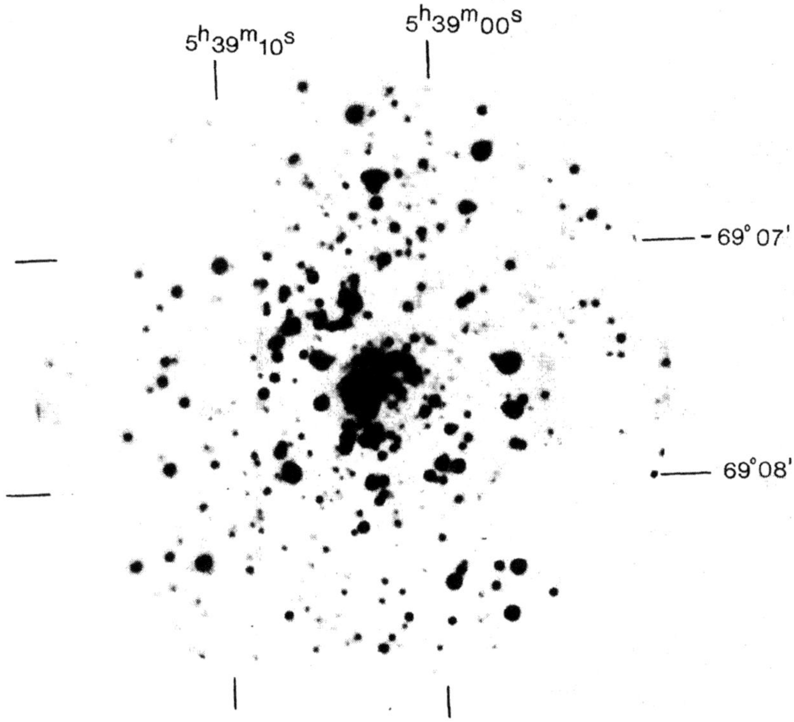

Figure 17 A continuum photograph of the central cluster surrounding R136.

report the detection of an exceptionally energetic ($\sim$3500 km s^{-1}) stellar wind with an expected emission of kinetic energy L_W of $\simeq 10^{39}$erg s^{-1} ($\equiv$ to that from 80-200 O4 stars - Table 3).

This wind, successive supernova explosions and the radiation pressure from R136 and the surrounding OB association (see Figure 17) must have important dynamical consequences to the interstellar material of 30 Doradus (see Section 5(iv)).

Table 3 Parameters of 30 Doradus Complex.

Dimensions (HII Region)	$\simeq$ 200 pc	See Figure 15
Dimensions (HI Region)	$\simeq$ 500 pc	McGee and Milton (1966)
HII Mass	$\simeq 10^5 M_\odot$	Mills et al (1978)
HI Mass	$\simeq 1.4 \times 10^7 M_\odot$	McGee and Milton (1966)
H_2 Mass	?	
Hot Dust Mass	$\simeq 1.4 \times 10^3 M_\odot$	Werner et al (1978)
Total Luminosity of OB Associations	$\simeq$ 0.5 to $1 \times 10^8 L_\odot$	Churchwell (1975) Israel and Koorneef (1979
Total Stellar Mass	$\simeq 4 \times 10^5 M_\odot$	Churchwell (1975)
Ionizing Photons	$\equiv$ 80-200 O4 Stars	Mills et al (1978), Churchwell (1975), Koorneef and Mathis (1980)
Local Electron Densities (central filaments shown in Figure 17)	$\simeq$ 100-1000 cm^{-3}	Canto et al (1980)
Extinction (core) A_v (radio cont/Hβ)	1.89	Israel and Koorneef (1979)

Large motions in the neutral and ionized material have been discovered in the close vicinity of R136. The latest (Meaburn, 1981) detailed comparison of these is given in Table 4. Here the Hα and [OII] components are from Meaburn (1981) and Cantó et al (1980). The separate velocity components in the CaII interstellar absorption line profile in R136 (obtained by the author) (see Cantó et al, 1980 and Blades and Meaburn, 1980) are from a 6 or 7 Gaussian fit by White (1981) to this (shown in Figure 18a). The values for components in the HI (21cm-emission) profile shown in Figure 18b of McGee and Milton (1966) are from a similar fit. The HI (21cm-absorption) are from Radharisan et al (1972) and the HI (21cm-emission) components are for the overlapping >3kpc HI sheets (see Figure 1) of McGee and Milton (1966). The stellar velocity of R136 is from Feast (1961).

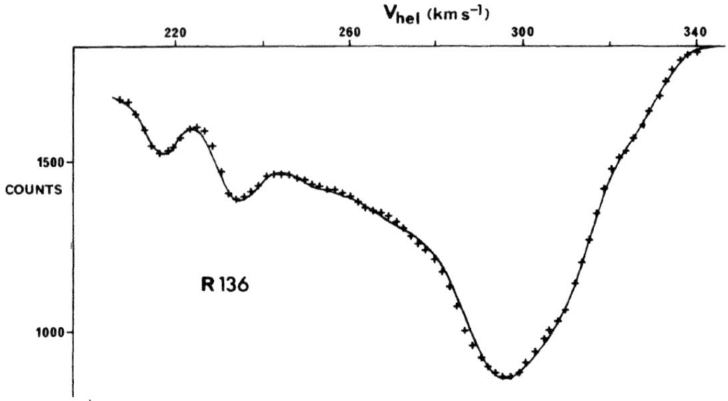

Figure 18a
CaII absorption profile for R136. The solid line is a 7 component fit.

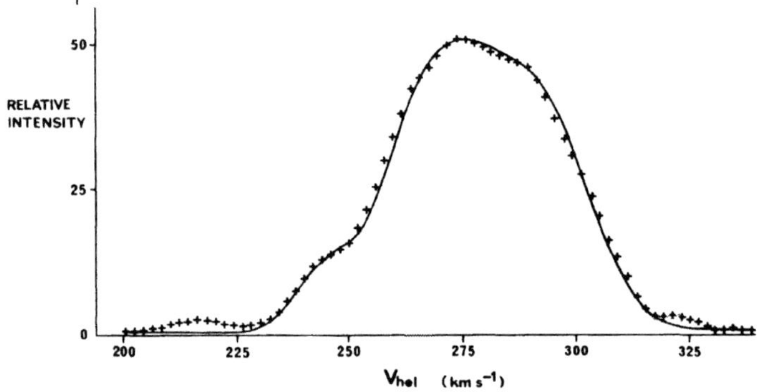

Figure 18b
HI 21cm emission profile over R136. The solid line is a 4 component fit.

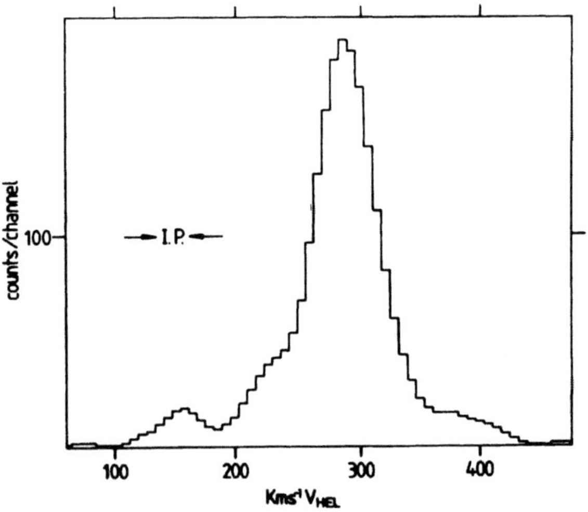

Figure 18c
An Hα profile from δ = -69° 10.5' in Figure 19.

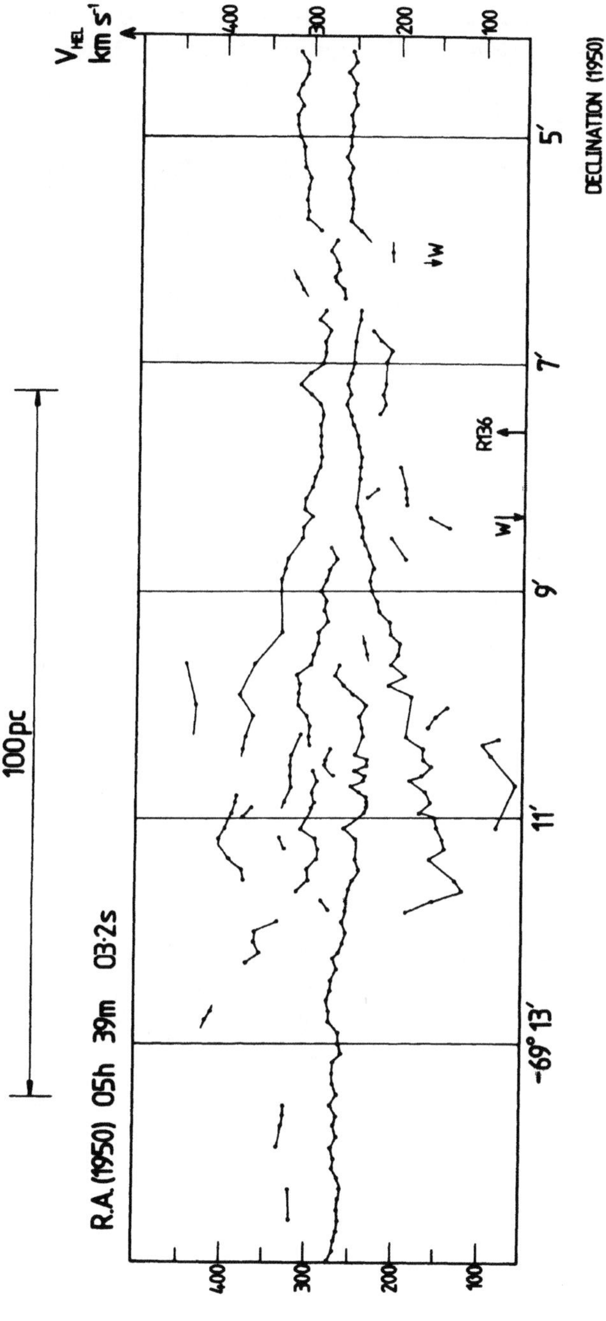

Figure 19 The trends in V_{HEL} for individual velocity components in the Hα profiles along the long line marked in Figure 15.

Table 4 Interstellar Motions in the Vicinity of R136 (dominant components are underlined).

Line	Comments		V_{HEL} (km s^{-1})							
Hα (6563 Å)	within ±30 arcsec of R136 (see Figure 19)		220	242		257		289		
[OII] (3727 Å)	within ±30 arcsec of R136		185	239		265		290		
CaII (3933)	in the light of R136 (see Figure 18a)									
	6 Gaussian fit	wing to 130	219	236		263		298	313	327
	7 Gaussian fit	wing to 130	218	234	250		274	296	311	326
HI (21cm)	in emission with 14.5 arcmin beamwidth on 30 Doradus (see Figure 18b)		217?		246	265	277	296		324?
HI (21cm)	absorption in the emission from the central radio sources in 30 Doradus core	wing to 220				260	280	292?		
HI (21cm)	in emission from the >3kpc sheets which overlap in the vicinity of 30 Doradus (see Figure 1)				248	260	273	298		
Stellar Velocity	R136 (see Figure 17)				251					

It is also important to note that de Boer et al (1980) have found similar very broad interstellar absorption line profiles in the ultra-violet light of R136 even in species with exceptionally high degrees of ionization such as CIV and SiIV.

Finally, the trends in V_{HEL} of distinctly separate velocity components in the Hα profiles (one from $\delta = -69°10.5'$ (in Figure 19) is shown in Figure 18c) obtained with the Anglo Australian Telescope (Meaburn, 1981) are presented in Figure 19. They were obtained along the longest north-south line marked in Figure 15. Extensive sheets of ionized material with distinctly separate values of V_{HEL} over a range of 80-410km s^{-1} are apparent. This is nearly symmetrical to negative and positive velocities when compared to the mean V_{HEL} of the bulk of the ionized material (between 250-300km s^{-1}). It is most important to note that systematic trends in V_{HEL} of individual components are present. For example, two components that have $V_{HEL} \simeq 250$ and 289km s^{-1} respectively over R136 progressively change to 150 and 400km s^{-1} at around $\delta = -69°11'$. Large-scale (>30pc) sheets of ionized material, flowing away from R136, are implied (see Section 5(iv)).

5. DISCUSSION

The observations, summarised here, illustrate the existence of intimate relationships between the creation of giant and supergiant shells, the young nucleus and young spiral arms of the LMC.

However, before considering these in more detail, it is essential to appreciate quantitatively the ways in which the interstellar medium can be disturbed during bursts of star formation.

i) Mechanisms

Newly formed, massive hot stars transmit kinetic energy and linear momentum to the surrounding interstellar medium by supernova explosions, energetic stellar winds and the pressure of their intense electromagnetic radiation. Sedov (1959) has calculated that a supernova explosion of energy E erg, of an isolated star in a uniform ambient cloud of HI density, n_o atoms cm^{-3}, produces an expanding volume of diameter D(t)pc after time t yr, where

$$D(t) = 4.12 \times 10^{-11} E^{0.2} n_o^{-0.2} t^{0.4} \text{ pc} , \quad (1)$$

which expands adiabatically up to a diameter of D_c pc, where,

$$D_c = 6.32 \times 10^{-14} E^{0.29} n_o^{-0.41} \text{ pc} , \quad (2)$$

after which it expands for constant momentum in its snow-plough phase (Chevalier, 1974; McKee and Ostriker, 1977; and also see Reynolds and Ogden, 1979) for $D(t) >> D_c$). Here,

$$E = 3.803 \times 10^{26}\, n_o^{\,1.12}\, V(t)^{4.52}\, t^{3.12}\,\mathrm{erg} \qquad (3)$$

and

$$D(t) = 6.36 \times 10^{-6}\, V(t)\, t\ \mathrm{pc} \quad , \qquad (4)$$

where D(t) and V(t) are the diameter and velocity (km s^{-1}) after time t yr.

This then reaches a maximum diameter D_{max} pc in time t_{max} yr when the internal pressure drops to equal the ambient pressure P_{04} for a medium of temperature T ($\simeq$80K) where from McKee and Ostriker (1977)

$$P_{04} = (n_o T)/10^4 \quad \mathrm{cm}^{-3}\ \mathrm{K} \quad ;$$

then

$$D_{max} = 5.26 \times 10^{-15}\, E^{0.32}\, n_o^{\,-0.16}\, P_{04}^{-0.2}\ \mathrm{pc} \qquad (5)$$

and

$$t_{max} = 1.29 \times 10^{-10}\, E^{0.31}\, n_o^{\,0.27}\, P_{04}^{-0.64}\ \mathrm{yr} \quad . \qquad (6)$$

In the adiabatic stages the isolated supernova remnant is characterized by an intense non-thermal radio source (like the Crab or Cassiopeia) whereas when approaching fossilization (Equations 2, 3 and 4) a non-thermal radio shell develops surrounded by collisionally ionized ([SII]/Hα high) filaments (Cygnus Loop, IC443 and Vela) and with a hot gas (10^6K), emitting X-rays, filling the central volume. When finally fossilized (Equations 5 and 6) the non-thermal radio emission, and collisional ionization, may have both declined, leaving only a nearly stationary shell of HI.

A stellar wind of power L_W erg s^{-1} from an isolated but hot star in a uniform cloud produces a similar expanding spherical shell (Pikelner, 1968; Dyson and de Vries, 1972; Castor et al, 1975; Dyson, 1977) where

$$L_W = 9.5 \times 10^{17}\, n_o\, V(t)^5\, t^2\ \mathrm{erg\ s}^{-1} \qquad (7)$$

and

$$D(t) = 3.5 \times 10^{-6}\, V(t)\, t\ \mathrm{pc} \quad . \qquad (8)$$

This 'bubble' is again full of hot gas (McCray, 1977), but now an emitter of thermal radio emission from the radiatively ionized (HII filament) inside surface of the shell. The radiative ionization dominates any of collisional origin in the shocked wind, etc. (Cantó, 1978).

In many instances, successive supernova explosions are expected (McCray and Snow, 1979) to have occurred in associations containing many OB stars. In this case the expansion of the ideally spherical bubble produced by this succession, each of kinetic energy E , spaced in time by T yr, is described (McCray, 1980 - private communication - and McCray and Snow, 1979) by a substitution of (3.2×10^{-8} E/T) for L_W in Equation 6 so that

$$3.2 \times 10^{-8}\ (E/T) = 9.5 \times 10^{17}\ n_o\ V(t)^5\ t^2\ \mathrm{erg\ s^{-1}} \qquad (9)$$

and Equation 7 remains the same. Here the inside surface of the expanding shell will be radiatively ionized by the remaining OB stars and perhaps the thermal radio source will dominate the non-thermal.

Furthermore, if in addition a strong stellar wind is present, particularly if Wolf-Rayet stars are included (Barlow et al, 1981), then to a reasonable approximation Equation 9 may become

$$3.2 \times 10^{-8}\ (E/T) + L_w = 9.5 \times 10^{17}\ n_o\ V(t)^5\ t^2\ \mathrm{erg\ s^{-1}} \qquad (10)$$

and Equation 6 remains the same.

In reality, complex interactions between the winds and the supernova ejecta are expected and their uneven energies and spacings in time and position are certain to lead to a highly irregular outer shell even in a uniform medium.

It is also relevant that Cantó (1980) has predicted the formation of cavities with flowing walls by the wind of mass loss rate $\dot{M}_*$ from a star which is embedded in the densest region of a neutral cloud where the ambient density decreases rapidly away from this position. To balance the pressure across the standing shock which defines the walls (for $T_e = 10^4$K) in the central regions where centrifugal pressure is unimportant

$$\frac{\dot{M}_*}{(10^{-6}\ M_\odot\ \mathrm{yr}^{-1})} \quad \frac{V_*}{1000\mathrm{km\ s}^{-1}} = \frac{5.2 \times 10^{-2} N_e R\ \mathrm{pc}^2}{\sin^2\theta} \qquad (11)$$

(Cantó, 1980) where θ is the obliquity of the wind at the shock at a distance R from the star and N_e is the electron density in the ionized flowing region. The tangential component of the linear momentum of the wind is transmitted to the flow as it refracts across the shock. If the star is hot, radiative ionization of the flowing region dominates the collisional ionization of the wind (Cantó, 1978).

It is worth considering if successive supernova explosions from a large OB association in similar circumstances (perhaps supplemented by stellar winds) could also produce flowing cavities (this has particular relevance to the discussion on 30 Doradus - see Section 5(iv)).

It is also possible that pressure of the electromagnetic radiation from the central massive OB stars may be important in the production of shell-like cavities, particularly where their inside surfaces are radiatively ionized. For instance, the linear momentum rate $\dot{P}_R$ of the electromagnetic radiation of a single O5 star with luminosity $L_* \simeq 10^6 L_\odot$ (Panagia, 1973) is 1.2×10^{29} gm cm s^{-2} for all wavelengths. This is comparable to $\dot{P}_w$ contained in the energetic particle wind expected from such a star. In this case for $\dot{M}_* = 10^{-5}$ to 10^{-6} $M_\odot$ yr^{-1} with $V_* = 2000$ km s^{-1}

$$\dot{P}_w = 1.3 \times 10^{29} \text{ to } 1.3 \times 10^{28} \text{ gm cm s}^{-2} .$$

Lyman photons transmit all their linear momentum to the ionized gas on the inside of an HI/HII shell, whereas only photons of longer wavelengths transmit theirs to the dust behind the ionization fronts. These particles drift and through collisions transmit their momentum to the larger (30 times) mass of neutral gas. These considerations can now be applied to the young phenomena of the LMC.

ii) Giant Shells

Several of the giant shells included in the histogram of Figure 8b are distinct non-thermal radio sources with surrounding collisionally ionized filaments (Mathewson and Clarke, 1973; Lasker, 1976) and must certainly be supernova remnants of isolated stars similar to the Cygnus Loop, IC443 and Vela in the Galaxy. These are approaching the snowplough phase described by Equations 2, 3 and 4. Furthermore, Mathewson and Clarke (1973) list six very young remnants in their early adiabatic phase whose diameters are <20pc, described by Equation 1.

On the other hand, Chu and Lasker (1980) have identified 8 of the giant shell-like nebulae with Wolf-Rayet stars in the LMC. These are then most likely wind-driven shells, for Barlow, Smith and Willis (1980) have shown that on average for 21 galactic Wolf-Rayet stars $L_w \simeq 5 \times 10^{37}$ erg s^{-1}, but only for the short lifetime of 4×10^5 yr (Chiosi et al, 1978). From Equations 7 and 8 and for $n_o \simeq 0.8$cm^{-3} (see Table 2), a spherical bubble of diameter 74pc will have formed with this wind, and after this time it will be expanding at 53 km s^{-1}. This will be a thermal radio source with radiatively ionized filaments like the galactic Wolf-Rayet nebula NGC 6888 (Johnson and Songsathaporn, 1981).

The majority of the other giant shells in the histogram in Figure 8b have a much more complex structure and origin. The parameters of those listed in Table 2 and whose structure and dynamics are shown in Figures 9 to 14 (inclusive) illustrate that this is certainly the case. In this they resemble the Gum Nebula, Barnards Arc, etc, in the galactic plane.

Consider N57 ideally as a spherical bubble expanding at 40 km s^{-1}. Then, from Equations 7 and 8, a wind of $L_w \simeq 66 \times 10^{36}$erg s^{-1} for $t = 10^6$yr, or from Equations 2, 3 and 4 a single supernova explosion of $E \simeq 3.3 \times 10^{51}$erg with $t \simeq 0.52 \times 10^6$yr could separately produce its present dimensions and internal motions (though not the complex velocity structure shown in Figure 13). Snow and Morton (1976) and McCray (1977) have shown that supergiant OB stars similar to the 13 in N57 each emit $L_w \simeq 1.3 \times 10^{36}$erg s^{-1} for $\lesssim 3 \times 10^6$yr. In which case there is $L_w \simeq 17 \times 10^{36}$erg s^{-1} available from this source (though one Wolf-Rayet would supplement this considerably). Also if all the giant shells in Figure 8b (and the 6 supernova remnants up to 20pc) are considered to be the consequence of supernova explosions, then the mean time between explosions $T \simeq 8300$ yr for the LMC (Meaburn, 1980). This is seventeen times the value of $T \simeq 500$ yr given by Mathewson and Clarke (1973) for the five adiabatic remnants (Equation 1) up to 10pc diameter where they assumed $E \simeq 10^{50}$ erg.

One explanaton of this discrepancy may be that many $E \simeq 10^{50}$ erg (McCray and Snow, 1978) successive explosions could have occurred within the OB associations within the majority of the giant shells (and the 30 Doradus Nebular complex). For instance, if Equations 9 and 8 are now applied to N57 which is approximated (Table 2) by an expanding bubble 160 x 105pc across, with $n_o \simeq 0.8$ cm^{-3} expanding at 40km s^{-1}, then $E/T \simeq 66 \times 10^{36}$ erg s^{-1} and $t \simeq 10^6$ yr; in which case for 17 explosions evenly spaced in time ($T \simeq 5.9 \times 10^4$ yr) each would have, on average, a value of $E \simeq 1.2 \times 10^{50}$ erg.

Many of the giant shells, which contain large OB associations, could then be formed by successive supernova explosions (as suggested by Reynolds and Ogden (1979) for Barnard's Arc, the Gum Nebula, etc.), supplemented by stellar winds and radiation pressure. The rapid compressions of the interstellar medium in their perimeters (by all these sources?) progressively trigger further bursts of star formation as the giant shell expands. In this way they 'chisel' larger supergiant structures (Section 5(iii)) out of the >3kpc sheets of HI shown in Figure 1.

Many giant shells may become cylindrical rings (rather than spherical bubbles) in two or more overlapping large HI sheets. This would explain the separate values of V_{HEL} on their edges (Table 2).

It is plausible to suggest that the three adjacent giant shells (one of which is N44) shown in Figure 9 have expanded until in one place their edges have collided. Here a dense ridge of HI has formed, stimulating the very active and most recent star formations (see Elliott et al, 1976).

iii) Supergiant Shells

It is demonstrably implausible (Meaburn, 1980) to suggest that the supergiant shells, shown in Figures 3 to 7 inclusive, could be approximately spherical remnants of individual supernova explosions or wind-driven bubbles. For example, even from Equations 5 and 6, for a fossilized remnant, single supernova explosions with $E \simeq 10^{53}$ erg with $t_{max} \simeq 5 \times 10^7$ yr are required. In any case, the >3kpc sheets of HI shown in Figure 1 are probably only $\lesssim$200pc thick.

The supergiant shells are then cylindrical (rather than spherical) in structure and again a mechanism which involves chiselling is strongly suggested. In this case giant shells (each formed by multiple events) occur in their expanding perimeters (see Figures 5, 6 and 7 for SGS

LMC 4 and 5). The process of stochastic, self-propagating star formation described by Gerola and Seiden (1978), Meaburn (1980) and Seiden (1981 - private communication) is very strongly substantiated by this evidence. This appears to be occurring in two distinctly separate 'cell' sizes ($\lesssim$220pc and $\simeq$1kpc) as shown in the histogram in Figure 8b.

Consider SGS LMC 4 shown in Figures 5, 6 and 7 within this interpretation. The most recent bursts of star formation in the inside edge of the neutral shell (Figure 6) are producing giant shells of $\lesssim$220pc diameter and of $\lesssim 10^6$yr old (from the parameters in Table 2 in Equations 4 and 8 as for N57) which have a total area of 9.5 percent of the whole area of SGS LMC 4. Then, at a constant rate of production of giant shells, SGS LMC 4 will have formed by this chiselling mechanism in $\simeq 10^7$ yr. The central OB stars which give rise to the UV emission in Figure 7 will then be from 10^6 - 10^7yr old within this argument and those around the perimeter $\lesssim 10^6$yr old.

iv) 30 Doradus and the Young Spiral Arms

Again within this interpretation the 30 Doradus complex has formed as the edges of several expanding supergiant shells, in separate >3kpc across sheets of HI (Figure 1), have collided as shown schematically in Figure 20. In this exceptionally dense and massive region (Table 3) extremely active star formation then occurred.

The localised ($\lesssim$100pc) motions shown in Figure 19 produced by the wind ($L_w \simeq 10^{39}$ erg s^{-1} in $\lesssim 10^5$yr with total linear momentum $P_w \lesssim 2\times10^{43}$ gm cm s^{-1}) and radiation pressure of Lyman photons ($P_R \simeq 4\times10^{43}$ gm cm s^{-1}) of R136, but supplemented by the successive 10^{50} erg explosions, winds and radiation pressure from the 29 (Feast, 1961 and Breysacher, 1980) Wolf-Rayet and multitude of OB stars, are superimposed on the motions of the >3kpc HI clouds (Figure 1 and Table 4) which existed prior to the formation of 30 Doradus.

At least one cavity (>30pc across) with flowing walls (see Figure 19 and Section 4) driven by all these sources in the densest regions near R136 (and perhaps described by the mechanisms which lead to Equation 11) is present. Incidentally, the high velocity CIV and SiIV interstellar absorption lines found by de Boer et al (1980) in the light of R136 (see Blades and Meaburn, 1980, for comparison with other motions) require the presence of a gas of 10^5K. De Boer et al (1980) therefore suggest that they originate in a hot halo of the LMC. Alternatively, the radiatively ionized ($T_e \simeq 10^4$K) high velocity sheets within 30 Doradus, which are revealed in Figure 19 may each have a collisionally ionized, high temperature ($T_e \simeq 10^5$K) component produced by successive supernova shocks.

As the cylindrical supergiant shells expand by the chiselling process towards the scale of the LMC they will be distorted by the overall differential rotation of that galaxy. Seiden (1981 - private communication) and Feitzinger et al (1981) have now shown most convincingly in a computer simulation that the continuation of this process of stochastic, self-prop-

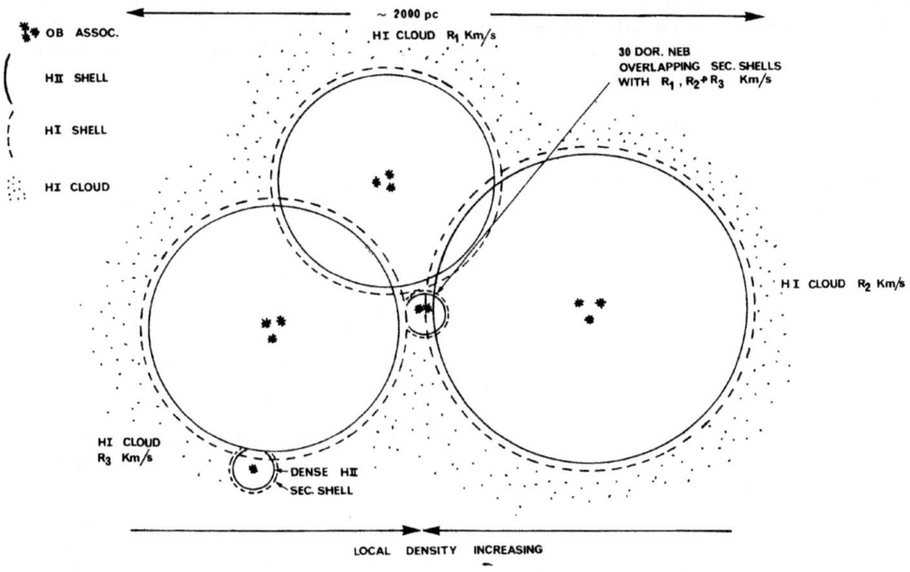

Figure 20 A schematic representation of the formation of supergiant shells in separate HI sheets, overlapping to form 30 Doradus.

agating, star formation, which first forms giant shells which proceed to chisel supergiant ones, can produce young spiral arms similar to those shown in Figure 2.

The displacement of 30 Doradus from the rotation centre (RC in Figure 2) is naturally (Seiden, 1981 - private communication) explained by this process. Here the shear (differential rotation) of the interstellar material is largest and the mechanisms causing star formation are replenished by new raw material at the maximum rate.

v) General Comments

A large number of questions concerning the young phenomena of the LMC remain unresolved. For instance, how do the separate $\gtrsim$3kpc HI sheets, shown in Figure 1, form?

Also, it will be most interesting to know whether R136 is a single 3500 $M_\odot$ star, or a compact group ($\equiv$ 100 O4 stars) - which is still most remarkable. In either case it may be comparable with the compact central 0.1pc diameter volume of our Galaxy where Oort (1971) gives a density of $2.6 \times 10^7\ M_\odot\ pc^{-3}$ with a mass of $3 \times 10^5\ M_\odot$.

The high angular resolution of the forthcoming space telescope will be required to attempt to resolve the angular extent to R136 or even the 'speckle' technique could be tried.

It is perhaps relevant that the spread in V_{HEL} of 400km s^{-1} shown in Figures 18c and 19 over 30 Doradus is comparable with the $\lesssim$500km s^{-1} range of motions found in Seyfert type 2 nuclei, as classified by Khachikian and Weedman (1971 and 1974).

The possibility of driving large cavities with flowing ionized walls, by stellar winds combined with successive supernova explosions and intense radiation pressure also requires exploring theoretically, as does the structure (and emissions from) a radiatively ionized HII/HI (mol?) interface which is also being subjected to supernova shocks.

ACKNOWLEDGMENTS

I joined the Astronomy Department in the University of Manchester as Professor Kopal's research student in 1961. Since then I have carried out my research, which has involved frequent visits to overseas observatories, from this firm base. I will always be grateful for the very lively atmosphere that Professor Zdenek Kopal generated within his Department and from which I have benefited so obviously. It has always been a great help to me in more recent years to know that he was in the background with his constant wisdom and support. I am hopeful that this will continue through his forthcoming long and happy Emeritus Professorship.

REFERENCES

Barlow, M. J., Smith, L. J. and Willis, A. J.: 1981, Mon. Not. Roy. Astron. Soc. (in press)
Blades, J. C., Elliott, K. H. and Meaburn, J.: 1980, Mon. Not. Roy. Astron. Soc., 192, p.101.
Blades, J. C. and Meaburn, J.: 1980, Mon. Not. Roy. Astron. Soc., 190, p.59.
Breysacher, J.: 1981, Astron. Astrophys. Suppl.,(in press).
de Boer, K. S., Koorneef, J. and Savage, B. D.: 1980, Astrophys. J., 236, p.769.
Cantó, J.: 1978, Astron. Astrophys., 70, p.111.
Cantó, J.: 1980, Astron. Astrophys., 86, p.379.
Cantó, J., Johnson, P. G., Meaburn, J., Mikhail, J. S., Terrett, D. L. and White, N. J.: 1979, Mon. Not. Roy. Astron. Soc., 187, p.673.
Cantó, J., Elliott, K. H., Goudis, C., Johnson, P. G., Mason, D. and Meaburn, J.: 1980, Astron. Astrophys., 84, p.167.
Cassinelli, J., Mathis, J. and Savage, B.: 1981, presented at American Astron. Soc. Meeting, Albuquerque, New Mexico.
Castor, J., McCray, R. and Weaver, R.: 1975, Astrophys. J., 200, L107.
Chevalier, R. A.: 1974, Astrophys. J., 188, p. 501.
Churchwell, E.: 1975, *HII Regions and Related Topics*, Springer Verlag, p.245.
Chiosi, C., Nasi, E. and Sreenivan, S. R.: 1978, Astron. Astrophys., 63, p.103.
Chu, Y. H. and Lasker, B. M.: 1981, Astrophys. J.,
Davies, R. D., Elliott, K. H. and Meaburn, J.: 1976, Mem. Roy. Astron. Soc., 81, p.89.
Dyson, J. E.: 1977, Astron. Astrophys., 59, p.161.
Dyson, J. E. and de Vries, J.: 1972, Astron. Astrophys., 20, p.233.
Elliott, K. H., Goudis, C., Meaburn, J., Pilkington, J. and Tebbutt, N. J.: 1976, Astrophys. Space Sci., 55, p.475.
Feast, M. W.: 1961, Mon. Not. Roy. Astron. Soc., 122, p.1.
Feitzinger, J. V., Schlosser, W., Schmidt-Kaler, Th. and Winkler, Chr.: (1980, Astron. Astrophys., 84, p.50.
Feitzinger, J. V., Glassgold, A. E., Gerola, H. and Seiden, P. E.: 1981, Astron. Astrophys. (in press).
Garner, W. V. and Meabrun, J.: 1979, A.A.S. Photo Bull., No. 20, p.3.
Gerola, H. and Seiden, P. E.: 1978, Astrophys. J., 223, p.129.
Goudis, C. and Meaburn, J.: 1976, Astron. Astrophys., 51, p.401.
Israel, F. P. and Koorneef, J.: 1979, Astrophys. J., 230, p.390.
Johnson, P. G. and Songsathaporn, S.: 1981, Mon. Not. Roy. Astron. Soc. (in press).
Khachikian, E. Y. and Weedman, D. W.: 1971, Astrofizika, 7, p.389.
Khachikian, E. Y. and Weedman, D. W.: 1974, Astrophys. J., 192, p.581.
Koorneef, J. and Mathis, J. S.: 1981, Astrophys. J. (in press).
Lasker, B. M.: 1976, in *The Galaxy and Local Group*, RGO Bull., 182, p.185.
Lasker, B. M.: 1980, Astrophys. J., 239, p.65.
Lucke, P. B. and Hodge, P. W.: 1970, Astrophys. J. Suppl., 75, p.171.
Lucke, P. B.: 1974, Astrophys. J. Suppl., 28, p.73

Malin, D. F.: 1978, A.A.S. Photo Bull., No. 16, p.10.
Malin, D. F.: 1978a, in *Modern Techniques in Astronomical Photography*, eds. West and Hendier, ESO, p.181.
Mathewson, D. S. and Clarke, J. N.: 1973, Astrophys. J., 180, p.725.
Mathewson, D. S., Clearly, M. N. and Murray, J. D.: 1974, Astrophys. J., 190, p.291.

McCray, R.: 1977, in *Topics in Interstellar Matter* (ed. H. van Woerden), ASSL Vol. 70, D. Reidel Publ. Co., Dordrecht, p.35.
McCray, R. and Snow, T. P.: 1979, Ann. Rev. Astron. Astrophys., 17, p.213.
McGee, R. X. and Milton, J. A.: 1966, Aust. J. Phys., 19, p.345.
McGee, R. X., Newton, L. M. and Butler, P. W.: 1978, Mon. Not. Roy. Astron. Soc., 183, p.799.
McKee, C. F. and Ostriker, J. P.: 1977, Astrophys. J., 218, p.148.
Meaburn, J.: 1976, *Detection and Spectrometry of Faint Light*, D. Reidel Publ. Co., Dordrecht.
Meaburn, J.: 1978, Appl. Optics, 17, p.127.
Meaburn, J.: 1980, Mon. Not. Roy. Astron. Soc., 192, p.365.
Meaburn, J.: 1981, Mon. Not. Roy. Astron. Soc. (in press).
Meaburn, J. and Blades, J. C.: 1980, Mon. Not. Roy. Astron. Soc., 190, p.403.
Meaburn, J. and Terrett, D. L.: 1980, Astron. Astrophys., 89, p.126.
Meaburn, J., Terrett, D. L. and Blades, J. C.: 1981, Mon. Not. Roy. Astron. Soc. (in press).
Meaburn, J. and Walsh, J.: 1981, Astrophys. Space Sci., 74, p.169.
Milne, D. K., Caswell, J. L. and Haynes, R. F.: 1980, Mon. Not. Roy. Astron. Soc., 191, p.469.
Mills, B. Y., Turtle, A. J. and Watkinson, A.: 1978, Mon. Not. Roy. Astron. Soc., 185, p.263.
Oort, J. H.: 1971, *Nuclei of Galaxies* (ed. D. J. K. O'Connell), North Holland Publ. Co., Amsterdam, p.321.
Page, T. and Carruthers, G. R.: 1978, NRL Report 8206, Naval Research Lab., Washington, D.C.
Pikelner, S. B.: 1968, Astrophys. J. Letts., 2, p.97.
Radharisan, V., Brooks, J. W., Goss, W. M., Murray, J. D. and Schwarz, U. J.: 1972, Astrophys. J. Suppl., 24, p.7.
Reynolds, R. J. and Ogden, P. M.: 1979, Astrophys. J., 229, p.942.
Schmidt-Kaler, Th.: 1977, Astron. Astrophys., 54, p.771.
Schmidt-Kaler, Th. and Feitzinger, J. V.: 1976, Astrophys. Space Sci., 41, p.139.
Sedov, L.: 1959, *Similarity and Dimensional Methods in Mechanics*, Acad. Press, New York.
Snow, T. and Morton, D.: 1976, Astrophys. J. Suppl., 32, p.429.
Walraven, J. H. and Walraven, Th.: 1975, Mon. Not. Roy. Astron. Soc., 34, p.56.
Werner, M. W., Becklin, E. E., Gatley, I., Ellis, M. J., Hyland, A. R., Robinson, G. and Thomas, J. A.: 1978, Mon. Not. Roy. Astron. Soc., 184, p.365.
White, N. J.: 1981, Astrophys. Space Sci. 78, p. 433.

HERBIG-HARO OBJECTS: RECENT OBSERVATIONAL AND THEORETICAL DEVELOPMENTS

Jorge Cantó
Instituto de Astronomia
Universidad Nacional Autonoma de Mexico

ABSTRACT

A brief summary of our present observational and theoretical knowledge of Herbig-Haro objects is given. Special emphasis is given to the recent observational results in the radio frequency range and on the most recent models advanced to explain these objects. It is concluded that HH-objects are most probably intrinsic emission nebulae produced by the cooling of shock-heated material. The source of energy is likely to be a strong wind from a new born star which is displaced ~ 0.1 pc typically from the optical object.

1. INTRODUCTION

The Herbig-Haro (HH) objects have been defined by Herbig (1969) as peculiar diffuse nebulosities, generally small in size (~ 0.01 pc.) with characteristic spectra dominated by strong emission in O I , [N II], [S II] and hydrogen lines. A weak continuum is sometimes present. These objects are always found projected against dark cloud complexes or closely related to them.

The first three HH-objects (HH 1, 2 and 3) were detected spectroscopically in the early fifties by Haro (1950, 1952) and Herbig (1951). Up to now more than 70 HH-objects have been identified.

Starting with Ambarzumian (1954) most authors have emphasized that there is a close relationship between HH-objects and the star formation process. This can be inferred, for instance, from the fact that they only occur in regions where T-Tauri stars are present (Herbig 1962, 1969) or from the high detection rate of high velocity masers in their vicinity (Rodriguez et al. 1978, 1980; Haschick et al.

F. D. Kahn (ed.), Investigating the Universe, 95–124.

1981). Therefore their study should provide valuable information on the birth of stars and their early evolution.

During the last few years these rather enigmatic objects have been the subject of extensive observations in a wide range of wavelengths, from UV to radio. These observations have provided unique empirical information about the objects themselves and their molecular environment and are the basis of several, quite different models which attempt to explain their peculiar nature. However, in spite of all these observational and theoretical contributions, there is no general concensus on the role of these objects in Nature.

Several review papers on HH-objects and related objects have appeared in the literature during the last decade (Strom et al. 1975; Haro 1976 and Böhm 1975, 1978b, 1978c). These papers were mainly concerned with the optical and IR characteristics of these objects. The purpose of the present paper is to give a brief summary of our observational and theoretical knowledge of the HH-objects, giving special emphasis to the most recent observational results in the radio frequency range.

2. OBSERVATIONAL RESULTS

2.1. Optical Properties

a) Lists of Herbig-Haro Objects. At the present time a large number of HH-objects have been identified. Herbig's (1974) catalogue contains a list of 42 objects for which accurate coordinates and finding charts are given. To them we have to add four objects discussed in Strom, Grasdalen and Strom (1974a); four more in the Orion Nebula observed by Gull et al. (1973), Münch (1977), Taylor and Münch (1978) and Cantó et al. (1980); sixteen new objects discovered by Schwartz (1977) in his survey of the southern dark clouds; one found by Elias (1978 in the Taurus dark cloud complex; two identified by Adams et al. (1979) near the centre of NGC 2264; and three HH-objects given by Cohen (1980) in his list of red and nebulous objects in dark clouds.

Parallel to the survey performed by Cohen (1980), Gyulbudaghian, Glushov and Denisyuk (1978) also searched the NGC-PO Sky Survey photographs looking for objects with external appearance similar to that of HH-objects (that is, red nebular objects with small angular dimensions). They ended up with a list of 37 new candidates (the GCD-objects) Preliminary spectroscopic observations by Gyulbudaghian et al. revealed the characteristic HH-spectrum in six of them (quoted by Rodriguez et al. 1978). In contrast, however, the spectra taken by Cohen (1980) of several of these objects showed an obvious continuum in three of them and a complete lack of emission in the spectra of other two. Therefore, whether the GGD-objects are HH-objects or not remains unclear; detailed optical observations of them are clearly needed.

In any case it has to be noted that detailed radio observations of the GGD-objects and their surrounding molecular environment by Rodriguez et al. (1978, 1980) reveal them to be extremely interesting by themselves and closely related to the star formation process.

Finally, we have the emission nebulae associated with T-Tauri stars, for instance, those around HL-XZ Tau and T Tau itself. It is well known that they show the typical HH-spectrum and are most probably excited by the same process (Schwartz, 1975).

b) Visual Appearance and Dimensions. The optical visual appearance of HH-objects is rather characteristic since they most frequently appear as small semistellar bright condensations or knots surrounded by a more diffuse nebulosity with a similar spectrum. They either occur as isolated condensations or in small groups containing several of them. This knotty appearance is particularly evident in HH2 (see Herbig's 1974 catalogue) in which at least nine bright and separated condensations of about 500 AU (a few arc sec) in radius form a tight group of 20 000 AU in extent. Such dimensions are typical for most HH-objects.

There are, however, a few exceptions, for instance, HH 100 in the R CrA region and HH 102, a diffuse nebulosity near XZ and HL Tau (Strom, Grasdalen and Strom 1974a). They are not semistellar but appear as large nebular complexes with highly irregular structure extending nearly 2' in the sky.

It is interesting to note here that the peculiar appearance of HH-objects, together with the fact that they occur only projected against or very near dark cloud complexes were the criteria used by Gyulbudaghian, Glushkov and Denisyuk (1978) in their search for new HH-objects.

c) Line Spectrum. The emission line spectrum shown by HH-objects is quite uncommon and actually constitutes the defining criterion of their class. As described by Herbig (1969) in HH-objects "..the H emission lines are strong and [O I] and [S II] are unusually intense. The [N II] lines are also strong ...". In the classical objects (HH 1 and 2) the H and K lines of CaII as well as forbidden lines of [O II] [N I], [Fe II] and [Ca II] are also found in appreciable strength.

After the first three objects had been discovered it became evident (Böhm 1956; Osterbrock 1958) that it is very unlikely that HH-objects are photoexcited by a normal star. Instead we know now that they most probably constitute the cooling region of moderately strong shock waves. This is evident from Table 2 of Mc Kee and Hollenbach (1980) which shows a comparison between the observed intensities of the OI, OII, NI, NII, SII and H_α lines in diffuse photoionized (HII and Planetary) nebulae, possible shocked regions (Supernova Remnants, HH and related objects) and those predicted by shock wave models (Shull and Mc Kee 1979). It follows from this table that while photoionized regions show quite weak lines from low excitation ions, like OI, OII, NI, NII, SII and CaII (compared with H_α) those lines are much stronger in shocked nebulae. Mc Kee and Hollenbach qualitatively explain this fact by pointing out that in photoionized nebulae the H_α line is mainly produced in the fully ionized region, while the low excitation lines come from the (much smaller) transition zone. In contrast, in shocked nebulae both sets of lines are only produced in the recombination region. Similarly, the temperature sensitive ratio of [O III], ($\lambda 4364$ / $\lambda 4959 + \lambda 5007$) (also included in this table) is roughly a factor of 4 greater in shocked nebulae than in photoionized gas. Obviously this is an indication of the high temperature ($> 10^4$ °K) reached by shocked material as compared with the rather constant temperature ($\lesssim 10^4$ °K) of photoionized nebulae.

Another interesting and useful comparison is given in Figure 1 between the relative line strengths in HH-objects and SNR and those in photoionized regions. In it, there are shown the relative intensities H_α/[N II], H_α/[S II] and the density (inverse) dependent ratio I(λ 6717) / I(λ 6731) of [S II] for a large number of HH-objects. [N II] indicates I(λ 6548) + I(λ 6583) and [S II] equals I(λ 6717) + I(λ 6731). In the same figure the general region occupied by H II Regic (HII), Planetary Nebulae (PN) and Supernova Remnants (SNR) (accordinc to Sabbadin, Minello and Bianchini 1977), are also exhibited. The right hand side scale of Figures 1a and 1b gives the electron density implied by the corresponding [S II] ratio when Pradhan's (1979) collision strengths are used.

The close similarity between SNR and HH-objects follows from Figure 1 and in turn supports their identification with the cooling region of a shock wave. It is also evident that the [S II] and H lines have comparable strengths in shocked nebulae, in contrast with photoionized material. This may suggest that Figure 1 can be used to indicate whether an emission nebula is mainly photoexcited or collisionally excited by a shock wave.

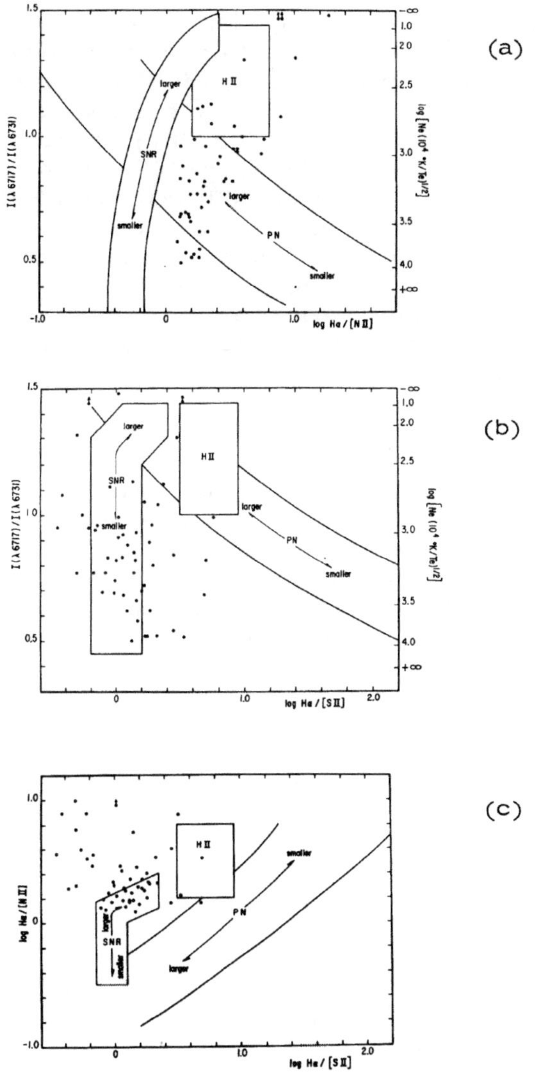

Figure 1. Diagrams showing the relative intensities of H_α , [S II]= I(λ6717) + I(λ6731), [N II] = I(λ 6548) + I(λ 6583) and the electron density dependent ratio I(λ 6717) / I(λ 6731). The right hand side scale of (a) and (b) gives the corresponding electron density (see text). Dots represent the HH-objects. The general regions occupied by H II Regions (H II), Planetary Nebulae (PN) and Supernova Remnants (SNR) are also exhibited. Data references are given in the text.

Besides the common features exhibited by HH-objects and SNR there are, however, two interesting differences which have to be mentioned. Firstly, we notice that the [S II] ratio indicates the electron density in the coding region of the shock wave, which in turn is proportional (or nearly) to $N_oV_s^2$ (Cox 1972; Raymond 1976, 1979; Dopita 1978; Shull and Mc Kee 1979) where N_o is the pre-shock particle density and V_s is the shock velocity. Although, the [S II] ratio in HH-objects indicates a wide range of electron densities (from less to 10 to nearly 10^4 cm^{-3}) most of them are similar to those of young (small radii) SNR, with high expansion velocities ($V_s \sim 1000$ km s^{-1}). Those facts imply that the pre-shock density in most HH-objects is much higher (roughly a factor of 100) than the typical hydrogen density in the interstellar medium.

The second interesting feature that follows from Figures 1a and 1c is the high $H\alpha$ /[N II] intensity ratio in some HH-objects in comparison with SNR. This feature can be qualitatively understood with the following arguments (Cantó et al. 1980): HI has an ionization potential of 13.6 eV whilst that for NI this is 14.5 eV. Therefore, at high enough shock velocities, both HI and NI will be fully ionized, giving rise to similar line intensities of $H\alpha$ and [N II] . However, for lower shock velocities the shock may be able to ionize HI but not NI. Therefore at low velocities ($\lesssim$ 50 km s^{-1}) an increase in the $H\alpha$ / [N II] ratio is expected. This may be the case for several HH-objects.

d) Radial Velocities and Line Widths. It is well known that the radial velocities of HH-objects are in general negative with respect to the surrounding cloud.

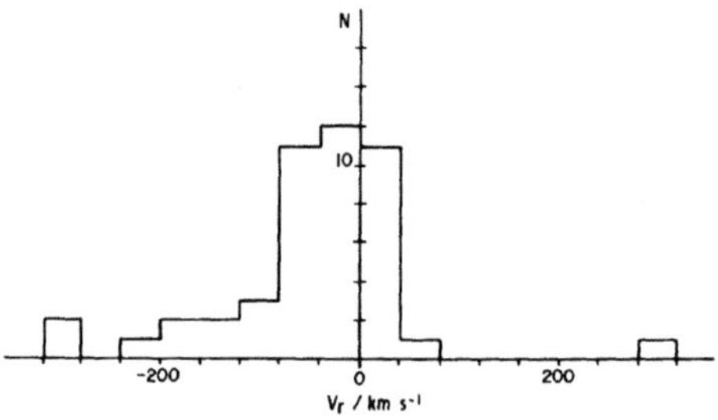

Figure 2. Histogram of the radial velocities of HH-objects. Data references are given in the text.

Figure 2 shows an histogram of the radial velocities of nearly 30 HH-objects. The data have been taken mainly from Strom, Grasdalen and Strom (1974a); Dopita (1978); Schwartz (1978); and Schwartz and Dopita (1980), complemented with those given by Böhm, Perry and Schwartz (1973); Münch (1977); Adams et al. (1979) and Cantó et al. (1980). Measurements of different nuclei within one object were taken as independent. For those objects with complex line profiles consisting of several (but distinguishable) components (see for instance Münch 1977 and Cantó et al. 1980), each component was also taken as independent. Finally, several data were not corrected for heliocentric motion since the errors in the measurements are expected to be of the same order of the correction itself. This effect is not likely to seriously affect Figure 2.

It follows from Figure 2 that, in fact, most ($\sim$ 75%) of the measured HH-objects possess negative radial velocities (with respect to the surrounding cloud), with an average value of $\sim$ -80 km/s. Since HH-objects are likely to be in the outer parts of the cloud (due to their low values of visual extinction; see Böhm 1975), this result implies that the emitting matter in those objects must be moving away from the related cloud. The average velocity for those objects which show positive radial velocities is about +40 km/s. In these cases we can always imagine some special geometry such that even if the HH-nebula is moving away from the cloud it would show a positive radial velocity to the observer.

Recently, Schwartz and Dopita (1980) showed that for a large sample of HH-objects the degree of excitation (defined by the ratio I(λ 5007) of [O III] over I(λ 6300) of [O I]) and the electron density (as given by the [S II] ratio, $T = 10^4$ °K) both correlate inversely with the modulus of the radial velocity. These tendencies were interpreted by Schwartz and Dopita (1980) as a confirmation of the shocked cloudlet model for HH-objects suggested by Schwartz (1978). However, they can also be understood simply as a consequence of oblique shocks (Cantó 1980).

Regarding the line widths, there is again no general tendency. While some objects show moderate values, $\sim$ 50 km s^{-1} (for instance in HH2; Schwartz 1978) others exhibit sizable ones, $\sim$ 150 km s^{-1} (Strom, Grasdalen and Strom 1974; Dopita 1978). However, the most interesting result concerning the line shapes in HH-objects is that discovered by Münch (1977) and later on measured in detail by Cantó et al. (1980) in M42-HH1. That is the presence of a long extended blue-shifted wing superimposed on a main strong component. This is shown in Figure 3.

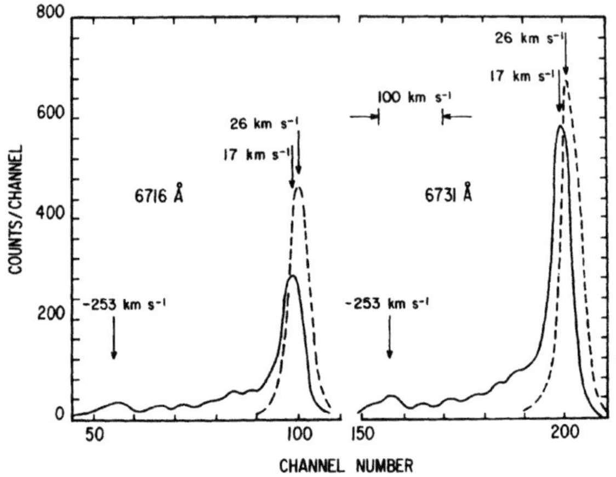

Figure 3. Profiles of the [S II], λλ 6717 and 6731 lines in the HH-object M42-HH1, (full line) and in the surrounding nebula (dashed line). Adapted from Figure 3 of Cantó et al. 1980.

It is striking that while the main component in the HH-object profile is only slightly (~ 9 km s^{-1}) blue-shifted with respect to the nebula, the wing extends by nearly 300 km s^{-1} toward the blue. Obviously such a profile has to be interpreted in terms of detailed theoretical line profiles in shock wave models which, unfortunately, are not available at present.

In this context it is interesting to notice that extended blue and red-shifted wings are also present in the spectra of the molecular surroundings of some HH-objects. This is the case for instance in the region containing HH 102, 28, 29, 30 (Snell, Loren and Plambeck 1980). This and other examples will be discussed in Section 2.3.a.

e) Proper Motions. In 1969, Cudworth and Herbig presented their proper motion measurements of three HH-objects (HH 28, 29 and 30) and other young objects in the L 1555 region. The large proper motions of HH 28 and 29, already discovered by Luyten (1963, 1970) were confirmed. For HH 30, however, no peculiar motion was detected within the observational errors. More recently Jones and Herbig (1980) have determined the proper motions of the knots in HH 1 and 2, and found them to be also quite large.

According to Cudworth and Herbig (1979) and Jones and Herbig (1980), the most probable interpretation of these motions is that they actually correspond to spatial displacements of the emitting material. This conclusion is backed, in the case of HH 28 and 29, by the fact that if the proper motion vectors of these two objects are extended backward they intersect very near an IR source discovered by Strom, Strom and Vrba (1976) in a 2.2 μ survey of the region. This IR source was later shown to be the center of a bipolar-type kinematic structure (detected in the CO lines by Snell et al. 1980) within which HH 28 and 29 are found (see Section 2.3.a for details).

At a distance of 140 pc (Elias 1978), the proper motions of HH 28 and 29 would correspond to about 145 km s^{-1} in tangential velocities. This value combined with the observed radial velocity of −40 km s^{-1} (Strom, Grasdalen and Strom 1974) for HH 29 gives a total spatial velocity of 150 km s^{-1} for this object, comparable with the radial velocities measured in other objects.

Similarly, for HH 1 and 2 the proper motions of both objects are directed away from the star which Cohen and Schwartz (1979) have suggested is their common exciting source.

At the distance of Orion the measured proper motions imply tranverse velocities as high as 400 km s^{-1} for these objects.

f) Variability. Although it appears that most, if not all, HH-objects are variable (Herbig, 1957, 1969, 1973), there is not enough information in this respect to make any statistical analysis. However, the variability, in both luminosity and spectra, of HH 1 and HH 2 has been studied for a long time (Herbig 1957, 1969, 1973; Böhm 1956; Böhm, Perry and Schwartz 1973; Böhm, Siegmund and Schwartz, 1976; Böhm and Brugel, 1979) and it is well documented. The main results of these studies seem to indicate (see Böhm 1978b for a detailed discussion) that 1) particular condensations within one object vary independently from each other on a time scale of the order of ten years, and 2) the observed brightness variations (about 2^m) are probably related to changes in physical parameters (say N_e and T_e) within the object by factors roughly less than two. Whether these two conclusions apply to all objects is not known; in any case they place strong restrictions on any model which may be advanced to explain the HH-objects.

g) Polarization and Continuum. Strom, Strom and Kinman (1974c) and Vrba, Strom and Strom (1975) undertook linear optical polarization measurements on several HH-objects. A rather high linear polarization was found in HH 24 ($\sim$ 24%) and HH 100 ($\sim$12%) with the property that the perpendiculars to the polarization vectors point toward a nearby IR source.

Strom and collaborators interpret their findings by assuming the HH-objects to represent the reflection nebulae of a heavily obscured T Tauri-type star (the IR source) which would have an emission circumstellar nebula similar to that around T Tauri. Surprisingly, however, no IR source and only weak ($\sim$ 3%) linear polarization were found in the strongest HH-objects 1 and 2. Later on Schmidt and Vrba (1975) showed this polarization to be interstellar in origin.

A simple explanation for the differences in linear polarization among HH-objects was first advanced by Schwartz (1975). It is well established that besides their characteristic emission line spectrum, HH-objects also show a rather weak continuous emission. Schwartz noticed that while the ratio of continuum to line intensity in HH 24 is rather high, this ratio appears to be very low in HH 1 and 2. It was therefore suggested that perhaps only the continuous emission is the result of a reflection process while the lines are formed in situ (probably as a result of a shock wave). This suggestion appears indeed to be correct at least in condensation A of HH 24.

Schmidt and Miller (1979) undertook spectropolarimetric observations of HH 24A. Their study revealed that this knot possesses a rich line spectrum typical of an HH-object superimposed onto a moderately strong continuum. While the lines are found to be essentially unpolarized, the continuum shows a strong polarization of 25%. Furthermore Schmidt and Miller commented on the sharp contrast between the spectrum of condensation A and that of HH 24E. In the latter the only line present is H_α (with a velocity dispersion of 160 km s^{-1}) and the continuum is much stronger, a spectrum quite similar to that of a T Tauri type star.

In their discussion, Schmidt and Miller interpret knot E as a pure reflection nebula scattering radiation of a deeply embedded T Tauri star. This star would also be the source of both the continuous emission (via reflection) observed in knot A and the outflowing wind which shocks in this condensation and produces the emission lines.

This simple and attractive interpretation of HH-objects however encounters some difficulties when detailed observations of the faint continuum are brought into the picture.

Böhm (1956); Böhm, Schwartz and Siegmund (1974) and Brugel, Böhm and Mannery (1980) have investigated in great detail the continuum emission in several HH-objects (HHl, 2A, 2G, 2H, 24A and 32A) in the spectral range 3200 Å $< \lambda <$ 8000 Å. Among their most interesting findings is the rapid increase in the continuum flux with decreasing wavelength in all observed HH-objects. Far UV observations of HH1 (Ortolani and D'Odorico 1980) show this blueward enhancement to extend up to ∼ 1200 Å. It also follows from the observations by Böhm and collaborators that the ratio of the total continuum flux to the H_β flux is almost the same (∼ 6) for all observed objects (with the exception of HH24A in which this ratio is about five times greater).

These two characteristics (the blue-UV excess and the nearly constant continuum/H_β ratio) lead to difficulties with the reflection hypothesis with a T Tauri star as the original source of radiation.

Brugel, Böhm and Mannery (1980) and Ortolani and D'Odorico (1980) conclude that if the continuum in HH-objects is really due to dust scattering the illuminating star must be hotter ($T_{eff} \simeq 40{,}000$ °K) than a T Tauri star. In this context it is also difficult to understand why the continuum to H_β flux ratio is nearly constant.

2.2. IR Observations

a) IR Searches and Photometry. Ever since their discovery, the central topic of controversy regarding HH-objects has been their energy source.

The searches for the exciting star in these objects started on deep direct photographs by Herbig (1951) and Haro and Minkowski (1960) of the HH 1 and 2 region in the visual and photographic IR. No stars up to the visual magnitude 21 were found to be directly involved with these objects.

Later on, Strom, Strom and Grasdalen (1974b), having noticed that all known HH-objects are within the boundaries of dark clouds, realised that the exciting star might be heavily obscured by cloud material and therefore invisible at visual wavelengths. Thus they searched for these sources in the IR. The first IR source associated with a HH-object was found in R CrA region, about 0.01 pc NE of HH 100. Strom et al. (1974b) obtained the IR spectrum of this source between 5 and 12.6μ and found it to be consistent with a T Tauri-like object but heavily obscured and reddened by about 15 mag in the visual. Since then, several other IR sources have been discovered in the vicinity of HH-objects (Strom, Grasdalen and Strom 1974a; Strom, Strom and Kinman 1974c; Strom, Strom and Vrbà 1976 and Cohen and Schwartz 1979, 1980). In general these sources do not coincide with the optical object but typically lie (in projection) displaced by up to 1' from it. In most cases these sources can (or may) be identified with a young T Tauri-like star reddened by 20–30 mag in the visual and therefore deeply embedded in the dark cloud.

Recently, Cohen and Schwartz (1979, 1980) undertook an extensive search for the exciting stars in a number of HH-objects by surveying their dark cloud environment at 2.3μ. Their limiting magnitude was ~ 11.4. Cohen and Schwartz (1980) also observed previously known IR sources close to HH-objects with the main purposes of verifying their existence, obtaining their positions accurately and, when possible, identifying their optical counterparts. Several IR sources were located near these HH-objects. However, in most cases, Cohen and Schwartz did not find enough evidence to be able to identify them with the exciting star of the optical object. They are most probably just reddened hot stars not associated with the HH-objects.

In the HH1 region, on the other hand, a bright source located about 30" SE of the optical object was detected (Cohen and Schwartz, 1979). The IR source coincides with a nebulous star, probably a G8 highly obscured T Tauri star, located just along a red faint nebulosity connecting HH 1 and 2 (see Herbig's 1974 catalogue). The star's spectrum was also taken by Cohen and Schwartz. It shows H_α, [N II], [O I], [S II], He I, H_β and [O III] emission lines superimposed on a very red continuum. Remarkably the forbidden lines emission is also present in the faint extended nebula which envelopes both HH1 and the star. Cohen and Schwartz (1979) conclude by argueing that the observational evidence indicates this star to be the long sought exciting source of HH1. This conclusion has been recently reinforced by the proper motion measurements of the knots in HH1 and 2 by Jones and Herbig (1980) and already described in Section 2.1.e.

b) Spectrophotometry. Molecular and Atomic Hydrogen. The search for molecular hydrogen line emission in the $2-2.5\mu$ range in the direction of HH-objects is quite recent. It started with the discovery of the v=1-0 S (1) line of H_2 around T Tauri and its associated emission (HH) nebula, Burnham's Nebula, by Beckwith et al. (1978). The emitting region was found to be centred on the star and probably larger than 5".

In their paper, Beckwith et al. briefly discuss the excitation mechanism for the H_2 in T Tau. They suggest shock excitation as the most viable one because T Tau is likely to possess a moderately strong wind (Kuhi 1964; Schwartz 1975). Assuming the emission to come from an optically thin, uniform spherical surface of 5" in diameter, a surface brightness in the v = 1 - 0 (S(1) line of 3×10^{-4} erg s^{-1} cm^{-2} sr^{-1} is derived. This value combined with Kwan's (1977) models for shock heated molecular material implies a minimum H_2 pre-shock density of 10^4 cm^{-3}. Furthermore, since a pressure of 10^{-7} dn cm^{-2} is derived as necessary to push the shock continuously and if the wind's velocity in T Tauri is taken to be 200 km s^{-1} (Kuhi 1964; Schwartz 1975) a mass loss rate of 5×10^{-8} $M_\odot$ yr^{-1} is then required to sustain the shock. It is interesting to notice that this result for the mass loss rate is quite similar to that derived by Cantó et al. (1980) for other T Tauri-type star, R Mon.

More recently Fischer et al. (1980) and Elias (1980) searched for H_2 emission in the direction of several HH-objects. The NGC 1333 region, surveyed by Fischer et al., showed no H_2 emission with flux in the S(1) v = 1 - 0 line greater than $6 - 7 \times 10^{-13}$ erg cm^{-2} s^{-1}. At the position of HH2, however, Fischer et al. detected the S(1) line and the blend of the Q-branch lines. The emission was found to extend over a region of at least 34", slightly greater than the optical image of this object (Herbig 1974).

The more complete search for H_2 emission in HH-objects is that by Elias (1980). Elias observed sixteen regions looking for hydrogen atomic and molecular lines; H_2 lines were detected in six of them (HH1, 2, 40, 46, 53 and 54) with a detection limit ($3 - \sigma$) of about 5×10^{-14} erg cm^{-2} s^{-1}. Among his main conclusions there are that: 1) There is no detectable ($3 - \sigma$ upper limit) excited atomic hydrogen emission in any observed object, implying that the hydrogen surrounding the HH-objects is mostly in molecular form. 2) There is no large amount of heavily obscured ionized gas in the HH-objects, since the upper limits on Brackett-γ emission are close to the fluxes expected from the visible Balmer line emission. 3) The absence of strong recombination line emission in the IR lines implies that the observed continuum is not mainly due to recombination. 4) The observed line intensity ratios of H_2 are very different from those predicted for emission following UV excitation. This result indicates shock heating. 5) Comparison of the observed surface brightness and line ratios of H_2 with shock models by Kwan (1977) and Shull and Hollenbach (1978) suggests shock velocities of no more than 20 km/s and pre-shock densities around 10^4 cm^{-3}.

2.3. Radio Observations

a) <u>Formaldehyde (H_2CO), Carbon Monoxide (CO) and Ammonia (NH_3)</u> Extinction measurements by Strom, Grasdalen and Strom (1974a) toward the 2.2 μ IR sources associated with HH-objects imply very high hydrogen densities ($\sim 10^4 - 10^5$ cm^{-3}) in those regions, high enough to excite the milimeter-wave lines of several molecules.

Lada et al. (1974) first included regions containing HH-objects in their study of star formation sites in dense molecular clouds. These were the NGC 1333 and M78 regions. Lada et al. reported detailed observations in the 140 GHz line of H_2CO, the 88.6 GHz line of HCN, the 1 cm lines of NH_3, the 147 GHz line of CS and the 115 and 110 GHz lines of ^{12}CO and ^{13}CO respectively. In both regions, Lada et al. found a broad, nearly circular, plateu of H_2CO emission of about 100 (arc min)^2 in area ($\sim$ 1.5 pc in diameter) with the HH-objects and IR sources near the peak of the molecular emission. Also kinetic temperatures and molecular hydrogen densities in the emission peaks were determined with values around 20°K and 2×10^4 H_2 cm^{-3} respectively. Those findings, therefore, suggested that HH-objects are closely associated with molecular density enhancements. Assuming a spherical cloud of 1.5 pc in diameter, a mass of 10^3 $M_\odot$ was derived.

Later on, Loren (1976) reported an extensive mapping of the NGC 1333 cloud in ^{12}CO and ^{13}CO lines. Based upon his observations, Loren suggests that the collision of two clouds has initiated the star formation in this region.

Further studies of the properties of molecular clouds containing HH-objects are presented in Loren et al. (1979). They observed 30 HH and similar objects in the 2 mm and 2 cm lines of H_2CO in order to determine the density of the clouds in which they are embedded. They also undertook CO observations to estimate kinetic temperatures.

The 2 mm line of H_2CO was positively detected in the direction of 12 objects. The density of neutral material around those objects was found to be quite high, ranging from 10^4 to 5×10^4 cm^{-3} (for the cloud near T Tauri a value of 9×10^4 cm^{-3} was obtained). Furthermore, in those regions where mapping was done, Loren et al. found that individual HH-objects are not associated with separate density enhancements, but are rather found in the vicinity of a common high density plateau. The results of Loren et al. also shows that in those cases where a bright IR source is associated with a HH-object, the position of the density enhancement is better correlated with the position of the IR source than with the position of optical HH-object. This result is not surprising since the extinction toward the IR source (typically 20 mag) is much greater than that toward the HH-object ($\sim$ 1 mag). Finally the kinetic temperature of the cloud material was found by Loren et al. to range from 9 to 34 °K; the core masses are between 18 to 1200 $M_\odot$.

Quite recently, Ho and Barrett (1980; see also Ho 1977) surveyed 29 HH-objects in the (1, 1) line of NH_3 with better angular resolution (85" beam) and better velocity resolution (0.21 km s^{-1}) than previous studies. The purpose of this survey was to confirm the association of HH-objects with high density material, to attempt to set limits on neutral molecular densities, to investigate the dynamics of the regions and to establish the spatial relationship between the HH-objects and the neutral material. They detected NH_3 in 15 objects (11 distinct regions).

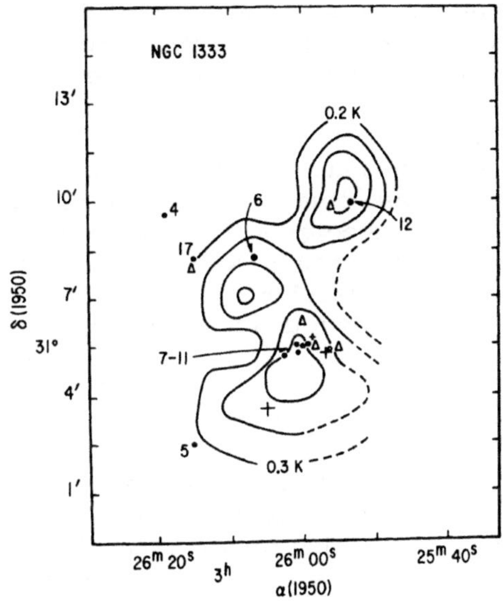

Figure 4. Contour map of the observed NH_3 antenna temperature in the NGC 1333 region. There are also included the HH-objects (filled circles), the IR sources (triangles), water masers (crosses) and a compact H II region (empty circle) found in the same region. Adapted from Figure 6.1 of Ho (1977) and Figure 2 of Haschick et al. (1980).

Figure 4 shows the contour diagram of the observed antenna temperatures of NH_3 in the NGC 1333 region. In this figure there are also included the HH-objects, the IR sources, water masers and a compact H II region also found in the region. It is noticeable that there is a close relationship between all these objects and molecular cloud enhancements, especially in the HH7-11 region. However, it is also evident that there is lack of spatial coincidence among them.

As noticed by Ho and Barret, the positions investigated were simply the positions of the optical objects, which a priori do not imply that these would be the densest areas in the cloud. Therefore, the fact that a large fraction of HH-objects can be detected in the NH_3 line definitely confirms the strong connection between the HH-objects and density enhancements within the cloud. The derived densities of molecular hydrogen were in general $\gtrsim 5 \times 10^3$ cm^{-3} in those regions where the (1,1) line of NH_3 was detected.

Ho and Barret also found that the observed line widths toward the optical HH-objects are quite narrow ($\lesssim$ 1 km s^{-1}), comparable to those in dark clouds. Therefore it was concluded that the positions of the optical object may not be the sites of current star formation since no peculiar movements are observed due to either collapse, expansion or enhanced turbulence of the neutral material (see below however).

Let us now briefly discuss some molecular line observations of the GGD-objects. Rodriguez et al. (1980) undertook CO and NH_3 observations in the direction of these objects. The CO measurements were used to determine the radial velocity of the associated cloud and in a few cases their kinematical distances; the NH_3 observations on the other hand, were used to estimate molecular hydrogen densities. The mean distance for the GGC-objects was found to be $\sim$ 1 kpc, which is about a factor of two greater than the mean distance of HH-objects ($\lesssim$ 500 pc).

Ammonia was detected in nine of the 24 regions surveyed with antenna temperatures in the range 0.2–0.7 $^{\circ}$k. A strong correlation was found between the sources that have H_2O masers and detectable NH_3 emission. In those regions the density is probably greater than 5×10^3 cm^{-3}. Rodriguez et al. also mention that the rate of detection and the line parameters obtained by them in GGD-objects are similar to those of the survey made by Ho and Barret (1980) of <u>bona fide</u> HH-objects.

On the other hand, perhaps, the most interesting feature discovered by CO observations in the direction of some HH-objects consists in the evidence of anisotropic (bipolar?) mass outflows in these regions.

Firstly, Knapp et al. (1976) detected a broad, low-velocity wing in the ^{12}CO emission line toward S 239 (HH 102; Strom, Grasdalen and Strom 1974a) in the dark cloud L1551. This cloud contains, apart from the extended HH-object 102, three other compact HH-objects, HH 28, 29 and 30, two T Tauri-type stars, HL and XZ Tau and an extremely reddened ($A_V \sim 20^m$) IR source. The latter lies in the dense ($\sim 10^5$ cm^{-3}) core of L1551 (Loren, Evans and Knapp 1979). The spatial distribution of all these objects is shown in Figure 5.

Recently, Snell, Loren and Plambeck (1980) reported extensive CO observations of the L1551 cloud. Surprisingly, their observations revealed broad and shifted velocity features superimposed to the narrow dark cloud line. But more striking, perhaps, was the fact that the broad feature appeared redshifted (with respect to the main line) NE of the IR source, while to the SW it appeared blueshifted. The CO emission line profile was found symmetrically broadened at the IR position. These features are also illustrated in Figure 5.

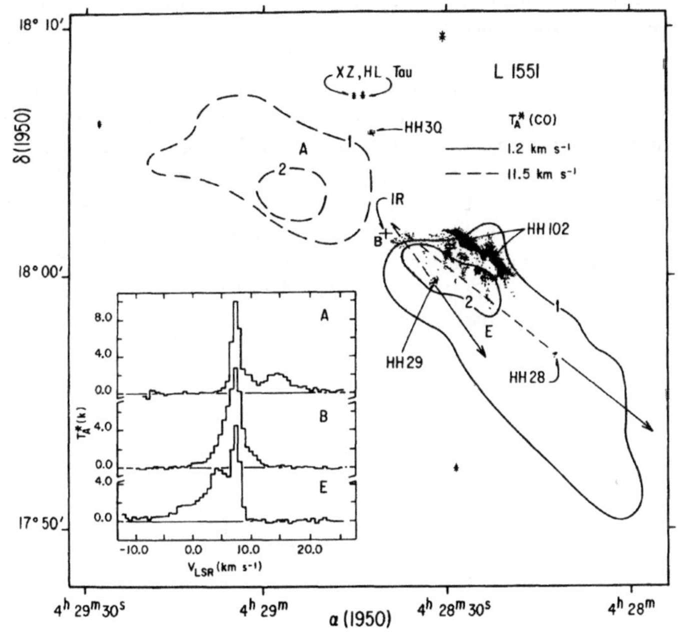

Figure 5. Contour map of the ^{12}CO corrected antenna temperature in the broad velocity components, superimposed on a sketch of the L1551 region. The cloud contains three compact HH-objects, HH 28, 29 and 30, an extended one, HH 102, two T Tauri-type stars, HL and XZ Tau and a strong IR source (cross). Letters A, B and E indicate the positions of the spectra in the insertion. The proper motions of HH 28 and 29 are also indicated; if they are extended backward, they intersect near the IR source. Adapted from Figures 2 and 3 of Snell et al. 1980.

If a map of the broad feature is made (Figure 5; adapted from Figure 2 of Snell et al. 1980) the presence of a symmetric double-lobed structure centered on the IR-source is revealed. The linear dimensions of each lobe are ∿ 0.5 pc. Also the HH-objects 102, 28 and 29 are found to lie within the blueshifted lobe.

The importance of these findings is enhanced when they are combined with the proper motion measurements of HH 28 and 29 by Cudworth and Herbig (1979) (Section 2.1.e.). Indeed, it is found that when the proper motions of HH 28 and 29 are extended backward, they intersect near the IR source which is also the center of the bipolar CO structure. This evidence strongly argues that the IR source is responsible for both the HH-objects and the double-lobed structure, probably via a strong stellar wind (Snell et al. 1980).

Cantó et al. (1981) also found a similar situation in the region containing RMon, NGC 2261 and HH 39. R Mon is a peculiar star located at the apex of the visible cometary nebula NGC 2261. If we extend the axis of symmetry of NGC 2261 toward the north, we find HH 39. Detailed CO observations of this region by Cantó et al. (1981) show an elongated ($\sim 0.8 \times 0.4$ pc) molecular cloud at whose centre lies R Mon. Furthermore, these observations also reveal a blueshifted CO component peaking to the north of R Mon (in the direction of HH 39) and a redshifted CO component peaking south.

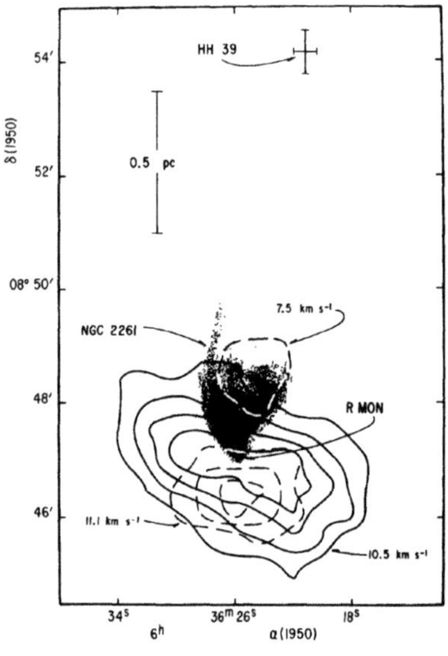

Figure 6. Contour map of the corrected antenna temperature of ^{12}CO at three different velocities, 7.7, 10.5 and 11.1 km s^{-1}, superimposed on a sketch of the R Mon, NGC 2261 and HH 39 region.

Figure 6 shows the brightness temperature of the ^{12}CO line contours at three different velocities, 7.7, 10.5 and 11.1 km s^{-1}, superimposed upon a sketch of the region. It appears that material moving at 10.5 km s^{-1} is spatially distributed in the form of a disk around R Mon. The major axis of the disk is perpendicular to the axis of symmetry of NGC 2261.

Cantó et al. interpret their observations by suggesting that R Mon has a strong stellar wind; this wind interacts with the disk-shaped cloud creating a bipolar cavity in it. In the visible, the northern cavity (NGC 2261) shines by reflecting light of R Mon. The southern cavity, on the other hand, is obscured in the visible by the tilted molecular cloud disk but can be detected in the CO observations. Moreover, following the work by Cantó (1980) Cantó and Rodriguez (1980) and Barral and Cantó (1981) it is proposed that the wind is channeled by the cavities into two antiparallel annular streams. The northern stream encounters another small molecular cloud, creating the HH-object 39.

There are further examples of anisotropic (perhaps bipolar) mass outflow in regions of star formation. Rodriguez et al. (1981) present CO observations of the molecular cloud associated with GGD-objects, 4, 12-15 and 27-28. The emission lines of GGD 12-15 show blueshifted and redshifted wings, each of them peaking at opposite sides with respect to a water maser also found in the region. Again, it is proposed that the stellar wind of a heavily obscured star is being channeled into two antiparallel streams by a surrounding disk. The wings in the emission lines are produced as those streams push the surrounding molecular gas. Furthermore, Rodriguez et al. point out that since this kind of anisotropic mass outflow has been detected in several regions of star formation, it could imply that the presence of disks (interstellar or circumstellar in size) around new-born stars is a common phenomenon.

b) Water Masers. H_2O masers are known to be associated with regions of current star formation, including those dark regions where HH-objects are present. The required molecular densities ($\sim 10^7$ - 10^{10} cm^{-3}), kinetic temperatures ($\sim$ 500-1000 ^oK) and dimensions ($\sim$ 10-100 AU) of masering regions are indeed similar to those expected to prevail in the circumstellar material around protostars.

The first detection of H_2O maser emission near an HH-object came in 1974. Dickinson, Kojoian and Strom (1974) undertook a search for the $J_{K^+ K^-} = 6_{16} - 5_{23}$ transition of water vapour at 22 GHz in the direction of 17 HH-objects. While they obtained negative results in 16 objects, they found a strong water maser near the cluster of HH-objects numbers 7 to 11. This water maser was independently detected by Lo, Burke and Haschick (1975) who also reported the discovery of H_2O maser emission in the direction of HH1. Several other HH-objects were also investigated for water emission by Lo et al. with negative results within $\sim$ 6 Jy of peak to peak noise.

Since the time of its discovery, the water maser near HH 7-11 has been recognized as variable, both in intensity and radial velocity.

Lo et al. (1976) monitored the maser over a 14-month period noticing peculiar variations on time scales as short as one day. The variations are such that two components appeared and dissappeared at a fixed radial velocity (~ −26 km s^{-1} and −13.8 km s^{-1}). Also, new components appeared randomly over a radial velocity range from −33 to +8 km s^{-1}, markedly blueshifted with respect to the molecular cloud which has a velocity of +8 km s^{-1} (Lada et al. 1974). Moreover, while the maser lies closest, in projected distance, to HH 11 (with radial velocity ~ −133 km s^{-1}) its radial velocity range is closer to the average velocity of HH 7, 8, 9 and 10 (~ −30 km s^{-1}). Later on Haschick et al. (1980) reported the monitoring of this maser at monthly intervals over a period of three years. In the course of the work, two additional water masers were discovered in the HH 7–11 region but displaced from the original maser (source A) by 0.5 (source B) and 2.5 arc minutes (source C) respectively (see Figure 4). Both masers have velocities similar to the cloud velocity. The position of the masers are determined to accuracies of 1, 10 and 15 arc sec for sources A, B and C respectively and are graphically given in Figure 4. Sources A and B have the same peak intensity of ~ 200 Jy and have displayed rapid amplitude variations in less than a week. Source C, on the other hand, is much weaker (~10 Jy) and it is not clear whether it is related to the cluster of HH-objects or not, since it is displaced ~10^{18} cm from them. All three masers, however, lie within equal brightness contour levels of the NH_3 emission (Figure 4).

Haschick et al. (1980) further report Very Long Baseline Interferometry (VLBI) observations of the H_2O maser source A. They conclude that the maser region is extremely compact, smaller than 7 x 10^{13} cm in diameter, ruling out the possibility of a circumstellar shell. Haschick et al. therefore suggest that the masing cloud is an isolated fragment with a mass about 10^{-5} $M_\odot$ which is currently confined to a diameter of 7 x 10^{13} cm but dispersing at 10 km s^{-1}. Whether this interpretation applies to other water masers associated with HH-objects is not known.

The most recent water maser associated with a HH-object was discovered by Haschick et al. (1981). This source is located ~ 1 arc min northeast of HH 23, near the reflection nebula NGC 2068. Its intensity (~ 20 Jy) and radial velocity (~ +20 km s^{-1}, LSR) are also variable in time.

Up to here we have discussed water masers located near genuine HH-objects. The GGD-objects, which morphologically resemble HH-objects, have also been searched extensively for water maser emission. Rodriguez et al. (1978, 1980) detected 10 H_2O masers in 24 regions studied. Additionally Rodriguez et al. (1980) report systematic radio observations of the GGD-object in the CO, NH_3, OH, SiO lines, together with continuum observations of ionized material at 4.9 GHz. The molecular line observations have already been discussed in Section 2.3.a. The main general results of Rodriguez et al. are the following: 1) There are large velocity differences between H_2O masers and the associated ambient cloud (typically $\gtrsim$ 20 km s^{-1}). Furthermore, in a statistical sense, water masers associated with GGD-objects show higher velocities than those related to HII regions ; 2) The masers are near but clearly not coincident with the optical objects ; 3) Compact deep embedded HII regions (probably early B-type stars) are usually found close ($\lesssim$ 30") to the high velocity water masers; 4) There is a good correlation between detectable NH_3 and water masers, since both emissions require high densities ; 5) No SiO masers were detected, indicating the youth of the GGD-objects. SiO masers are generally associated with evolved stars.

Again, the lack of spectroscopic confirmation of the nature of the GGD-objects (that is whether they are HH-objects or not) prevents us from assigning these general characteristics to the HH-object class. Moreover, the high detection rate of water masers in GGD-objects ($\sim$40%) compared with the modest rate found in HH-objects ($\sim$ 15%) may suggest a different kind of phenomenon.

3. MODELS

Once Herbig (1951) presented the spectra of the newly discovered HH 1 and 2, their peculiar characteristics were immediately recognized. Especially striking was the strong emission in the [O I] and [S II] lines as compared with diffuse H II regions and planetary nebulae. This fact led Osterbrock (1958) to conclude that radiative processes could not be the dominant ionization mechanism. Furthermore, a simple photoionization model would require a "central star" equivalent to a black body with radius 1 $R_\odot$ and temperature $\sim$ 25,000 ^oK (Böhm 1956; Osterbrock 1958). No such star is seen in any known HH-objects.

Beside this, Böhm (1956) explored tne possibility that the source of ionizing radiation might be accretion onto a protostellar object. This mechanism was found to fail by several orders of magnitude.

Magnan and Schatzman (1965) on the other hand, suggested that the ionization and excitation in HH-objects might be due to the streaming of high energy protons ($\sim$100 MeV) through a neutral medium. Also, Gurzadyan (1974, 1975) advanced the idea that the ionization is the result of transition radiation between fast, 1 MeV electrons, and dust particles within the nebula. However, Gurzadyan's idea requires the stream of electrons to have a very large energy and the model by Magnan and Schatzman is unable to account for the dimensions of HH-objects (Schwartz 1974).

Following extensive search for the exciting star in several HH-objects, Strom, Grasdalen and Strom (1974a); Strom, Strom and Grasdalen (1974b), and Strom, Strom, Kinman (1974c) reported the discovery of several IR-sources not within but displaced ($\sim$ 30 arc sec typically) from the optical object. Strom and collaborators also observed a large and peculiar optical polarization in HH 24 and HH 100. These findings would be naturally explained if the HH-objects were reflection nebulae and not intrinsic emission ones. Thus, Strom et al. proposed a model in which an extremely young star, similar to a T Tauri but surrounded by an active line emitting envelope, is located deep into the cloud. The light from the star and envelope propagates through "holes" or "channels" in the non-uniform surroundings in such a way that the obscured star is able to illuminate patches of cloud material available to the observer. These patches would represent the optical HH-object.

However, recent spectropolarimetric observations of HH 24 by Schmidt and Miller (1979) have shown that while the underlying continuum is indeed strongly polarized, the emission lines show a very weak polarization. This fact thus suggests that the lines are not emitted in a circumstellar envelope but within the optical object itself. In addition to this difficulty the reflection model is unable to explain other observational evidence, for instance, the quite different spectra among knots of a single HH-object or the large proper motions measured by Cudworth and Herbig (1979) and Jones and Herbig (1980) in HH 29, 28, 1 and 2. It therefore appears that a pure reflection model for HH-objects is not completely adequate.

At present, most observational evidence favours the interpretation that HH-objects are the result of the interaction of a strong stellar wind with its ambient medium. Let us now summarize the main arguments that back this interpretation.

Ever since Schwartz (1975) drew attention to the great similarities between the spectra of HH-objects and the SNR N49 in the LMC, a great amount of evidence has been accumulated to show that HH-objects derive their energy from the cooling of shock-heated material. In addition to the geometrical arguments given by Münch (1977), the strongest support for this hypothesis comes from theoretical calculations of slow plane-parallel shock waves by Raymond (1976, 1979), Dopita (1978) and Shull and Mc Kee (1979). These calculations are able to account for the principal features in the line spectra of HH-objects by assuming preshock densities and shock velocities of about 100 cm^{-3} and 100 km s^{-1} respectively. The source of mechanical energy for the supersonic flow would be a pre-main sequence object, probably similar to a T Tauri star.

Indeed, the strong similarities in the spectra of HH-objects and the nebula around T Tauri was first pointed out by Herbig (1951). Later Osterbrock (1958) and more recently Schwartz (1975) also noticed the observational evidence which indicates the presence of high velocity outflow motions in T Tauri (see also Kuhi 1964). Therefore, it was suggested by Schwartz (1975) that HH-objects could be formed by the interaction of a strong stellar wind with the ambient material around the star. Several detailed models have been advanced based on this idea.

First there is the cloudlet model studied by Schwartz (1978) in which there is assumed to be a strong stellar wind from an embedded recently formed star. The wind is taken to be non-ionized (since the star is proposed to be a cold T Tauri-type one) and isotropic, emerging radially from a dense circumstellar envelope. The ambient material is thought to consist of discrete dense cloudlets surrounded by a nearly uniform and less dense material. As the wind expands through this environment it sweeps up the tenuous medium leaving behind the dense clouds. The wind, now able to collide directly with the cloudlets, forms bow shocks at their front surfaces. The HH-objects are identified, with the cooling of the shock-heated wind through these shocks. A dynamical consequence of this interaction is the acceleration of the cloudlets to velocities comparable with the wind itself.

This model provides natural explanations for several observational features, for instance, the lack of spatial coincidence between the optical object and its source of energy; and the inverse correlation between the object's radial velocity and either its shock pressure or its degree of excitation (Schwartz and Dopita 1980). Also the proper motions of some HH-objects (Herbig and Cudworth 1979; Jones and Herbig 1980) fit naturally into this model.

Application of this model to HH 1 and HH 2 requires a wind with velocity $\sim$ 100 km s^{-1} and mass loss rate of $\sim 10^{-5}$ to 10^{-6} $M_{\odot}$ yr^{-1}. These values give a pre-shock density of $\sim$ 300 cm^{-3} at a typical distance of about 10^{17}cm. As noticed by Schwartz (1978) the required mass loss rate is quite high, in the sense that a star could not sustain it for a long time (more than a few thousand years). Therefore, Schwartz concluded that the HH-phenomenon must be short-lived.

However the most serious objection to the model is that it requires the star to be obscured visually by the wind itself since the cloud material in which the star was embedded has already been removed. Schwartz holds that this can be the case if only 10^{-3} of the mass in the wind consists of dust grains with radii of $\sim 10^{-5}$ cm and specific gravity around 3. The optical depth at the star would be ~ 35.

An interesting variation of the cloudlet model was recently advanced independently by Norman and Silk (1979) and Rodriguez et al. (1980). This new model (called the "interstellar bullet" model) was constructed in order to explain not only the optical HH-objects but also the water masers sometimes found in their vicinity.

Norman and Silk (1979) start by considering the early stages of protostar formation. In their view, there is a central core generating radiation pressure or a T Tauri-like wind which interacts with the still infalling matter. Rayleigh Taylor instabilities occur at the interface of the wind-cloud region and produce infalling clumps. These clumps are eventually swept up by the wind and blown outward. The clumps are maintained at high density by the wind's ram pressure. Norman and Silk identify these cloudlets with the H_2O masers, while the heat input for the masering is provided by the Kelvin-Helmholtz instability which is generated at the interface between the clumps and the supersonic wind. The maser phase terminates when the fast moving clumps (with terminal velocity ~ 100 km s^{-1}) emerge beyond the cloud interface and their molecular hydrogen density drops below that required to maintain maser conditions (between 10^8 and 10^{11} cm^{-3}). Subsequently the expanding clumps interact collisionally with the ambient cloud and a bow shock preceding the condensations is formed. The ambient gas is shock-heated and accelerated through these shocks, to velocities similar to those of the condensations. Radiation emitted by the shocked gas become observable as the HH-object.

This suggestive model is strongly supported by the proper motions of some HH-objects (Cudworth and Herbig 1979; Jones and Herbig 1980) and it also explains, naturally, the observed mean spatial distribution of HH-objects and related water masers by Rodriguez et al. (1980) in the sense that H_2O maser are generally closer to the exciting source than HH-objects.

It is worth noticing here that both the "interstellar bullets" model and Schwartz' (1978) cloudlet model assume that wind-accelerated dense clouds are responsible for the HH-phenomenon, but there is a key difference. In Schwartz' model the HH-objects represent the cooling of shocked wind material, while, in the model just described it is the shocked ambient material which radiates as the optical HH-object. The energy requirements are about the same for both models: wind velocities

100 - 200 km s^{-1} and mass loss rates $\sim 10^{-6} - 10^{-5}$ $M_\odot$ yr^{-1}, that is winds with total mechanic power $\sim 1 - 40\, \mathcal{L}_\odot$.

This high value for the wind's mechanical power constitutes the main difficulty for these models since it is expected that the wind carries only a very small fraction of the star's total luminosity. For instance, for a wind driven by radiation pressure the ratio between the power of the wind ($\mathcal{E}_\omega$) and the star's luminosity ($\mathcal{L}$) is restricted to be

$$\frac{\mathcal{E}_\omega}{\mathcal{L}} \lesssim \frac{1}{2} \left(\frac{V^*}{c} \right) = 1.67\; 10^{-4} \; \frac{V_*}{100 \text{ km s}^{-1}}$$

where V_* is the wind's terminal velocity and c the speed of light. Therefore, the star in the cloudlets model should be a very luminous one, luminous enough to be detected in the radio continuum. Unfortunately in most HH-objects such a star is absent.

A model which greatly eases the power requirement for the exciting star is that proposed by Cantó and Rodriguez (1980) (see also Cantó 1979, 1980; Cantó et al. 1980, and Barral and Cantó 1981). In it a wind producing star is assumed to be located at the edge of a molecular cloud. There, the ambient pressure is not uniform but decreases in the outward direction. In such conditions the configuration of the cavity produced by an isotropic wind is not spherical but ovoid. The shock has a larger radius of curvature in the direction of the less dense outer regions of the cloud.

Cantó (1979, 1980) and Cantó and Rodriguez (1980) analyse the static configuration maintained by the flow. The ram pressure of the wind plus the centrifugal pressure generated by the shocked wind stream balance the gas pressure in the molecular cloud. The flow configuration thus consists of an ovoid cavity filled by unshocked stellar wind and bounded by a shock in the wind. As the wind crosses the shock it is refracted, retaining an appreciable fraction of its momentum and kinetic energy, and slides along the walls of the ovoid toward the distant extreme of the configuration. The resulting annular stream of shocked wind narrows down as it moves away from the star until it shocks for the second time, on this occasion, against itself. Radiation of this second shock is identified with the optical HH-object. A schematic representation of this stellar-wind focusing effect is given in Figure 7.

Cantó and Rodriguez (1980) are able to show that the luminosity per unit area of the second shock (see Figure 7) is much larger than that of the first shock. This fact illustrates the high efficiency of the focusing mechanism advanced by them. For example the application of

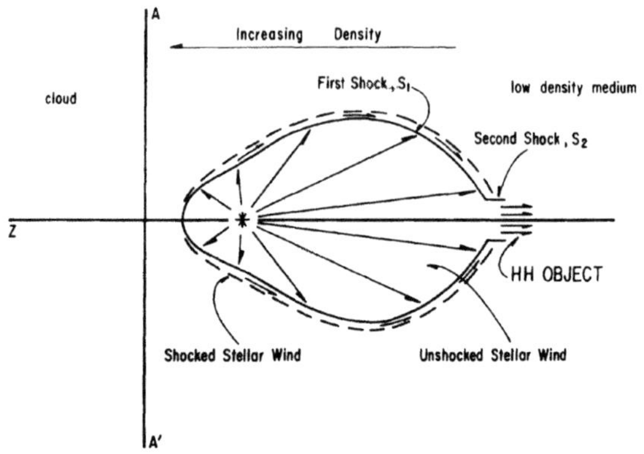

Figure 7. A schematic representation of the stellar-wind focusing model for HH-objects (Cantó and Rodriguez 1980). See text for details.

the model to HH 1 requires only about 0.3 per cent of the power required for an isotropic distribution of the wind energy.

In brief, the stellar wind focusing model is capable of explaining the small solid angles subtended by the HH-objects with respect to the exciting source without the need for unrealistically powerful stellar winds. Also, since the star is required to be in a denser region than the HH-object, the large extinction needed optically to obscure the exciting star follows naturally.

But on the other hand the focusing model, as advanced by Cantó and Rodriguez (1980), cannot account for the large proper motions recently measured in some HH-objects (Cudworth and Herbig 1979; Jones and Herbig 1980). A more detailed analysis of this model is therefore required.

4. CONCLUSIONS

Our present observational and theoretical knowledge of HH-objects was briefly summarized in Sections 2 and 3. These results seem to indicate the following conclusions,

1) The HH-objects are closely related to the star formation process and the early evolution of young stars.

2) Their emission line spectrum is produced <u>in situ</u> by the cooling

of shocked material with pre-shock densities 100 cm^{-3} and shock velocities 100 km s^{-1}.

3) The exciting source is not within the optical object but displaced from it 0.1 pc typically, and deeply embedded into the surrounding cloud.

4) The weak continuum which underlies the emission lines rapidly increases toward the blue and UV, and is likely to be the result of a reflection process.

5) A strong stellar wind is the most likely agent by which the embedded energy source is able to excite the optical object.

6) This stellar wind has to be focused in preferential directions in order to explain the small solid angles subtended by the HH-objects with respect to the exciting source.

Even though all these conclusions are well founded and represent a great advance in the understanding of HH-objects, it is clear that much more work remains to be done.

ACKNOWLEDGMENTS

I would like to thank J. E. Dyson, F. D. Kahn, J. Meaburn, L. F. Rodriguez and M. Tapia for their careful reading of the manuscript and their valuable suggestions. This work was partly carried in a visit in February 1981 to the Astronomy Department, Manchester University, sponsored by the S.R.C. I am grateful to Professor Z. Kopal for the hospitality of his Department.

REFERENCES

Adams, M. T., Strom, K. M. and Strom, S. E.: 1979, Ap. J. Letters, 230, L183.

Ambarzumian, V. A.: 1954, Comm. Burakan Obs., No.13.

Barral, J. F. and Cantó, J.: 1981, Rev. Mexicana Astron. Astrof., 5, xxx.

Beckwith, S., Gatley, I., Matthews, K. and Neugebauer, G.: 1978, Ap. J. Letters, 223, L41.

Böhm, K. H.: 1956, Ap. J., 123, p.379.

Böhm, K. H.: 1975, "Problems in Stellar Atmospheres and Envelopes" (B. Baschek, W. H. Kegel and G. Traving, eds.) p.205, Heidelberg - New York: Springer-Verlag.

Böhm, K. H.: 1978a, Astron. Astrophys., 64, p.115.

Böhm, K. H.: 1978b, "Protostars and Planets" (T. Gehrels, ed.) p.632, The University of Arizona Press.

Böhm, K. H.: 1978c, "The Interaction of Variable Stars with their Environments", IAU Coll. No.42, ed. R. Kippenhahn, J. Rahe, and Strohmeier, Ver. Remeis-Sternwarte Bamberg 11, No.121 p.13.

Böhm, K. H. and Brugel, E. W.: 1979, Astron. Astrophys., 74, p.297.

Böhm, K. H., Perry, J. F. and Schwartz, R. D.: 1973, Ap. J., 179, p.149.

Böhm, K. H., Siegmund, W. A. and Schwartz, R. D.: 1976, Ap. J., 203, p.399.

Brugel, E. W., Böhm, K. H. and Mannery, E.: 1980, Ap. J., preprint

Cantó, J.: 1979, Ph.D. Thesis, University of Manchester.

Cantó, J.: 1980, Astron. Astrophys., 86, p.327.

Cantó, J., Goudis, C., Johnson, P. G. and Meaburn, J.: 1980, Astron. Astrophys., 85, p.128.

Cantó, J. and Rodriguez, L. F.: 1980, Ap. J., 239, p.982.

Cantó, J., Rodriguez, L. F., Barral, J. F. and Carral, P.: 1981, Ap. J., 244, xxx.

Cohen, M.: 1980, Ap. J., 85, p.29.

Cohen, M. and Schwartz, R. D.: 1979, Ap. J. Letters, 233, L77.

Cohen, M. and Schwartz, R. D.: 1980, Mon. Not. R. astr. Soc., 191, p.165.

Cox, D. P.: 1972, Ap. J., 178, p.143.

Cudworth, K. M. and Herbig, G. H.: 1979, Ap. J., 84, p.548.

Dickinson, D. F., Kojoian, G., and Strom, S. E.: 1974, Ap. J. Letters, L93.

Dopita, M. A.: 1978, Ap. J. Suppl., 37, p.117.

Elias, J. H.: 1978, Ap. J., 224, p.857.

Elias, J. H.: 1980, Ap. J., 241, p.728.

Fischer, J., Righini-Cohen, G. and Simon, M.: 1980, Ap. J. Letters, 238, L155.

Gull, T. R., Goad, L., Chiu, H-Y., Maran, S. P. and Hobbs, R.: 1973, Publ. Astron. Soc. Pacific, 85, p.526.

Gurzadyan, G. A.: 1974, Astron. Astrophys., 33, p.307.

Gurzadyan, G. A.: 1975, Astrofizika, 11, p.531.

Gyulbudaghian, A. L., Glushkov, Yu. I. and Denisyuk, E. K.: 1978, Ap. J. Letters, 224, L137.

Haro, G.: 1950, Ap. J., 55, p.72.

Haro, G.: 1952, Ap. J., 115, p.572.

Haro, G.: 1976, Bol. Instituto Tonantzintla, 2, p.3.
Haro, G. and Minkowski, R.: 1960, Ap. J., 65, p.490.
Haschick, A. D., Moran, J. M., Rodriguez, L. F., Burke, B. F., Greenfield, P. and Garcia-Barreto, J. A.: 1980, Ap. J., 237, p.26.
Haschick, A. D., Rodriguez, L. F., Ho, P. T. P. and Moran, J. M.: 1981, (in preparation).
Herbig, G. H.: 1951, Ap. J., 113, pp.697.
Herbig, G. H.: 1957, in IAU Symp. No.3. (G. H. Herbig ed.) p.3., Cambridge, University Press.
Herbig, G. H.: 1962, Advances Astron. Astrophys., 1, p.75.
Herbig, G. H.: 1969, in "Nonperiodic Phenomena in Variable Stars", (L. Detre, ed.) p.75, Dordrecht: Reidel.
Herbig, G. H.: 1973, Comm. 27, IAU Inf. Bull. Var. Stars, No.832.
Herbig. G. H.: 1974, Lick Obs. Bull., No.658.
Ho, P. T. P.: 1977, Ph.D. Thesis, Massachusetts Institute of Technology.
Ho, P. T. P. and Barret, A. H.: 1980, Ap. J., 237, p.38.
Jones, B. P. and Herbig, G. H.: 1980, Bull. AAS., 12, p.843.
Knapp, G. R., Kuiper, T. B. H., Knapp, S. L. and Brown, R. L.: 1976, Ap. J., 206, p.443.
Kuhi, L. V.: 1964, Ap. J., 140, p.1409.
Kwan, J.: 1977, Ap. J., 216, p.713.
Lada, C. J., Gottlieb, C. A., Litvak, M. M. and Lilley, A. E.: 1974, Ap. J., 194, p.609.
Lo, K. Y., Burke, B. F. and Haschick, A. D.: 1975, Ap. J., 202, p.81.
Lo, K. Y., Morris, M., Moran, J. M. and Haschick, A. D.: 1976, Ap. J., 204, L21.
Loren, R. B.: 1976, Ap. J., 209, p.466.
Loren, R. B., Evans, N. J. and Knapp, G. R.: 1979, Ap. J., 234, p.932.
Luyten, W. J.: 1963, Harvard Annu. Card., No.1589.
Luyten, W. J.: 1970, in "The Hyades", University of Minnesota, Minneapolis, Min.
Magnan, C. and Schatzman, E.: 1965, C. R. Acad. Sci. Paris, 260, p.6289.
Mc Kee, C. F. and Hollenbach, D. J.: 1980, Ann. Rev. Astron. Astrophys., 18, p.219.
Münch, G.: 1977, Ap. J. Letters, 212, L77.
Norman, C. and Silk, J.: 1979, Ap. J., 228, p.197.
Ortolani, S. and D'Odorico, S.: 1980, Astron. Astrophys., 83, L8.
Osterbrock, D. E.: 1958, Publ. Astron. Soc. Pacific, 70, p.399.
Pradhan, A. K.: 1978, Mon. Not. R. astr. Soc., 183, p.89.
Raymond, J. C.: 1976, Ph.D. Thesis, University of Wisconsin, Madison.

Raymond, J. C.: 1979, Ap. J. Suppl., 39, p.1.
Rodriguez, L. F., Carral, P., Ho, P. T. P. and Moran, J. M.: 1981a (preprint).
Rodriguez, L. F., Ho, P. T. P. and Moran, J. M.: 1981b, Ap. J. Letters, 240, L149.
Rodriguez, L. F., Moran, J. M., Dickinson, D. F. and Gyulbudaghian A. L.: 1978, Ap. J. Letters, 226, L21.
Rodriguez, L. F., Moran, J. M., Ho, P. T. P. and Gottlieb, E. W.: 1980, Ap. J., 235, p.845.
Sabbadin, F., Minello, S. and Bianchini, A.: 1977, Astron. Astrophys 60, p.147.
Schmidt, G. D. and Miller, J. S.: 1979, Ap. J. Letters, 234, L191.
Schmidt, G. D. and Vrba, F. J.: 1975, Ap. J. Letters, 201, L33.
Schwartz, R. D.: 1974, Ap. J., 191, p.419.
Schwartz, R. D.: 1975, Ap. J., 195, p.631.
Schwartz, R. D.: 1977, Ap. J. Suppl., 35, p.161.
Schwartz, R. D.: 1978, Ap. J., 223, p.884.
Schwartz, R. D. and Dopita, M. A.: 1980, Ap. J., 236, p.543.
Shull, J. M. and Mc Kee, C. F.: 1979, Ap. J., 227, p.131.
Shull, J. M. and Hollenbach, D. J.: 1978, Ap. J., 220, p.525.
Snell, R. L., Loren, R. B. and Plambeck, R. L.: 1980, Ap. J. Letters, 239, L17.
Strom, S. E., Grasdalen, G. L. and Strom, K. M.: 1974a, Ap. J., 191, p.111.
Strom, K. M., Strom, S. E. and Grasdalen, G. L.: 1974b, Ap. J., 187, p.83.
Strom, S. E., Strom, K. M. and Grasdalen, G. L.: 1975, Ann. Rev. Astron. Astrophys., 13, p.187.
Strom, K. M., Strom, S. E.and Kinman, T. D.: 1974c, Ap. J. Letters, 191, L93.
Strom, K. M., Strom, S. E. and Vrba, F. J.: 1976, Ap. J., 81, p.320.
Taylor, K. and Münch, G.: 1978, Astron. Astrophys., 70, p.359.
Vrba, F. J., Strom, S. E. and Strom, K. M.: 1975, Publ. Astron. Soc. Pacific, 87, p.337.

THE DYNAMICAL EFFECTS OF HYPERSONIC STELLAR WINDS ON INTERSTELLAR GAS

J. E. Dyson
Astronomy Department, University of Manchester.

ABSTRACT

Early type stars have energetic winds which impinge hypersonically on interstellar gas. We review this interaction with particular emphasis on the resulting dynamical effects. The basic flow pattern and physical state of the gas are first briefly discussed and the results then applied to the competition between a wind driven shell and the ionization front produced by the stellar radiation field. We then examine a variety of topics: the dynamical structure of galactic and extragalactic HII regions, the production and maintenance of density fluctuations within nebulae, and the formation of large scale loop structures in the plane of the galaxy. The emphasis throughout the review is on simple semi-qualitative models of various wind related phenomena.

1. INTRODUCTION

The physical and dynamical state of interstellar matter in the spiral arms of galaxies is largely determined by its interaction with Population I stars. This interaction occurs in a variety of ways. The most violent - and probably most important - interaction occurs when a star becomes a supernova. But the more massive hot stars have two other options open to them. They can interact through their UV radiation fields and - the topic of this review - through continuous ejection of matter in the form of a wind. Many different types of star possess winds but massive young stars have particularly energetic ones. They can eject several solar masses of material at velocities of many thousand kilometres per second over their lifetimes. We here concentrate on a particular aspect of this interaction, namely the dynamical effects produced when such a violent wind impinges on interstellar gas. There are other effects of importance, for example the heavy element enrichment of interstellar gas which occurs through mixing with these winds. However these will not be discussed here.

F. D. Kahn (ed.), Investigating the Universe, 125–150.

2. OBSERVATION OF MASS LOSS

The first observational evidence for the existence of strong stellar winds in early type stars came from UV observations made during unmanned rocket flights (Morton, 1967). The profiles of resonance lines (e.g., Si IV) showed the P Cygni characteristics associated with extended absorbing stellar atmospheres. The wavelength shifts of the absorption edges showed that the velocity of the absorbing material always exceeded the stellar escape velocity. The conclusion that the stars were losing matter was unavoidable. Since these pioneering observations were made considerable effort has gone into the observation of mass loss from hot stars and its interpretation. We briefly review some of the details pertinent to a study of the interaction of this mass loss with interstellar gas and refer the interested reader to the Proceedings of I.A.U. Symposium No. 83 (1979) for more extensive discussion.

Essentially all stars of bolometric magnitude $M_{BOL} \leqslant -6$ have spectral features indicating mass loss. The terminal wind velocities $\bar{V}_*$ range from about 1000 to 4000 km s^{-1}. Abbott (1978) has proposed the empirical relationship $V_* \simeq 3\ V_{esc}$ where V_{esc} is the stellar escape velocity. Quite clearly, such velocities are hypersonic with respect to almost any conceivable surrounding interstellar medium. They are also hypersonic with respect to the winds themselves. A steady wind has a constant velocity (equal to V_*) and an associated inverse square fall-off of density with distance from the star.

The rates of mass loss have been determined by a wide variety of methods (Conti and McCray 1980). For O stars, the inferred rates lie in the range $10^{-8} - 10^{-5}\ M_{\odot}\ yr^{-1}$. Their relationship to the stellar properties is far from unambiguous, but they do seem correlated with stellar luminosity (though, see Conti and Garmany (1980) for a caution to this statement). This correlation argues strongly that the winds are driven by radiation pressure (Abbott et al. 1980). There is also evidence that the mass loss rate depends on the evolutionary status of the star. Generally, stars to the right of the ZAMS in the H-R diagram have the highest mass loss rates. Particularly large mass loss rates are associated with Of stars. Note that the uncertainty in the measured mass loss rate for a given star is typically a factor 3 (Conti and Garmany 1980).

Of all the stars observed, the largest mass loss rates are associated with Wolf-Rayet stars (of both W-N and W-C subgroups). Barlow, Smith and Willis (1981) have observed a selection of W-R stars and find average mass loss rates of about $3.10^{-5}\ M_{\odot}\ yr^{-1}$ for W-N types and $5.10^{-5}\ M_{\odot}\ yr^{-1}$ for W-C types. It seems very likely that Of stars evolve into W-R stars (Conti 1976). Further, it is thought (Willis and Wilson 1978) that the W-N types may evolve into W-C types. A general trend towards increasing mass loss rate during the stellar evolution seems indicated.

A very important parameter from the gas dynamical standpoint is the stellar mechanical luminosity $\dot{E}_*$ ($\equiv \frac{1}{2} \dot{M}_* V_*^2$). Values inferred from observation lie in the range 10^{34} - 10^{37} erg s^{-1} for O and Of stars and up to about 10^{38} erg s^{-1} for the most vigorous W-C stars. In extreme cases a star can inject $\sim 10^{50}$ erg in the form of mechanical energy into its surroundings. As has been noted many times, this total injection is comparable with the energy released in a Type II supernova. The ratio of stellar mechanical to radiative luminosity is considerably less than 1:100 for main sequence O stars, about 1:100 for Of stars, and up to 1:5 for W-R stars. However, the efficiency of conversion of stellar UV radiation into kinetic energy is extremely low (<< 1%). A much higher ($\sim$ 20% efficiency obtains (in the idealised case) for the conversion of stellar mechanical energy output and this compensates for the disparity in the two forms of luminosity). Note that the spectra of HII regions stirred up by winds will still be essentially radiative. However, the spectra will be influenced by the winds indirectly through their effects on such parameters as dilution factors, densities etc.

It is extremely unfortunate that so little is known for certain at present about the way in which stars switch on their winds and how the winds vary during the course of stellar evolution. Hydrodynamic calculations demand the specification of initial and boundary conditions which clearly cannot be done in any unambiguous way. There are, of course, additional uncertainties in the specification of conditions in the surrounding interstellar gas. It is again regrettable that many phenomena which seem ascribable to stellar wind activity are associated with stars or stellar clusters whose mass loss rates have not been observationally determined.

3. BASIC PHYSICAL CONSIDERATIONS

3.1 The Flow Pattern

A hypersonic wind incident on a uniform distribution of interstellar gas produces the characteristic two shock pattern first described by Pikelner (1968) (Fig. 1). An outwards facing shock S accelerates ambient gas and an inwards facing shock S^1 decelerates the wind. Regions (a), (b), (c) and (d) (Fig. 1) respectively contain unshocked wind shocked wind, shocked ambient gas and unshocked ambient gas. A contact discontinuity C separates regions (b) and (c). The physical conditions in the various regions will be briefly discussed in the appropriate context later.

3.2 The Very Early Stages of the Flow

Assume that the star switches on its wind luminosity suddenly and maintains it thereafter at a steady rate $\dot{E}_{36}$ ($\equiv \dot{E}_*/10^{36}$). The ambient density is taken as uniform (n_0 cm^{-3}). A similarity flow is set up at a time of the order of that required for the wind to sweep up its own mass in ambient gas. This occurs at radius R_d and time t_d given by

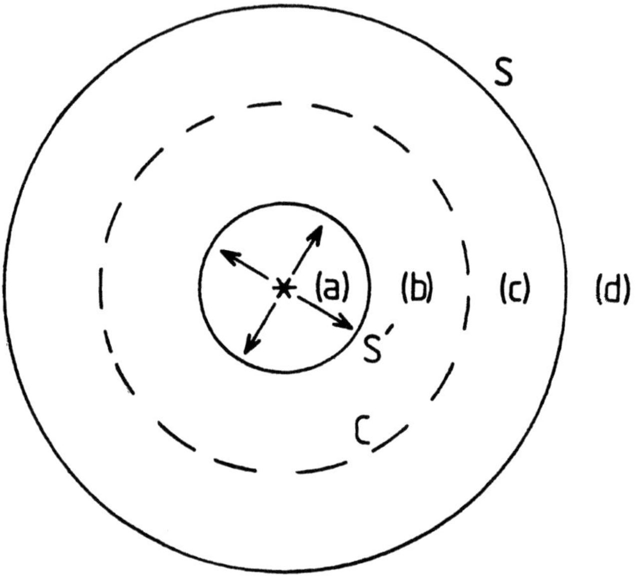

Fig. 1 The flow pattern resulting from the interaction of a high speed wind with surrounding interstellar gas. S and S^1 are shocks and C is a contact discontinuity.

$$R_d \simeq 0.25 \ (\dot{E}_{36}/V_8{}^3 \ n_0)^{\frac{1}{2}} \ \text{pc},$$
$$t_d \simeq 200 \ (\dot{E}_{36}/V_8{}^5 \ n_0)^{\frac{1}{2}} \ \text{yr}. \qquad (1)$$

Here and afterwards, V_8 is the wind terminal velocity in units of 1000 km s^{-1}. Falle (1975) noted that the contact discontinuity in a decelerating flow is Rayleigh-Taylor unstable if the gas in region (b) is denser than that in region (c). The density in region (b) decreases rapidly with time because of the geometrical dilution of the wind. The instability dies out when the outer shock has a radius comparable to R_d. It is thus reasonable to ignore this instability and, since t_d is much less than the dynamical timescale of the flow, to assume that the similarity flow is set up instantaneously.

3.3 The Early Stages of the Flow

Initially regions (b) and (c) behave adiabatically. The time dependences of the outer shock radius R_s and velocity V_s can be found from dimensional considerations. Taking power law dependences of the for

$$\dot{E}_* = e_* \, t^{\alpha},$$

$$\rho_0 = \rho_* r^{\beta}, \tag{2}$$

for the mechanical luminosity and the ambient density respectively, then

$$R_s \propto (e_*/\rho_*)^{\frac{1}{(5+\beta)}} t^{\frac{3+\alpha}{5+\beta}}, \tag{3}$$

$$V_s \propto (e_*/\rho_*)^{\frac{1}{(5+\beta)}} t^{\frac{\alpha-\beta-2}{5+\beta}},$$

In principle, both accelerating and decelerating solutions are possible. The contact discontinuity is always Rayleigh-Taylor unstable in an accelerating flow. To examine consequences of the flow we generally adopt $\alpha = \beta = 0$ giving the well known dependences $R_s \sim t^{3/5}$, $V_s \sim t^{-2/5}$. Note that it is here implicitly assumed that S and S^1 are strong shocks.

3.4 The Breakdown of Adiabaticity

(i) Region (b). The post-shock gas temperature in this region is of order $T_b \sim 10^7 V_8^2$ K and the gas cools by bremsstrahlung. Because of geometrical dilution, the gas density is low and the radiative cooling time is long. In fact, provided that conduction effects at the interface C are neglected, the cooling time is much longer than the overall evolution time of the flow. Region (b) will then act as an adiabatically expanding bubble whose pressure drives the outer shock S.

Weaver et al. (1977) have considered in some detail the effects produced by allowing conduction at the interface C. They conclude that the mass of region (b) is dominated by material evaporated from region (c). The mean density and temperature of region (b) are respectively increased and decreased thereby decreasing the radiative cooling time in this gas. For moderate ambient densities ($n_0 \lesssim 10^3$ cm^{-3}), the dynamical evolution of the system is relatively unaffected. The gas at the conductive interface may contribute to the production of interstellar O VI absorption lines seen in the spectra of many hot stars (Jenkins and Meloy 1974). At high ambient densities ($n_0 >> 10^4$ cm^{-3}) however, the cooling time in region (b) becomes considerably less than the dynamical timescale (Shull, 1980). Region (b) collapses and the outer shock is driven by the wind momentum. Equations (3) are then of course, no longer applicable.

However, for the conductive mass flux across C to assume such significance, suppression of conduction by magnetic fields must not be important. It seems very unlikely that the ambient gas will not contain a magnetic field with a strength of the order of microgauss. Compression across S amplifies the tangential field component and tangential fields of the order of microgauss will be present in the region (c). They will strongly inhibit conduction across them. We adopt the viewpoint that conduction is unimportant and that S is driven by an adiabatically expanding bubble.

(ii) Region (c). Adiabaticity breaks down quickly in this gas for two reasons. Firstly the density is always relatively high. Secondly, as the flow decelerates, the gas temperature behind S decreases and the shocked gas moves towards temperatures lying in the thermally unstable section of the interstellar cooling function (Raymond, Cox and Smith 1976). Strong cooling sets in behind S at a time t_{sg} given by (cf. Dyson, Falle and Perry 1980)

$$t_{sg} \simeq 3\ 10^3\ (\dot{E}_{36}^{3}\ n_0^{-8})^{1/11}\ \text{yr} \tag{4}$$

The shock S has radius R_{sg} and velocity V_{sg} at this time which are given by

$$R_{sg} \simeq (\dot{E}_{36}^{4}\ n_0^{-7})^{1/11}\ \text{pc},$$
$$V_{sg} \simeq 150\ (\dot{E}_{36}\ n_0)^{1/11}\ \text{km s}^{-1} \tag{5}$$

These results are derived using the approximate cooling law given by Kahn (1976). The insensitivity of V_{sg} to the parameters has been noted elsewhere (Dyson et al. 1980).

The evolution of the swept-up gas in the phase after $t = t_{sg}$ is complex and extremely difficult to follow numerically even in very idealised cases (Falle 1975). The difficulties arise because in essentially all cases of interest, the cooling time in the swept up gas becomes much less than the sound crossing time in region (c). The computations show that initially the driving pressure on S falls and V_s therefore decreases. Steep local pressure gradients occur in region (c) and they produce secondary shock waves. Falle (1975) showed that S eventually is reaccelerated by an outwards moving secondary shock. Eventually a thin shell of compressed ambient gas forms at time $t_{sh} \sim 3\ t_{sg}$.

3.5 The Flow After Thin Shell Formation

At times $t \gtrsim t_{sh}$, the flow morphology is that of a thin shell surrounding a bubble (at least in this idealised case). The shell thickness $\Delta R_s \sim R_s\ M^{-2}$, where M is the mach number of S relative to the sound speed in the shell gas. As long as the shell is optically thin in the Lyman continuum, the sound speed is ≈ 10 km s^{-1}. The shell gas may partially recombine (section 4) and $\Delta R_s/R_s$ becomes much reduced.

The time dependences of the outer shock radius and velocity are still those of relationships (3) although S is now an isothermal shock. Useful approximations for this thin shell phase are ($\alpha = \beta = 0$),

$$R_s \simeq 1.7\ (\dot{E}_{36}/n_0)^{1/5}\ t_4^{3/5}\ \text{pc},$$

$$V_s \simeq 100\ (\dot{E}_{36}/n_0)^{1/5}\ t_4^{-2/5}\ \text{km s}^{-1} \qquad (6)$$

Here t_4 is the time in units of 10^4 yr. The kinetic energy of the swept up gas $E_K \overset{\sim}{\sim} 0.2\ E_* t$ (Weaver et al., 1977).

It should be noted that the sound speed in region (b) is always much greater than V_s. Region (b) is spatially - but not temporally - isobaric. This would still be the case even if the mass loss from the central star were non-isotropic. Departures of the outer shock shape from sphericity occur as a result of non-uniformities in the ambient gas.

We note finally that there is evidence for short timescale (less than a few months) variability in some winds. This variability is most likely related to instabilities occurring in the winds near the stellar surface (Kahn, 1981). This timescale is always far less than the dynamical timescale and the variability can be ignored as far as the interaction dynamics is concerned.

4. THE COMPETITION BETWEEN THE SHELL AND THE IONIZATION FRONT PRODUCED BY AN ISOLATED STAR

We assume that the star switches on its Lyman continuum output, S_* photons s^{-1}, and steady wind simultaneously and instantaneously. The ambient density is taken as uniform. With these assumptions, the ionization front (I-F) moving into the ambient gas is initially weak-R type. An I-F of this type heats the gas it overruns but does not set it into motion or change its density. The initial Strömgren radius R_0 ($\equiv (3\ S_*/4\ \pi\ n_0^2\ \gamma)^{1/3}$) is reached by the I-F in a time $t_0 \overset{\sim}{\sim} (\gamma\ n_0)^{-1}$ in the absence of a wind. In the numerical estimates to follow, γ, the recombination coefficient, takes the value 2.10^{-13} cm^3 s^{-1}.

The ratio of the time when strong cooling sets in behind a wind driven shock to that when the I-F reaches R_0 is $t_{sg}/t_0 \overset{\sim}{\sim} 0.02\ (\dot{E}_{36} n_0)^{3/11}$. Therefore if $\dot{E}_{36} n_0 \overset{<}{\sim} 2\ 10^6$, $t_{sg}/t_0 \overset{<}{\sim} 1$ and strong cooling sets in before the initial Strömgren sphere can be set up. In this case, Equations (6) adequately describe the shell dynamics for a comparison with those of the I-F.

The shell reaches a radius R_s (t_0), at the time when the initial

Strömgren sphere is established, given by

$$R_s\ (t_0) \simeq 0.3\ (\dot{E}_{36}\ n_0^{-2/3})^{1/5}\ S_{48}^{-1/3}\ R_0 \tag{7}$$

where $S_{48} \equiv S_*/10^{48}$ Under all realistic circumstances, $R_s(t_0) << R_0$ i.e. the I-F moves out to the Strömgren radius well ahead of the shell.

The recombination rate in the swept up shell, $\dot{N}_R$, is given approximately by

$$\dot{N}_R \simeq \frac{4}{3}\pi\ R_s^{\ 3}\ V_s^{\ 2}\ n_0^2\ \frac{\gamma}{C_i^{\ 2}} \tag{8}$$

where C_i ($\simeq$ 10 km s^{-1}) is the sound speed in the ionized gas. At $t = t_0$, the fraction of the stellar Lyman continuum absorbed in the shell is

$$\dot{N}_R/S_* \simeq 0.2\ \dot{E}_{36}/S_{48} \tag{9}$$

From the observational data referred to in Section 2, we find that this fraction is << 1 for O stars and generally $\lesssim$ 0.1 for Of stars. Thus the shell is absorbing an insignificant fraction of the stellar radiation field when the initial Strömgren sphere is set up.

At time t_0 the wind driven shell velocity is

$$V_s\ (t_0) \simeq 20\ (\dot{E}_{36}\ n_0)^{1/5}\ \text{km s}^{-1} \tag{10}$$

Thus $V_s\ (t_0) > C_i$, which is required for the consistency of the strong shock approximation.

The I-F changes character from weak-R type to weak-D type (via a strong-D phase) at a time $t \simeq t_0$. The weak-D front is preceded by a shock moving directly into neutral gas. In the frame of reference of the star, the I-F and the shock move at essentially the same velocity which rapidly becomes less than C_i. Provided therefore that the shell is being pushed supersonically (with respect to C_i), the I-F is approximately stationary whilst the shell tries to catch up with it. We can therefore neglect changes in the gas density behind the I-F (i.e. ahead of the shell) in the estimates which follow.

Equation (8) shows that the recombination rate is increased by sweeping ambient gas into a thin shell. If $\dot{N}_R/S_* \gtrsim 1$ before the shell reaches a radius equal to that of the initial Strömgren sphere, R_0, the

I-F becomes trapped within the shell. Clearly this must occur if the shell velocity is greater than C_i when it reaches a radius equal to R_0. When $R_s = R_0$, the shell velocity is

$$V_s(R_0) \simeq 14\ (n_0/S_{48}{}^2)^{1/9}\ \dot{E}_{36}{}^{1/3}\ \text{km s}^{-1} \tag{11}$$

For trapping, $V_s(R_0) > C_i$ giving the trapping criterion

$$n_0 \gtrsim n_{crit} \simeq 0.4\ S_{48}{}^2/\dot{E}_{36}\ \text{cm}^{-3} \tag{12}$$

We briefly consider the two extreme cases.

(a) $n_0 \gg n_{crit}$. The I-F becomes trapped when $\dot{N}_R/S_* \simeq 1$ which occurs at time $t = t_R$ given by

$$t_R \simeq 100\ S_{48}/\dot{E}_{36}\ n_0\ 10^4\ \text{yr} \tag{13}$$

The shell then has radius $R_s(t_R)$ and velocity $V_s(t_R)$ given by

$$R_s(t_R) \simeq 0.9\ \dot{E}_{36}{}^{-2/5}\ n_0{}^{-2/15}\ S_{48}{}^{4/15}\ \text{pc},$$

$$V_s(t_R) \simeq 15\ \dot{E}_{36}{}^{3/5}\ n_0{}^{1/5}\ S_{48}{}^{-2/5}\ \text{km s}^{-1} \tag{14}$$

The gas ahead of the wind driven shell recombines in a time $\sim (\gamma\ n_0)^{-1}$. This is much shorter than the dynamical timescale for the subsequent evolution of the flow and the recombination effectively is instantaneous. Clearly, when recombination occurs, the shell velocity $V_s(t_R)$ is supersonic with respect to neutral gas (sound speed $\lesssim 1$ km s^{-1}) and a strong shock is driven into neutral gas.

Mazurek (1980) has proposed that the classical expansion of HII regions around early type stars may be responsible for the dispersal of the molecular clouds in which they are formed. Since the ambient density in these clouds is generally high enough to be greater than n_{crit}, the inclusion of stellar wind effects on the expansion should follow the above general outline. Suppose that the main sequence lifetime and wind emitting period of the star are identical and equal to t_{m4} (in units of 10^4 yr). It can be shown that provided $t_{m4} \gtrsim (S_{48}{}^5/\dot{E}_{36}{}^7\ n_0{}^3)$, and $n_0 \gtrsim n_{crit}$, the effect of the inclusion of the wind is to attain a greater radius (and velocity) at t_{m_4} than is realised if the wind is neglected. These inequalities can readily be satisfied at densities characteristic of molecular clouds ($n_0 \gtrsim 10^4$ cm^{-3}). The wind assisted HII regions should be even more efficient at dispersing molecular clouds than the classical expanding HII regions. This further strengthens Mazurek's (1980) conclusions regarding the importance of HII regions in dispersing molecular clouds.

(b) $n_0 \lesssim n_{crit}$. In this situation the I-F never becomes trapped inside the wind driven shell. It is, in fact, possible for the outer shock S to degenerate into a sound wave even before reaching a radius equal to R_0. Degeneration occurs after a time $t_{s0} \approx 2.10^6$ $(\dot{E}_{36}/n_0)^{\frac{1}{2}}$ yr. provided t_{s0} is less than the time for which the star blows off the wind. The overall evolution of the HII region is very little affected by the wind inclusion although the region has a somewhat different morphology.

Considerable refinement needs to be made to the above treatment. Shull (1980) notes that the time for a main sequence O star to turn on its radiation field may be considerably greater than t_d and t_0. Dust opacity can affect the propagation of the I-F. Further, early type stars most likely reach their main sequence luminosities whilst still surrounded by a cocoon of dust and gas (Kahn, 1974). The propagation of both the I-F and the shell will surely be affected by the surrounding structure. Perhaps the major uncertainties lie in the relationship between, and the temporal behaviour of, the stellar radiative and mechanical luminosities. Considerable work is needed to clarify the various possibilities.

5. STELLAR WINDS AND THE DYNAMICS OF HII REGIONS

Many HII regions contain several exciting stars which are distributed inside the region and presumably were not formed simultaneously. Problems such as those relating to the trapping of ionization fronts then become very difficult to treat with any certainty. We here consider only very simple models and ignore many such complications.

5.1 The Rosette Nebula

Very few HII regions clearly exhibit the spherical cavity appearance predicted by the theory of Section 3. An exception though is the Rosette Nebula. The internal velocity structure of the ionized gas has been studied in great detail by Smith (1973) and Fountain, Gary and O'Dell (1979). The gas velocities in the inner region of the nebula are about 20 km s^{-1}. Fountain et al. (1979) show that the observations can be fitted with a velocity distribution in which the gas velocity decreases outwards from the centre of the region. This distribution is consistent with a model in which there exists a wind driven shell in the interior of the nebula and in which the outer regions expand at about the sonic velocity. The central cavity in the Rosette Nebula is precisely the morphology expected on the wind model.

The shell characteristics can be obtained from Equations (6). An estimated cavity radius ($\equiv R_s$) of 12 pc and an ambient density of 10 cm^{-3} (Menon, 1962) lead to a shock age $t_4 \simeq 30$ and a required mechanical luminosity $\dot{E}_{36} \simeq 4$. The stellar cluster in the Rosette Nebula contains six O stars, the two most important are spectral types O4 ((f)) and O5 ((f)) (Pottasch, Wesselius and van Duinen, 1979). Conti and Garmany (1979) quote an estimated mechanical luminosity for the O4 ((f)) star of $\dot{E}_{36} \gtrsim 3$, which is in excellent (though fortuitous) agreement with that deduced from the dynamical model.

In order that the expanding shell make a significant contribution to the line profiles, the shell must clearly absorb a substantial fraction of the Lyman continuum output of the exciting stars. This fraction, ϕ say, is given by

$$\phi \simeq \dot{E}_{36}\, n_0\, t_4/100\, S_{48} \tag{15}$$

A comparison of the parameters of the two main exciting stars (Pottasch et al. 1979) with the results of Panagia (1973) gives the estimate $S_{48} \simeq 50$ for the Lyman continuum output from the cluster. From the previously derived parameters and Equation (15) we find $\phi \simeq 0.24$. This is high enough (in view of the uncertainties) to make the general model very plausible.

5.2 General Velocity Structure in Nebulae

(a) Galactic Nebulae. The observed velocity structure of diffuse nebulae is often extremely complex (see, for example, the article by J. Meaburn in this volume). Here we pick out some general features and examine them in the context of the stellar wind interaction.

The spherically symmetric HII region formed by the sudden turn on of an early type star in a uniform medium is largely a figment of the theorists' imagination. Many nebulae (for example M8, M17, M42) appear to be located at the front edge (with respect to the observer) of extensive neutral clouds (Zuckerman, 1973; Gull and Balick, 1974; Elliott and Meaburn, 1975). This morphology finds a natural explanation in the so called 'blister' models of HII regions. An early type star forms near the edge of a molecular cloud and its radiation field ionizes through the cloud edge into the surrounding low density (and pressure) medium. The dynamical evolution of the high density ionized gas as it flows into the low pressure surroundings has been investigated numerically (Bodenheimer, Tenorio-Tagle and Yorke, 1979). Quite clearly a spherically symmetric wind interaction is no longer valid since the wind must blow into ionized gas which is far from uniform.

Dyson (1977a) has analysed a very simple model for this situation using the geometry depicted in Fig. 2. Ionized gas is assumed fed continuously into the nebula by the erosion of the neutral cloud by the stellar UV. This gas expands away from the neutral cloud into the much lower pressure surroundings and the ionized gas density therefore decreases with distance from the cloud. (Of course, this flow itself will produce a velocity pattern). The wind from a star within the ionized region blows into gas containing pronounced density gradients. The interior pressure driving the shell swept up by the wind remains uniform (Section 3.5) but the shock velocity at any point on the shock surface depends on the gas density immediately ahead of the shock. The outer shock (or the shell) becomes distorted. In the context of Fig. 2 the portion of shell moving towards the neutral cloud (i.e. into regions of increasing density) slows down more rapidly than that moving away.

The shape and velocity structure of these distorted bubbles have been calculated by Dyson (1977a) using a variant of the well known Kompaneets' (1960) theory developed to deal with explosions in media of variable density. The results have been applied to the particular problem of generating low intensity high velocity structures observed in the line profiles of various diffuse nebulae. Meaburn (1971) and Elliott and Meaburn (1975) reported blue shifted features moving at velocities of about 70 km s^{-1} with respect to the main emission line peaks in various nebulae (e.g. M8, M17, M42). The geometry of Fig. 2 provides a natural explanation of the asymmetry of these features. The shell section moving towards the cloud produces a red-shifted component but this portion slows down in a relatively short (typically $\sim 10^4$ yr) timescale. The blue shifted features result from the shell section moving away from the cloud and firstly have much higher velocity and secondly persist much longer. Qualitative models for various nebulae (Dyson 1977a) show that blue shifted features with velocities consistent with the observations could exist for timescales in excess of 10^5 yr provided that the winds are powerful enough ($\dot{E}_{36} \gtrsim 10$). Interestingly, the O4 ((f)) star 9 Sgr associated with M8 has a very large estimated mechanical luminosity $\dot{E}_{36} \simeq 100$ (Abbott et al. 1980).

A stellar wind bubble in a nebula with the geometry of Fig. 2 will affect the gas flow in the region between the I-F (on the cloud surface) and the bubble provided that the flow is subsonic there. The flow would deflect past the bubble. If the flow ahead of the bubble is supersonic, the bubble provides an obstacle which would produce a stand off shock in the streaming gas, again causing a deflection. This situation has been invoked by Pankonin, Walmsley and Harwit (1979) to explain the velocity structure observed in H 76 α recombination line measurements on the Orion Nebula.

A variant of the model involving density gradients has been suggested by Weaver et al. (1977). They suggest that high velocity blue shifted features are produced when the wind driven shell itself breaks out of the cloud edge and accelerates because of the sudden drop in the ambient density. On acceleration the shell becomes Rayleigh-Taylor unstable and breaks up. (Note that the density distributions used by Dyson (1977a) did not produce accelerating shells). It seems inescapable that geometrical effects as described above are of considerable importance in determining the effects of the winds on the dynamics of nebulae.

(b) Nebulae in External Galaxies. The widths of Hα line profiles observed in giant HII regions in the spiral galaxies M33 and M101 show clearly that the ionized gas is in a state of supersonic motion (Smith and Weedman 1970, Melnick 1977). The observed profiles are gaussian and the widths correspond to velocities of, typically, 25 km s^{-1} in the emitting gas. Although it is possible to produce velocities of this magnitude by the flow of ionized gas off the surfaces of neutral intrusions within nebulae, there are problems with advancing this mechanism for these regions. In particular, since the [OI] 6300 A line is not observed, there may be little neutral gas within them.

The nebulae have a core-halo structure and the exciting stars are located in the cores. The excitation parameters of the HII region are extremely high and imply that large numbers of early type stars must be present to maintain the ionization. For example the nebulae NGC 595 (in M33) and NGC 5461 (in M101) need a UV radiative input equivalent to that of 18 and 1700 O5 supergiants respectively. A priori then we expect a considerable input of energy by stellar wind activity into the nebular gas. Melnick (1980) has shown that in at least one of these nebulae the gas velocity is greatest in the core region where the exciting stars are situated. This supports the possibility that the motions are generated by some stellar induced activity.

Dyson (1979) has investigated a simple model in which the line broadening is produced by the emission from shells of material swept up by stellar winds from background gas. The line profile is produced by contributions from both shell and background. On general grounds, the velocity dispersion due to the background material alone is about equal to the sound velocity ($\approx$ 10 km s^{-1}). The line profile produced by integrating over a single thin expanding shell is non-gaussian. However the observations integrate over the entire core region, i.e. over a distribution of shells of varying velocities, radii etc. Further we would anticipate that the shells themselves would be non-uniform (because of density irregularities etc.). These effects would tend to produce gaussian line profiles. Since there is not enough information available to make it meaningful to integrate over some particular shell distribution, the combined effects of shell and background material can be fitted (reasonably) to the observed gaussians to yield the properties of an "average" shell.

The significant - and hardly surprising - point is that the wind driven shells must contribute about half the observed Hα intensity. Alternatively stated, the shells must absorb about the same fraction of the stellar Lyman continuum output.

An acceptable fit to the observational data is obtained if the stellar mechanical luminosity driving a typical shell is $\dot{E}_{36} \approx 50$ for nebulae in M33 and up to $\dot{E}_{36} \simeq 500$ for those in M101. The maximum shell

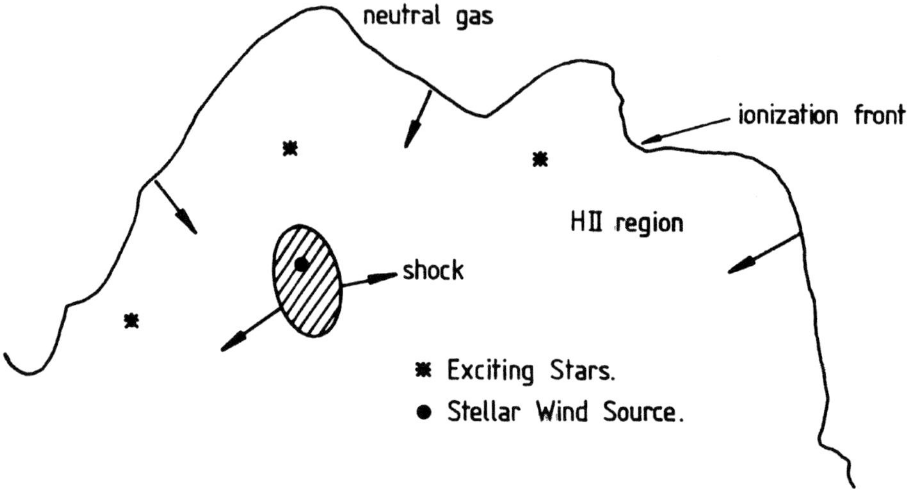

Fig. 2. A sketch of an HII region located at the periphery of a neutral cloud.

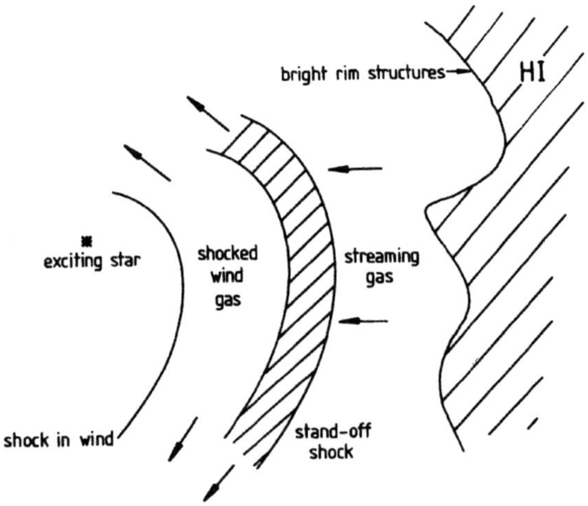

Fig. 3. A sketch of the arc-like structure in S206 based on its appearance in the 5007A line of O^{++}

ages range from about 10^5 yr (M33) to about 10^6 yr (M101). The turn on of these rather powerful winds must occur throughout the core regions over a timescale of this order otherwise many shells would be moving too slowly to broaden the lines appreciably.

We now consider the possible origin for these high mechanical luminosities. Dyson (1979) supposed that each group contained N_* stars and that each star in a group gave some average contribution to the mechanical and radiative luminosities. It turned out that the shells must be driven by groups of stars with numbers ranging up to $N_* \sim O(10)$ (M33) and up to $N_* \sim O(100)$ (M101). A serious objection to this particular interpretation is that these groups of stars must form and turn on their winds throughout the cores of the nebulae over the very short timescale of 10^5-10^6 yr.

An alternative and perhaps more plausible interpretation is to note that the typical mechanical luminosities required to drive a shell are comparable to those produced by the very energetic Of or W-R stars. These high loss rates presumably cannot be sustained for timescales much longer than 10^5-10^6 yr. Suppose that there exists a distribution of O stars (with main sequence lifetimes of several million years on average) which can provide sufficient photons to maintain ionization of the nebular gas. If a proportion of these stars evolves into a more vigorous mass loss phase over a timescale of 10^5–10^6 yr, a continuous creation of new shells might occur. As the older shells slow to below the velocity required to broaden the line profiles ($\simeq$ 25 km s^{-1}), new fast moving shells are produced by the turn on of the more energetic winds. Roughly 10% of the stars need, at only one time, be in (or on their way to) a more vigorous mass loss phase. This evolution would, of course have to occur fairly uniformly through the core regions.

5.3 Specific Structures Associated With Galactic Nebulae

We now consider specific structures associated with galactic nebulae whose forms can be explained in terms of a stellar wind interaction with interstellar gas. These structures generally are non-spherical and appropriate modification must be made to the theory of Section 3.

The nebula S206 contains an arc of high excitation nebulosity apparently centred on the exciting star of the nebula (Deharveng, Israel and Maucherat, 1976). This structure is visible on radio maps of S206 and so is not produced by some quirk of obscuration. The presence of bright rim structures pointing towards the star shows that the nebula itself is formed at the edge of a neutral cloud. It has been proposed (Dyson, 1977b) that this structure is produced by the interaction of the stellar wind from the exciting star with ionized gas streaming away from the ionization front towards the star. (This is somewhat similar to the situation discussed by Pankonin et al. (1979) in the Orion Nebula). The arc of nebulosity is interpreted as a layer of shocked nebular gas behind a stand-off shock produced by the interaction of nebular gas with the wind (Fig. 3). The shape of this shock (basically the shape of the layer) is determined by the balance between the ram pressures of the wind and the streaming gas. The required stellar mass loss rate can be estimated by balancing the ram pressures at the stagnation point (i.e. the head of the arc), and is given by

$$\dot{M}_* \simeq 4 \pi r_0^2 f \rho_0 V_0^2/V_*, \tag{16}$$

where r_0 is the distance from the star to the stagnation point, ρ_0 is the ambient density, and V_0 the streaming velocity of the ambient gas relative to the star. The factor f ($\simeq$ 0.2) allows for the finite thickness of the layer of shocked wind (Fig. 3) and is calculated on the assumption that this gas can expand freely out of the layer (Dyson 1977b). The mass loss rate derived is $\dot{M}_* \approx 1.5\ 10^{-7}\ V_8^{-1}\ M_\odot\ yr^{-1}$, which seems consistent (Conti and Garmany, 1980) with the stellar spectral type O5 V. The calculated shape of the arc is in very reasonable agreement with that observed (Dyson 1977b).

Weaver et al (1977) have given a general discussion of the distortion of a wind driven shell resulting from the motion of the star relative to the interstellar gas. Significant distortion sets in when the shell velocity V_s becomes about equal to the velocity of the star, V_T, relative to the gas. This occurs at a time $t_4 \approx 56\ (\dot{E}_{36}/n_0)^{1/2}\ V_T^{-5/2} 10^4 yr$ (where V_T is measured in units of 20 km s^{-1}). At times t >> t_4, the trailing part of the bubble becomes elongated but the frontal part of the shock (i.e. ahead of the moving star) remains a fixed distance from the star. Its shape is fixed by the ram pressure balance discussed previously.

Gull and Sofia (1979) have explained an extended arc-like structure seen in the vicinity of the runaway star ζ Oph in terms of this distortion effect. Determinations of the wind velocity in this star range from $V_* = 1590$ km s^{-1} (Morton, 1976) to 2730 km s^{-1} (Abbott et al. 1980). Ram pressure balance (Equation 16) then requires a mass loss rate in the range $3\ 10^{-9} - 2\ 10^{-8}\ M_\odot\ yr^{-1}$. Abbott et al. (1980) give $\dot{M}_* \lesssim 3.8\ 10^{-7}\ M_\odot\ yr^{-1}$ for the mass loss rate from this star. Gull and Sofia (1979) estimate $\dot{M}_* \simeq 2\ 10^{-8}\ M_\odot\ yr^{-1}$ for the star by scaling its mass loss rate to that of τ Sco using an empirically determined mass loss luminosity relationship. The agreement is certainly close enough to make this interpretation very plausible. The minimum timescale t_4 required for significant bubble destortion is a few thousand years using the parameters above, which seems not unreasonable.

It is particularly tempting to associate the bubble nebula, NGC 7635, with stellar wind activity. Icke (1973) analysed this object in some detail and fitted its structure to a model in which the shell of the bubble is formed from interstellar matter swept up by the shock wave caused by a point explosion. The nebula is rather asymmetric and the interstellar gas distribution must contain a steep pressure gradient to account for this. Icke (1973) also noted that continuous mass loss from the star might produce much the same effects. We here examine this suggestion but will neglect the problems introduced by a non-uniform ambient gas density. We also assume that the strong shock formulation of Section 3 is applicable although this is a marginal assumption (Icke, 1973).

The mean age derived for the shock is 6.10^{12}s which follows from Icke's (1973) age estimate of 9.10^{12}s for an explosion ($R_s \propto t^{3/5}$ for a wind, $R_s \propto t^{2/5}$ for an explosion). Using an ambient density of 10 cm^{-3}, the average pressure driving the shell is presently 9.10^{-11} dyne cm^{-2} (Icke, 1973). We equate this to the interior pressure of the hot bubble P_I, which is given by

$$P_I \simeq 1.4\ 10^{-10}\ \dot{E}_{36}^{\ 2/5}\ n_0^{\ 3/5}\ t_4^{\ -4/5}\ \text{dyne cm}^{-2} \tag{17}$$

The required mechanical luminosity of the star is then $\dot{E}_{36} \simeq 4$. The spectral type of the exciting star of NGC 7635 is O6.5 III f (Conti and Alschuler, 1975). A plausible estimate of the mechanical luminosity of such a star based on the data of Conti and Garmany (1980) and Abbott (1978) is $\dot{E}_{36} \simeq 2$ (although this should be treated with caution since the exciting star may be binary). Direct observation of mass loss from this star would strengthen the case for this interaction considerably.

There are many other nebulae in which bubble or arc-like structures have been observed. IC 1805 contains a relatively large shell of radius 37pc (Valleé, Hughes and Viner, 1979) which can be interpreted as wind driven provided that a mechanical luminosity $\dot{E}_{36} \gtrsim 13$ is available. Balick, Boeshaar and Gull (1980) note the existence of a small bubble in the nucleus of NGC 3603. The central star cluster appears to contain W-R stars which are known to have high mass loss rates. Other examples are cited by J. Meaburn (this volume).

We note here that it is usually extremely difficult to rule out supernovae as the formation mechanism for shell like structures. Even if a particular feature can definitely be associated with a star or stellar cluster, the possibility of its being caused by a supernova remains. It is effectively impossible to differentiate the two mechanisms from the dynamics alone. The absence or presence of non-thermal emission may not be significant since the expanding shell around a wind emitting star will sweep up the interstellar particles and magnetic field in the same way as would a supernova blast wave. The most likely way of differentiating them may be through their spectral properties. Although in both cases the interstellar gas is heated by a shock wave, gas swept up by a wind will also be exposed to a UV radiation field, (though of course, this would also be true for a supernova remnant which was formed inside a cluster of early type stars).

Considerable work remains to be done on this problem.

5.4 Stellar Winds and Density Fluctuations in Nebulae

Considerable observational evidence points to the presence of fully ionized and neutral cored density fluctuations within diffuse nebulae. We now briefly consider some of the ways in which stellar winds may maintain and perhaps cause such fluctuations.

Pikelner and Sorochenko (1974) pointed out that fully ionized density irregularities overrun by the wind driven shell would be maintained in pressure equilibrium with the shocked wind gas. Since the interior pressure in the bubble is uniform, their density would equal that in the shell. Dyson (1977c) considered this problem further and concluded that fully ionized fluctuations having surface densities greater than that of the impinging shell could develop neutral cores by the compression. The mechanical luminosities driving the shell need to be fairly high ($\dot{E}_{36} \gtrsim 10$). As the shell travels outwards, the interior pressure decreases and partially ionized condensations expand and become fully ionized in timescales short compared to the dynamical timescales associated with the shell motion. Any partially ionized condensations produced in this way would tend to lie fairly close behind the shell. It is worth noting here that small clouds left behind in the interior of the bubble would not be accelerated by the outwards flow of the shocked wind gas. There is far too little momentum in this flow. Any outwards motion of these clouds is essentially due to that given by the shell impact. This is in distinction to the situation when small clouds are overrun by supernovae blast waves and accelerated outwards by the flow in the remnant (McKee and Cowie, 1975).

If a driven shell accelerates, the contact discontinuity C becomes Rayleigh-Taylor unstable and the shell will fragment. There are various possibilities for producing this acceleration. The shell could pass suddenly into a region of lower ambient density. Alternatively, the mechanical luminosity driving the shell could increase if, for example, a star evolves to a more vigorous mass loss phase, or perhaps another star switches on a wind. Whether this fragmentation produces fully or partially ionized fragments depends on the physical conditions in the shell when fragmentation occurs and also on the cause of the shell acceleration. If the interior pressure in the bubble decreases continuously with time, partially ionized fragments can be formed only if the I-F has been trapped inside the shell at the time of fragmentation. If the acceleration is due to an increase in the mechanical luminosity, the bubble pressure may actually increase with time and compress fully ionized fragments until they develop neutral cores.

Schneps, Ho and Barrett (1980) have suggested that the elephant trunk structures and globules lying at the periphery of the Rosette Nebula (Herbig, 1974) may be a consequence of a Rayleigh-Taylor instability induced by the shell moving into a low density region. These globules have a space motion away from the stellar cluster and their measured velocity is almost identical to that of the ionized gas (≈ 20 km s^{-1}). Since stellar wind activity seems to account rather well for the morphology and velocity structure of the ionized gas (Section 5.1), the connection seems a logical one to attempt.

Schneps et al. (1980) assumed that the instability was caused simply by a density decrease ahead of the shell. But the estimate of ϕ in Section (5.1) suggests that the I - F is not presently trapped inside the shell (the presence of the bright rim structures in the nebula support this). Therefore, if the shell fragmented fairly recently, it is necessary for there to have been an increase in the interior bubble pressure in order to compress the fragments until their interiors could recombine. (There could, of course, have also been fragmentation effects due to a density drop).

It is difficult to estimate the sizes of fragments which can be formed by the instability but the upper limit should be comparable with the shell thickness at the time of fragmentation. Suppose that a fully ionized shell fragments when its radius is R_s. Let M_i and M_n respectively be the mach numbers of the outer shock relative to the sound speeds in ionized and neutral gas respectively. Then on the assumption that a pressure increase occurs after fragmentation sufficient to cause recombination, an upper limit on the size of a fragment is $d_u \approx R_s/(M_n M_i^2)^{2/3}$. At present $M_i \approx 2$ and we take $M_n \simeq 60$ (assuming a neutral gas temperature of 10K (Schneps et al. 1980). Then $d_u \approx 0.02\ R_s \approx 0.2$ pc. This is comparable with the sizes of larger globules, but is an upper limit. We can estimate a lower limit for their sizes by assuming the shell fragmented recently with the I - F trapped (i.e. ignore the calculated ϕ). This lower limit is $d_\ell \approx R_s/M_n^2 \sim 0.002$ pc which is far too small. We conclude that the hypothesis is an attractive one but that the sizes of fragments may only be large enough to account for the smaller globules.

Falle (1975) has suggested another possible way of producing shell break up. The strengths of the secondary shocks produced during the thin shell formation period depend sensitively on the ambient density ahead of the main shock. Even slight ($\sim$ 20%) fluctuations in ambient density lead to large differences in secondary shock strength. This effect would certainly produce shell break up.

Although shell break up has the desirable effect of explaining certain phenomena, it has important consequences for the explanation of other effects. Firstly, shell breakup, when coupled with fluctuations in ambient density, could well allow the hot bubble gas to escape through holes in the shell, thus reducing the driving pressure and decreasing the dynamical importance of the wind. Secondly, it is usually necessary for the shell to absorb a substantial fraction of the stellar Lyman continuum flux in order to contribute significantly to, say, Hα line profiles. Break up of the shell will make this more difficult to achieve. Quantitative investigation of such problems is highly desirable.

We finally emphasise that there are probably several different dynamical effects occurring simultaneously within diffuse nebulae and attempts to explain their velocity structure in terms of any one mechanism are unlikely to succeed.

6. THE EFFECTS OF STELLAR WINDS ON THE EVOLUTION OF SUPERNOVA REMNANTS

It is plausible to suppose that at least a proportion of massive stars with energetic winds are the progenitors of Type II supernovae. The supernova remnant's evolution will be affected because of the disturbances on the interstellar gas caused by the wind. Specifically, a supernova might occur inside a cavity which has been blown in the interstellar gas by the wind. We discuss here some of the gross modifications that will be made to the remnant's evolution.

We firstly consider the case of a single star immersed in a homogeneous low density medium where the wind driven shock has degenerated into a sound wave (Section 4). During its main sequence lifetime, the star will be surrounded by an expanding HII region containing an interior bubble of hot shocked wind gas. The two regions will have approximately equal pressures.

Neglecting the effects of the HII region expansion, the mean density in the bubble is $n_B \approx (T_0/T_B)n_0$. Here T_B is the mean temperature in the bubble and T_0 the HII region temperature. For numerical estimates we take $T_0 = 10^4$K, $T_B \simeq 5\ 10^7$K. If the wind blows for a time t_{ω_4} (in units of 10^4 yr), the bubble radius after time t_{ω_4} is $R_B \simeq 7\ (\dot{E}_{36}\ t_{\omega_4}/n_0)^{1/3}$ pc. We assume this configuration remains unchanged whilst the star then evolves to the supernova stage, thus ignoring such interesting possibilities as the effects of a W-R phase. We assume that the supernova injects 10^{51} erg (i.e. about $5M_\odot$ ejected at about $V_{ej} \simeq 5000$ km s^{-1}). The associated momentum is $\approx 5\ 10^{42}$ gm cm s^{-1}.

We now consider a specific case. Conservatively we adopt $n_0 = 1$ cm^{-3}, $t_{\omega_4} = 500$, $\dot{E}_{36} = 1$, giving $n_B \simeq 2\ 10^{-4}$ cm^{-3}, $R_B \simeq 58$ pc. The mass of gas ejected in the wind phase will be comparable with that ejected by the supernova. Momentum considerations imply that the shock impacting on the contact discontinuity separating ambient and shocked wind gas will have a velocity $\approx V_{ej}/2$ ($\simeq 2500$ km s^{-1}). A shock is transmitted into the ambient gas at a velocity $V_{TR} \approx (n_B/n_0)^{\frac{1}{2}}\ (\varepsilon\ V_{ej}/2)$. The factor ε depends on the density contrast n_0/n_B, and here is about equal to 2.5 (Sgro, 1975). Thus $V_{TR} \simeq 90$ km s^{-1}. The cooling time behind this transmitted shock is very short ($\sim 10^4$ yr). We therefore assume that momentum conservation determines the subsequent dynamics (though this neglects any push due to the very hot interior of the bubble). The time elapsed between the supernova and this snowplough phase is short ($\lesssim 10^4$ yr). There is now no classical Sedov phase in the evolution of the remnant. The shell loses its identity when its velocity is about equal to the mean velocity of interstellar clouds. Taking this as 5 km s^{-1}, the shell (which is neutral) becomes indistinct after a time $\sim 2\ 10^5$ yr when its radius is about 62 pc. The chief result of including the wind has been to remove the Sedov phase of evolution.

There are clearly other possibilities even with a single star. For example, the supernova could occur whilst there exists a well defined dense shell of swept up interstellar gas driven by the bubble. The supernova blast wave might then produce a shell of shocked hot gas which would cool. The supernova blast wave could reaccelerate once it had driven through the wind produced shell. There would then be a fast shock with a high temperature immediately behind it and interior to this could be a shell of optically emitting gas (the wind produced shell). This interaction could produce structures rather similar to the Cygnus loop (McCray and Snow, 1979).

A perhaps more realistic situation involving joint stellar wind - supernova activity would occur in a group of early type stars. Bruhweiler et al. (1980) have suggested that the very large radii ($\gtrsim$ 100 pc) shells of neutral hydrogen found by Heiles (1979) result from this process since several of the large shells show a possible spatial correlation with OB associations. We adopt another simple model based on Bruhweiler et al.'s (1980) discussion.

Suppose that an OB association contains 30 stars each having a mechanical luminosity $\dot{E}_{36} \simeq 1$ blowing into ambient gas of density 1 cm^{-3}. Each star blows its wind for a time $t_{\omega 4} \simeq 500$ and the winds commence simultaneously (though a spread in turn-on time would be more realistic). After time $t_{\omega 4}$, the shell has radius $R_{SH} \gtrsim 130$ pc, velocity $V_{SH} \simeq 17$ km s^{-1} and an associated momentum of 8 10^{44} gm cm s^{-1}. The maximum density in the shell depends on whether or not the I-F is trapped inside it. The results of Section 4 shows that this will be the case provided that the average UV photon output rate from the cluster stars is $S_{48} \lesssim 5$. This corresponds to about O7 on the ZAMS (Panagia, 1973). We therefore assume that this gas recombines and the maximum shell density is therefore $n_s \simeq 300$ cm^{-3} with an assumed neutral gas temperature of 100K. Supernova blast waves will again produce transmitted shocks into the shell but the cooling will be very strong. Consequently, as for the single star, the evolution of the shell after the stars become supernovae is determined by momentum conservation. However, even if all 30 stars were to simultaneously switch off their winds and become supernovae, they could have little effect on the shell dynamics. The momentum input by the 30 supernovae is $\sim$ 1.5 10^{44} gm cm s^{-1}, i.e. about 20% of that already in the shell. Thus the momentum of the shell in the subsequent motion is essentially generated by the wind activity. This shell loses its identity ($V_{sh} \simeq 5$ km s^{-1}) at a radius of about 200 pc, in encouraging accord with observation. Shells of even larger radii are produced if there is a second wave of supernovae produced by lower mass stars (Bruhweiler et al. 1980). However under most circumstances it is really the stellar wind activity which produces these large scale structures.

Bruhweiler et al. (1980) pointed out that the characteristic parameters of the shells (velocities, radii etc.) depend upon their position relative to the galactic centre since the ambient gas distribution varies along the galactic plane. In the solar neighbourhood, the shell radii become comparable to the scale height of the neutral gas even before the supernovae occur. There are two possible effects of this on the shell dynamics (if, for the moment, we ignore gravitational forces). Firstly, the shells ought to elongate in the direction perpendicular to the galactic plane. Secondly, most of the common approximations (gaussian, exponential etc.) used for the ambient gas distribution lead to a blow-out effect in which the shell section moving perpendicular to the plane eventually accelerates rapidly. This section of shell would become Rayleigh-Taylor unstable and break up. If the shell has not reached escape velocity when this occurs the shell fragments would fall back towards the galactic plane. Further, the hot gas in the bubble would escape and other sections of the shell would lose their driving pressure. They might also be destroyed by expansion waves moving through the bubble.

To quantify these ideas, we take the following simple picture. Up to about 250 pc from the galactic plane, the gravitational acceleration in the solar neighbourhood is approximately linear in z (Spitzer, 1978). The gas density distribution is therefore gaussian (for an isothermal gas). If we neglect gravity in the dynamics, the motion of the shell driven by the winds from a group of stars can be followed using the method developed by Dyson (1977a). We then find that the shell section moving perpendicular to the plane initially decelerates and then starts to accelerate at a characteristic time t_{acc} and characteristic R_{acc} given by

$$t_{acc} \simeq 10^7 \left(n_0 \Delta_{100}{}^5/\dot{E}_{36}\right)^{1/3} \text{ yr,}$$

$$R_{acc} \simeq 1.2\ \Delta_{100} \text{ pc.} \qquad (18)$$

Here, n_0 is the density in the plane (z = 0), and Δ_{100} is the scale height of the gaussian distribution in units of 100 pc. In the numerical estimates we take, $\Delta_{100} = 1$ and $n_0 = 1$ cm^{-3}. The stellar group is located at z = 0 and has a total mechanical luminosity $\dot{E}_{36}$. For blow out to occur, t_{acc} must be less than the wind emitting duration. If as before, we assume a wind emitting phase lasting 5.10^6 yr, blow out occurs if $\dot{E}_{36} \gtrsim 8 n_0 \Delta_{100}{}^5$. This is a modest requirement and would be satisfied for the group of 30 stars discussed previously (for which $E_{36} \simeq 30$).

The neglect of the gravitational field on the shell dynamics is justified provided $g_z \lesssim V_s{}^2/R_s$, where g_z is the z component of the gravitational acceleration. If $dg_z/dz \simeq -2.10^{-11}$ cm s^{-2} pc^{-1} (Spitzer, 1978), then gravity is not important up to $t = t_{acc}$ if $\dot{E}_{36} \gtrsim 4\ n_0\ \Delta_{100}{}^5$. Both criteria will be satisfied for the group discussed and shell fragmentation starts at $t \gtrsim t_{acc}$. With the parameters as given, $t_{acc} \simeq 3\ 10^6$ yr, $R_{acc} \simeq 120$ pc. The shell velocity at time of fragmentation is

$V_{acc} \simeq 40$ km s^{-1} and the fragments rise to a few hundred pc above the plane before they fall back. The surface densities of the falling fragments would, at most, be comparable with the shell surface density at fragmentation, ($\simeq 5\ 10^{-4}$ gm cm^{-2} with the parameters used). The interaction of these fragments with ensuing supernovae and the effect on the supernova remnant evolution of producing low density channels merit further investigation. The production of shells and their subsequent break up may contribute to the 'bubble-bath' structure of the galaxy discussed by Brand and Zealey (1975).

7. WOLF-RAYET STARS AND THEIR WINDS

W-R stars have often been invoked as the driving mechanism behind particularly energetic phenomena because of their very high mechanical luminosities. Usually, Equations (6) (or their equivalent) have been applied directly using values of $\dot{E}_{36}$ appropriate to W-R winds. Since it appears that W-R stars result from the evolution of Of stars (Conti, 1978), mass loss from the progenitor Of stars will have produced a cavity before the vigorous W-R wind turns on.

We here consider only the very simplest possibility in which an isolated Of star evolves effectively instantaneously into a W-R phase. In order that the sudden increase in mechanical luminosity appreciably affect the shell produced during the Of phase, the internal pressure driving the shell must be substantially increased. This requires a W-R mechanical luminosity $\dot{E}_{WR} \gtrsim \dot{E}_{Of}\ (t_{Of}/t_{WR})$, where $\dot{E}_{Of}$ is the mechanical luminosity in the Of phase (assumed constant), and t_{Of} and t_{WR} are the wind duration phases of the Of and W-R phases respectively. Since we expect t_{Of} and t_{WR} to be comparable with the stellar hydrogen and helium burning lifetimes respectively, $t_{Of}/t_{WR} \gtrsim 10$. Thus the mechanical energy output rate must be increased by at least a factor ten by the transition to have much effect. Numerical calculations based on the thin shell model show that the shell radius and velocity at the end of the W-R phase are increased relative to their values at the end of the Of phase by about 5% and 8% respectively if $\dot{E}_{WR}/\dot{E}_{Of} = t_{Of}/t_{WR} = 10$. As a more extreme case, the shell radius and velocity are increased by about 8% and 85% respectively if $\dot{E}_{WR}/\dot{E}_{Of} = 100$, $t_{Of}/t_{WR} = 10$. We see that for the case of an isolated star, the dominant dynamical effects are most likely produced in the Of phase prior to the W-R phase. This may not be the case if a star undergoing this evolution is part of an extended stellar group. There winds from other stars might move interstellar gas into the neighbourhood of the W-R star where wind could then directly interact with interstellar gas. (In fact the concluding discussion in Section 5.2 (b) implicitly assumes that this effect can occur).

An interesting possibility is that winds in the W-R phase are responsible for the formation of the ring like nebulosities observed around several isolated W-N type stars. The masses of these nebulae are uncertain and estimates for the various nebulae give a possible range of less than 1 $M_\odot$ to several hundred $M_\odot$. If the masses are of the order of a hundred

$M_{\odot}$, the nebulae must presumably be produced by interstellar matter swept up jointly in the Of and W-R phases. This explanation is unlikely in the specific case of the best observed object NGC 6888. The relative abundance of nitrogen is high in the nebular gas (Wendker et al. 1975), and this suggests that it is composed of stellar ejecta rather than swept-up interstellar matter. Various stellar evolution calculations (e.g. Chiosi, Nasi and Sreenivasan, 1978) predict that the transition Of $\rightarrow$ W-R can take place via a red giant phase. The star may eject several solar masses of material in the form of a cool slowly moving ($\simeq$ 20 km s^{-1}) shell during the red giant phase. This is consistent with the estimated mass ($\simeq$ 4 $M_{\odot}$) of NGC 6888 (Wendker et al., 1975). The ring nebulae (or at least some of them) may be formed by the interaction of the W-R wind with the red giant ejecta and not with interstellar matter. Interestingly, no ring-like nebulae seem yet to have been observed around W-C type stars. If W-C stars are at a later stage of evolution than W-N stars (e.g. Willis and Wilson, 1978), this may be due to a decrease in surface brightness of the shell during its evolution.

8. DISCUSSION

The very simple models presented in this review suggest that stellar winds from early type stars may be the agency responsible for a great many interesting dynamical phenomena observed in the interstellar medium. There are however (at least) two areas of potential difficulty which have been carefully skirted. The effects of the interstellar magnetic field on either the shell dynamics or its structure have not been discussed. Falle (1975) has shown that the main effect of a moderate interstellar field is to reduce the compression in the swept up shell since the magnetic field dominates in the cooled shocked gas. This reduces considerably the recombination rate in the shell and thus reduces the fraction of the stellar Lyman continuum output absorbed in the shell. This has serious implications for several of the problems discussed in Section 5 where it is necessary that a significant fraction of the stellar UV output must be absorbed by moving shells in order to produce observable effects. Secondly, we have noted that the shells may break up under a variety of circumstances and the dynamics of fragments are not the same as that of uniform shells. This breakup compounds the problem of the stellar UV absorption. We have already noted other uncertainties relating to such problems as the duration of the wind phase, the time dependence of the winds etc. Because of these various omissions and uncertainties, the explanations offered in this review of various phenomena are at best plausibility arguments. It is highly desirable to have observations of both dynamical phenomena and the stars presumably the origin of these phenomena in order to resolve the ambiguities.

ACKNOWLEDGEMENTS

My first acquaintance with Professor Kopal came through his articles and books. Since then, first as a research student and then as a staff member in his department, I have benefitted considerably from his great knowledge of and enthusiasm for astronomy. It is a great pleasure for me to dedicate this article to him with gratitude for his help and support over the years.

REFERENCES

Abbott, D. C., 1978. Astrophys. J., 225, p. 893.
Abbott, D. C., Bieging, J. H., Churchwell, E. and Cassinelli, J. P., 1980. Astrophys. J., 238, p. 196.
Balick, B., Boeshaar, G. O. and Gull, T. R., 1980. Astrophys. J., 242, p. 584.
Barlow, M. J., Smith, L. J. and Willis, A. J., 1981. Mon. Not. Roy. Ast. Soc., (in press).
Bodenheimer, P., Tenorio-Tagle, G. and Yorke, H. M., 1979. Astrophys. J., 233, p. 85.
Brand, P. W. J. L. and Zealey, W. J., 1975. Astron. Astrophys., 38, p. 363.
Bruhweiler, F. C., Gull, T. R., Kafatos, M. and Sofia, S., 1980. Astrophys. J., 238, L27.
Chiosi, C., Nasi, E. and Sreenivasan, S. R., 1978. Astron. Astrophys., 63, p. 103.
Conti, P. S., 1976. Mem. Soc. Roy. Sci. Liege IX, p. 193.
Conti, P. S. and Alschuler, W. R., 1971. Astrophys. J., 170, p. 325.
Conti, P. S. and Garmany, C. D., 1979. Bull. AAS. 11, p. 631.
Conti, P. S. and Garmany, C. D., 1980. Astrophys. J., 238, p. 190.
Conti, P. S. and McCray, R., 1980. Science, 208, p. 9.
Deharveng, L., Israel, F. P. and Maucherat, M., 1976. Astron. Astrophys., 48, p. 63.
Dyson, J. E., 1977a. Astron. Astrophys., 59, p. 161.
Dyson, J. E. 1977b. Astrophys. Space Sci., 51, p. 197.
Dyson, J. E. 1977c. Astron. Astrophys., 62, p. 269.
Dyson, J. E. 1979. Astron. Astrophys., 73, p. 132.
Dyson, J. E., Falle, S. A. E. G. and Perry, J. H., 1980. Mon. Not. Roy. Ast. Soc., 191, p. 785.
Elliott, K. H. and Meaburn, J., 1975. Astrophys. Space Sci., 35, p. 81.
Falle, S. A. E. G., 1975. Astron. Astrophys., 43, p. 323.
Fountain, W. F., Gary, G. A. and O'Dell, C. R., 1979. Astrophys. J., 229, p. 971.
Gull, T. P. and Balick, B., 1974. Astrophys. J., 192, p. 63.
Gull, T. P. and Sofia, S., 1979. Astrophys. J., 230, p. 782.
Heiles, C., 1979. Astrophys. J., 229, p. 533.
Herbig, G. H., 1974. Publ. Ast. Soc. Pac., 86, p. 604.
Icke, V., 1973. Astron. Astrophys., 26, p. 45.
Jenkins, E. B. and Meloy, D. A., 1974. Astrophys. J., 193, L121.

Kahn, F. D., 1974. Astron. Astrophys., 37, p. 149.
Kahn, F. D., 1976. Astron. Astrophys., 50, p. 145.
Kahn, F. D., 1981. Mon. Not. Roy. Ast. Soc., (in press).
Kompaneets, A. S., 1960. Soviet Phys. Doklady., 5, p. 46.
Mazurek, T. J., 1980. Astron. Astrophys. 90, p. 65.
McCray, R. and Snow, T. P. Jnr., 1979. Ann. Rev. Astron. Astrophys., 17, p. 213.
McKee, C. F. and Cowie, L. L., 1975. Astrophys. J., 195, p. 715.
Meaburn, J., 1971. Astrophys. Space Sci., 13, p. 110.
Melnick, J., 1977. Astrophys. J., 213, p. 15.
Melnick, J., 1980. Astron. Astrophys., 86, p. 304.
Menon, T. K., 1962. Astrophys. J., 135, p. 394.
Morton, D. C., 1967. Astrophys. J., 150, p. 535.
Morton, D. C., 1976. Astrophys. J., 203, p. 386.
Panagia, N., 1973. Astron. J., 78, p. 929.
Pankonin, V., Walmsley, C. M. and Harwit, M., 1979. Astron. Astrophys., 75, p. 34.
Pikelner, S. B., 1968. Astrophys. Lett. 2, p. 97.
Pikelner, S. B. and Sorochenko, R. L., 1974. Soviet Ast., 17, p. 443.
Pottasch, S. B., Wesselius, P. R. and van Duinen, R. J., 1979. Astron. Astrophys., 77, p. 189.
Raymond, J. C., Cox, D. P. and Smith, B. W., 1976. Astrophys. J., 204, p. 16.
Schneps, M. H., Ho, P. T. P. and Barratt, A. H., 1980. Astrophys. J., 240, p. 84.
Sgro, A., 1975. Astrophys. J., 197, p. 621.
Shull, J. M., 1980. Astrophys. J., 238, p. 860.
Smith, M. G., 1973. Astrophys. J., 182, p. 111.
Smith, M. G. and Weedman, D. W., 1970. Astrophys. J., 161, p. 33.
Spitzer, L. Jr., 1978. *Physical Processes in the Interstellar Medium*, Wiley (New York).
Vallee, J. P., Hughes, V. A. and Viner, M. R., 1979. Astron. Astrophys., 80, p. 186.
Weaver, R., McCray, R., Castor, J., Shapiro, P. and Moore, R., 1977, Astrophys. J., 218, p. 377.
Wendker, H. J., Smith, L. F., Israel, F. P., Habing, H. J. and Dickel, H. R., 1975. Astron. Astrophys., 42, p. 173.
Willis, A. J. and Wilson, R. 1978. Mon. Not. Roy. Ast. Soc., 182, p. 559.
Zuckerman, B., 1973. Astrophys. J. 183, p. 863.

THE DISCOVERY AND OBSERVED PROPERTIES OF QSOs AT LARGE REDSHIFTS - AN UPDATE

Malcolm G. Smith
The Royal Observatory, Edinburgh, Scotland

CONTENTS

F. D. Kahn (ed.), Investigating the Universe, 151–206.

1. TECHNIQUES, SURVEYS AND SELECTION EFFECTS

Apart from studies of the microwave background, direct exploration of the universe beyond the established redshifts of ordinary galaxies ($z \sim 1$) has, at least so far, had to rely on the detection and study of high redshift QSOs. This review covers work, principally on quasi-stellar objects with large redshifts, that has come to my attention in the interval May 1978 to February 1981. I have assumed, except where otherwise noted, that redshift is an indicator of distance in a zero-pressure Friedmann cosmology with zero cosmological constant, and zero deceleration parameter. I have taken the Hubble parameter as $H_o = 100$ km s^{-1} Mpc^{-1}. In this first section I shall give a rather detailed discussion of some of the techniques used to discover high-redshift QSOs, recent major surveys for high-redshift objects and efforts to quantify selection effects.

In May 1978, I gave a detailed review (Smith, 1978) of the development of the slitless-spectrum search technique - the most efficient method known for discovering QSOs at large redshifts. Basically it depends on the direct detection of spectral features - usually emission lines - in survey spectra. These spectra have generally been taken with Schmidt telescopes through an objective prism, or with large reflectors using a grism (see, e.g., Hoag and Schroeder 1970, Smith 1975, Hoag and Smith 1977). Lyα, being the strongest emission line in QSO spectra, is the one normally detected. It turns out that the slitless-spectrum search technique, being strongly biassed towards high redshifts ($z > 1.8$), is complementary to the ultraviolet-excess (UVX) search method, which is best suited for redshifts $z < 2.3$ (see, e.g., Bolton and Savage 1978). Though use of the technique has proved extremely successful in the discovery of large numbers of high redshift QSOs, it is so plagued with selection effects that cynics are already referring to it as "subjective-prism spectroscopy". Nevertheless, as its widespread application has now significantly affected all branches of observational work on QSOs, an updated review seems timely. The reader more interested in the results, rather than the technique should, after reading warnings about selection effects, pass directly to section 2. For a catalogue of the basic optical data on all 1491 QSOs and 58 BL Lacertae objects identified up to September 1979, see Hewitt and Burbidge (1980).

1.1 Details of recent applications of the slitless survey technique

In Table 1.1 I have reproduced and slightly extended Osmer's (1980a) summary table of recent slitless surveys.

Osmer has compared the different surveys in order to check their consistency. He concludes that the internal agreement between different surveys using the same technique - measured by the total surface density of quasars in each survey - is excellent. His main questions (Osmer 1977, 1980a,b, 1981) centre around the steepness of the increase from the Curtis Schmidt value, σ(apparent) = 0.34 deg^{-2}, to the 4m value,

σ(apparent) = 13 deg^{-2}, and "the lack of agreement between the two surveys at magnitude 18, where they might be expected to give similar results". Vaucher and Weedman (1980) also state, in their summary: "Comparisons of the surveys indicate serious and ill-understood incompleteness effects. The most puzzling such effect is the discontinuity in counts, such that surveys with 4m telescopes give sixty times as many quasars deg^{-2} with $17.5 \leq$ mag ≤ 18.5 as are found from Schmidt telescope surveys in the interval $16.5 \leq$ mag < 17.5. Results for independent surveys with the same techniques agree however".

Table 1.1 Major surveys using the slitless spectrum technique

CTIO

Curtis Schmidt (thin prism) 1360Å/mm (Hγ) 97 arcsec/mm
$m \leq 20.1$ (1" seeing)

MacAlpine and Lewis (1978)* IIIaJ
Osmer and Smith (1980) [OS] IIIaJ

4m (grism) 1580Å/mm 18.5 arcsec/mm $m \leq 21.4$ (1" seeing)

Hoag and Smith (1977) [HS] IIIaF, blue blazed (see also Osmer 1980b)
Bohuski and Weedman (1979) IIIaJ, blue blazed
Osmer (1980a) IIIaF, red blazed

KPNO 4m (grism)		UK Schmidt 2480Å/mm (Hγ) 67 arcsec/mm $m \leq 20\frac{1}{2}$?	
Hoag, Burbidge and Smith (1977)	IIIaJ	Bolton and Savage (1978)	IIIaJ
Sramek and Weedman (1978)	IIIaJ	Smith (1978)	IIIaJ
Koo and Kron (1980)	IV-N	Hazard (1979)	IIIaJ
	& CCD	Clowes et al (1980)	IIIaJ

Note that in general the Schmidt plates are, at best, only very poorly calibrated. Most of the grism plates have broad-band sensitometric calibration - see, e.g., Vaucher et al 1980 - but this may not be enough for accurate spectrophotometry - Clowes et al 1980. Limiting magnitudes are given only as a very approximate guide and depend on a variety of parameters (see, e.g. Fig 1.1); limiting magnitudes are typically ∿ 0.7 mag brighter for 2" seeing. Osmer's IIIaF survey reaches emission-line fluxes 3 times fainter than the survey by Hoag and Smith; the values given for the grism are to be taken only as representative of the various different surveys.

*One of a series of papers by MacAlpine and collaborators - see final reference list.

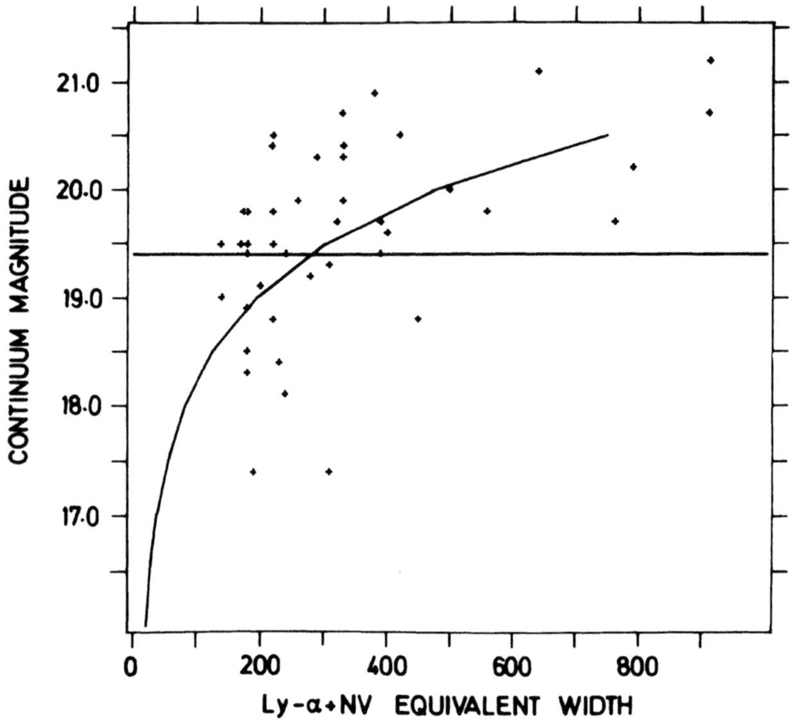

Figure 1.1: 4-metre grism plates yield fainter, weaker-lined quasars than those detectable on Curtis Schmidt plates. This diagram (Clowes 1981) illustrates selection lines for the CTIO Curtis Schmidt survey (Osmer and Smith 1980) drawn on the data for the CTIO grism survey (Hoag and Smith 1977; Osmer 1980b). The horizontal line represents the limiting magnitude for continuum detection, the curved line the limit for detection of emission lines. Quantitative criteria, based on idealised machine-based searches, are given in Clowes' paper (the CTIO plates were not searched by machine). Of objects picked up in the grism survey, about half would have been missed with the Curtis Schmidt in this particular overlap region. In fact the Curtis Schmidt technique has worked so well only because there is an excess of bright, strong-lined QSOs in this overlap region compared with the rest of the Curtis Schmidt survey. See discussion in section 1.1.

Two obvious causes for this apparent discrepancy are (i) the 4m allows one to pick up many weak-lined QSOs that were missed on the Curtis Schmidt plates and/or (ii) the small-area 4m samples happened to be taken in regions of locally enhanced surface density of QSOs. The first effect has led to a detailed quantitative study by Clowes of the effects governing the detectability of emission lines on slitless spectra. The second effect (which I confess I did not believe would be important because of the good agreement between the various grism surveys in both hemispheres, even using different emulsions) led to a general study by Osmer (1981) of the degree and significance of clustering in the slitless survey lists, and by Clowes (1981) of the overlap region in the surveys by Osmer and Smith (1980) - OS - and by Hoag and Smith (1977) - HS. These surveys were used because they are presently the only ones available for which every object has absolute spectrophotometry (Osmer and Smith 1980, Osmer 1980b).

Osmer and Clowes' figures for apparent surface density σ (deg^{-2}) on the sky are given in Table 1.2, which summarises the problem. (Class I QSOs are those with strong emission lines, as defined later in terms of Clowes' selection curves).

Table 1.2 Detection rates in the −40° zone averaged over the entire respective survey areas − 5.1 deg^2 for the grism survey and ∿ 340 deg^2 for the Curtis Schmidt survey (data from Osmer (1981) and Clowes (1981))

Grism − all QSOs	13 deg^{-2}
Grism − class I QSOs (strong liners − see text)	2.5 deg^{-2}
Curtis Schmidt − all QSOs	0.34 deg^{-2}
Curtis Schmidt − class I	0.2 deg^{-2}

Thus the ability of the grism to detect weak lines (and reach fainter limiting magnitude − Clowes 1981) increases its detection efficiency per unit area of sky by a factor up to ∿40. However we shall now see that, vis-a-vis the mean for the whole Curtis Schmidt survey, there happen to be more than 10 times as many strong-lined (class I) QSOs in the −40° zone of overlap between the grism and prism surveys. It is only because the amplitude of this particular clumping of prominent (class I) QSOs is so small in terms of total numbers, that independent surveys with the same (grism) technique from observatories in both hemispheres still arrived at similar overall numbers of QSOs (∿ 10 deg^{-2}).

Figure 1.1 is an adaptation of a diagram in Clowes' thesis − see also Clowes (1981) − which demonstrates the ability of the CTIO −4m plates to show weaker lines at a given continuum magnitude than the Curtis Schmidt plates. Only those CTIO −4m quasars with $m_\nu < 19.4$ and to the lower right of the selection curve could, with 2 arcsecond seeing, also be detected with the Curtis Schmidt. For the purposes of this discussion, it is these relatively bright, strong-lined objects which I am calling "class I" QSOs. The surface densities of these class I QSOs are what must be compared between the two basic survey techniques.

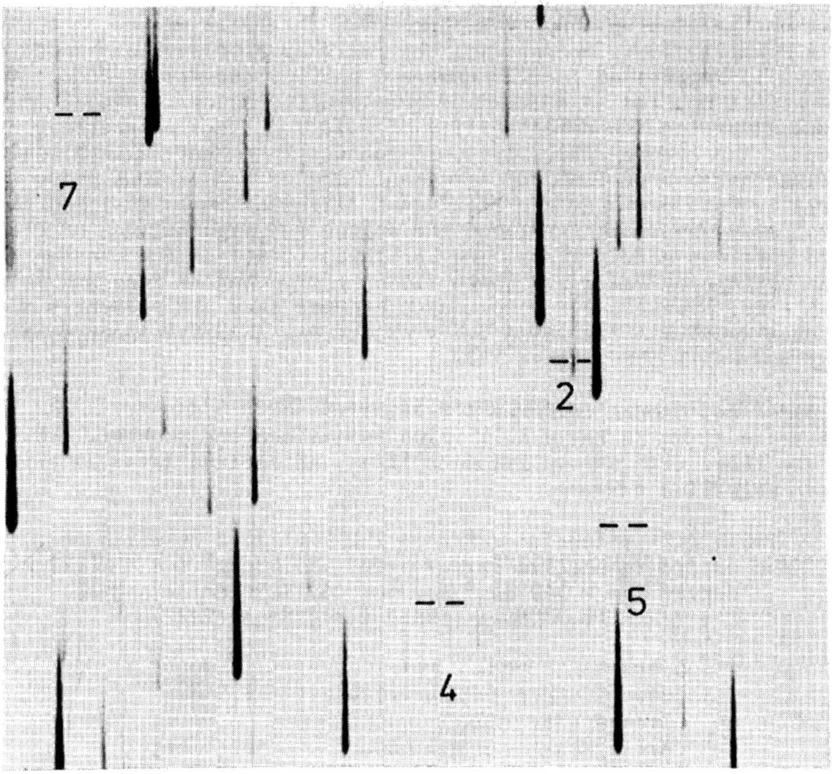

Figure 1.2: Some QSOs in the region of overlap between the CTIO grism and prism surveys. This photograph is taken from the original −40° Curtis Schmidt plate and can be compared with an illustration (Smith 1978) based on a UK Schmidt plate of the same overlap region; the numbering of QSOs is the same in both illustrations.

#2	Q0130-403	; $z = 3.02$;	$m_\nu = 17.4$;	$W = 190$,	Lyα + NV;	061
#4	Q0131-404	; $z = 1.48$,	$m_\nu = 19.0$;	$W = 130$,	CIV ;	064
#5	Q0130.7-4025;	$z = 2.16$;	$m_\nu = 19.2$;	$W = 280$,	Lyα + NV;	060
#7	Q0131.6-4017;	$z = 2.11$;	$m_\nu = 20.3$;	$W = 170$,	Lyα + NV;	071

Numerical data are taken from Osmer (1980b). m_ν is the continuum magnitude at 1475Å in the rest frame of the quasar. W is the equivalent width, in Angstroms, in the observer's frame. The final numbers (060 to 071) are the designations in Osmer's (1980b) paper. See discussion near the end of section 1.1; the object marked #5 is one which Clowes (1980) considers should have been detected with the Curtis Schmidt. It, and #7, were not noticed until the grism was used (Hoag and Smith 1977; Osmer and Smith 1980).

In the 3.5 deg^2 of sky in common to the two surveys (OS and HS), Clowes finds that the local density of class I QSOs, 1.7 deg^{-2}, is substantially higher than the average value (0.2 deg^{-2}) for the whole Curtis Schmidt sample of class I QSOs. However, in this small area of sky common to both surveys, 3 class I QSOs found with the 4-metre telescope were missed with the Curtis Schmidt, so the observed density of such objects in the common 3.5 deg^2 area is in fact 2.6 deg^{-2}, more than ten times the average. (The figure of 2.5 deg^{-2} given in Table 1.2 refers to the full 5.1 deg^2, not the 3.5 deg^2 overlap zone.) Clowes identifies the three missing Type I QSOs as numbers 12, 26 and 60 in Osmer's list. He adds: "Number 12 lies on the continuum detection limit at $m_\nu = 19.4$; number 26 has Lyα beyond the range of IIIaJ so it would indeed be missed in the Curtis Schmidt survey; number 60 should have been detected". I have searched retrospectively for number 60 on the CTIO plates and include an illustration of its spectrum in Figure 1.2. Thus one can see that, although one can indeed derive interesting selection lines, they have only qualitative value until applied to proper machine-based surveys; the completeness limit for eyeball searches is not a sharp boundary, but a very broad, fuzzy line. I think completeness could only be claimed for regions very far to the lower right of Clowes' relation.

Clowes (1980, 1981) concludes from his work that, in the region of overlap (on the sky, on the magnitude m_ν - equivalent width W_λ plane, and in wavelength λ), the Curtis Schmidt and grism techniques are equally effective. There is no "puzzling discrepancy". The prism gives much greater sky coverage, whereas the grism allows greater coverage of m_ν (fainter objects), W_λ (weaker lines) and to some extent z (higher redshifts). Clowes cautions that the CTIO-4m and even the Curtis Schmidt survey do not obviously represent the whole sky.

1.2 Some other problems

It is clear from Clowes' work, and indeed from the most cursory examination of plates of the same field taken in different seeing conditions, that the detectability of QSOs is a strong function of seeing (see, e.g., Clowes et al 1980, Clowes 1981). A simpler, less rigorous version of Clowes' argument might run as follows: if a seeing degradation of 1 arc second produces, say, a 0.7 mag loss in limiting magnitude on a IIIaJ emulsion, and if the luminosity function for QSOs is guessed to be 6 per magnitude increment, then a Curtis Schmidt plate taken in 2 arc second seeing may yield ∿ 12 or 13 QSOs, while one taken in 3 arc second seeing yields only ∿ 3. This illustration, though not rigorous, serves to illustrate the need for very careful calibration. I had not expected the Curtis Schmidt plates to be used for clustering tests. They are not calibrated (see footnote to Table 1.1). Indeed power-spectrum analysis applied to the -40° zone (Webster 1981) shows that the detection efficiency for quasars is half, at the edges, compared to what it is at the centre of each individual plate. It is therefore unlikely that Arp's (1980) arguments, concerning the astrophysical significance of the different apparent surface densities

Table 1.3
Curtis Schmidt Fields

Plate No.	Plate Centers (1950)		No. of Quasars
16985*	$21^h53^m53^s$	−30°39'13"	(2)
18610*	22 17 30	−40 03 30	(11)
16408	22 42 00	−39 48 00	8
16465	23 10 08	−40 00 00	8
16414	23 30 00	−38 54 00	9
16458	00 01 42	−40 07 00	10
16466	00 22 07	−40 31 00	13
17001*	00 54 47	−39 41 34	(4)
16467	01 14 14	−39 58 00	15
16459	01 40 02	−40 21 00	7
17049	02 03 22	−39 50 29	3
17050	02 30 11	−39 46 50	6
17051*	02 54 12	−39 50 30	(3)
16995*	03 20 59	−39 39 31	(4)
17052*	03 44 22	−39 42 16	(5)

Fields Arranged by Plate Number

Plate No.	Spectrum Width	Diffuse density of background	No. of Quasars	Plate No.	Spectrum Width	Diffuse density of background	No. of Quasars
16408	–	–	8	16985*	2"	1.1	(2)
16414	2"	0.7	9	16995*	4"	0.9	(4)
16458	2½"	0.5	10	17001*	3½"	0.9	(4)
16459	2½"	1.0	7	17049	2½"	1.0	3
16465	2"	0.7	8	17050	2½"	0.9	6
16466	2"	0.7	13	17051*	3½"	1.0	(3)
16467	2"	0.7	15	17052*	3"	0.9	(5)
				18610*	3½"	0.8	(11)

*These plates were recognised as substandard from straightforward eyeball inspection at the time of the original survey; they should be repeated.

of QSOs in different parts of the Curtis Schmidt -40° zone, are valid. Osmer (1981) goes so far as to tabulate the order in which the Curtis Schmidt plates were taken so as to indicate how smaller numbers of QSOs were detected on plates taken closer together in time (when, presumably, weather conditions or plate sensitivities were systematically poorer). I give, in Table 1.3, a reproduction of Osmer's (1981) Table 1, but have added rough measures, from the original discovery plates, of diffuse densities (to ±0.1) and spectrum widths (to ±½ arcsecond). It is clear that plates numbered up to 16467 are generally better than later ones and yield more quasars. Indeed, many of us believe that counting quasars on a plate gives a very accurate measure of seeing! However, I would find it hard to explain on this basis alone (and in the absence of calibration) why plate 17049, with sky background density (0.95) only slightly above average, and with a spectrum width of 2½ arc seconds yielded only 3 quasars, while plate 18610, background density 0.8, spectrum width 3½ arcseconds (immediately obvious to the eye as a substandard plate) yielded 11! It appears that a sufficiently low background density may also be a critical factor for success in these eyeball searches even though the higher densities are more favourable for machine measurement. Copy plates with carefully controlled (low) backgrounds <u>may</u> be more suitable for this kind of work. Were it not for the fact that Cecilia Aguirre searched each of the plates three times after I had searched them once, I could advance all sorts of reasons why I may have been distracted while searching one plate; anyone who has tried QSO hunting this way will realise how poor the method is, and <u>should realise how careful one has to be in applying any kind of quantitative statistical test to this material</u>. Continuing the litany of problems, it should be noted that sufficiently bright guide stars were not too easy to find with such a small telescope, so Mario Zemelman and I produced a less than ideal strip of sky, as can be seen by examining the co-ordinates of the field centres.

We turn to another particularly nasty selection effect mentioned by Arp (1980), when he announced in the Astrophysical Journal that quasars in the Curtis Schmidt survey violate requirements for quasars at their redshift distances. This conclusion was based on the correct observation (e.g. Arp's Figure 2) that the average apparent continuum magnitude, m(1475Å) given by Osmer and Smith (1980) for the Curtis Schmidt QSOs is about 1 magnitude <u>brighter</u> for QSOs at redshift > 2.5 than for QSOs at redshift $1.0 < z \leq 2.5$, contrary to what one might expect if redshift was an indicator of distance; the volume sampling effect is likely to be small. The blending of the forest of absorption lines shortward of the Lyα emission (section 5.2) blankets the continuum and the short wavelength side of the Lyα emission line, particularly at higher redshifts.

The continuum at 1475Å is directly observable on IIIaJ plates only up to redshifts $z \sim 2.5$. Beyond that redshift, the detection of a QSO (which depends, usually, on seeing emission lines <u>and</u> at least a hint of some continuum) is related to the visibility of the blanketed continuum region shortward of Lyα emission, not on the magnitude, m(1475Å)

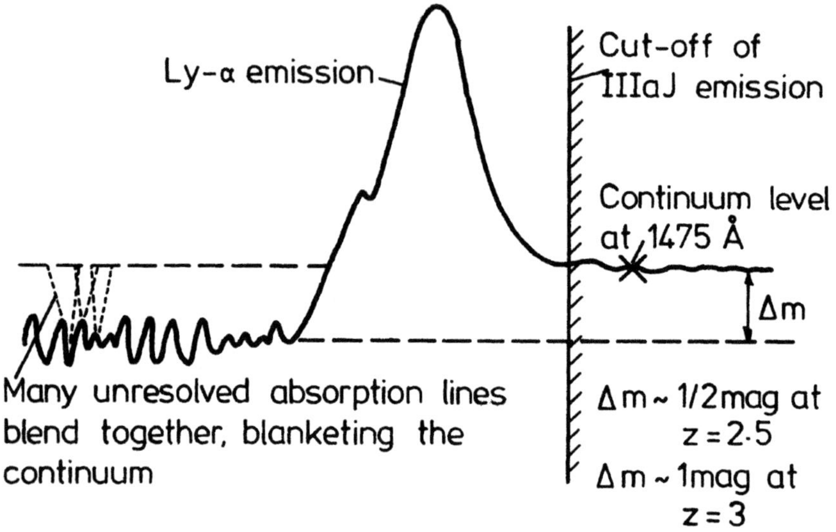

Figure 1.3: The mean apparent magnitude (at 1475Å in the rest frame) for high redshift QSOs ($z > 2.5$) found with the Curtis Schmidt is brighter than the corresponding magnitude (at 1475Å in the rest frame) for the lower-redshift QSOs (Arp 1980). However, this does not necessarily provide evidence for non-cosmological redshifts. Line blanketing in the ultraviolet makes the continua of QSOs harder to see on IIIaJ plates at higher redshifts (where the continuum level on the long-wavelength side of Lyα is beyond the emission cutoff; that region, near 1475Å in the rest frame, is where the monochromatic magnitudes of high-redshift QSOs are usually defined). Incompleteness can set in for high-redshift objects ($z \sim 3$) at a level as much as 2½ magnitudes brighter than the magnitude limit of the survey. As a result, nearly all of the QSOs in this redshift range found with the Curtis Schmidt (e.g. Smith 1976) have proved suitable for more detailed study at higher resolution (see, e.g., Sargent et al 1979). See discussion in section 1.2.

measured later with a sufficiently red sensitive photoelectric slit spectrophotometer on a large reflector. It is also dependent on the visibility of the increasingly blanketed emission-line feature itself. These effects are seen in Plate 1(c) of Clowes et al. (1980) and the result is seen in Fig 5 of Smith and Wright (1980) in the form of a selection line which can be drawn to exclude objects from the faint, high-redshift corner of the diagram. A schematic illustration of this effect is given here as Figure 1.3. Arp illustrates the limit of detectability near 20 mag for QSOs with $1.0 < z < 2.5$ in the lower panel of his Figure 2. He then raises the apparently reasonable objection that the peak of the histogram for objects with $z > 2.5$ is fully $2\frac{1}{2}$ magnitudes brighter than this limit. Examination of the histogram for the lower-redshift group suggests that incompleteness sets in near $18\frac{1}{2}$ magnitude; as pointed out in my earlier review (Smith 1978), the distribution is so broad (Osmer 1980b; Arp 1980 - see also Figure 1.1) because both lines and continuum have to be detected (Clowes 1981). For the higher redshift objects, the whole histogram is shifted about 1 magnitude brighter. Thus, regrettably, in the Curtis Schmidt survey overall incompleteness at $z \sim 3$ does set in at magnitudes as bright as $17\frac{1}{2}$ or so. That is why I am not enthusiastic either about conclusions using this material to support non-cosmological redshifts (or about V/V_m tests applied blindly to this kind of material without reference to a diagram of apparent magnitude vs redshift).

In table 1.1, we see details of some very recent surveys for QSOs of redshift $z > 4$. One is Osmer's most recent, as yet unpublished work using a red-blazed grism and IIIaF plates (Section 2.1). Hoag and Smith (1977) used a blue-blazed grism, which was particularly well suited for the detection of objects near redshifts $z \sim 2$. However, a red-blazed grating should be much more sensitive for the detection on IIIaF plates of Lyα in quasars (or primeval galaxies) with $3.5 < z < 4.7$. Koo and Kron (1980) have employed, for the first time, a CCD detector behind a grating-prism assembly. For Lyα detection, the CCD system is open to a range of redshift $4 < z \lesssim 6.4$. Line detection limits estimated by Koo and Kron were 1.1×10^{-14} erg s^{-1} cm^{-2} (IIIaJ), 4×10^{-15} erg s^{-1} cm^{-2} (IV-N) and 7×10^{-16} erg s^{-1} cm^{-2} (CCD). Phillips (personal communication) has pointed out that the availability of IV-N emulsion for this kind of work should allow detection of low redshift QSOs from their Hα emission (which will be roughly as strong as Lyα), up to $z \sim 0.34$.

1.3 Filter and Fabry-Pérot search techniques

Slitless spectroscopy is of course only one of many possible ways to search for QSOs. This - as we have seen from the nastier selection effects involved - is fortunate.

Broad-band imaging yields optimum detection limits for sources with unknown continuum profiles (see, e.g., Braccesi et al, 1970; Savage and Bolton 1979; Steppe, Véron and Véron 1979; Koo and Kron 1980). A colour-colour diagram allows one to isolate objects with unusual spectra. For detection of broad emission lines with large equivalent widths, a series of intermediate band filters have been ordered for the UK Schmidt telescope (Hazard and Cannon, personal communication). The first 5 filters received have passbands spaced at 200Å intervals over the range 5800Å-6600Å. Koo and Kron suggest a further improvement in detectability might be achieved with slit spectroscopy, using the slit to suppress the sky background. As they themselves point out, the slit will of course strongly cut down the accessible solid angle of sky, indicating the need for a long slit (see, e.g., Spinrad 1977, Hogan and Rees 1979). A Lyα luminosity limit of $\sim 3 \times 10^{43}$ erg s^{-1} at $z = 5$ is implied. Greater solid angle can also be obtained while retaining sky suppression, using (in combination with a grism) a mask having holes corresponding to the position of direct (stellar) images in the field. Such experiments are, I believe, underway at the Steward Observatory in Arizona and at the AAT (Atherton and Taylor, personal communication).

An even more interesting possibility is offered by the imaging Fabry-Pérot system, TAURUS (Taylor and Atherton 1980; Atherton et al. 1980), at the Anglo-Australian Telescope (AAT). This device, when combined with the Boksenberg Image Photon Counting System (IPCS - Boksenberg 1972), allows one to survey individual fields $\sim$ 9 arc minutes in diameter with a spatial resolution $\sim$ 1.3 arc sec. The Fabry-Pérot behaves as a tunable filter, of bandpass $\sim$ 50Å for this application. In one night's observation, a single field can be surveyed down to a visual magnitude limit of $\sim 24\frac{1}{2}$, over a wavelength range of, say, 400Å. In view of the fact that no published redshifts exist for QSOs fainter than 22nd magnitude (see, e.g., Richer and Olsen, 1980), TAURUS will permit the first spectroscopy ever made on such faint objects. Proposals have been made, by Drs Keith Taylor, Paul Atherton and collaborators, to survey several QSO fields with the following scan ranges; note that each 50Å setting of the tunable Fabry-Pérot filter gives a two-dimensional picture of the field in the 50Å passband selected.

	Wavelength Scan range (Å)	Redshift zones Lyα + NV	CIV	OVI
1.	4250-4620	2.5-2.8	1.7-2.0	-
2.	5420-5890	3.5-3.8	2.5-2.8	-
3.	4600-5000	2.8-3.1	1.9-2.2	3.5-3.8

Using observed surface densities of QSOs in these redshift intervals and a luminosity function which scales as $\sim$ 3 per magnitude interval, Taylor et al expect $\sim$ 6 new QSOs in the redshift zone $2.5 < z < 2.8$ in the 9 arcminute field to a limiting magnitude of 24.5. There has been considerable speculation, both from the work on the X-ray background as well as from straight number counts of star-like objects that there will be a distinct drop in the rate of increase of QSO numbers as a function of magnitude (e.g. Setti and Woltjer 1979; Bahcall and Soniera 1980; Vaucher and Weedman 1980; Zamorani et al 1981). The importance of direct measurement of the luminosity function of QSOs at very faint magnitudes becomes obvious from a perusal of these papers.

1.4 Variability as a discriminant for discovery of QSOs

Usher (1978), Usher and Mitchell (1978) have demonstrated that searches for variable star-like objects with ultraviolet excess yield candidate lists which may have $\geq$ 60% QSO content. Original searches (based on UVX criteria alone) were extended by Usher and Mitchell to ensure completeness, in two Palomar Schmidt fields, to an average photographic magnitude $\bar{B}_{pg}$ = 18.6 mag. Photometry of a series of Palomar 1.2m Schmidt plates yielded their final sample of (variable) UVX QSO candidates. Note, however, that of the 96 candidates only 16 have published redshifts, and one other has an unidentified line at λ4161Å (Schmidt 1974).

Hawkins (1981) has assembled a sample of 27 variable objects in the range $14 < B < 21$ from a series of 8 UK Schmidt plates of a single field using the COSMOS measuring machine (Pratt 1977); Hawkins expects the sample to be complete to B = 21, and to an r.m.s. variation $\geq$ 0.25 mag. No attempt was made to preselect blue or ultraviolet objects. Objective prism classifications include 3 quasars, 1 object with an apparently featureless blue spectrum, 4 objects with flat continua and sharp cutoffs (not at a common observed wavelength) and 6 RR Lyrae stars. Observing time has been allocated for follow-up slit spectroscopy and polarimetry later this year.

QSOs selected on the basis of variability seem likely to differ from most other optically selected QSOs, as it is believed that optically selected QSOs are usually not variable at the level of 0.25 mag (Grandi and Tifft 1974; Hayman, Hazard and Sanitt 1979). An important preliminary investigation of the physical nature of this new class of objects has recently resulted in completion of a survey with the VLA, at 4885 MHz (Condon et al 1980a) of the 96 candidates selected by Usher and Mitchell. The idea was to test the simple 2-component model for QSOs and BL Lacertae objects which holds that the optical continuum consists of a steady component and a variable component in different proportions. The steady component, which is unpolarised (Moore and Stockman 1981), not strongly variable (de Veny, Osborn and James 1971) and has no associated radio emission dominates the continuum emission from radio quiet QSOs. The optical and infrared continua could be primarily thermal emission from hot plasma and dust. According to this model, there is a second variable component, which has strong optical polarisation (independent of wavelength), which is more prominent in radio-loud QSOs and BL Lacertae objects. During large outbursts, the radio and optical emission may be correlated (e.g. Rieke et al 1976). Most violently variable QSOs and BL Lacertae objects contain strong nuclear radio sources. It thus seems likely that the optical radiation in this second component is synchrotron emission.

According to the simple two-component model, one would have expected that, with moderate variability, the second (radio-emitting) component would be fairly readily observable with the VLA in most, if not all cases. In fact, only 3 candidates were detected, to a 5σ upper limit of 2.0 mJy. Most UVX optically variable QSOs are radio quiet.

1.5 Searches at radio wavelengths

Radio search techniques, the original method used to identify QSOs, are obviously of continued importance. Work in this area has been well reviewed recently by a number of authors (see, e.g. Wall 1977, Bolton 1977, Wall, Pearson and Longair 1980). A brief summary of some recent major surveys is given in Table 1.4.

Table 1.4 Some major radio surveys

High flux density	Medium flux density	Low flux density
3CR 178MHz (Ref 1)	6C 151MHz (Unpublished)	5C1-5C13 408MHz 1407MHz (Refs 8,9)
408MHz Molonglo (Ref 2)	4C 178MHz (Ref 6)	WSRT 1415HMz (Ref 10)
2.7GHz Northern Hemisphere (Ref 3)	Bologna 408MHz (Ref 7)	WSRT 5GHz (Ref 11)
2.7GHz Southern Hemisphere (Parkes - Ref 4)		
5GHz Northern Hemisphere (Ref 5)		
5GHz Southern Hemisphere (Ref 4)		

References:
1. Bennett, A.S., 1962; Jenkins, Pooley and Riley 1977; see also Smith and Spinrad 1980; Laing, Riley and Longair (in preparation).
2. Molonglo Reference Catalogue - Large *et al* 1981.
3. Peacock and Wall 1981; see also Peacock *et al*, 1981.
4. Parkes Catalogues - summarised by Wall 1977.
5. Pauliny-Toth 1977.
6. Pilkington and Scott 1965; Gower, Scott and Wills 1967; see also Caswell and Crowther 1969; Lynds and Wills 1972.
7. Colla *et al* 1973; see also Fanti *et al* 1975 and Fanti *et al* 1979.
8. Pearson and Kus 1978 (5C6 and 5C7).
9. Benn (5C12 unpublished).
10. Willis, Oosterbaan and de Ruiter 1976.
11. Willis and Miley (1979*a*, *b*).

The major conclusions from all this work, regarding quasars, are usually that the most powerful sources of either the steep or flat-spectrum type undergo strong evolution (see, e.g., Peacock and Gull 1981; also Schmidt 1968, 1970, 1972, 1976; Masson and Wall 1977; Wall, Pearson and Longair 1980; Peacock *et al* 1981). There is little or no evolution for intrinsically weak sources of either type. However, the critical luminosity dividing evolving sources from non-evolving sources is much higher in the flat-spectrum case. On the other hand, Barnothy and Barnothy (1968, 1972), Turner (1980) and others have shown that large values of V/V_m and steep slopes to number-magnitude counts need not necessarily imply such strong source density evolution if lensing events are common. Turner (1980) has suggested that "brightness amplifications by undetected gravitational lenses could be responsible in part for the apparent evolution of quasars, particularly for those which appear to be of high luminosity". There is no evidence from the radio data alone for any directly observed turnover in the comoving space density of QSOs (Peacock and Gull 1981; see also section 2.1).

It has long been assumed that the reason that so few QSOs with $z > 3$ have been discovered by radio techniques is probably associated with bias introduced by the UVX selection procedure needed with low-accuracy radio positions. In an effort to overcome this bias, extensive optical studies have been made of high-latitude neutral and red coloured stellar sources which coincide with accurate radio positions (e.g. Hoskins et al 1972; Browne and McEwan 1972; Browne, Crowther and Adgie 1973). A complete sample of identified objects based on radio positions good to better than 1 arc second has recently been published by Peacock and Wall (1981). Although some redshifts have still to be obtained, the general lack of objects with $z > 3$ remains. Wills and Wills (1976), Condon et al (1980b) and Baldwin, Wampler and Burbidge (1981), have also found no new very high redshift objects. The latter authors, from a set of 21 mainly neutral coloured QSO candidates, found a maximum redshift of $z = 2.43$. Condon et al found 20 objects with $0.82 < z < 3.30$. In a recent observing run at the AAT, Bolton, Savage, Wright and I had similar experiences with a set of red and neutral stellar objects identified with flat-spectrum radio sources; these objects represent about 1% of an initial sample of 1500 weak radio sources ($0.35 > S_{11} > 0.25$ Jy) from the Parkes 11 cm catalogue. In this case, the situation was made worse because the Parkes positions alone are not sufficiently accurate; subsequent selection of red objects then favours high latitude K stars more than QSOs and BL Lac objects. (Ward, Griffiths and I had a rather similar experience earlier in the year with attempts to observe serendipitous sources in Einstein Observatory IPC fields; it seems that because the x-ray counts from sources in these fields were generally lower than in the survey by Chanan, Margon and Downes (1981), the associated optical identification was likely to be very faint, and was usually swamped in a brighter population of G and K stars having $m_B \leq 20$ mag).

1.6 X-ray searches, the X-ray background and luminosity evolution

As expected, the successful launch of HEAO-2 on 13th September 1978 (thus becoming the "Einstein Observatory") has had a significant impact on many fields of astrophysics, including the study of quasars. Now upwards of 200 QSOs are known x-ray sources. Prior to launch, only four other quasars had been detected at x-ray wavelengths. The two major surveys were carried out by the Columbia University group (Ku, Helfand and Lucy 1980) and by a large consortium headed by the x-ray group at the Center for Astrophysics, Harvard (e.g. Giacconi et al 1979, Tananbaum et al 1979, Zamorani et al 1981, Griffiths et al 1981). As a result, quasars immediately became a clear favourite for the dominant component of the x-ray background. The x-ray luminosities of quasars, in the range 0.5 to 4.5 keV, are typically between 10^{43} and 10^{47} erg s^{-1}; the lower luminosities overlap with the values obtained for Seyfert type 1 galaxies.

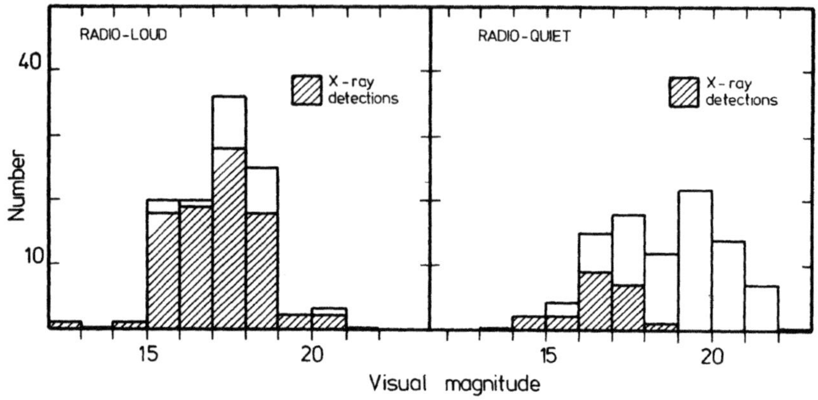

Figure 1.4: The x-ray detectability of QSOs (from Kembhavi and Fabian 1981). Optically faint, radio-quiet QSOs are very hard to detect with the Einstein satellite. See discussion in section 1.6.

Chanan, Margon and Downes (1981) have recently reported a 50% success rate in discovering quasars or Seyfert galaxies by means of their x-ray emission (i.e. half their x-ray sources at high galactic latitude turn out to have active galactic nuclei). A list of x-ray selected QSOs has also been published recently by Grindlay et al (1980). A striking feature of these lists is that nearly all the objects have very low redshifts. Only one object in the Chanan et al list has $z > 1$ (0302-223, $z = 1.40$), and the mean redshift is only 0.36. Because the survey of very bright optical QSOs by Green (1976) also yields objects of low mean redshift, one might expect that deeper x-ray surveys (e.g. Giacconi et al 1979; Griffiths et al 1981) will yield more high redshift QSOs. However, the success rate so far in identifying faint x-ray sources is much lower than that obtained on bright sources by Chanan et al., even though the x-ray flux limit of $\sim 2 \times 10^{-13}$ erg cm^{-2} s^{-1} chosen by Chanan et al is certainly low enough to detect at least some high-redshift objects (as can be seen by comparing with the x-ray fluxes listed by Zamorani et al (1981) for QSOs discovered by other means). It is also interesting to note that radio selection effects are important (Ku, Helfand and Lucy 1980); it seems that x-ray emission is more likely to be detected from radio loud QSOs than radio quiet QSOs.

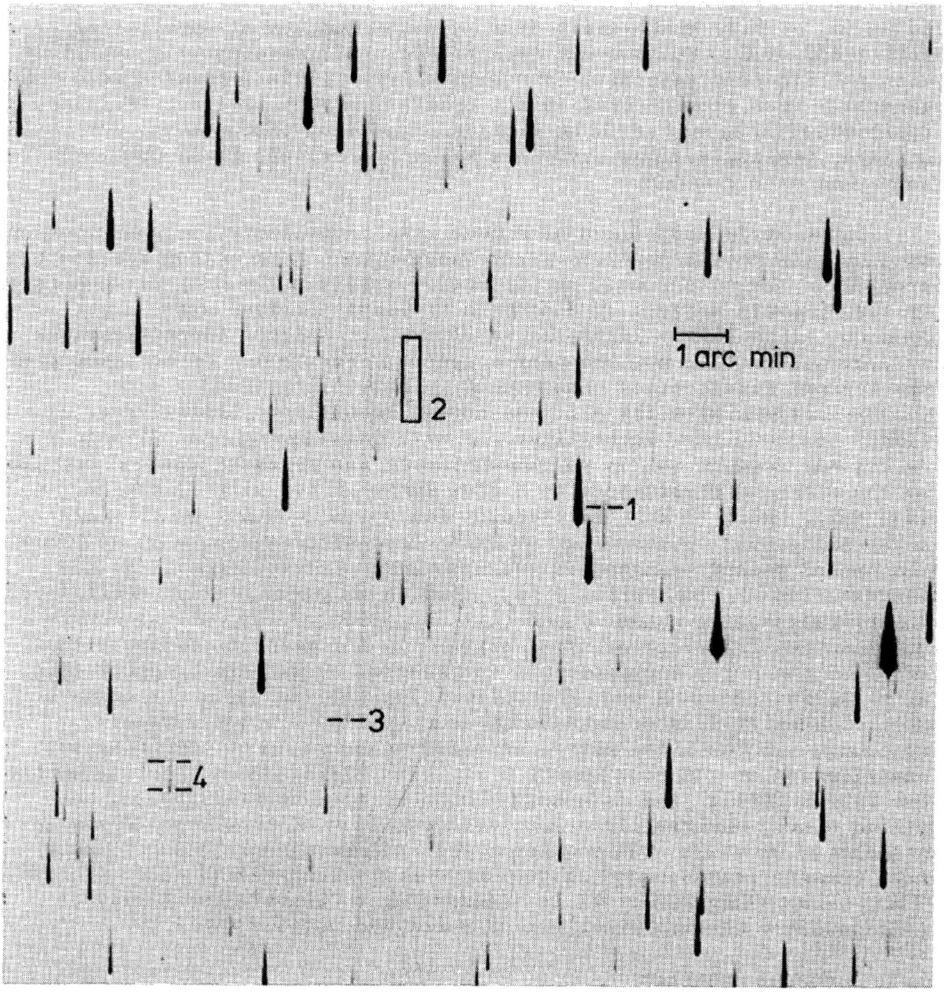

Figure 1.5: Objective-prism photograph of part of the Eridanus Einstein Deep Survey field (Giacconi et al 1979). The original plate was taken, and objects numbered 1, 3 and 4 were identified, prior to launch of the satellite. Object #1, 1E043811-1638.2, is one of the few high-redshift QSOs found in the deep x-ray surveys (Giacconi et al 1979; Griffiths et al 1981). The object marked 2 was detected as an x-ray source (1E043826-1635.6) and identified as a low-redshift QSO at the Anglo-Australian Observatory by Wright, Peterson and Jauncey. It was not noticed on the objective-prism plate. Objects 3 and 4 are emission-line galaxies in the field, but do not seem to have been detected as x-ray emitters. See discussion in section 1.6.

Kembhavi and Fabian (1981) point out that only one object - out of 55 faint ($m_v > 18.0$) radio-quiet QSOs from the samples of Zamorani et al (1981) and of Ku, Helfand and Lucy (1980) - was successfully detected with the Einstein satellite (see Figure 1.4). This important result has generally been glossed over in the original survey papers. As faint, radio-quiet QSOs appear to be a major proportion of the total set of quasars, generalisations about the x-ray properties of all QSOs have to be viewed with caution.

Three basic techniques have been used to estimate the contribution made by QSOs to the observed x-ray background. Each method has its own problems. The first, most direct method involves the Deep Surveys (such as the Einstein surveys of fields in Eridanus and Pavo - see e.g., Giacconi et al 1979; Griffiths et al 1981). Optical identifications are, unfortunately, nowhere near complete; Figure 1.5 is an example of the typical difficulties encountered in attempting to make such identifications with the slitless survey technique. Griffiths et al (1981) conclude that identifications with QSOs having photographic J < 24 mag may account for up to two-thirds of the x-ray background resolved by the Einstein Observatory to a flux limit of 1.3×10^{-14} erg cm^{-2} s^{-1} (1-3 keV), which in turn may account for up to $\sim$35-40% of the total x-ray background (Giacconi et al 1979; see also Cavaliere et al 1981). The second method is based on two uncertain distributions: (i) the distribution of flux ratios S_x/S_{opt} (which is particularly uncertain for faint, radio-quiet quasars) and (ii) the slope of the optical counts (see section 2.5). Zamorani et al (1981), for example, estimate that $\sim$30% of the x-ray background is contributed by quasars brighter than $m_B = 20$, and conclude that the optical log N-m_B relation for quasars must flatten at fainter magnitudes (see section 2.5 for further discussion). The third method is based principally on combining the distribution of the flux ratios S_x/S_R with radio source counts (Kembhavi and Fabian 1981); the data here should be more secure, though, as stated above, uncertainty arises from our lack of knowledge of the x-ray properties of faint, radio-quiet QSOs. Kembhavi and Fabian estimate that quasars probably contribute between about 25 and 40 percent of the 2 keV x-ray background. It is interesting to compare these various post-launch estimates with that of Setti and Woltjer (1979).

1.7 Infrared searches

Rieke and collaborators (Rieke, Lebovsky and Kinman 1979, Rieke 1980) have recently discovered a number of objects with steep infrared/optical spectral indices (α = d ln flux/d ln frequency) $\alpha_{KB} \sim -3$ in optically blank fields of flat spectrum radio sources (Condon, Hicks and Jauncey 1977; Condon, Jauncey and Wright 1978). Because the remaining properties of these objects were similar to those of violently variable QSOs or BL Lac objects, they suggested that these new steep-spectrum infrared sources represent a new class of quasar with larger infrared excess than any known extragalactic object. They also noted that at the faintest blue magnitudes α_{KB} tended to be highest.

Impey and Brand (1981) have shown, however, that these red sources are just the tail of the normal distribution of quasar colours, that they are relatively common, and that a considerable fraction of flat spectrum sources from a complete sample will have properties similar to the BL Lac objects. They base these conclusions on K(2.2 μm) and J(1.25 μm) filter observations of a sample of 21 Parkes flat spectrum sources (α(6-11 cm) > -0.5, S(11 cm) > 0.5 Jy). Of these 21, 20 were detected at K and 19 at J. The sample is reasonably free of obvious selection effects as it is a radio sample which is completely identified optically and almost completely identified at 2.2 μm. Comparing their sample with the Rieke et al objects, Impey and Brand conclude that the Rieke et al objects do not come from a separate population at better than the 1.5σ level. Impey and Brand find in their sample a distribution of indices α_{KB} covering the full range $0 \geq \alpha_{KB} \geq -3$ and show in addition that a trend of α_{KB} with apparent blue magnitude m_B is a selection effect. Finally they find that a simple composite object spectrum consisting of the energy distribution of a giant elliptical added to a power law spectrum (which flattens off at some wavelength) is consistent with the range of spectral indices found in a plot of α_{KJ} vs α_{KB} for their sample.

1.8 Proper-motion search

Sandage and Luyten (1967) pioneered the use of proper motions in efforts to locate QSO candidates, but worked only on samples of objects already selected by colour. Kron and Chiu (1981) have recently completed a survey of star-like objects in SA57 which lack detectable proper motion; they did not appeal to colours in the process of isolating their sample.

As all of the alternative search techniques suffer to a greater or lesser degree from selection effects, the renewed application of this proper-motion search technique is most welcome. I confess to having not realised that proper motion data on such faint objects could be obtained to sufficient accuracy that most Galactic stars are observed to move (Chiu 1980). In addition to the proper motion data, they report on information gleaned from red and blue grism plates of SA57, multichannel photometry and a variability study over five epochs spanning 14 years. With all this data and slit spectra of nine objects, they classified all 23 of the objects they studied as either definitely a Galactic star or definitely a QSO. The UVX criterion would have failed to discover a QSO in only one out of ten cases. The variability criterion was successful in eight out of ten cases, but more than two epochs are an essential requirement for such high success rates. Allowing for the brighter limiting magnitude, the grating-prism survey failed in one instance to find a QSO. The proper-motion technique was in fact fairly successful, but all the 5 QSOs found with this method would have been found with one or other of the alternative techniques. Deep x-ray and VLA surveys of this zone would be interesting. In summary, Kron and Chiu discovered, or rediscovered a total of ten QSOs brighter than B = 21.4 in 0.1 deg^2; this figure strengthens and supersedes Becker's value (Sandage and

Luyten 1969) of ∿ 4 QSOs in a field of 0.04 deg^2, brighter than B = 21.4. Some uncertainty remains, as apparent clumping of QSOs by factors of ten over the average background has already been observed (Clowes 1981 - see also section 1.1). Redshifts for these ten objects would be most valuable; such objects are now within reach of modern spectrometers (see, e.g., Richer and Olsen 1980).

2. THE DISTRIBUTION OF QSOs ON THE SKY

2.1 Is there an observed turnover in the comoving space density of quasars at high redshift?

Rapid density evolution models are observed not to be valid out to redshifts z ∿ 4.7; indeed, evidence has been obtained very recently for an apparent rapid downturn near z ∿3.5 in the comoving space density of quasars.

Longair (1976) has summarised the reasons for believing that significant information about the birth of galaxies and QSOs may be obtained by probing to redshifts z ∿ 5. Several times in the past, there has been a wide consensus that the 'switch-on' epoch for quasars had been directly observed in the form of a downturn at high redshift in the comoving space density of QSOs, $\rho(z)$. The most recent of these was about a decade ago, when it became widely assumed that very few QSOs would be found beyond redshift $z > 2.5$ (see, e.g., Schmidt 1968, 1970, 1972, Sandage 1972). These assumptions had to be changed following the discovery of OH471 (Carswell and Strittmatter 1973) and OQ172 (Wampler et al 1973) and the subsequent success of the slitless spectrum search technique (Smith 1976; Osmer and Smith 1977b). For the interval $1.9 < z < 3.2$, the space density of luminous quasars was at least constant with redshift and was consistent with Schmidt's exponential density law, $\rho(z) \propto \exp(-10t/t_o)$, where t/t_o is the cosmic time in units of the present epoch. However, OQ172 (1442+101), discovered in 1973, still has the highest redshift (z = 3.53) known. Although perhaps as many as 100 QSOs have been found with $z > 3$ by means of the slitless spectrum technique, the largest published redshift for an optically selected QSO is z = 3.45 (Smith et al 1977). Why have no larger redshifts been discovered when so many objects with $z > 3$ have been found?

All of the published slitless surveys have been done with IIIaJ except for the grism survey on IIIaF emulsion by Hoag and Smith (1977) and the recent CCD and IV-N surveys by Koo and Kron (1980) - see Table 1.1. Thus, only these two latter surveys were sensitive to Lyα emission beyond z ∿ 3.3, and these surveys only covered a few square degrees of sky. Carswell and Smith (1978) showed (with semi-quantitative illustrations) that the 5 degrees covered by Hoag and Smith was not enough, at least with a blue-blazed grism. The failure to detect any QSO of redshift $z > 3.45$ was not sufficient evidence in itself for a turn down in the comoving space density beyond z ∿ 3.2; indeed, Carswell and Smith's models were only marginally inconsistent with continued density evolution of the form $\rho(z) \propto \exp(-10t/t_o)$. They did manage however to rule out very rapid density evolution models such as $\rho(z) \sim (1 + z)^6$ extending out to z = 4.7.

Koo and Kron (1980) may have advanced a little further by means of their prime focus slitless spectroscopy with CCD and photographic detectors. They looked for evidence of emission lines or continuum breaks in the first order spectra. With the CCD, about 200 arc mins was surveyed with a bandpass λλ6100-9100Å (Lyα redshift $4 < z < 6.5$) to a level ∿1% of sky. With IV-N plates, approximately 4000 arcmin^2 was surveyed with a bandpass λλ6100-8800 (Lyα redshift $4 < z < 6.2$). Koo and Kron find that "Even though both the CCD and the IV-N data are of high quality, no object of clearly unusual nature has yet been confirmed". They had expected, even in the absence of evolution, to have detected ∿3 QSOs with $z > 4$ in their 200 arc min^2 CCD survey and up to ∿ 2 more in their 4000 arc min^2 IV-N survey. With evolution laws such as $\rho(z) \propto \exp[-10t/t_o]$, they would have expected ∿ 3 times more objects. However, because the flux limit of their survey is not well determined, these estimates must be regarded as tentative, especially as no positive QSO detections were made.

The most significant work on the redshift cutoff is, in my opinion, a very recent, as yet unpublished survey by Osmer (preprint) using a red blazed grism and IIIaF plates (see e.g. Osmer 1980a). He has surveyed ∿ 5 deg^2 with a bandpass 5700Å-6900Å (Lyα redshift 3.7-4.7). He discovered 5 hitherto unknown QSOs with $z > 2.7$ .... from their CIV lines! Though he discovered these very high redshift QSOs the hard way, he saw no evidence for any object with Lyα emission, which would seem to be suggestive of a sharp cutoff in the numbers of QSOs beyond $z \sim 3.7$. However the radio results remain inconclusive about the reality of any redshift cut off near $z \sim 3.5$ (Peacock and Gull 1981). Should one still worry that we may yet have overlooked some subtle selection effect? QSOs are very easy to find (see e.g. Smith 1976, 1978) just to the point ($z \sim 3.3$) where Lyα reaches the long-wavelength limit of IIIaJ spectral sensitivity. In the words of Dr R.F. Carswell - "How did Eastman Kodak know"? More to the point, evidence not unlike some of that found recently by Osmer has been published for the window 4500Å $< \lambda <$ 4900Å, yet no-one now believes there is a redshift cutoff there. This wavelength range for Lyα emission corresponds to redshifts $2.70 \leq z \leq 3.03$; for CIV λ1549 emission, the redshift range is $1.91 \leq z \leq 2.16$. The wavelength range was chosen a priori, for another purpose (Fabry-Pérot spectroscopy - see section 1.3). Hoag and Smith surveyed an area of 5.1 deg^2. They reported a number of independently discovered objects with $m_B \leq 19\frac{1}{2}$ and $1.91 \leq z \leq 2.16$ with the comment "α4"; "α4" means that Art Hoag and I clearly detected Lyα and CIV on the original survey spectra. 5 of these objects were confirmed by Osmer's (1980b) spectrophotometry as QSOs falling within these redshift and magnitude constraints (numbers 7, 12, 46, 48 and 60 in Osmer's list). Interestingly enough, we found only two new objects within the redshift range $2.70 \leq z \leq 3.03$ (numbers 37 and 42) but both were 20th magnitude objects (Osmer 1980b). Thus, looking through the window 4500Å $\leq \lambda \leq$ 4900Å at objects brighter than 19½ magnitude, we saw 5 objects with detectable CIV emission on the survey plates yet none with Lyα emission in 5.1 deg^2. Now Art Hoag and I clearly spotted most, if not all these 5 objects from their Lyα emission at shorter wavelengths. Nevertheless, this observation serves as a cautionary note to anyone advocating the reality of a cutoff based on the non-detection of Lyα through a fixed wavelength window when small numbers of objects are involved.

The strength of Osmer's case, however, is that it rests not only on the failure to detect Lyα when CIV (and other weak lines) were readily seen. By showing that his equipment would be sensitive to known QSOs from the HS survey with $2.5 < z < 3.5$, if such QSOs were moved out to $3.7 < z < 4.7$, he has based his expectations on the known properties of quasars in the redshift bin immediately below $z = 3.5$, and so the least possible extrapolation is involved. His calculations yield a sharp turn down in the comoving space density; his 95% upper limit is that the space density must be at least a factor 3 less at $3.7 < z < 4.7$ than at $2.5 < z < 3.5$. Osmer concludes his summary of future prospects: "The real questions are how steep is the turn down in density, and where in the interval $3.2 < z < 3.6$ does it occur .... The UK Schmidt offers .... a possibility. With its large field, a suitable objective prism and filter ...... could yield excellent plate material".

Delivery of a higher dispersion prism for the UK Schmidt is scheduled for the middle of this year (a discussion of its specification and advantages for work at very high redshifts was given in my earlier review article - Smith (1978)). We plan first of all to photograph, on red-sensitive emulsions, a number of fields containing high redshift QSOs of differing apparent magnitudes and emission-line strengths, in order to calibrate the searches. You can be sure that a number of groups will be busy with IIIaF and IV-N objective-prism plates from the UK Schmidt later this year.

2.2 Clustering of QSOs

There is still no unambiguous evidence for clustering of QSOs.

In another very important paper Osmer (1981) has used the data from the 174 high-latitude quasars from the HS (grism) and OS (prism) surveys at CTIO to take a first look at the spatial (three-dimensional) distribution of redshift $z \sim 2$ quasars on scales of order 50-1500 Mpc in present epoch co-ordinates (with $H_o = 100$ km s^{-1} Mpc^{-1}). In the HS (grism) survey, he finds, by visual inspection, a number of close groups and pairs with small Δz; note that at $z = 2$, 1° on the sky corresponds to 70 Mpc, or $z = 0.07$. He uses three statistical techniques (binning analysis, the nearest neighbour test and the correlation function) to show that these groups and pairs turn out to be chance fluctuations. He finds no evidence for clustering or other deviations from randomness on the 50-1500 Mpc scale. It is particularly important to notice that Osmer has taken at least some precautions to allow for selection effects such as the strong redshift bias, edge effects etc.

The HS (grism) survey yields a high surface density of objects, but covers only a small total area of sky, itself split into (widely separated) regions. Thus only limited ranges of spatial frequencies are sampled. Though the OS (Curtis Schmidt) survey covered a much larger region of sky, its lower surface density of objects reduced the sensitivity to small-scale clustering.

It is on the smaller scale that most of the astrophysical interest and controversy centres; scales up to 40 Mpc (at the present epoch) are of interest for investigation of groups, clusters (see, e.g. Peebles 1974) and superclusters (e.g. Hauser and Peebles 1974; Gregory, Thompson and Tifft 1981). Oort, Arp and de Ruiter (1981) make a rough estimate of 20 Mpc for the diameters of superclusters at $z = 2$. They have found 12 pairs of neighbouring quasars (from a number of different samples) "whose redshifts are so nearly equal that they have a high probability of being physically associated", at separations between 5 and 30 Mpc, i.e. of the order of separations one might expect if quasars were thinly represented in superclusters of galaxies. They state that "in order to exclude spurious pairs a special effort has been made to determine the probability of accidental coincidences". Though the statistics are poor they estimate that roughly 20% of the powerful superclusters contain one quasar brighter than 20m while only 3% have 2 quasars.

It seems to me that, in contrast to Osmer (1981), Oort et al (1981) have not properly accounted for samples in which no close pairs can be found. It is not sufficient to make a complete search in a very small area and produce statistics yielding low probabilities of occurrence, unless reasonably large numbers (10 or 20) of similar control areas are searched. What we all want to know is how often, when you search a section of sky, do you not find pairs. Osmer has attempted his statistics on a sample sufficiently large to make at least a preliminary assessment of the noise in addition to running Monte Carlo simulations to provide further checks. He concludes: "a major lesson is that the eye is good at picking out associations ...." "....... but poor at evaluating the noise. Beware of redshifts written on the sky!" De Ruiter and Zuiderwijk (1981), following some elegantly simple mathematics, produce a clumpy quasar sky from a quasar-generating supercluster model. Though I would not accept their assertion "observations now suggest that quasars are not randomly distributed in space but instead are located in superclusters" as an established fact, it is equally true that the number of corrections that Osmer (1981) was forced to make to the rather unsatisfactory Tololo surveys could easily mask the subtle effects de Ruiter and Zuiderwijk are searching for. Webster (1981) has also made tests for clustering in the Curtis Schmidt -40° zone, using power-spectrum analysis, and finds no evidence for clustering, even though he searched various subsamples in an attempt to enhance the contrast of any potential clustering. He concludes that substantially larger samples will be necessary to investigate the very weak associations being postulated (1 QSO brighter than 20^m per 5 major superclusters, and 1 QSO pair in 3% of such superclusters - Oort, Arp and de Ruiter 1981). In summary, there is not enough data yet to give sufficient signal-to-noise ratio to evaluate the significance of clustering on scales less than ~ 50 Mpc.

The study of quasars has been bedevilled by 'a posteriori' statistics on small numbers of objects in the presence of large selection effects. A vast and apparently impressive bulk of research papers thus has to be tackled by one brave enough to attempt to enter this field alone. Many observational effects have had to be tested. Quasars on great circles? Quasars at the galactic poles? Quasars associated with galaxies at vastly differing redshifts? Quasars in pairs, triples, clusters, lines? Non-cosmological redshifts, line locking, gravitational lenses? The difficulty is to avoid being so disgusted with the whole business that one misses the really important discoveries - e.g. that which arose when Walsh and Carswell (i) happened to notice that the optical counterpart to the Jodrell Bank radio source 0957+561 was in fact resolved into 2 blue star-like objects on the Palomar sky survey but (ii) did not assume that the second object was just a foreground star or a chance association (Walsh, Carswell and Weymann 1979; see also section 6). One must avoid being blinded to the possibility that special, but rare phenomena can exist even when no evidence can be found from general statistical tests. Arp (1981b), in his characteristically entertaining way, illustrated this point at a recent meeting thus: "This talk of statistics reminds me of a story of two people walking along in the jungle. They remarked on an interesting striped pattern in the undergrowth, and while they were discussing the probability that the stripes could arise from random leaves and twigs, the tiger leaped out and ate one of them".

2.3 Lines of QSOs

Lines of QSOs in the sky probably represent chance associations.

Hazard, Arp and Morton (1979) - see also Arp and Hazard (1980) - have found a particularly interesting grouping of QSOs at 11^h46^m, +11° 30' from searches of UK Schmidt objective-prism plates followed up with slit spectroscopy at the 5-meter telescope on Palomar mountain. Two straight lines can be drawn on their photograph (Arp and Hazard 1980, Figure 3) passing through the seeing discs of two sets of 3 QSOs. Two more such triplets have been found in another QSO association a few degrees away.

The idea that the triplets are associated would imply a physical relation between the objects, which, in view of their disparate redshifts, implies a large, non-cosmological component to at least some of the redshifts. Various tests are in progress to check this:

(a) Edmunds and George (1981) have produced a series of randomly generated points, sprinkled over the typical area of a UK Schmidt plate; using a mean surface density $\sigma = 6$ QSO deg^{-2}, they "expect some eleven alignments good to 2.4 arc seconds and of overall size less than a degree to appear on a typical Schmidt plate". A more restricted alignment, of length $d < 30$ minutes of arc and good to less than 1 arc second, would occur once every other Schmidt plate. They conclude that triple alignments of quasars, of the kind found by Arp and Hazard (1980), are not unexpected.

(b) Various UK Schmidt objective-prism plates are being searched, by Mr Arthur Trew, at Edinburgh, following definition of selection criteria for such QSO alignments. Various alignments have been found, but as their statistical significance appears to be marginal, a fair amount of careful work has still to be done to complete the project.

(c) Proposals have been made (e.g. by Clube, Arp, Brand and Trew) to search for common absorption-line systems in front of two of the QSOs in a line which have $z > 2.1$ and are brighter than 19 mag. Lyα absorption lines have been found in all QSOs of appropriate redshift, once sufficient resolution has been employed. However, there has never been any particular reason to invoke non-cosmological redshifts for those QSOs for which high-resolution observations have so far been made. Thus Clube et al propose to observe such specially selected QSOs to look for any radical departures from the commonly observed distribution of sharp lines. [Similar efforts that were made a few years ago by Boksenberg, Carswell, Whelan and myself to observe common absorption lines in closely-spaced QSOs (not expected to have large non-cosmological redshifts) were frustrated by the fact that, with the AAT and the RGO/IPCS spectrometer combination, exposures of more than six hours are needed to detect lines of equivalent width 0.7Å with confidence in individual objects of $\sim$ 19 mag even in fairly good seeing.]

2.4 Quasar-galaxy associations

An important sample of quasar candidates very close to galaxies has recently been reported, which would appear to be grossly inconsistent with commonly accepted estimates for the background density of quasars.

Arp has continued to present examples of highly redshifted quasars close to galaxies as evidence for non-cosmological redshifts (e.g. Arp 1977, Arp and Sulentic 1979, Arp, Sulentic and di Tullio 1979, Arp 1981a). G. Burbidge (1979) on the one hand and Wills (1978) and Weedman (1980) on the other have come to essentially opposite conclusions concerning the statistical significance of these apparent associations. The latest independent check of Arp's work, by Weedman, compares the position of 259 quasars (discovered with the Curtis Schmidt) with 62 bright galaxies. Weedman finds that the number of close ($2' < r < 15'$) quasar-galaxy pairs predicted from a random distribution lies within 1 standard deviation of the number actually found. He also finds that quasar-quasar pairs agree with the number expected by chance. By using QSOs found in surveys as reference points for the associations rather than galaxies, the controversy concerning magnitude limits and the actual surface density of QSOs is not brought in.

Weedman (1980) then goes on to Burbidge's 10 best pairs on a case-by-case basis. Weedman concludes "It is tempting to feel that, while a few of these associations can be explained away, surely they cannot all be dismissed. Yet we see arguments can be made that, indeed, they all should be. Note that these arguments were not forced upon us simply by prejudice in favour of cosmological redshifts. They arose because large samples of quasars now exist, such as those in the Tololo survey, that show no evidence of quasar-galaxy associations. If these associations are physically real, they have to show up in objective samples of quasars. No doubt, further interesting examples of quasars near galaxies will continue to turn up. All should be considered as honestly as possible." (One should, as I have mentioned earlier, realise that special cases, of potentially great physical importance, can occur, but one must also realise that statistical tests of the type being considered here will not isolate them. Who can say, at the moment, what is the significance of the quasars grouped near NGC1097? One is quite near the end of one of the jets (Wolstencroft, personal communication). Is the group near M82 important? Obviously Burbidge et al (1980) think so, but again have used a small sample and 'a posteriori' statistics. Independent lines of enquiry are needed, but telescope time to do this is often hard to get, given the number of failures in the past).

True to Weedman's predictions, Arp (1981*a*) has just produced a new list of 22 quasars close to 34 "companion" galaxies. However, on this occasion the results are presented in a form quite different from most of his previous work, and in this case, the argument of uncertain background density cannot be so easily invoked (though it is instructive to note that a change of only a factor 16 in this background density produces a change from 10^{-17} to 10^{-5} in Arp's estimated probabilities for these associations). Arp states that "If the objects tabulated .... are confirmed to be quasars, it seems necessary to conclude that these quasars are at the distances of bright galaxies". Note that the thrust of Arp's argument is slightly different here; although he uses an almost complete sample of bright galaxies for his tests, he is actually invoking physical association between quasars and companion galaxies to the bright galaxies (see also Sulentic, Arp and di Tullio 1978). He does make a useful advance on his earlier, almost purely 'a posteriori' statistics, by analysing how many companion galaxies do not show associated quasars; he also defines his sample of main and companion galaxies and uses a background density of ~ 10 deg^{-2}. He mentions that some of the companions have considerably larger redshifts than the central galaxy, but points out that "the companions as used here merely furnish a defined sample of moderate to low redshift galaxies around which densities of quasars may be measured and compared to quasar densities in the general field". In essence, his companion galaxies can be taken as points in the sky around which nearest neighbour distances have been measured. A histogram of such a distribution shows an overwhelming preference for much smaller separations than would be predicted for global background densities anywhere near 10 deg^{-2} (Zuiderwijk, personal communication); QSO densities nearer 200 deg^{-2} would be needed to explain the small separations (Arp 1981*a*).

Barnothy (1974) has suggested that in at least some quasar-galaxy associations, the quasar is in fact the image of a distant, bright, compact object with spectral characteristics of quasars, intensified through the gravitational lens effect of one of the globular clusters expected to surround the associated galaxy; thus, large differences in redshifts between galaxy and quasar would not necessarily be evidence of non-cosmological redshifts.

It is of interest to note, in a different context, that these close quasar-galaxy configurations may be ideal subject matter for study of the absorption of QSO light in various galaxies.

2.5 The slope and evolution of the luminosity function

The slope of the luminosity function remains as uncertain as ever, in the range 3 to 9 per magnitude interval. Star counts strongly suggest the existence of pronounced evolution of the intrinsic luminosity function with redshift when using standard Friedmann models.

The slope of the number counts of bright QSOs appears to be remarkably steep. Green and Schmidt (1979) and Braccesi et al (1980) quote factors $\sim$ 7-9 per magnitude interval for the rate of numerical increase in the surface density of QSOs. Bahcall and Soniera (1980) conclude, from models of the halo population of our galaxy, and star counts to faint magnitudes (see also Kron 1978, 1980; Peterson et al 1979; Tyson and Jarvis 1979; Bohuski and Weedman 1979; Shanks, Phillips and Fong 1980; Gilmore 1981), that (i) extrapolation of such steep slopes to faint magnitudes ($m_v \geq 22$ mag) predicts that virtually all optically unresolved objects at the galactic pole are QSOs yet (ii) models of our Galaxy can already account for most of these observed stellar objects. Continuation of such steep slopes much beyond 20th magnitude could potentially violate the observed x-ray background (see e.g., Setti and Woltjer 1979; Tananbaum et al 1979; Ku, Helfand and Lucy 1980; Bònoli et al 1980; Cavaliere et al 1980, 1981). Thus, the number counts of QSOs must flatten at faint magnitudes. Vaucher and Weedman (1980) conclude that at faint magnitudes ($m \geq 18$ mag), the differential number counts of quasars do not increase by more than about 3.5 times per magnitude interval; they quote similar results for the nuclei of local class 1 Seyfert galaxies. (A similar slope was also obtained from a sample of unconfirmed UKSTU QSO candidates - Smith 1978). Indeed Bǒnoli et al (1980) have discovered that many of the fainter UVX objects in the Braccesi et al (1980) counts are in fact extended and are not quasars; the proportion of such objects increases with magnitude. A representation of the Braccesi-Bònoli counts (e.g. Kembhavi and Fabian 1981) is $\log N(< m_B) = 0.86\ (m_B-18.33)$ for $m_B \leq 19$, and $\log N(<m_B) = 0.36\ (m_B-17.40)$ for $m_B > 19$. These slopes correspond to $\sim$ 7.3 per magnitude for the brighter objects, 2.3 per magnitude at the faint end. Other surface-density estimates are given, for example, in papers by Sandage and Luyten (1969), Bolton and Savage (1978), Osmer (1978), Steppe, Véron and Véron (1979), Lewis, MacAlpine and Weedman (1979), Kron and Chiu (1981) - (see also the summary by Wills 1978).

Some degree of luminosity evolution seems inescapable, when using standard Friedmann models (see, e.g., Cheney and Rowan-Robinson 1981). Segal, Loncaric and Segal (1980) have shown that the chronometric cosmology (Segal 1976) has the potential for non-evolutionary

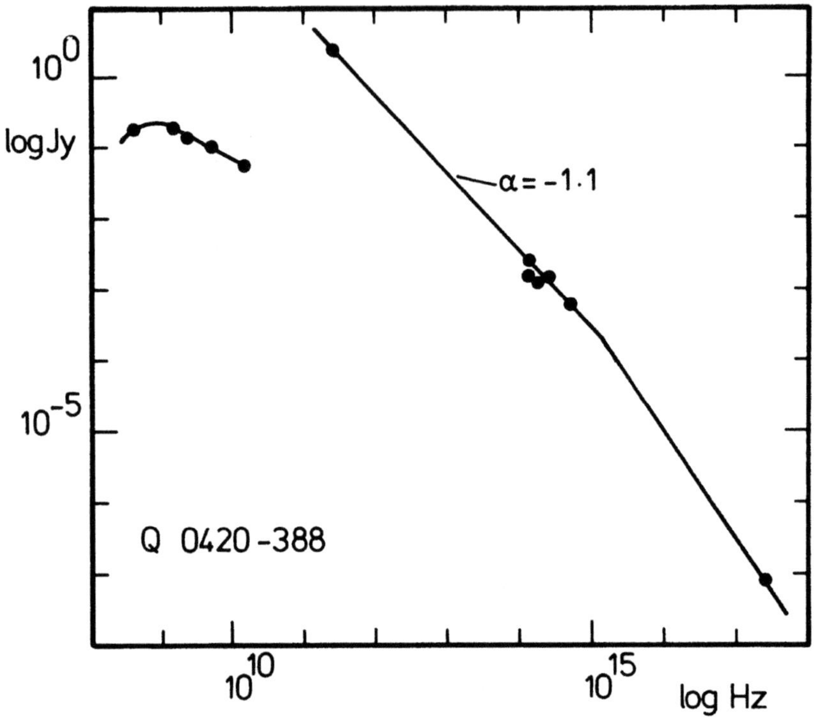

Figure 3.1: The continuous spectrum of the high redshift (z = 3.13) optically selected QSO, Q0420-388 (Sherwood et al 1981; see also Glass 1980).

reconciliation of this problem; as they point out, it will be interesting to compare its prediction of an extreme paucity of very faint quasars with observations (for objects of realistic spectral index it predicts an infinite $\partial \log N/\partial m$ at finite values of m, beyond which N is constant). Regardless of which cosmology is adopted, better observational data is badly needed before more reliable quantitative statements can be made. We have seen earlier (section 1.3) how the use of the TAURUS Fabry-Perot interferometer may enable one, for the first time, to do spectroscopic observations and surveys for QSOs fainter than $m_B \sim 22$ and thus begin to make direct observational tests of the evolution of the luminosity function. The faintest QSO with a published redshift has B = 21.8, z = 1.92 (Richer and Olsen 1980), so that, near redshift $z \sim 2$, QSOs have been found over the six-magnitude range 16-22 mag.

3. THE CONTINUOUS SPECTRUM OF QSOs, AND THE NATURE OF THE CENTRAL SOURCE OF ENERGY

Improvements in sensitivity at all wavelengths of the electromagnetic spectrum from radio to x-ray have enabled very full coverage of the electromagnetic spectrum of QSOs to be extended to fainter objects and higher redshifts. X radiation beyond redshift z = 3 was reported (from Q0420-388 z = 3.13), very shortly after launch of the Einstein satellite (see e.g., Zamorani et al 1981). Detailed IUE observations of various bright QSOs have recently been published - e.g. 3C273 (Ulrich et al 1980), Q1101-264 (Boksenberg and Snijders 1981). Infrared (Glass 1980) and millimetre (Sherwood et al 1980, 1981) detections of Q0420-388 and other QSOs with z > 3 have also been reported. At lower redshifts, the continuous spectrum of 3C273 is being explored in even greater detail (see e.g. Ulrich et al 1980; Cutri et al 1981, and references therein). It is becoming increasingly clear that the continuum in most QSOs is not a simple power law (see e.g. Figs 3 and 4 in Ulrich et al 1980; however there are reasons for believing that their reddening corrections and consequent interpretation of the continuum may not be correct - much of the enhancement may be due to FeII emission). Figure 3.1 shows the spectrum of Q0420-388, as given by Sherwood et al 1981 (c.f. Glass 1980; see also Kreysa et al 1980). More detailed observations of QSOs in the ultraviolet, optical and infrared (e.g. Baldwin 1975; Neugebauer et al 1979; Richstone and Schmidt 1980; Grandi and Phillips 1980; Ulrich et al 1980; Puetter 1980; Glass 1981; Puetter et al 1981; Canfield and Puetter 1981; Cutri et al 1981; Gaskell 1981) reveal considerable lumps and bumps in the spectrum, suggestive of a multicomponent origin, perhaps involving a number of different compact sources, thermal emission from dust, Balmer, Paschen and Brackett continua, FeII emission, two-photon continuum in a high-density emission-line region with very high Lyα optical depth, reprocessed emission from dense clouds or filaments near the central source, thermal radiation from the dense inner part of an accretion disk surrounding a massive black hole ....; see Ulrich et al (1980), O'Dell (1979) and Shields and Wheeler (1978) for further discussion. Hard x-ray spectra of Seyfert galaxies, in the range 5 keV up to several hundred keV, exhibit a remarkably smooth power law, of slope ~ -0.7 (e.g. Mushotsky et al 1980; Mushotsky 1981).

Most QSOs have very low levels of polarisation, $P \leq 1\%$, but those with high, variable optical polarisation, the HPQs (Moore and Stockman 1981) have provided many of the most significant clues as to the origin of the continuous spectrum. If one assumes no bulk relativistic motion, then light-travel arguments and the timescale for variability set upper limits to the size of the region emitting the radiation. There is a strong correlation between high polarisation and rapid variability. In the case of the high-polarisation quasars (and BL Lac objects) it is likely that we are observing a region only a few light days across, quite close to the central energy source. Moore and Stockman have suggested that spectropolarimetric monitoring of high polarization QSOs may, during faint phases, reveal the presence of a lower-polarisation, less variable background continuum with a concave spectrum and with polarisation decreasing to the blue. In the majority of QSOs it seems that the radiation from the central source has been obscured or reprocessed, so as to yield low polarisation, flatter optical spectra and less dramatic variability.

Blandford and Rees (1978) and many others have shown that incoherent synchrotron theory is hard pressed to explain the observations of the HPQs unless some kind of continuous energy generation process is invoked (as discussed below). In order to avoid excessive inverse - Compton losses, very high magnetic fields are required, which in turn result in rapid synchrotron cooling of the relativistic electrons. In the absence of an appropriate source of energy the resulting high column density of non-relativistic electrons would depolarise the radiation, either by Faraday rotation, or by electron scattering. The reader interested in more extensive discussion of the fundamental radiation processes in quasars is referred to a number of articles on the subject by Rees (1977a,b; 1978a,b; see also Ozernoy 1974). The standard, but unproven model for quasars is that proposed by Lynden Bell (1969, 1978). According to such models, a quasar luminosity, $\sim 10^{47}$ erg s^{-1}, can be derived by accretion of $\sim 10 M_{\odot}$ y^{-1} of material onto a massive black hole. Because the surrounding material has significant angular momentum, some sort of disk-like configuration of material will form around the black hole (e.g. Shields and Wheeler 1978). This has led observers to attempt to explain some of their emission-line observations in terms of rotating disks (see, e.g. Osterbrock 1978).

Very rapid x-ray variability (timescale $\sim$ 100 seconds, amplitude $\Delta L \sim 10^{43}$ erg s^{-1}) has recently been reported in the Seyfert 1 galaxy NGC 6814 (Tennant et al 1981). There was no indication of spectral variability in the x-ray region during these rapid variations. Tennant et al use these observations to set constraints on the radiation mechanisms and conclude that "unless reacceleration or continuous injection of particles occurs, neither the synchrotron nor the synchrotron self-Compton processes can be the origin of the x-ray emission. However, we do find that our data are consistent with the general predictions of thermal-Compton models". Nevertheless, from a sample of over 30 active galactic nuclei, NGC 6814 was the only object in which Tennant et al. found rapid x-ray variability.

The radio luminosity function for QSOs is observed to be very broad (see, e.g., Sramek and Weedman 1978; Smith and Wright 1980; Strittmatter et al 1980). Any theoretical model of the radio emission process has to account for this. The most common explanations depend on preferred alignment of relativistic beams - see, e.g., Blandford and Rees (1978), Blandford and Königl (1979), Scheuer and Readhead (1979), Marscher (1980). By invoking relativistic enhancement, the rest-frame luminosities are lower and the rest-frame variability timescales are longer than in the observed frame, thus easing some of the difficulties mentioned earlier. Recent observations of relativistic velocities in SS433 have lent credence to such models as have VLA maps of the radio emission from many sources. Further possible support has come from some recent, very exciting world-wide polarisation observations of BL Lac itself (Moore et al 1980). The Stokes parameters Q and U have short and longer-term periodicities. Both Q and U appear to vary on a timescale ∿1 day, while no corresponding variation in the total flux was observed. The longer-term variability is of larger amplitude and of order 10 days. Rotation of the polarisation vector suggests large geometrical changes (beaming?) very close to the dragon's mouth. This is particularly interesting for the massive-black-hole pundits, who point out that while you can only expect to probe the immediate environs of a stellar mass black-hole via millisecond timescale x-ray studies, corresponding regions around a $10^9 M_\odot$ black hole may expect to produce observable effects in the optical and infrared regions with timescales of hours and days ($t \sim 3M/M_\odot c$ seconds, where c is in km s^{-1}). The rotation of position angle could be produced by the rotation of, say, an accretion disk and a magnetic field around the black hole.

Preferred alignment of the relativistic beams has also been offered as an explanation why most QSOs are radio quiet. The basic idea is that the radio emission is only detected when the beam is pointed more or less directly at the observer. However, Smith and Wright (1980), Strittmatter et al (1980) and Condon et al (1981) have all given a number of reasons why flux enhancements by relativistic beaming of intrinsically indistinguishable objects cannot be the principal cause of the observed radio power differences between radio-loud and radio-quiet QSOs. In addition, recent observations of millimetre radiation from optically selected QSOs by Sherwood et al (1980, 1981 - see e.g. Figure 3.1), may lend further support for the belief that most QSOs are radio quiet because their radio radiation is self absorbed at frequencies below ∿ 300 GHz. It is clearly of great importance to follow up these observations to define the shape of the continuum at several frequencies in this region and attempt to distinguish between synchrotron self absorption and free-free absorption. The new He^3 cooled (0.3°K) bolometric 1, 2 and 3 mm receiver to be installed on the NRAO 11 m telescope at Kitt Peak should prove outstandingly suitable for this kind of research particularly in view of the high transmission and selectivity of its filter. RMS flux densities of ∿ 2 Jy are expected to be reached in 1 second at 230 GHz (depending on the transparency of the atmosphere).

4. THE EMISSION-LINE SPECTRUM

4.1 Introduction

The reader persistent enough to have come across an article on QSOs in a rather unusual book like this will probably have no trouble in locating the "Review of Modern Physics". If you wish to obtain a detailed account of the state of modern research into QSO emission-line regions, you cannot do better than to go at once to the excellent but necessarily lengthy review by Davidson and Netzer (1979) [you will also find in that article some very good illustrations of typical low-resolution spectra of high and low redshift quasars]. If however you are content with a relatively brief, and therefore superficial paraphrasing of the article by Davidson and Netzer, along with a few more recent observations, read on ....

In a qualitative sense, our description for the emission-line region of a quasar has not changed greatly during the last ten years. What has been gained is a much deeper understanding of what happens to various photoionisation models as parameters are varied and also the extent to which models can be used in interpretation of the observed spectra. Key observational data still needed to clarify the models are measurements of a number of weak emission lines (e.g. Shields 1976, Osmer 1980c).

These models (see discussion in section 4.3) usually involve electron densities in the range $10^7 < N_e < 10^{10}$ and electron temperatures $T_e \sim 10{,}000°K - 25{,}000°K$. The dominant source of ionising energy appears to be photoionisation rather than shock heating or ionization by fast particles. The region emitting the broad permitted lines (the BLR) consists of only $\sim 100\ M_\odot$ of ionised gas (and an unknown amount of neutral gas). Abundances, remarkably enough, appear to be approximately solar in most QSOs, and so processing of the BLR material within stars seems to be indicated. Indirect measurements of the size of the BLR suggest values < 1pc. This needs to be checked by monitoring the variability of QSO emission lines. The ratio of radiation to gas pressure is ~ 0.1 and the ratio of the flux of ionising particles to the electron density is $\sim 10^8$ cm s^{-1}. The relative rôle of soft x-ray flux (½-1 keV) in controlling the enhancement of features at the red end of the spectrum versus the rôle of dust in depressing emission-line strengths in the ultraviolet is still rather ill-defined. The BLR seems most likely to consist of very large numbers of tiny, optically thick clouds, (or perhaps a system of sheets and filaments) having a small covering factor, $\Omega/4\pi < 0.1$, as viewed from the central source of ionising radiation. Observations of emission-line profiles in QSOs suggest that the profiles are quite symmetric, which appears to indicate (Section 4.4) that radial flow is not the dominant mechanism causing the line broadening.

4.2 General problems

Davidson and Netzer give three basic rules for the newcomer to modelling QSO emission-line regions:

(i) Simplified models of quasar's emission regions are usually wrong, in the sense that many complicated effects (discussed with great clarity and detail by them) can be shown to modify the physical parameters of the situation.

(ii) Because observational data are limited (e.g. on the faint lines or the x-ray and ultraviolet continuum shapes), in practice one may as well use simplified models for many purposes. The most useful observational data would be (Davidson 1977, Baldwin 1979) (a) the slope of the ionising continuum, (b) the rough spatial and density distribution of the gas envelope surrounding the central source (c) the optical depth of the gas clouds to the ionising continuum and (d) the relative abundance of key elements (H, He, C, N, O).

(iii) A specific explanation of just one or two spectral features - without considering the other features - is usually worthless.

Some basic observational programmes come to mind immediately in support of these cautionary remarks. The first set of data illustrates what can happen when spatial resolution is inadequate - a situation likely to be true for all QSO observations. The objects in question are the "hot-spot" nuclei described by Sersic and Pastoriza (1965); some spiral and barred-spiral galaxies have peculiar nuclei which appear on photographs as a cluster of bright spots (see, e.g., Fig 4 in the paper by Osmer, Smith and Weedman 1974). Early speculations (e.g. Tovmassian 1966) suggested that these were in fact active components of a fragmented galaxy nucleus; galaxies similar to these were referred to as having "split nuclei". These speculations appeared to have been laid to rest following the work by Osmer, Smith and Weedman, which led to the widely-held view that the "hot spots" were short-lived, massive HII regions, photoionised by OB stars, themselves created in a recent burst of star formation.

Osmer, Smith and Weedman (1974) used large circular diaphragms (up to 40 arc seconds diameter) for their observations. The signal received by the CTIO scanner was then the integrated light from all of the spots in each nucleus. Osmer, Smith and Weedman made the apparently reasonable assumption that all the spots in a given nucleus were similar. This seemed even more reasonable as the spectra of the different nuclei were qualitatively similar to each other - as is the case for quasars ......

More recent studies by three different groups, using the high spatial and spectral resolution provided at the Anglo-Australian Telescope (AAT), have led to a radical revision of this picture. Though the integrated light from these nuclei may indeed be dominated by giant

HII regions, some of the individual hot spots show characteristics of active (Seyfert) galaxy nuclei. Perhaps the most exciting result of this work so far is the discovery, by Edmunds and Pagel (personal communication), of (i) a Seyfert type 1 hot spot in the nuclear region of NGC 1365, (ii) a surrounding highly-ionised narrow emission-line region showing evidence of excitation by a non-thermal source, and finally, (iii) a set of nearby conventional low-excitation HII regions. The Seyfert characteristics of another "hot spot" galaxy, NGC 1097, have recently been discussed by Meaburn et al, 1981.

These observations demonstrate the need for high spatial resolution in discussion of the physical properties of galaxy nuclei; by implication, caution is needed in accepting the results of uniform models for quasi-stellar objects.

The second set of data which may help to warn the unwary newcomer derives from what many consider is the most significant recent paper on observations of QSO emission lines - a fairly short communication by Baldwin (1977) which has demonstrated that not even the atomic spectrum of hydrogen is understood in quasars. From a composite spectrum (necessary in the days preceding satellite data) made up from ground-based spectra of high- and low-redshift QSOs, he found a Lyα/Hβ intensity ratio of about 3, surprisingly smaller than the values of around 40 produced by the standard models then in vogue. (These observations have since been generally confirmed, at least to within an inessential factor of two or three, by observations from space - e.g. Wu 1977; Davidsen, Hartig and Fastie 1977; Boksenberg et al 1978; Green et al 1980 - and, in the infrared, from the ground - e.g. Hyland, Becklin and Neugebauer 1978; Soifer et al 1979; Puetter, Smith and Willner 1979. A review of the data and interpretations is given by Davidsen 1980).

4.3 Models involving dust and a strong soft-x-ray flux

Because (i) the apparent intensity ratios of the OI λ1302/λ8446 and HeII λ1640/λ4686 lines are, like the Lα/Hβ intensity ratio, lower than expected, and (ii) the general curved shape of QSO continua resembles the general form of reddened power laws, Davidson and Netzer (1979) conclude - after reviewing a long series of possible complications - that: (i) there is probably an overall ultraviolet/visual discrepancy between the predictions of models then current vs observations and (ii) this provides evidence for interstellar "smoke", or dust, along the line of sight to each quasar. MacAlpine, Feldman and Weedman (1981) also conclude, from their Lyman-limit observations coupled with consideration of the equivalent width of HeII λ4686, that intrinsic spectral indices for quasar continua, as seen by the line-emitting material, do not decline toward higher energies as rapidly as observations would appear to indicate (see also Shuder and MacAlpine 1979). Some confirmation of this is provided by the power-law form of the hard x-ray spectra of Seyfert galaxies and objects such as 3C273 and Cen A (Mushotsky 1981). The observation that the narrow-line-emitting gas in 3C 390.3 exhibits normal line ratios (perhaps slightly reddened by dust in our galaxy) whereas the broad lines show low UV to optical line ratios, also led Ferland et al (1979) to conclude that external reddening is unlikely to account for the generally reduced Lyα/Hβ ratios observed. The most probable location for most of the dust appears to be in the galaxy-sized region around the quasar, at distances between 100 and 10^5 light years from the radiation source, i.e. outside the broad emission-line region itself. A nearby example is perhaps 1C4329A (Wilson and Penston 1979). Non-neglible internal reddening is possible from dust grains within the line-emitting region itself (at temperatures $\sim$ 1000°K); however, if that were invoked to account for all the Lyα/Hβ discrepancy, i.e. Lyα were strongly suppressed by dust during resonance scattering, then HeII λ1640 and other ultraviolet lines would appear much stronger, relative to Lyα, than is in fact observed. Note that accurate JHK photometry of about 40 QSOs by Hyland and Allen (1981) indicates the presence of an infrared excess appearing longward of 1.1 μm in the rest frame; see also Neugebauer et al (1979). They suggest an origin in thermal re-radiation from dust largely outside the broad-line region (BLR). Further evidence for dust comes from the detection of 3.28 μm emission in 3C273 (Allen, 1980).

Although dust may be responsible for a general depression of the ultraviolet end of the rest-frame spectrum of quasars, an intense soft-x-ray flux around 200eV will tend to enhance many features (Netzer 1981; see also Halpern and Grindlay 1980; Weisheit, Shields and Tarter 1981). For a discussion of far ultraviolet dust opacity, see Martin and Ferland (1980). Important indication of how the x-ray flux governs the optical emission-line spectrum is given by correlations between x-ray continuum flux and various line fluxes, line flux ratios and line widths over a wide range of intrinsic luminosities. It has been suggested that such correlations apply not only to x-ray selected QSOs (Grindlay et al 1980) but also to both types of Seyfert galaxy (Kris, Canizares and Ricker 1980). Strong soft-x-ray photoionisation models appear to be remarkably successful in accounting for the observed emission lines, particularly OVI λ1034 and NV λ1240, which are observed to be very bright. Such models typically involve the following parameters:-

Element abundances - generally similar to solar values, but perhaps with some slight overabundance in C, N and O. The observational data make it difficult to obtain sufficient cooling if the carbon abundance were less than about half the solar value. Heavy elements are perhaps unlikely to be more than a few times more abundant than in the Sun (see arguments on p736 of Davidson and Netzer's article). A detailed account of abundances in Q0353-383 is given by Osmer (1980c), who concludes that in this very unusual QSO (which has the lowest measured value of CIV/(Lα + NV) emission of 80 objects in which he measured it), the N/C abundance is enhanced by a factor $\sim$ 10. This large factor results from a low abundance of carbon rather than a high nitrogen abundance. Other physical conditions such as the gas density and ionising spectrum seem normal in this quasar - which is what led Osmer to appeal to element abundance to explain the unusual spectrum.

Ionisation parameter - Davidson and Netzer define an ionisation parameter U_H in terms of radiation flux density/electron density. $U_H = Q(H)/4\pi D^2 n_e$, where Q(H) is the number of ionising photons emitted by the central object per second, and D is the distance between the central object and the clouds. U_H has velocity-like dimensions. In a pure hydrogen nebula $n(H^+)/n(H^\circ) \approx (U_H/10^{5.3}\ \mathrm{cm\ s^{-1}})$ with a reasonable value of the slope of the spectrum of the incident energy flux. Under similar conditions, but with a more realistic nebula having helium, $n(He^{++})/n(He^+) \approx (U_H/10^{7.2}\ \mathrm{cm\ s^{-1}})$. In a typical QSO model, $U_H \sim 10^8\ \mathrm{cm\ s^{-1}}$.

Electron density - After dealing with the usual wealth of potential complications, Davidson and Netzer conclude (mainly on the basis of the observed Balmer line ratios) that values of $10^9 < n_e < 10^{10}\ \mathrm{cm^{-3}}$ are most likely. The most quoted initial constraint, $10^7 \leq n_e \leq 10^{10}\ \mathrm{cm^{-3}}$, is obtained from arguments based (i) on the absence of broad forbidden lines, (ii) the presence of the intercombination CIII] λ1909 line as a broad feature (intercombination lines involve spin changes), and (iii) on an assumption of near-solar abundances. However, one should note, for example, that the upper limit of $n_e \sim 10^{10}\ \mathrm{cm^{-3}}$ cannot be proved absolutely, and no self-consistent models are available for higher densities. Bergeron and Kunth (1980) consider that the extreme FeII emission from PHL1092 and particularly the relative weakness of the Balmer and MgII lines is indicative of higher electron densities ($n_e \sim 10^{12}\ \mathrm{cm^{-3}}$).

The ionising continuum - Davidson and Netzer used, for illustrative purposes, a continuum shape of the form $F(\varepsilon) \sim \varepsilon^{1/3}\, e^{-\varepsilon/\varepsilon_o}$, with $\varepsilon_o = 200\mathrm{eV}$. This reflects a fundamental difference between the classical models to describe HII regions (where the ionising flux comes from hot stars) and models for quasar spectra. The wide range of ions represented in quasar spectra, particularly those with high ionisation thresholds such as N^{4+} (77.5 eV) and O^{5+} (ionisation threshold 113.9 eV), require substantial fluxes of high-energy photons, which cannot be provided by models based on hot stars as the ionising source. Stars and gaseous nebulae exhibit wide ranges of ionisation from object to object. The fact that quasars do not, (i.e. their emission-line spectra do not differ greatly from object to object in most cases) appears to lend some validity to the attempt to derive "typical" models (the potential dangers of even such an innocent assumption as this have, of course, just been highlighted in section 4.2). It is because of the apparently restricted range of physical conditions in the QSO envelope that Baldwin's deductions based on composite spectra were so successful.

Electron temperatures - The temperature limit of $\sim$25,000°K suggested by the generally low value of the CIII λ977/CIII] λ1909 intensity ratio (e.g. Osmer and Smith 1976) provides fairly good evidence in favour of photoionisation and against strong additional heating by, say, shocks or fast particles (Davidson and Netzer 1979). However, in the words of Davidson and Netzer, "the reader will not be surprised to learn" that complications can again be introduced to reduce the certainty of this temperature limit. Davidson's (1977) dictum remains true - apart from the crude limits discussed above, "gas densities and temperatures in the permitted-line regions associated with quasars cannot be estimated directly".

A real photoionised gas is thermally stable because the cooling/heating ratio rises quickly with temperature. The input of kinetic energy into free electrons by means of photoionisation is rather easily dominated by the cooling effect associated with collisional excitation of heavy ions (such as C^{3+}). Temperatures range from 12,000°K in the H^+/H° transition zone (sufficiently hot to cause most hydrogen lines to be strongly enhanced by collisional excitation) to 25,000°K in the most highly ionised zones. In effect, because the CIII λ977/CIII] λ1909 ratio indicates $T_e \leq 30,000$°K, the ratio also favours photoionisation (rather than other forms of heating - e.g. shocks or fast particles) as the dominant mechanism; only photoionisation produces doubly and triply ionised species at equilibrium temperatures below 30,000°K. Further evidence in favour of photoionisation as the main form of energy input to the broad-line region is provided by the fact, noticed by Searle and Sargent (1968), that the ratio of the Hβ recombination line intensity to the underlying continuum brightness is roughly the same (20-100Å) for many Seyfert galaxies and quasars, over a wide range of absolute luminosities. The Hβ flux is an indirect measure of the number of absorbed ionising photons and the range of Hβ equivalent widths is consistent with photoionisation by a moderate power-law continuum spectrum (see also Osterbrock 1980).

The geometry and optical depth of the emission-line material - Here we shall go into a little more detail than in other areas (though still glossing over much of Davidson and Netzer's discussion), so as to ensure that the reader is left with no doubt how complicated the subject can become! First let us consider the overall size of the broad emission-line region. The observed state of the emission-line material obviously depends on the ionising luminosity of the central continuum source and the light travel times from source to emission-line region and from that region to us. The ensemble of emission-line material associated with the quasar will have a minimum possible timescale for variability which will be a measure of its total size. Emission-line profiles in Seyfert galaxies have been observed to vary on timescales of a few months or less, and show global intensity variations by factors up to ∿4 (e.g. Andrillat and Souffrin 1967; Pastoriza and Gerola 1970; Andrillat 1971; Tohline and Osterbrock 1976; Boksenberg and Netzer 1977; Phillips 1978b; Snijders 1981). Size scales for the Seyfert broad-line region ∿0.1pc may be indicated. For quasars, variations of continuum flux appear to be concurrent with non-variations of emission-line intensities, suggesting at least that the line emitting gas is not distributed compactly [see, e.g., Sandage, Westphal and Strittmatter 1966; Kinman, Lamla and Wirtanen 1966; Oke 1967; Shields, Oke and Sargent 1972; Netzer et al 1979; note however that the often quoted report by Osmer and Smith (1977a) of continuum variations in Q0231-410 is incorrect. Such variations probably did not occur (Osmer and Smith 1980)]. From the preferred values of U_H, Ne and ionising luminosities given in the model by Davidson and Netzer that we are considering, the likely size scales for the (broad-line) emission region [roughly $(Q(H)/(4\pi U_H n_e))^{\frac{1}{2}}$] range from about a light year (for a moderate Seyfert) to tens of light years (for a luminous quasar).

Unfortunately, ionised gas at $N_e \sim 10^9$ cm^{-3} filling a spherical volume ∿ 10 parsecs across would yield emission-line intensities far greater than those actually observed. Furthermore, it turns out that a single cloud of electron density $N_e \geq 10^9$ cm^{-3} around a compact continuum source would be optically thick to electron scattering, and the value of the radiation/gas density parameter $U_H \sim 10^8$ cm s^{-1} used in the above quasar models would be much too low to keep all the gas ionised (Davidson and Netzer 1979, p729). Attempts to increase U_H by decreasing the distance D between the central object and the ionised gas would also tend to make the emission-line cloud too small in comparison to observed timescales for emission-line variability. There are two fairly straightforward ways out of these difficulties. The first is to draw an analogy with the inhomogeneous distribution of gas in galactic nebulae and supernovae - or for that matter with the distribution of galaxies in the universe. The gas in quasars is probably inhomogeneously distributed in cloudlets, filaments or sheets, so that even if the gas in such condensations were to be optically thick, the light from the central continuum source may be seen directly, or with lower attenuation, through spaces between the condensations (see, e.g., Schmidt 1964; Oke 1974). The fraction of the overall volume actually occupied by high density gas, the so-called "filling factor", is small. The second way to explain why the observed emission-line intensities may be so low is to suggest that lines such as Lyα are suppressed in some way, e.g. by reddening which is internal or external to the ionised gas clouds, resonance trapping of Lyα (at high optical depths) on dust etc (see the detailed discussion by Davidson and Netzer 1979, which suggests that most of these effects are probably operating, but on the model we are considering, they are unlikely to be sufficient). To choose between these possible approaches, we must make an independent appeal to observational data to elucidate more information about the cloud geometry.

First we must decide whether or not the condensations are optically thick. It is as well to be clear that in an optically thick cloud with normal chemical composition, most of the ionising photons are absorbed by hydrogen and helium; these ions control the positional dependence of the ultraviolet flux in three major ionisation zones (see, e.g., Davidson and Netzer 1979):

(a) A highly ionised inner zone, where, if U_H is high enough (as in the model we are considering where $U_H \sim 10^8$ cm s^{-1}) most of the helium is doubly ionised (and most of the hydrogen is in the form of H^+). Most photons with energy greater than 54.4 eV are absorbed in maintaining the supply of doubly-ionised helium.

(b) An intermediate, singly-ionised zone in which He^+ and H^+ predominate, and most photons have energies less than 54.4 eV. For typical model parameters, the H^+/H° and He^+/He° ionisation fronts occur almost together. Mg° and Fe°, with ionisation potentials of $\sim$ 7.7 eV, cannot survive in this region.

(c) A neutral zone with mainly H° and He° and photons having energies below 13.6 eV. Some high-energy x-ray photons ($\varepsilon \geq 200$ eV) may, however, penetrate into this zone and cause some ionisation and heating. This warm neutral gas may be an amenable site for molecule formation; the consequences of efficient molecule formation in the presence of dust have recently been examined by Ferland (1980).

Where accessible, MgII λ2800 emission is seen as a strong feature in the spectra of quasars. Optical FeII emission is often found in Seyfert 1 galaxies and in some low - z QSOs, and ultraviolet resonance lines of FeII have recently been recognised as occuring in the spectra of virtually all moderate - redshift QSOs (Wills *et al* 1980; Grandi and Phillips 1980, Wills, Netzer and Wills 1980). As we have seen, the presence of these lines indicates an environment of photons with energies less than 13.6 eV, i.e. MgII and FeII indicate an important component of the gas which is optically thick at the Lyman limit (912Å in the rest frame of the quasar); MgII and FeII emission originate in the H^+/H° transition region. The amount of neutral gas involved must be substantial, because otherwise the strength of the transition-zone ions (CIII] λ1909, MgII λ2800) would be very sensitive to the actual outer edge of the gas. Not only is no correlation between line profile and ionisation potential observed for the strong lines in QSOs, including MgII, but also MgII appears to be strong in quasar spectra whenever the λ2800Å rest-wavelength region is accessible to observation. For discussions of these points see, e.g., Osterbrock (1978), Baldwin and Netzer (1978), Grandi and Phillips (1979), Kwan and Krolik (1979), Netzer (1980), Carswell and Ferland (1980) and Weisheit, Shields and Tarter

(1981). Independent evidence for large optical depths in the H^+/H° transition region is found in the relative intensities of FeII lines within optical multiplets and also between optical and ultraviolet multiplets in Seyfert 1 galaxies and low-redshift QSOs (Phillips 1978a, b).

However, when high-redshift quasars are observed, the majority of them show no evidence for a cutoff at 912Å (in the rest frame of the quasar). Even where a cutoff is seen near 912Å, there is no unambiguous case of absorption which must be attributed to optically thick material which is actually in the emission-line region and along the line of sight between us and the continuum source (see, e.g., Wright *et al* 1978, Osmer 1979; Green *et al* 1980; Smith *et al* 1981; MacAlpine, Feldman and Weedman 1981). Indeed, there appears to be no correlation between the Lyα emission-line equivalent width and the strength of the Lyman continuum cutoff (seen normally at rest wavelengths less than 912Å); this is taken to imply that the cutoff is usually, (if not always) associated with material outside the QSO emission-line region.

As very large optical depths are considered necessary in order to explain the Balmer line ratios, Carswell and Ferland (1980) have investigated what can happen if a cloud of HI column density $\sim 10^{23}$ cm^{-2} is photoionised by a power-law continuum extending to the x-ray region. In these clouds of such large optical depth, a CI absorption edge at $\sim$ 1100Å may be observed. No trace of such an edge is present in the hundred or so sufficiently high redshift QSO spectra that I have seen.

In summary, the presence of MgII and FeII emission indicates the presence of optically thick material in the broad-emission-line region yet the Lyman-limit and CI edge data yield no unambiguous evidence for optically thick material inside the emission-line region and covering the continuum source. As Lyman-limit observations have been performed for ≥ 50 quasars, the line-emitting, optically thick gas condensations appear likely to cover $< 2\%$ of the total solid angle around a hypothetical point source of ionising radiation. On the other hand, though the interpretation is rather ambiguous, many cases are known where a sharp continuum drop *is* observed at wavelengths near (though usually less than) 912Å in the emission frame. Thus, a more conservative upper limit for the covering factor of the "sky" seen from the central source is most likely to be in the range $\Omega/4\pi < 0.02$ to 0.1 (see, e.g., the discussion by Smith *et al* 1981). Independent evidence favouring a small covering factor (at least for Seyfert galaxies) comes from the lack of an x-ray cutoff in most Seyfert galaxies observed so far (e.g. Fabian 1981) and the smooth power-law form of the hard x-ray continuum (e.g. Mushotsky et al 1980; Mushotsky, 1981) in Seyferts.

Now the fact that both the filling factor and the covering factor are small need not imply that the condensations of optically thick, line-emitting gas would completely cover the continuum source if they happened to get into the line of sight. To draw on an analogy from our solar system, the situation may be much more like a transit of Mercury than a total eclipse by the Moon (Mushotsky 1981). Carswell and Ferland (1980) have investigated the possibility that the clouds are small compared with the continuum source, and many lie along a line of sight. If these clouds are thermally bound, the individual clouds will have thermal widths of, for example, $\sim$ 4 km s^{-1} for carbon. The observed carbon lines in quasars have widths $\geq 10^4$ km s^{-1}, and the profiles appear to be rather smooth when sufficient resolution is used to correct for the presence of absorption lines. To produce such smooth profiles, Carswell and Ferland estimate that at least 10^4 such clouds are required; see also the discussion by Capriotti, Foltz and Byard (1981). With many such clouds along the line of sight, but effectively covering only 10% or less of the continuum source, then, for a number of technical reasons, direct observational verification of their effect on the continuum is difficult. Among these problems is the existence of a dense forest of absorption lines (section 5.2, Type 'D') which prevents one obtaining, from the ground, a direct look at the continuum at sufficiently high signal-to-noise ratio. Space Telescope observations of low-redshift objects, (which according to the conventional wisdom set out in section 5 may lack such a forest of lines), should allow one to obtain sufficient signal-to-noise ratio to detect the small changes in the continuum level at 912Å or $\sim$ 1100Å. It is also worth noting that MR2251-178 exhibits variable x-ray absorption consistent with the passage of a dense cloud across the line of sight to the nucleus (Halpern and Elvis 1980). NGC 4151 also shows apparently variable absorption, near the emission-line redshift, in HeI λ3889 and in several ultraviolet lines (e.g. CIV, MgII - Phillips personal communication). Gilmore (1981b) has recently proposed using variations in the double QSO 0957+561 as a direct probe of the size and number of clouds in the emission-line region.

4.4 Profiles of the emission lines

The profiles of the emission lines in quasars depend not only on the velocities of cloud motion (line widths $\sim 10^3$-10^4 km s^{-1} have to be explained) but also on the details of geometry and optical depth discussed above. Davidson and Netzer (1979) discuss a number of general statements that can perhaps be made if one assumes that the observed velocity dispersion is related to gravitationally controlled speeds in the emission regions. For example, from a consideration of the statistics of line widths for Seyfert galaxies, Osterbrock (1977 , 1980) has derived a simple observational model in which the dense gas forms a rotating disc with typical rotational velocity 5000 km s^{-1}, and typical turbulent velocity 2000 km s^{-1}. The characteristic dimension of such a disc in Seyfert galaxies, as we have seen earlier, is $\sim$0.1 pc. Osterbrock points out that not only is a rotational velocity $\sim$ 5000 km s^{-1} at a radius $\sim$ 0.05 pc. consistent with the presence of a central black hole of mass 3 x 10^8 $M_\odot$, but the observed continuum luminosities from Seyfert galaxies are also consistent with the presence of such an object (Rees 1977a,b). Quasar data are consistent with, though do not prove (Davidson and Netzer 1979), the existence of a massive central continuum source $\sim 10^{8}-10^{12}\ M_{\odot}$; a graph giving the distribution of emission-line widths for a complete sample of quasars is given by Baldwin (1979).

Seyfert galaxies often exhibit asymmetric, sometimes variable emission-line profiles. Most quasars exhibit rather symmetric profiles, with the broad lines usually having essentially the same shape and relative width. There are a few exceptional quasars, however (see discussion by Davidson and Netzer 1979).

In general, if the broad emission-line region is made up of individual clouds, then the similarity of the broad-line profiles in a given quasar indicates that with regard to excitation and ionisation each velocity bin should have the same population. Examples of how geometry, dynamics and optical depth become intertwined in any discussion are provided by Capriotti, Foltz and Byard (1979 see particularly their Figure 2) and Smith et al (1981). Because the individual clouds are optically thick to the Lyman continuum, most of the recombination flux should arise on the illuminated side of the cloud; because Lyα optical depths could be as high as 10^{8} or 10^{9} (Weisheit, Shields and Tarter 1981), the multiply "scattered" Lyα photons will leak out preferentially from the illuminated side (i.e. towards the continuum source at the centre of the quasar). In Lyα, the inner sides of the clouds are bright, the outer sides opaque. The occurrence of asymmetries in the line profiles therefore indicates radial flow, the direction (inwards or outwards) being determined by the sense of the asymmetry - see discussion by Capriotti, Foltz and Byard (1979), Ferland, Netzer and Shields (1979), Davidson and Netzer (1979) Heckman *et al* (1980), Wilkes (1980) and Smith *et al* (1981).

The apparent symmetry in line profiles of QSO emission lines (when observed with sufficient signal-to-noise ratio and resolution to correct for the effect of absorption lines) suggests that we should, for most quasars, discount models in which most of the material in the emission-line region is flowing radially with respect to a central continuum source at speeds comparable to the line widths. Such models in combination with rotation and dust may however be applicable to Seyfert galaxies (see, e.g., Capriotti, Foltz, and Byard).

5. ABSORPTION LINES IN QSO SPECTRA

5.1 Introduction

Once again my task is eased - this time by the existence of a review in preprint form (Weymann, Carswell and Smith 1981 "WCS") which should appear in "Annual Review of Astronomy and Astrophysics" at about the same time as this article appears. I shall therefore do little more than summarise some of the main points of that review and refer you to its reference list. Considerable progress has been made in the last five years towards understanding the location of the material responsible for the absorption lines in QSOs: it appears that most of it is not associated with the QSOs themselves. The biggest single factor in producing the improvement in the quality (and to some extent the quantity) of QSO absorption-line data necessary for such an understanding has been the development of high-resolution photon-counting detectors. For an outstanding application of the best known detector system in this class (the IPCS - Boksenberg 1972), incorporating a very detailed discussion of many of the most significant observational results obtained so far from QSO absorption lines, see the article by Sargent et al (1980). It is equally clear that the Space Telescope will play an even more dominant role in attacking most of the present problems in this field.

5.2 Classification of the absorption-line systems and their basic physical properties

Four classes of absorption-line system may be distinguished (Weymann 1980, see especially his Table 1); all types can occur in a single QSO (see e.g., Fig 1 of WCS), though the significance of Type 'B' (as discussed below) has been questioned very recently.

Type 'A': Absorption systems consisting of extremely broad features which appear to be consistent with ejection velocities reaching or even exceeding 0.1c. Detailed observational criteria for membership of this class are given in WCS, the most important being that the broad lines must be observed to occur in a line other than Lyα (to eliminate those objects with strongly damped Lyα absorption where the Doppler widths of the lines in the system are relatively small). Table 3 of WCS includes a list of such objects and references to observations of them. Considerable substructure is often observed, particularly in the Si IV profile. The ability to find such objects directly from examination of slitless spectra (Smith 1979; Clowes et al 1979) has contributed to a large recent increase in the numbers of objects known in this class. Uncertainty regarding selection effects makes it difficult to be sure of the total fraction of all QSOs that this class represents. It is likely to be in the range 1-10% of all high-redshift QSOs. A universal characteristic of these objects is their high level of ionisation. Where it is observable, OVI is very strong, and NV is often much stronger than Lyα . Christiansen, Scott and Weymann (1980) have proposed that the broad-absorption-line QSOs may simply be QSOs in edge-on-spiral galaxies. In this context, it is interesting to note that the edge-on, very luminous Seyfert (or quasar - Wilson and Penston 1979) IC4329A shows interstellar velocity dispersion similar to that found along sight lines through spiral arms in our own galaxy. There is no sign of disruption to the interstellar medium in IC4329A. The most important uncertainty in a description of these objects occurs in our lack of knowledge, even to within several orders of magnitude, of the distance between the absorbing material and the central source of ionisation.

Type 'B': These are sharp-absorption-line systems having $Z_e - Z_a$ / $1 + Z_e \leq 0.01$, i.e. the velocity differences between the emission and absorption line system do not exceed $\sim$ 3,000 km s^{-1}. A histogram of the observed frequency of CIV systems vs $(Z_e - Z_a)/(1 + Z_e)$ can exhibit a prominent peak distributed symmetrically about zero velocity difference (see Fig 2 in WCS). Such systems would seem to arise from material associated somehow with the QSO; the most obvious explanation, in view of the symmetry about zero, is that the CIV systems arise from clouds in or near galaxies bound to a cluster which also contains the QSO. This apparently straightforward explanation runs into the problem that, at low redshifts, QSOs appear only to be associated with groups of galaxies whose velocity dispersion is much too low to explain the width of the peak. Though observational work to resolve this dilemma is already underway, at present we are faced with the unsatisfying 'ad hoc' postulate that some mysterious evolutionary effect causes QSOs to avoid associating with rich clusters at redshifts $z \leq 0.4$ and yet to reside in them quite happily at $z \geq 1.2$; see, however, the discussion by Stocke and Perrenod (1981). (I suppose one could say that city environments are generally more appealing to the young. However, just before I sent in this manuscript, Boksenberg and Sargent both informed me that they have recently observed a similar sized, but independent sample to that used by WCS, and yet they find no evidence for the supposed peak near zero! Thus the reality of type 'B' systems appears to be in question).

Type 'C': These are sharp-lined systems containing 'metal' lines (in this context, lines in addition to hydrogen lines), which have $(Z_e - Z_a)/(1 + Z_e) > 0.01$. Most (but not all) workers in the field now accept that these type C systems are likely to correspond to absorbing material in extensive, low density haloes of normal galaxies at great (cosmological) distances from the QSO. This idea stems from the observed frequency of occurence of such systems (Burbidge et al 1977; Sargent et al 1979; Weymann et al 1979b). Effective halo radii for absorption may be $\gtrsim$50-100 kpc (depending on one's assumed value for the Hubble constant, what types of galaxy can produce absorption, the luminosity function of such galaxies etc). Bahcall and Spitzer (1969) pointed out the extreme sensitivity of UV lines to small amounts of gas, so that these large values may not present a serious problem. Several workers have found (e.g. Savage and Jeske 1981) that comparisons of high dispersion UV spectra of hot stars in the Magellanic Clouds with representative quasar absorption-line systems yield remarkable similarities between the Milky Way halo and quasar absorption spectra, viz: "mixed ion absorption", a general absence of excited fine-structure levels, similar column densities and similar multi-component line profiles.

The fact that galaxies can absorb the light from background quasars has been demonstrated directly on several occasions - most notably by Boksenberg and Sargent (1978 - 3C232/NGC 3067), but also, since then, by Boksenberg et al (1980 - PKS2020-370/Klemola 31) and by Blades, Hunstead and Murdoch (1981 - Q0446-208/Anon SO galaxy). However, IUE spectra of 3C232 show absorption from neutral or singly-ionised atoms; the SiIV and CIV resonance lines are weak or absent (Snijders 1980) in this system, unlike in most QSO absorption systems or those in sight-lines through the halo of our own galaxy. Furthermore, in the three cases cited above, the CaII lines from the intervening galaxy are somewhat stronger than those from our own Galaxy (along the same line of sight). For gas in our Galaxy, a typical value of N(CaII)/N(NaI) is ~ 1 for material in the disc and ≥ 8 for halo material (Cohen and Meloy 1975), so this ratio provides a useful diagnostic. In our Galaxy, though there is evidence for absorbing material in the inner halo (scale height ~ 2 kpc), the equivalent width, W, of the CaII K line appears to be independent of galactic latitude above $|b| \sim 30°$, lying in the range 0.1Å or less up to $\sim$ 0.4Å, except in the direction of the Magellanic Clouds (here the sight-lines may not be typical as interactions between our Galaxy and the Magellanic Clouds may be responsible for the larger amounts of intervening material in this direction - e.g. Blades, Hunstead and Murdoch 1981; Blades 1981. Thus, the evidence for general, strong absorption in the halo of our Galaxy is, shall we say, thin).

Because the three QSOs cited above were selected to be near galaxies, one can adopt the point of view that the physical properties of the absorbing clouds in the intervening galaxies may not be typical. To take what is perhaps an extreme example, suppose in fact say only 1 in 100 absorption-line systems in QSOs were really produced in galaxies, and that the remainder were produced in something else. Searching close to galaxies only demonstrates the plausibility of the intervening-galaxy hypothesis; it doesn't prove anything. What is important now is to make the converse observation, starting with, say, a typical low-redshift absorption-line system, and then try to find the galaxy which produced it (or at least, galaxies in the same cluster). Low redshifts are required in order to resolve the galaxy. Then only MgII λ2800 is accessible until the Space Telescope is launched. So far, the various ground-based attempts to locate such galaxies have either given ambiguous results and/or been dogged by bad luck with the weather. Some sceptic is sure to comment, anyway, that this doesn't prove anything about the majority of known absorption systems, namely those at high redshifts!

The physical conditions in sharp-lined, metal-containing systems are not well defined, as the spectral resolution has usually been insufficient to resolve out potential blends; because of the shape of the curve of growth, metal abundance may be systematically underestimated if an optically thick component is present in a system with a complex velocity structure. Similarly, because neutral hydrogen column densities are based on profile fitting, strong lines which are in fact hidden blends (with velocity structure) yield overestimates of neutral hydrogen column densities.

The assumption used in deriving column densities, namely that the velocity distribution in the absorbing region is gaussian, is almost certainly not correct for these poorly resolved lines. Finally individual components probably have different relative ionisation. Large numbers of physical analyses have, nevertheless, been attempted and published; these yield estimates for velocity dispersions in the range 20–100 km s^{-1}, and metal column densities typically in the range 10^{13} to 10^{16} cm^{-2}, implying total hydrogen column densities $\sim 10^{17}$ to 10^{21} cm^{-2}. High-resolution observations of B2 1225+31 (Pettini, West and Boksenberg, in preparation) and Q1101-264 (Turnshek, personal communication) have allowed very tight upper limits (≤ 1 cm^{-3}) to be set on the electron density. These limits are derived from the absence of the excited fine-structure line CII* λ1335.7 and the strong presence of CII λ1334.5 in these objects. The line is occasionally (but rarely) detected in other QSOs. Photoionisation is considered a much more likely explanation than collisions for the ionisation equilibrium in the absorbing regions (see, e.g., the discussions by Arons 1972; Röser 1975; Sargent et al 1979). The metal-lined systems appear to be clumped, i.e. small velocity separations between metal-line systems are common. Sargent et al (1980) find a peak in the two-point correlation function at $\sim$ 150 km s^{-1}, though observations have revealed correlations on scales from 40 km s^{-1} to thousands of km s^{-1}.

Type 'D': The fourth obvious class of absorption lines involve the "Lyα forest" seen shortwards of the Lyα emission line in all QSO spectra in which this region has been studied extensively. Cross-correlation analyses by Young et al (1979) and Chen et al (1980), showing strong peaks at the separation of Lyα/Lyβ, tend to confirm Lynds' (1971) suggestion that most lines in the 'forest' are Lyα at different redshifts. Nearly all the useful results on this class of systems are derived in the paper by Sargent et al (1980).

Bahcall (1980) has shown that if these absorbers exist in galactic haloes, then the haloes must have sizes of at least $\sim$ 200 kpc. However, Sargent et al (1980) find no clumping on velocity scales in the range 300–30,000 km s^{-1}, suggesting an intergalactic origin (in view of the lack of clustering on velocity scales one might associate with clusters of galaxies). From profile fitting to very high resolution data for Q1101-264, Carswell et al (in preparation) deduce HI column densities generally $< 10^{14}$ cm^{-2} and a Doppler b-parameter in the range 20–40 km s^{-1} for the Lyα-only systems. Oort (1981) suggests that these Lyman-alpha lines arise from uncondensed gas in superclusters, on the grounds that the spacing between the lines agrees with what one might assume is a reasonable distribution of superclusters in space. Epstein and Fransson (1980) speculate that absorption by neutral hydrogen in dwarf-galaxy winds could be responsible for these type "D" absorption lines. No obvious bunching around values of $z_{abs} \sim z_{em}$ is detected.

Sargent et al (1980) find, from a sample of 5 objects, that the number density of lines per unit redshift, $N(z) \propto (1 + z)^{0.5\pm0.5}$. They underline the importance of ensuring that the sample of lines is selected to include only those with (rest) equivalent width greater than some value. A uniform co-moving distribution of absorbers yields $N = N_o(1 + z)(1 + 2q_o z)^{-\frac{1}{2}}$, where N_o is the (extrapolated) zero-redshift value of $N(z)$. For $q_o \simeq 0$, Sargent et al obtain $N_o = 17.7 \pm 1.3$; a similar value is obtained by Carswell et al (1981), suggesting that there are no obvious systematic differences between the various QSOs in these samples.

6. GRAVITATIONAL LENSES

Two examples of probable gravitational lensing have been observed. They have been so extensively discussed in recent literature that only the briefest summary is given here.

The discovery* of the twin QSO 0957+561 A,B by Walsh, Carswell and Weymann (1979) - see also Weymann et al 1979a - led them to suggest that because the two images have essentially identical observed spectra, the images could have been formed as a result of imaging by some intervening mass. Young et al (1980, 1981b) provided the most convincing evidence in favour of this interpretation with (i) their observation of a cD galaxy near image B and (ii) their demonstration that plausible gravitational imaging models can reproduce the observed configuration. A series of observations at wavelengths in the radio (Greenfield, Roberts and Burke 1980; Greenfield, Burke and Roberts 1980; Porcas et al 1981), infrared (Soifer et al 1980; Lebofsky et al 1980) optical (Wills and Wills 1980, Stockton 1980) and ultraviolet (Gondhalekar and Wilson 1980) regions of the spectrum generally support this interpretation. Variability in this object has recently been reported (Miller, Antonucci and Keel 1981; Young et al 1981b). An interesting semi-popular account of the observational discovery of gravitational lenses has been given by Chaffee (1980). The long-standing discussions by the Barnothys (e.g. Barnothy and Barnothy 1968, 1972) of the potential effect of gravitational lenses on source-count models have received much greater attention recently as a result of these observations (see, e.g., Turner 1980).

A second case of probable gravitational-lens imaging has been discovered by Weymann et al (1980). PG1115+08 contains at least three images. Bourassa and Kantowski (1975) and more recently Young et al (1981a) have shown that when realistic density distributions of matter in the deflector are considered, a rich variety of brightness ratios and

---

*Bob Carswell provided me with an interesting anecdote on the discovery. Prior to the original Kitt Peak observing run, Dennis Walsh had a bet with Derek Wills; Derek was to pay 25¢ if one of the star-like images was a QSO, $1.00 if both were quasars. When both images turned out to be quasars, Derek paid up. It seems that "Derek refused to accept 75¢ change when it was established that we were, in fact, looking at only one quasar".

image displacements are possible (see also Dyer and Roeder 1980). I shall never forget an even richer combination presented on the TV console of Keith Hege's intensified imaging system when he and Richard Green were searching for more examples of gravitational lensing. They were using the MMT (and its already renowned good seeing) to form separate images with each of the six mirrors; we were thus presented with the spectacle of 18 dancing images on the console screen each with the same redshift!

Acknowledgements: I should particularly like to thank Bob Carswell, Ray Weymann, John Peacock, Pat Osmer, Alan Wright, Peter Brand, Chip Arp, Roger Clowes, Arthur Trew, Alec Boksenberg, Gerry Gilmore, Paul Atherton, Gordon MacAlpine, Dan Weedman, Keith Taylor, Andy Fabian, Martin Ward, Ann Savage, Mike Edmunds, Ed Zuiderwijk, John Bahcall, Hagai Netzer, Wal Sargent, Mike Hawkins, Victor Clube, Ian McLean, Russell Cannon, Malcolm Longair, Jack Baldwin, Paul Hewett and Riccardo Giacconi for advice, discussions and preprints.

REFERENCES

Allen, D.A.: 1980, Nature, 284, p.323.
Andrillat, Y.: 1971, Astrophys. Lett., 8, p.161.
Andrillat, Y., and Souffrin, S.: 1967, Astrophys. Lett., 1, p.111.
Arons, J.: 1972, Ap.J., 172, p.553.
Arp, H.C.: 1977, "L'Evolution des Galaxies et ses Implications Cosmologique", Colloque IAU No.37, Colloque CNRS263, Paris, Sept, 1976, p.377.
Arp, H.: 1980, Ap.J., 239, p.463.
Arp, H.: 1981a, preprint "Quasars near companion galaxies".
Arp, H.: 1981b, Observatory, in press (proceedings of RAS conference).
Arp, H.C., and Hazard, C.: 1980, Ap.J., 240, p.726.
Arp, H., and Sulentic, J.W.: 1979, Ap.J., 229, p.496.
Arp, H., Sulentic, J.W., and di Tullio, G.: 1979, Ap.J., 229, p.489.
Atherton, P.D., Taylor, K., Pike, C.D., and Hook, R.N.: 1981, M.N.R.A.S., (submitted), "Taurus: A Wide-Field Imaging Fabry-Perot Spectrometer for Astronomy".
Bahcall, J.: 1980, IAU Colloquium No. 54 "Scientific Research with the Space Telescope", (Washington: U.S. Gov. Print. Off.), p.215..
Bahcall, J.N., and Soniera, R.M.: 1980, Ap.J.(Suppl.), 44, p.73.
Bahcall, J.N., and Spitzer, L., Jr.: 1969, Ap.J., 156, L63.
Baldwin, J.A.: 1975, Ap.J., 201, p.26.
Baldwin, J.A.: 1977, M.N.R.A.S., 178, 67P.
Baldwin, J.A.: 1979, Proc. NATO Conference, "Active Galactic Nuclei", eds. C. Hazard and S. Mitton (Cambridge : University Press), p.51.
Baldwin, J.A., and Netzer, H,: 1978, Ap.J., 226, p.1.
Baldwin, J.A., Wampler, E.J., and Burbidge, E.M.: 1981, Ap.J., 243, p.76.
Barnothy, J.M., and Barnothy, M.F.: 1968, Science, 162, p.348.
Barnothy, J.M., and Barnothy, M.F.: 1972, Ap.J., 174, p.477.
Barnothy, J.: 1974, Bull. A.A.S., 6, p.212.
Bennett, A.S.: 1962, Mem. R.A.S., 68, p.163.
Bergeron, J., and Kunth, D.: 1980, Astron. and Ap., 85, L11.

Blades, J.C.: 1981, Observatory, in press (proceedings of RAS conference).
Blades, J.C., Hunstead, R.W., and Murdoch, H.S.: 1981, M.N.R.A.S., 194, p.669.
Blandford, R.D., and Königl, A.: 1979, Ap.J., 232, p.34.
Blandford, R.D., and Rees, M.J.: 1978, "Pittsburgh Conference on BL Lac Objects", ed. A.M. Wolfe (Pittsburgh: University of Pittsburgh), p.328.
Bohuski, T.J., and Weedman, D.W.: 1979, Ap.J., 231, p.653.
Boksenberg, A.: 1972, in "Auxiliary Instrumentation for Large Telescopes", Proc. of the ESO-CERN Conference, ed. S. Laustsen and A. Reiz, p.295.
Boksenberg, A., Danziger, I.J., Fosbury, R.A.E., and Goss, W.M.: 1980, Ap.J., 242, L145.
Boksenberg, A., and Netzer, H.: 1977, Ap.J., 212, p.37.
Boksenberg, A., and Sargent, W.L.W.: 1978, Ap.J., 220, p.42.
Boksenberg, A., and Snijders, M.A.J.: 1981, M.N.R.A.S., 194, p.353.
Bolton, J.G.: 1977, I.A.U. Symp., No. 74, "Radio Astronomy and Cosmology", ed. D.L. Jauncey (Dordrecht: Reidel) p.85.
Bolton, J.G., and Savage, A.: 1978, I.A.U. Symp., No.79, "The Large Scale Structure of the Universe", eds. M.S. Longair and J. Einasto (Dordrecht: Reidel), p.295.
Bònoli, F., Braccesi, A., Marano, B., Merighi, R., and Zitelli, V.: 1980, Astron. and Ap., 90, L10.
Bourassa, R.R., and Kantowski, R.: 1975, Ap.J., 195, p.13.
Braccesi A., Formiggini, L., and Gandolfi, E.: 1970, Astron. and Ap., 5, p.264.
Braccesi, A., Zitelli, V., Bònoli, F., and Formiggini, L.: 1980, Astron. and Ap., 85, p.80.
Browne, I.W.A., Crowther, J.H., and Adgie, R.L.: 1973, Nature, 244, p.146.
Browne, I.W.A., and McEwan, N.J.: 1972, Nature Phys. Sci., 239, p.101.
Burbidge, E.M., Junkkarinen, V.T., Koski, A.T., Smith, H.E., and Hoag, A.A.: 1980, Ap.J.(Letters), 242, L55.
Burbidge, G.: 1979, Nature, 282, p.451.
Burbidge, G., O'Dell, S.L., Roberts, D.H., and Smith, H.E.: 1977, Ap.J., 218, p.33.
Canfield, R.C., and Puetter, R.C.: 1981, Ap.J., 243, p.390.
Capriotti, E., Foltz, C., and Byard, P.: 1979, Ap.J., 230, p.681.
Capriotti, E., Foltz, C., and Byard, P.: 1981, Ap.J., 245, p.396.
Carswell, R.F., and Ferland, G.J.: 1980, M.N.R.A.S., 191, p.55.
Carswell, R.F., and Smith, M.G.: 1978, M.N.R.A.S., 185, p.381.
Carswell, R.F., and Strittmatter, P.A.: 1973, Nature, 242, p.394.
Carswell, R.F., Whelan, J.A.J., Smith, M.G., Boksenberg, A., and Tytler, D.R.: 1981, M.N.R.A.S., submitted.
Caswell, J.L., and Crowther, J.H.: 1969, M.N.R.A.S., 145, p.181.
Cavaliere, A., Danese, L., De Zotti, G., and Francheschini, A.: 1980, Astron. and Ap., 85, L9.
Cavaliere, A., Danese, L., De Zotti, G., and Francheschini, A.: 1981, Astron. and Ap., 97, p.269.
Chaffee, F.H., Jr.: 1980, Scientific American, 243, p.60.
Chanan, G.A., Margon, B., and Downes, R.A.: 1981, Ap.J., 243, L5.
Chen, J.-S., Morton, D.C., Peterson, B.A., Wright, A.E., and Jauncey, D.L.: 1981, M.N.R.A.S., in press.

Cheney, J.E., and Rowan-Robinson, M.: 1981, M.N.R.A.S., 195, p.497.
Chiu, L.-T.G.: 1980, Ap.J. (Suppl.), 44, p.31.
Christiansen, W., Scott, J., and Weymann, R.: 1980, Bull. A.A.S., 12, p.808.
Clowes, R.G.: 1980, Ph.D. thesis, University of Edinburgh.
Clowes, R.G.: 1981, M.N.R.A.S., in press "Selection Effects in Spectral Searches for Quasars".
Clowes, R.G., Emerson, D., Smith, M.G., Wallace, P.T., Cannon, R.D., Savage, A., and Boksenberg, A.: 1980, M.N.R.A.S., 193, p.415.
Clowes, R.G., Smith, M.G., Savage, A., Cannon, R.D., Boksenberg, A., and Wall, J.V.: 1979, M.N.R.A.S., 189, p.175.
Cohen, J.G., and Meloy, D.A.: 1975, Ap.J., 198, p.545.
Colla, G., Fanti, C., Fanti, R., Ficarra, A., Formiggini, L., Gandolfi, E., Gioia, I., Lari, C., Marano, B., Padrielli, L., and Tomasi, P.: 1973, Astron. and Ap. (Suppl.), 11, p.291.
Condon, J.J., Condon, M.A., Mitchell, K.J., and Usher, P.D.: 1980, Ap.J., 242, p.486.
Condon, J.J., Condon, M.A., Jauncey, D.L., Smith, M.G., Turtle, A.J., and Wright, A.E.: 1981, Ap.J., 244, p.5.
Condon, J.J., Hicks, P.D., and Jauncey, D.L.: 1977, Astron.J., 82, p.692.
Condon, J.J., Jauncey, D.L., and Wright, A.E.: 1978, Astron.J., 83, p.1036.
Condon, J.J., Ledden, J.E., Burbidge, E.M., Koski, A.T., and Smith, H.E.: 1980, Bull. A.A.S., 12, p.808.
Cutri, R.M., Aitken, D.K., Jones, B., Merrill, K.M., Puetter, R.C., Roche, P.F., Rudy, R.J., Russell, R.W., Soifer, B.T., and Willner, S.P.: 1981, Ap.J., 245, p.818.
Davidsen, A.F.: 1980, I.A.U. Symp., No. 92, "Objects of High Redshift", eds. G.O. Abell and P.J.E. Peebles (Dordrecht : Reidel), p.235.
Davidsen, A.F., Hartig, G.F., and Fastie, W.G.: 1977, Nature, 269, p.203.
Davidson, K.: 1977, Ap.J., 218, p.20.
Davidson, K., and Netzer, H.: 1979, Rev. Mod. Phys., 51, p.715.
de Ruiter, H.R., and Zuiderwijk, E.J.: 1981, preprint "Quasar-generating Superclusters : an explanation for a clumpy quasar sky?".
De Veny, J.B., Osborn, W.H., and Janes, K.: 1971, P.A.S.P., 83, p.611.
Dyer, C.C., and Roeder, R.C.: 1980, Ap.J., 238, L67.
Edmunds, M., and George, G.H.: 1981, preprint "The Random Alignment of Quasars".
Epstein, R.I., and Fransson, C.: 1980, Bull. A.A.S., 12, p.853.
Fabian, A.C.: 1981, Observatory, in press (proceedings of 1981 Herstmonceux conference).
Fanti, C., Fanti, R., Ficarra, A., Formiggini, L., Giovannini, G., Lari, C., and Padrielli, L.: 1975, Astron. and Ap. Suppl., 19, p.143.
Fanti, R., Feretti, L., Giovannini, G., and Padrielli, L.: 1979, Astron. and Ap., 73, p.40.
Ferland, G.J.: 1980, Bull. A.A.S., 12, p.853.
Ferland, G.J., Netzer, H., and Shields, G.A.: 1979, Ap.J., 232, p.382.
Ferland, G.J., Rees, M.J., Longair, M.S., and Perryman, M.A.C.: 1979, M.N.R.A.S., 187, 65P.
Gaskell, C.M.: 1981, Ph.D. thesis, University of California, Santa Cruz.
Giacconi, R., et al.: 1979, Ap.J. (Letters), 234, L1.

Gilmore, G.: 1981a, M.N.R.A.S., 195, p.183.
Gilmore, G.: 1981b, preprint "The Double Quasar 0957+561 A,B as a Probe of Quasar Structure".
Glass, I.S.: 1980, M.N.R.A.S., 192, 37P.
Glass, I.S.: 1981, M.N.R.A.S., 194, p.795.
Gondhalekar, P.M., and Wilson, R.: 1980, Nature, 285, p.461.
Gower, J.F.R., Scott, P.F., and Wills, D.: 1967, Mem. R.A.S., 71, p.49.
Grandi, S.A., and Phillips, M.M.: 1979, Ap.J., 232, p.659.
Grandi, S.A., and Phillips, M.M.: 1980, Ap.J., 239, p.475.
Grandi, S.A., and Tifft, W.G.: 1974, P.A.S.P., 86, p.873.
Green, R.F.: 1976, P.A.S.P., 88, p.665.
Green, R.F., Pier, J.R., Schmidt, M., Estabrook, F.B., Lane, A.L. and Wahlquist, H.D.: 1980, Ap.J., 239, p.483.
Green, R.F., and Schmidt, M.: 1978, Ap.J. (Letters), 220, L1.
Greenfield, P.E., Burke, B.F., and Roberts, D.H.: 1980, Nature, 286, p.865.
Greenfield, P.E., Roberts, D.H., and Burke, B.F.: 1980, Science, 208, p.495.
Gregory, S.A., Thompson, L.A., and Tifft, W.G.: 1981, Ap.J., 243, p.411.
Griffiths, R.E., and 14 collaborators: 1981, preprint "The Optical and Radio Identification Context of the Einstein Observatory Deep X-ray Survey of a Region in Pavo".
Grindlay, J.E., Steiner, J.E., Forman, W.R., Canizares, C.R., and McClintock, J.E.: 1980, Ap.J., 239, L43.
Halpern, J.P., and Elvis, M.: 1980, Bull. A.A.S., 12, p.795.
Hawkins, M.R.S.: 1981, Nature, (submitted) "The Automated Detection of Variable Objects on Schmidt Plates".
Hayman, P.G., Hazard, C., and Sanitt, N.: 1979, M.N.R.A.S., 189, p.853.
Hazard, C.: 1979, Proc. NATO Conference, "Active Galactic Nuclei", eds. C. Hazard and S. Mitton (Cambridge : University Press), p.1.
Hazard, C., Arp, H.C., and Morton, D.C.: 1979, Nature, 282, p.271.
Hauser, M.G., and Peebles, P.J.E.: 1973, Ap.J., 185, p.757.
Heckman, T.M., Miley, G.K., van Breugel, W.J.M., and Butcher, H.R.: 1980, Bull. A.A.S., 12, p.809.
Hewitt, A., and Burbidge, G.: 1980, Ap.J. (Suppl.), 43, p.57.
Hoag, A.A., Burbidge, E.M., and Smith, H.E.: 1977, "L'Evolution des Galaxies et ses Implications Cosmologiques", Colloque IAU No. 37, Colloque CNRS 263, Paris, Sept. 1976, p.521.
Hoag, A.A., and Schroeder, D.J.: 1970, P.A.S.P., 82, p.1141.
Hoag, A.A., and Smith, M.G.: 1977, Ap.J., 217, p.362.
Hogan, C.J., and Rees, M.J.: 1979, M.N.R.A.S., 188, p.791.
Hoskins, D.G., Murdoch, H.S., Hazard, C., and Jauncey, D.L.: 1972, Aust. J. Phys., 25, p.559.
Hyland, A.R., and Allen, D.A.: 1981, QJRAS, 22, p.61 (Section 1.4).
Hyland, A.R., Becklin, E.E., and Neugebauer, G.: 1978, Ap.J. (Letters), 220, L73.
Impey, C., and Brand, P.W.J.L.: 1981, Nature, submitted "Infrared Photometry of Flat Spectrum Radio Sources".
Jenkins, C.J., Pooley, G.G., and Riley, J.M.: 1977, Mem. R.A.S., 84, p.61.
Kembhavi, A.K., and Fabian, A.C.: 1981, preprint "X-ray Quasars and the X-ray background".

Kinman, T.D., Lamla, E., and Wirtanen, C.A.: 1966, Ap.J., 146, p.964.
Koo, D.C., and Kron, R.G.: 1980, P.A.S.P., 92, p.537.
Kreysa, E., Pauliny-Toth, I.I.K., Schultz, G.V., Sherwood, W.A., and Witzel, A.: 1980, Ap.J. (Letters), 240, L17.
Kriss, G.A., Canizares, C.R., and Ricker, G.R.: 1980, Ap.J., 242, p.492.
Kron, R.G.: 1978, Ph.D. thesis, University of California, Berkeley.
Kron, R.G.: 1980, Ap.J. Suppl., 43, p.305.
Kron, R.G., and Chiu, L.-T.G.: 1981, preprint "Stars with zero proper motion and the number of faint QSOs".
Ku, W.H.-M., Helfand, D.J., and Lucy, L.B.: 1980, Nature, 288, p.323.
Kwan, J., and Krolik, J.H.: 1979, Ap.J., 233, L91.
Large, M.I., Mills, B.Y., Little, A.G., Crawford, D.F., and Sutton, J.M.: 1981, M.N.R.A.S., 194, p.693.
Lebofsky, M.J., Rieke, G.H., Walsh, D., and Weymann, R.J.: 1980, Nature, 285, p.385.
Lewis, D.W., MacAlpine, G.M., and Weedman, D.W.: 1979, Ap.J., 233, p.787.
Longair, M.S.: 1976, Q.J.R.A.S., 17, p.422.
Lynden-Bell, D.: 1969, Nature, 223, p.690.
Lynden-Bell, D.: 1978, Physica Scripta, 17, p.185.
Lynds, C.R.: 1971, Ap.J. (Letters), 164, L73.
Lynds, C.R., and Wills, D.: 1972, Ap.J., 172, p.531.
MacAlpine, G.M., Feldman, F.R., and Weedman, D.W.: 1981, Ap.J., to be submitted.
MacAlpine, G.M., and Lewis, D.W.: 1978, Ap.J. Suppl., 36, p.587.
MacAlpine, G.M., Lewis, D.W., and Smith, S.B.: 1977, Ap.J. (Suppl.), 35, p.203.
MacAlpine, G.M., Smith, S.B., and Lewis, D.W.: 1977, Ap.J. (Suppl.), 34, p.95.
MacAlpine, G.M., Smith, S.B., and Lewis, D.W.: 1977, Ap.J. (Suppl.), 35, p.197.
Marscher, A.P.: 1980, Ap.J., 235, p.386.
Martin, P.G., and Ferland, G.J.: 1980, Ap.J. (Letters), 235, L125.
Masson, C.R., and Wall, J.V.: 1977, M.N.R.A.S., 180, p.193.
Meaburn, J., Terrett, D.L., Theokas, A., and Walsh, J.R.: 1981, M.N.R.A.S., 195, p.39.
Miller, J.S., Antonucci, R.R.J., and Keel, W.C.: 1981, Nature, 289, p.153.
Moore, R.L., and Stockman, H.S.: 1981, Ap.J., 243, p.60.
Moore, R., McGraw, J., Angel, R., Duerr, R., Lebofsky, M., Rieke, G., Wiesniewski, W., Axon, D., Bailey, J., Hough, J., Breger, M., Clayton, J., Martin, P., Miller, J., Schmidt, G., Schutz, H., and Thompson, I.: 1980, Bull. A.A.S., 12, p.809.
Mushotsky, R.F., Marshall, F.E., Boldt, E.A., Holt, S.S., and Serlemitsos, P.J.: 1980, Ap.J., 235, p.377.
Mushotsky, R.F.: 1981, Observatory, in press (proceedings of 1981 Herstmonceux conference).
Netzer, H.: 1980, Ap.J., 236, p.406.
Netzer, H., Wills, B.J., Uomoto, A.K., Rybski, P.M., and Tull, R.G.: 1979, Ap.J. (Letters), 232, L155.
Neugebauer, G., Oke, J.B., Becklin, E.E., and Matthews, K.: 1979, Ap.J., 230, p.79.
O'Dell, S.L.: 1979, Proc. NATO Conference "Active Galactic Nuclei",

eds. C. Hazard and S. Mitton (Cambridge : University Press), p.95.
Oke, J.B.: 1967, Ap.J., 147, p.901.
Oke, J.B.: 1974, Ap.J., 189, L47.
Oort, J.H.: 1981, Astron. and Ap., in press.
Oort, J.H., Arp, H., and de Ruiter, H.: 1981, Astron. and Ap., 95, p.7.
Osmer, P.S.: 1978, Physica Scripta, 17, 357.
Osmer, P.S.: 1979, Ap.J., 227, p.18.
Osmer, P.S.: 1980a, I.A.U. Symposium No.92, "Objects of High Redshift", eds. G.O. Abell and P.J.E. Peebles, (Dordrecht : Reidel), p.77.
Osmer, P.S.: 1980b, Ap.J. (Suppl.), 42, p.523.
Osmer, P.S.: 1980c, Ap.J., 237, p.666.
Osmer, P.S.: 1981, Ap.J., in press.
Osmer, P.S., and Smith, M.G.: 1976, Ap.J., 210, p.267.
Osmer, P.S., and Smith, M.G.: 1977a, Ap.J., 215, L47.
Osmer, P.S., and Smith, M.G.: 1977b, Ap.J., 217, L73.
Osmer, P.S., and Smith, M.G.: 1980, Ap.J. (Suppl.), 42, p.333.
Osmer, P.S., Smith, M.G., and Weedman, D.W.: 1974, Ap.J., 192, p.279.
Osterbrock, D.E.: 1977, Ap.J., 215, p.733.
Osterbrock, D.E.: 1980, Ann. N.Y. Acad. Sci., 336, p.22.
Ozernoy, L.M.: 1974, "Galaxies and Relativistic Astrophysics", eds. B. Barbanis and J.D. Hadjidemetriou (Springer), p.65.
Pauliny-Toth, I.I.K.: 1977, I.A.U. Symp., No.74, "Radio Astronomy and Cosmology" ed. D.L. Jauncey (Dordrecht: Reidel) p.63.
Peacock, J.A., and Gull, S.F.: 1981, M.N.R.A.S., in press.
Peacock, J.A., Perryman, M.A.C., Longair, M.S., Gunn, J.E., and Westphal, J.A.: 1981, M.N.R.A.S., 194, p.601.
Peacock, J.A., and Wall, J.V.: 1981, M.N.R.A.S., 194, p.331.
Pearson, T.J., and Kus, A.J.: 1978, M.N.R.A.S., 182, p.273.
Peebles, P.J.E.: 1974, Astron. and Ap., 32, p.197.
Peterson, B.A., Ellis, R.S., Kibblewhite, E.J., Bridgeland, M.T., Hooley, T., and Horne, D.: 1979, Ap.J., 233, L109.
Phillips, M.M.: 1978a, Ap.J. (Suppl.), 38, p.187.
Phillips, M.M.: 1978b, Ap.J., 226, p.736.
Pilkington, J.D.H., and Scott, P.F.: 1965, Mem. R.A.S., 69, p.183.
Pratt, N.M.: 1977, Vistas in Astronomy, 21, p.1.
Porcas, R.W., Booth, R.S., Browne, I.W.A., Walsh, D., and Wilkinson, P.N.: 1981, Nature, 289, p.758.
Puetter, R.C.: 1980, Bull. A.A.S., 12, p.836.
Puetter, R.C., Smith, H.E., Willner, S.P., and Pipher, J.L.: 1981, Ap.J., in press.
Puetter, R.C., Smith, H.E., and Willner, S.P.: 1979, Ap.J., 227, L5.
Rees, M.J.: 1977a, Q.J.R.A.S., 18, p.429.
Rees, M.J.: 1977b, Ann. N.Y. Acad. Sci. 302, p.613.
Rees, M.J.: 1978a, Physica Scripta, 17, p.193.
Rees, M.J.: 1978b, Observatory, 98, p.210.
Richer, H.B., and Olson, B.I.: 1980, P.A.S.P., 92, p.573.
Richstone, D.O., and Schmidt, M.: 1980, Ap.J., 235, p.361.
Rieke, G.H., Grasdalen, G.L., Kinman, T.D., Hintzen, P., Wills, B.J., and Wills, D.: 1976, Nature, 260, p.754.
Rieke, G.H.: 1980, I.A.U. Symposium No.96, "Infrared Astronomy", eds. C.G. Wynne-Williams and D.P. Cruikshank (Dordrecht: Reidel) p.317.

Rieke, G.H., Lebofsky, M.J., and Kinman, T.D.: 1979, Astrophys. J., 232, L151.
Röser, H.-J.: 1975, Astron. and Ap., 45, p.329.
Sandage, A.R.: 1972, Q.J.R.A.S., 13, p.282.
Sandage, A.R., and Luyten, W.J.: 1967, Ap.J., 148, p.767.
Sandage, A., and Luyten, W.J.: 1969, Ap.J., 155, p.913.
Sandage, A.R., Westphal, J.A., and Strittmatter, P.A.: 1966, Ap.J., 146, p.322.
Sargent, W.L.W., Young, P.J., Boksenberg, A., Carswell, R.F., and Whelan, J.A.J.: 1979, Ap.J., 230, p.49.
Sargent, W.L.W., Young, P.J., Boksenberg, A., and Tytler, D.: 1980, Ap.J. (Suppl.), 42, p.41.
Savage, A., and Bolton, J.G.: 1979, M.N.R.A.S., 188, p.599.
Savage, B.D., and Jeske, N.A.: 1980, Ap.J., 244, p.768.
Scheuer, P.A.G., and Readhead, A.C.S.: 1979, Nature, 277, p.182.
Schmidt, M.: 1964, "Quasars and High-Energy Astronomy" (New York : Gordon and Breach 1969), p.55.
Schmidt, M.: 1968, Ap.J., 151, p.393.
Schmidt, M.: 1970, Ap.J., 162, p.371.
Schmidt, M.: 1972, Ap.J., 176, p.273.
Schmidt, M.: 1974, Ap.J., 193, p.509.
Schmidt, M.: 1976, Ap.J., 209, L55.
Searle, L., and Sargent, W.L.W.: 1968, Ap.J., 153, p.1003.
Segal, I.E.: 1976, "Mathematical Cosmology and Extragalactic Astronomy" (New York : Academic Press).
Segal, I.E., Loncaric, J., and Segal, W.: 1980, Ap.J., 238, p.38.
Sersic, J.L., and Pastoriza, M.: 1965, P.A.S.P., 77, p.287.
Setti, G., and Woltjer, L.: 1979, Astron. and Ap., 76, L1.
Shanks, T., Phillips, S., and Fong, R.: 1980, M.N.R.A.S., 191, 47P.
Sherwood, W.A., Kreysa, E., and Schultz, G.V.: 1981, Astronomische Gesellschaft Mitteilungen, in press.
Sherwood, W.A., Schultz, G.V., and Kreysa, E.: 1980, contributed (unpublished) paper at I.A.U. Symposium No.96, "Infrared Astronomy", eds. C.G. Wynne-Williams and D.P. Cruikshank (Dordrecht: Reidel).
Sherwood, W.A., Schultz, G.V., and Kreysa, E.: 1981, preprint "Millimetre Observations of the Quasar Q0420-388".
Shields, G.A.: 1976, Ap.J., 204, p. 330.
Shields, G.A., Oke, J.B., and Sargent, W.L.W.: 1972, Ap.J., 176, p.75.
Shields, G.A., and Wheeler, J.C.: 1978, Ap.J., 222, p.667.
Shuder, J.M., and MacAlpine, G.M.: 1979, Ap.J., 230, p.348.
Smith, H.E., and Spinrad, H.: 1980, P.A.S.P., 92, p.553.
Smith, M.G.: 1975, Ap.J., 202, p.591.
Smith, M.G.: 1976, Ap.J. (Letters), 206, L125.
Smith, M.G.: 1978, Vistas in Astronomy, 22, p.321.
Smith, M.G.: 1979, I.A.U. Colloquium No.47, "Spectral Classification of the Future", McCarthy, Philips, Coyne (eds.), Vatican Observatory, p.117.
Smith, M.G., Boksenberg, A., Carswell, R.F., and Whelan, J.A.J.: 1977, M.N.R.A.S., 181, p.67P.
Smith, M.G., Carswell, R.F., Whelan, J.A.J., Wilkes, B.J., Boksenberg,

A., Clowes, R.G., Savage, A., Cannon, R.D., and Wall, J.V.: 1981, M.N.R.A.S., 195, p.437.
Smith, M.G., and Wright, A.E.: 1980, M.N.R.A.S., 191, p.871.
Snijders, M.A.J.: 1980, Proc. 2nd European IUE Conf., Tubingen, p.71.
Snijders, M.A.J.: 1981, Observatory, in press (proceedings of 1981 Herstmonceux conference).
Soifer, B.T., Oke, J.B., Matthews, K., and Neugebauer, G.: 1979, Ap.J., 227, L1.
Soifer, B.T., Neugebauer, G., Matthews, K., Becklin, E.E., Wynn-Williams, C.G., and Capps, R.: 1980, Nature, 285, p.91.
Spinrad, H.: 1977, in "Evolution of Galaxies and Stellar Populations", eds. B.M. Tinsley and R.B. Larson (Yale : New Haven), p.301.
Sramek, R.A., and Weedman, D.W.: 1978, Ap.J., 221, p.468.
Steppe, H., Véron, P., and Véron, M.P.: 1979, Astron. and Ap., 78, p.125.
Stocke, J.T., and Perrenod, S.C.: 1981, Ap.J., 245, p.375.
Stockton, A.: 1980, Ap.J. (Letters), 242, L141.
Strittmatter, P., Hill, P., Pauliny-Toth, I.I.K., Steppe, H., and Witzel, A.: 1980, Astron. and Ap., 88, L12.
Sulentic, J.W., Arp, H.C., and di Tullio, G.A.: 1978, Ap.J., 220, p.47.
Tananbaum, H., Avni, Y., Branduardi, G., Elvis, M., Fabbiano, G., Feigelson, E., Giacconi, R., Henry, J.P., Pye, J.P., Soltan, A., and Zamorani, G.: 1979, Ap.J. (Letters), 234, L9.
Taylor, K., and Atherton, P.D.: 1980, M.N.R.A.S., 191, p.675.
Tennant, A.F., Mushotsky, R.F., Boldt, E.A., and Swank, J.H.: 1981, preprint "Rapid X-ray Variability in NGC 6814".
Tohline, J.E., and Osterbrock, D.E.: 1976, Ap.J., 210, L117.
Tovmassian, G.M.: 1966, Astrofizika, 2, p.419.
Turner, E.L.: 1980, Ap.J., 242, L135.
Tyson, J.A., and Jarvis, J.F.: 1979, Ap.J., 230, L153.
Ulrich, M.H., and 19 collaborators: 1980, M.N.R.A.S., 192, p.561.
Usher, P.D.: 1978, Ap.J., 222, p.40.
Usher, P.D., and Mitchell, K.J.: 1978, Ap.J., 223, p.1.
Vaucher, B., Thomas, N., Kreidl, T., and Hoag, A.: 1980, Bull. A.A.S., 12, p.836.
Vaucher, B.G., and Weedman, D.W.: 1980, Ap.J., 240, p.10.
Wall, J.V.: 1977, I.A.U. Symposium No.74, "Radio Astronomy and Cosmology", ed. D.L. Jauncey, Reidel, Holland, p.55.
Wall, J.V., Pearson, T.J., and Longair, M.S.: 1980, M.N.R.A.S., 193, p.683.
Walsh, D., Carswell, R.F., and Weymann, R.J.: 1979, Nature, 279, p.381.
Wampler, E.J., Robinson, L.B., Baldwin, J.A., and Burbidge, E.M.: 1973, Nature, 243, p.336.
Webster, A.S.: 1981, Observatory, in press (proceedings of RAS conference).
Weedman, D.W.: 1980, Ap.J., 237, p.326.
Weisheit, J.C., Shields, G.A., and Tarter, C.B.: 1981, Ap.J., 245, p.406.
Weymann, R.J.: 1980, I.A.U. Symposium No. 92, "Objects of High Redshift", G.O. Abell and P.J.E. Peebles (eds.), p.107.

Weymann, R.J., Carswell, R.F., and Smith, M.G.: 1981, Ann. Rev. Astr. and Ap., 19, in press.
Weymann, R.J., Chaffee, F.H., Davis, M., Carleton, N.P., Walsh, D., and Carswell, R.F.: 1979a, Ap.J. (Letters), 233, L43.
Weymann, R.J., Williams, R.E., Peterson, B.M., and Turnshek, D.A.: 1979b, Ap.J., 234, p.33.
Weymann, R.J., Latham, D., Angel, J.R.P., Green, R.F., Liebert, J.W., Turnshek, D.A., Turnshek, D.E., and Tyson, J.A.: 1980, Nature, 285, p.641.
Wilkes, B.: 1981, Ph.D. thesis, University of Cambridge.
Willis, A.G., and Miley, G.K.: 1979a, Astron. and Ap. (Suppl.), 37, p.397.
Willis, A.G., and Miley, G.K.: 1979b, Astron. and Ap., 76, p.65.
Willis, A.G., Oosterbaan, C.E. and de Ruiter, H.R.: 1976, Astron. and Ap. (Suppl.), 25, p.453.
Wills, B.J., Netzer, H., Uomoto, A.K., and Wills, D: 1980, Ap.J., 237, p.319.
Wills, B.J., Netzer, H., and Wills, D.: 1980, Ap.J. (Letters), 242, L1.
Wills, B.J., and Wills, D.: 1980, Ap.J., 238, p.1.
Wills, D.: 1978, Physica Scripta, 17, p.333.
Wills, D., and Wills, B.J.: 1976, Ap.J. (Suppl.), 31, p.143.
Wilson, A.S., and Penston, M.V.: 1979, Ap.J., 232, p.389.
Wright, A.E., Peterson, B.A., Jauncey, D.L., and Condon, J.J.: 1978, Ap.J., 226, L61.
Wu, C.-C.: 1977, Ap.J., 217, L117.
Young, P.J., Sargent, W.L.W., Boksenberg, A., Carswell, R.F., and Whelan, J.A.J.: 1979, Ap.J., 229, p.891.
Young, P., Gunn, J.E., Kristian, J., Oke, J.B., and Westphal, J.A.: 1980, Ap.J., 241, p.507.
Young, P., Deverill, R.S., Gunn, J.E., Westphal, J.A., and Kristian, J.: 1981a, Ap.J., 244, p.723.
Young, P., Gunn, J.E., Kristian, J., Oke, J.B., and Westphal, J.A.: 1981b, Ap.J., 244, p.736.
Zamorani, G., Henry, J.P., Maccacaro, T., Tananbaum, H., Soltan, A., Avni, Y., Leibert, J., Stocke, J., Strittmatter, P.A., Weymann, R.J., Smith, M.G., and Condon, J.J.: 1981, Ap.J., 245, p.357.

THE ALGOL-TYPE SYSTEMS

Alan H. Batten,
Dominion Astrophysical Observatory
Herzberg Institute of Astrophysics
Victoria, B.C.

ABSTRACT. The development of our ideas about Algol-type systems over the past quarter of a century is discussed with special reference to the hypothesis of mass-transfer between the components, and to the role of circumstellar matter in the evolution of these systems.

I. INTRODUCTION

My connection with Professor Kopal and the Department of Astronomy in the University of Manchester extends back almost to the time that he arrived there. Being, like Kopal himself, a student of the late E. Finlay-Freundlich (Figure 1), I was introduced to Kopal and visited Manchester even before I enrolled as a graduate student in September 1955. My interest in binary systems had been stimulated by Finlay-Freundlich, and my decision to come to Manchester had been largely based on Kopal's presence there and my wish to learn more about close binary stars. At that time, Kopal himself was just beginning to explore the evolutionary problems posed by Algol-type systems, and it was natural that he should direct my own attention to them. This direction he gave was reinforced by friendships made in Manchester (particularly with Mirek Plavec) and has determined a large part of my own research ever since. During the quarter of a century, or somewhat more, that has elapsed since I began my studies in Manchester, mass-transfer between the components of Algol-type systems has progressed from being a speculative and barely credible idea (Crawford 1955) to being a respectable scientific theory which, whether it is ultimately judged to be right or wrong, is capable of making predictions that can be tested by observation. Although I played very little part in effecting this transformation, circumstances enabled me to watch it from a ring-side seat. I have chosen to write about these events, now history, in a somewhat more personal vein than is usual in scientific writing, and I hope that Professor Kopal will accept what follows as thanks for the start he gave me in my own career, even though I shall endeavour to persuade my readers to look at these events from a point of view somewhat different from that which he himself recently urged (Kopal 1978).

F. D. Kahn (ed.), Investigating the Universe, 207–225.

Figure 1: Some members of the Manchester Department of Astronomy during a visit by Prof. E. Finlay-Freundlich in the spring of 1957. Left to Right: (Back Row) Alan Batten, John Owen, Dr. Rudolf Kurth, Bill Olle, John Hazlehurst, Michael Elphick; (Front Row) Welwyn Davies, Muriel Hodgkinson, Prof. E. Finlay-Freundlich, Ellen Finlay, Prof. Zdeněk Kopal, Wallace Sargent. (Photograph by T.W. Olle).

Although, as Kopal has been careful to point out, the first suggestion that Algol-type systems contain one component near the limit of stability appears to have been made by Walter (1931), the idea did not win general acceptance until some time after the end of the second world war, and it was particularly promoted by Crawford in the paper already cited and by Kopal (1955, 1956, 1957). The last of these papers was in fact written first and presented at the Dublin General Assembly of the I.A.U., just before I came to Manchester. In it, Kopal introduced the ideas developed in the other two. That is, he drew attention to the central problem of Algol-type systems, that the less-massive star appeared to have evolved first and to have reached its limit of dynamical stability; he introduced the terminology of detached, semi-detached and contact systems now generally adopted; finally, he published one of the earliest attempts to compute trajectories of particles ejected from one component into the system.

At the time, our detailed knowledge of the evolution even of single stars was rudimentary. Haslegrove (then in Manchester but not in the Astronomy Department) and Hoyle (1956) were attempting to follow the early evolution of massive stars off the main sequence. It was accepted that massive stars should evolve first, and this of course created the paradox of the apparent situation in Algol-type systems. Crawford's attempt to resolve this was to suggest that the most massive star <u>had</u> expanded first and transferred the greater part of its mass to its companion. Although this idea is now a commonplace, it was not generally accepted. Detailed computations of mass-transfer, even under artificially restricted assumptions, were quite beyond the capacities of the computers then available, and arguments on both sides were necessarily largely qualitative. Kopal was sceptical, and justifiably so. One obvious question to ask was: why do we see no systems actually transferring mass? The answer that the process is so quick (astronomically speaking) that very few systems are going through it at some particular time, although now generally accepted, was not fully convincing then. A nova is a very transitory phenomenon, even on the time scale of a human life, but most of us have seen at least one, and any amateur who tries hard enough has a reasonable chance of discovering one. Yet a nova expels a very small amount of mass - 10^{-5} m$\odot$ or 10^{-4} m$\odot$ at most, while it was proposed that several solar masses can be shot across from one member of a binary to the other, without anyone noticing. This criticism was made by Kopal at the time and is implicit in his latest critique. I shall, therefore, return to it later. Morton (1960) showed that mass-transfer, if it occurred at all, would take place on a Kelvin-Helmholtz time scale and is thus, indeed, astronomically a rapid process, although, as we later came to believe, the Algol systems themselves have passed this rapid phase and are transferring mass much more slowly (on a nuclear time-scale). Morton's work, confirmed by more detailed calculations by Smak (1962), thus partially answered the foregoing criticism and introduced a quantitative element into the discussion. Nevertheless, it left untouched the central questions of whether or not mass-transfer can occur at all, or, if it does, whether it could produce anything like the observed Algol systems.

II. MASS-TRANSFER AS A SCIENTIFIC THEORY

Until August 1966, talk of mass-transfer in close binaries was little more than "hand-waving", to use the term that Kopal has recently applied to it. At the end of that month it became a scientific theory, fulfilling K.R. Popper's (1959) criterion of making quantitative predictions that are statistically improbable, by which the theory may be falsified. This change of status can be dated so precisely because it occurred at a conference held in Brussels and devoted primarily to the study of visual double stars. The conference included a session in which Paczynski (1967) and Kippenhahn and Weigert (1967) presented the first results of calculations of mass-transfer. That meeting has always seemed to me to have been rather an ironic presentation of a breakthrough. The interest of most of the audience was centred on binary systems quite different from the Algol systems, while Weigert - who, as it happened, had proceeded closer to the goal of producing such a system - openly admitted that his aim had not been to compute a theoretical Algol system, but to find a means of producing white dwarfs. Nevertheless, Kippenhahn and Weigert had shown the way and other groups soon followed.

The next few years might be called the halcyon days of mass-transfer. Three possible cases were quickly distinguished: case A in which mass was transferred from one component to the other before hydrogen had stopped burning in the core of the mass-losing star; case B in which the transfer takes place while hydrogen is burning only in a shell; and case C in which no mass is transferred until helium has been ignited in the core of the mass-losing star. A fourth case, AB, was not recognized at first but is now considered important. In this case, mass-transfer begins before all the hydrogen in the core is burnt but does not end until afterwards (Ziółkowski 1969, Horn 1971). A great variety of model systems was computed during the years 1969-72. Although Kopal has strongly criticized the artificial assumptions on which many of these calculations were based, particularly that of conservative mass-transfer (i.e. no mass lost to the system) and that the mass-gaining star could adjust immediately to its new mass, yet these computations served to show that, even under such restrictive assumptions, a wide variety of systems could be produced. Moreover, careful perusal of the original papers shows that most, if not all, of the investigators were aware of the importance of these assumptions and were properly cautious in applying their results to real systems. Nevertheless, it must be admitted that there was some euphoria in those years, which indeed I shared. At the time I wrote my own book (Batten 1973), which went to press in 1971, it seemed very likely that mass-transfer in conservative case A could come close to explaining the origin of Algol-type systems. At that time we still knew the dimensions of almost all such systems only very approximately, and it is as much the improvement in the observational data as any theoretical considerations that led to a revision of our opinions.

Indeed, the history of this subject in the last decade or two offers an interesting study of the interaction between observation, theory and technology. The data available in the 1950's were sufficient to define the problem of the Algol systems, but the computers available then were simply inadequate for the necessary theoretical calculations to be attempted. (This point should be remembered when we criticize the assumptions made in earlier work: computer technology has moved so fast that it is easy to judge the methods of yesterday by the facilities available today.) When such computations became possible, it was clear that the observational data were inadequate to test them properly. It was particularly important to attempt to detect more secondary spectra of Algol systems. This meant working in the red region of the spectrum. Thus, D.M. Popper extended his fundamental studies of stellar masses in this way, with results that he has recently summarized (Popper 1980) but an even more significant improvement in our data has come from the application of solid-state detectors characterized by a high signal-to-noise ratio. These were essential in the detection of the secondary spectrum of Algol itself (Tomkin and Lambert 1978) and very useful for several other systems (e.g. Tomkin 1981). It is the combination of these interwoven developments that enables us now to look critically at the state of our knowledge of Algol-type systems. Had any one of them been missing, the subject might not yet have developed as far as it has.

About the end of this early period of computation, Popper (1973) completed his thorough investigation of the Algol-type system AS Eri and showed that its properties were consistent with theoretical results for case B mass-transfer (Refsdal and Weigert 1971), a conclusion confirmed by more refined calculations (Refsdal et al. 1974). At the same time, Plavec (1973) looked more carefully at the theoretical predictions and the available data, and declared unequivocally "... I have little doubt now: I think that conservative case A cannot explain the Algol systems". Although much of the improvement in our observational data has come since Plavec reached that conclusion, the better data have confirmed it and it is now generally accepted.

III. SUBDIVISION OF THE ALGOL CLASSIFICATION

In his very first paper on mass-transfer (1967), Paczynski drew attention to two sub-types amongst the Algol systems. The distinction appears very obvious in Popper's (1980) recent list of masses of these systems. In that list RS Vul and U Sge stand out as each having total masses between 5 m$\odot$ and 10 m$\odot$ and secondaries more massive than the Sun, although they are roughly of solar type. These two secondaries, although subgiants, are not greatly over-luminous for their masses. Other systems could be added to this group if one was prepared to relax Popper's high standards for inclusion. Recent results on U CrB (Batten and Tomkin 1981), unavailable to Popper, would qualify, I believe, even by his standards. I would be prepared to add U Cep, while admitting that my velocity-curve for the primary star is very uncertain (Batten 1974). My results for the secondary component, however, have been fairly closely confirmed (Tomkin 1981). All the other secondaries listed in Popper's table have masses less than that of the Sun, the extreme case (AS Eri B, m = 0.2 m$\odot$) being the best determined. Such a star, estimated by Popper to have an absolute visual magnitude of 3.9, is exceedingly over-luminous if judged by its mass rather than its spectral type. The total mass of nearly all these systems is less than 5 m$\odot$. It is perhaps still questionable whether or not there is a real dichotomy between massive and less massive Algol-type systems, but it is undeniable that there is a very wide spread between extreme members of the group. This has naturally led to the question of whether or not more than one process may have operated to produce Algol-type systems.

This question has been considered by several investigators, but it is perhaps still too early to reach a decision. The latest study by de Grève (1980) indicates that few if any Algol-type systems can be explained by conservative mass-transfer in either case A or B, and leaves open the question whether or not a unique process has produced the observed variety of systems. Basing his comparison with observations partly on Popper's list of data, he finds that the theory can be made to agree with observations of most systems only if more than half the mass of the original primary is lost to the system in the process. He is careful, however, to point out that there is still a need for a more systematic grid of computations, covering all possibilities.

IV. U CORONAE BOREALIS

At this point I would like to recapitulate the development of my own thinking on this topic of mass-transfer by discussing a system that I have studied in some detail - U CrB. An analysis of a visual light-curve of this system figured in the first paper I published - under Kopal's direction (Batten 1956). Spectroscopic observations by Plaskett (1920) and Pearce (1936) appeared to show that the secondary component was an early-type star comparable in luminosity to the primary star. This result, which was not confirmed by Sahade and Struve (1945), contradicted what was known of the light-curve. The observations I analyzed (obtained by Dugan), unlike any other group, showed an appreciable secondary minimum that could almost be reconciled with the results of Plaskett and Pearce. At almost the same time as my work, however, D.B. Wood (1958) published a photoelectric light-curve that showed a very shallow secondary minimum (since confirmed by Catalano et al. 1966) and it became impossible to maintain that the secondary star is of early type. Dugan's results must represent either serious errors in reduction, or a transient stage in the system. Despite some evidence of errors in the reduction of phases, I am inclined still to believe the latter possibility - partly because we now know how much the light-curve of the similar system U Cep can be affected.

Some years after completing that work, I moved from Manchester to Victoria and was fortunate to learn spectroscopy from the late R.M. Petrie, for whose scientific and personal qualities I still retain the highest respect. Like Kopal, Petrie was sceptical both of Otto Struve's ideas of gas streams within binary systems and of the theory of mass-transfer. Petrie did not live to see the paper by Kippenhahn and Weigert and, to his dying day, maintained that binaries showing the phenomena that Struve ascribed to gas streams were few compared with the vast number of "normal" spectroscopic binaries. He was right, in one sense, since Algol-type systems are intrinsically rare. They must be, because early-type massive stars, such as are found in these systems, are themselves rare. Algol-type systems seem more common than they are because the combination of a large cool star with a hot one of comparable size favours the occurence of conspicuous deep eclipses. On the other hand, within this group of systems, the phenomena that we now interpret as caused by gaseous streams are virtually universal.

At Victoria I had the chance to study Plaskett's and Pearce's spectrograms of U CrB and became convinced that they genuinely showed the features that Plaskett and Pearce had interpreted as the secondary spectrum. That interpretation was no longer tenable, so the hypothesis of a non-stellar source of absorption in the system appeared (and still does appear) inescapable. This thought led me to look again at F.B. Wood's (1950) idea of giant prominences exerting an effect on the orbital motion of the late-type component and thus changing the period. Clothed in more mathematical form by Huang (1956) the idea no longer seemed as incredible as I had once thought. At the time it seemed to fit quite well with the history of U CrB, although I would no longer defend all the details of my second attempt to interpret this system (Batten 1964). About the same time, Piotrowski (1964) published an important paper on period changes which also pointed to a possible connection between them and the evolution of binaries by the transfer of mass between their components.

Recently, I have returned to the study of U CrB in collaboration with J. Tomkin. I believe that we have found reliable orbital elements and absolute dimensions for the system (Batten and Tomkin 1981). The secondary spectrum, first detected by Plavec and Polidan (1976), has now been measured to give us individual masses of 4.8 $m_\odot$ and 1.4 $m_\odot$ for the two components. There is no longer any room for doubt that the secondary spectrum is of approximately solar type. The secret of measuring the primary velocity-curve has been to measure the weak lines of ionized "metals" visible in the spectrum. Velocities derived from them lead to a circular orbit, in accord with the light-curve but in contradiction to velocity-curves based on measures of either the hydrogen or helium lines. We can therefore determine the residuals of velocities determined from these latter lines with respect to what we may reasonably assume to be the true velocity curve. The residuals for the hydrogen lines are shown in Figure 2, reproduced from the paper by Batten and Tomkin. Those for the helium lines follow a similar pattern but are not so large and might not be thought significant if we considered them in isolation. The trend shown in Figure 2 is remarkably like that shown schematically by Struve (1949). The residuals are close to zero (perhaps slightly negative) at all phases except between about $0^P\!.7$ and primary mid-eclipse (zero phase). From around $0^P\!.72$ to $0^P\!.83$, the residuals increase almost linearly to a maximum value of around + 30 km s^{-1} and then decrease, again approximately linearly and symmetrically until close to zero phase. The absence of any strong rotation effect in U CrB helps to show this trend more clearly than it can be seen in other systems, but even U Cep would show a similar trend somewhat later in phase. The obvious interpretation is essentially that first put forward by Struve. There is a stream of gas, presumably originating in the secondary component, descending onto the primary star from the direction defined by our line of sight to the system at or about phase $0^P\!.83$. The state of excitation in this stream is similar to that of the B-type atmosphere since both helium and hydrogen absorption lines occur in it (even the dilution-sensitive Mg II line, λ 4481, may be weakly present) but nevertheless not quite the same since the helium lines are weaker relative to the hydrogen lines in the stream spectrum than they are in the stellar spectrum. From the range of phases over which the velocity-curve is distorted, one shoud be able to deduce the approximate cross-section of the stream.

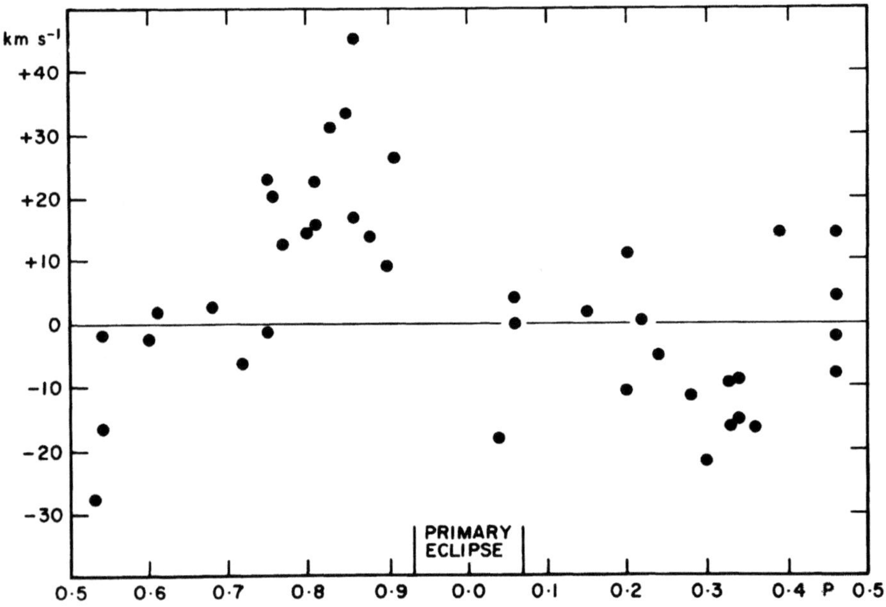

Figure 2. Residuals of hydrogen-line velocities from the velocity-curve of U CrB defined by measurements of "metallic" lines.

V. AN ALTERNATIVE INTERPRETATION

In two recent papers, Kopal (1980a, 1980b) has suggested an alternative interpretation of the distortion of the velocity-curves of Algol-type systems. Although he applies his theory to U Cep rather than to U CrB, the systems are in many respects similar. If I understand him correctly, Kopal accepts that the velocity-curves are indeed distorted and that the distortion arises because the spectral lines are formed in matter moving with respect to the centre of mass of the star. He suggests, however, that this matter is still confined within the atmosphere of the primary star, moving at less than the velocity of escape, and that the distortion of the velocity-curve is thus caused by strong radial currents within the atmosphere. He derives a formula (equation 3.18 in 1980b) for ΔV - the difference between the observed and true velocities of the primary star and shows that even for U Cep, with its highly distorted velocity curve, ΔV never exceeds 20 km s^{-1} to 40 km s^{-1}. As I understand him, he equates ΔV to the speed of the atmospheric currents themselves.

Now ΔV is precisely the quantity plotted for U CrB in Figure 2, and we may note in passing that it is of the same order of magnitude for this system as for U Cep. On Kopal's interpretation, therefore, we would say that around phase $0^P\!.83$ we are observing a strong downward current in the atmosphere, but at other phases we do not see any strong radial currents. One might perhaps argue that such a downward current is the resultant of heating by the companion star and rotation of the primary star, and that there is no need to postulate infall of matter from outside the primary. I suspect that a proper mathematical treatment of such an idea would prove as intractable as, so far, has a proper treatment of the motion of gas streams. One might wonder, too, why the distortion of the velocity-curve of the rapidly rotating primary of U Cep is so similar to that of the more slowly rotating primary of U CrB. There is, moreover, one situation in which we know that we are looking at matter between the two components of the system: when during the total phase of primary eclipse in systems like U Cep and RW Tau we see hydrogen emission lines. These emissions must originate in regions lying above the B-type star as Joy (1942) first pointed out. Observations at Hα have recently shown such "emission disks" to be a common property of Algol-type systems (Peters 1980).

The chief difficulty, however, is that it is almost impossible to separate out from any observed line profile the stellar and (for want of a better word) stream components. Struve used to write of a sharp stream component superimposed on a typical stellar line. I have measured plates of both U Cep and U CrB that present this appearance, but I believe it to be illusory, arising from the eye's sensitivity to changes of intensity gradient or chance grain effects, and only to a minor degree corresponding to the actual profile. One can only proceed by assuming some model and trying to interpret the observations within its limitations. Of course the model may be wrong, but if it is, that will sooner or later be made obvious by the inconsistencies between theory and observation. Figure 3a shows schematically how a shallow broad "stream" line, centred on a quite different velocity from the stellar line may have only a small effect on the velocity measured for the latter. I think that such broad shallow lines are more common than the sharp stream lines Struve wrote of because, as Kopal (1978) has urged, the flow in any gas streams that may exist is probably highly turbulent. Figure 3b illustrates another situation: the stream is seen in emission when it is moving almost across the line of sight. This time it distorts the line profile in such a way as to change the observed velocity of the star even though the stream has only a very small radial velocity with respect to us. This situation must exist in Algol systems at phases near quadratures. There is, in fact, no simple or unique relation between the distortion of the velocity-curve ΔV and the speed of matter in any streams that may be producing the

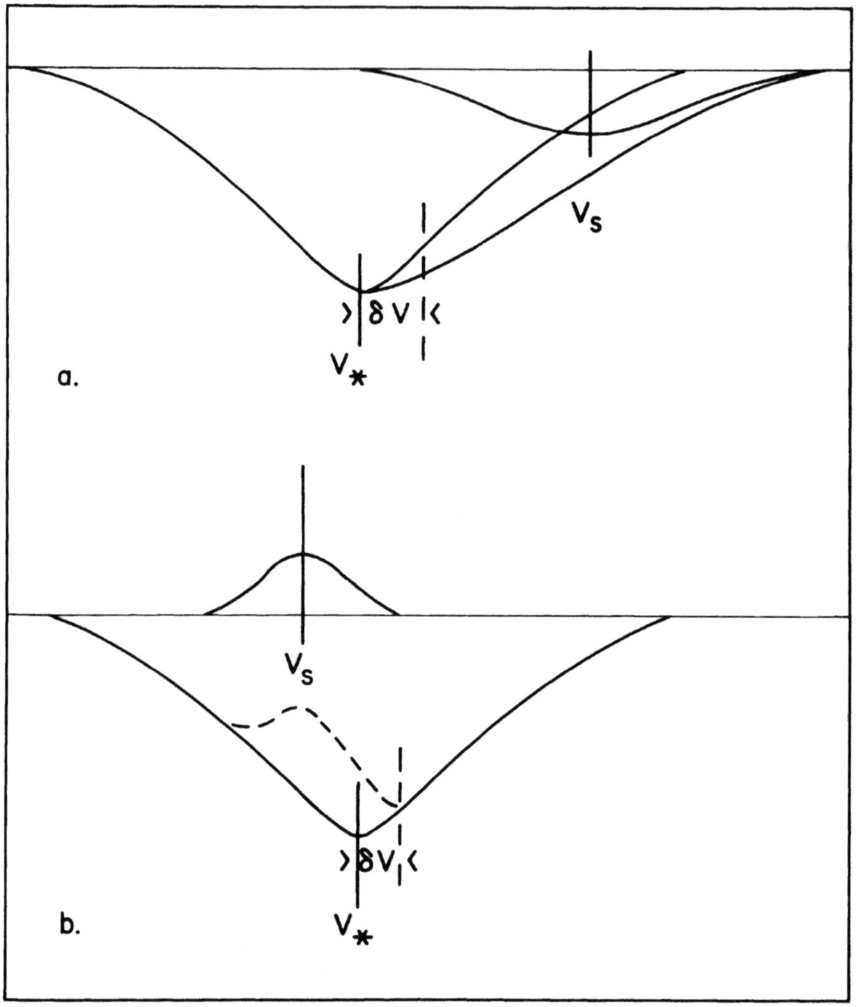

Figure 3. These schematic diagrams illustrate that there is no simple relation between the distortion of a velocity curve (ΔV), the true velocity of the star (V*) and the radial velocity of the stream (V_S). In (a) the stream is seen as a broad absorption at high velocity, making the stellar line slightly asymmetric. Viewed from a different angle (b), the same stream may appear as a relatively narrow emission of low velocity, but it still makes the main part of the stellar line seem to have a more positive velocity than it should have. (Longer wavelengths are to the right in both sketches).

distortion. Streams moving at velocities of several hundred kilometres per second may produce values of ΔV that are much smaller. Again, in the only situation in which we can measure stream emission-lines directly, that is during total eclipse of the primary star, we find velocities of precisely that order of magnitude.

VI. OTHER EVIDENCE OF CIRCUMSTELLAR MATTER

At this point we should remind ourselves that spectroscopic evidence, still less the distortion of velocity-curves, is not the only evidence we have for circumstellar matter in binary systems. Much of the resistance (which only the older workers in the field seem to remember) to Struve's ideas arose from the fact that there appeared to be little or no corroborative evidence for them independent of spectroscopy. It was even questioned whether or not such streams as he postulated could affect the light-curve (see the discussion after a paper on pp. 173-4 in the reference cited as Kopal 1957). The gross distortions of the light-curve of U Cep observed in recent years by several investigators must now have removed all doubt on that score. Olson (1976, 1978, 1980a, 1980b) has made the most thorough study of these distortions and finds them fully consistent with an optically thick stream between the components. Accompanying spectroscopic phenomena (Batten et al. 1975, Plavec and Polidan 1975) and a period change (Crawford and Olson 1979) are well attested. This outburst also demonstrated that the UV excesses observed in so many Algol systems must be partly due to circumstellar matter. As late as 1972 it was possible to argue that these excesses might be caused by a deficiency of metals in one of the component stars (Hall 1973) but, during the outburst, the UV excess of U Cep varied by a factor of two in a few days - which can hardly be explained by a changing metal content. Plavec (1980) has found a strong UV continuum in the spectrum of SX Cas, already known to show shell-like features. This strongly suggests a connection between UV excesses and circumstellar matter.

Neither should we forget the radio observations of Algol and β Lyrae (Wade and Hjellming 1972). The radiation from Algol was shown to be variable and sometimes non-thermal in origin (Hjellming, Wade, and Webster 1972). It is not, unfortunately, possible to pinpoint just where it comes from in the system, but it must come from an extended region (Hjellming, Webster and Balick 1972). In some ways it is a pity that the RS CVn group of systems has proved so interesting as radio sources, since it has deflected attention from Algol itself. It is also a pity that most Algol-type systems are so far away. No certain radio radiation was detected from U Cep during the 1974 outburst, although I am convinced that had this system been no further from us than Algol is, we would have learned much from radio observations.

Since this discussion concerns Algol-type systems, I have deliberately avoided introducing arguments from observations of X-ray binaries containing compact components. It is relevant, however, that

Algol itself sometimes gives off X-rays (Schnopper et al. 1976). I have also deliberately avoided discussing Be and shell stars, in which spectroscopic evidence for gas streams is much clearer, although in some of these systems the streams may be largely removing matter from the system (Polidan and Peters 1980). Much of the attraction of mass-transfer theory is precisely that it seems to be able to explain a wide variety of systems that would otherwise all appear anomalous in different ways.

VII. IS MASS TRANSFERRED?

There is no need to labour further the point of the last section: little doubt can remain that the Algol-type systems are one of several groups of binaries that show evidence of containing appreciable amounts of interstellar matter. The only questions to ask are: what is this matter doing and is it sufficient to account for mass transfer?

As Kopal (1978) has pointed out, the matter, once in the region between the component stars, can do one (or more) of only four things: (i) leave the system, (ii) return to the star from which it came, (iii) move over to the other star, or (iv) remain where it is in some kind of quasi-equilibrium. Quite probably some matter does each of these things; we are concerned to know something about the proportions. Most people will probably agree that (ii) and (iv) are rather unimportant. Smak (1980) has pointed out that one reason why mass-transfer calculations reproduce observed systems tolerably well is that the evolution of the mass-losing star is unaffected by the ultimate fate of the mass it loses. This would not obviously remain true if a substantial portion of the mass the star ejected eventually returned. Although I have suggested some matter does this in U Cep (Batten 1974) it is very unclear how much is involved. As for (iv), the speed with which emitting rings are observed to collapse after an outburst suggests that the storage of any substantial amounts of mass between the components is a strictly temporary measure. This conclusion finds some theoretical support in the work of Gorbatskij (1969). It seems likely, then, that (i) and (iii) are the most probable fates of circumstellar matter in Algol-type systems. That some matter escapes is virtually certain. Kondo et al. (1980) have published evidence of mass leaving U Cep; whether or not the amount is anything like the 50 per cent or more required by de Grève's theoretical work is still (and likely to remain) an open question.

Nevertheless, matter escaping from a star that fills its Roche lobe is presented with a fairly large target area by the companion star and some of it, at least, must be picked up by that star. We can, of course, argue whether the Roche lobe represents the shape of any real star, but I believe this argument is a red herring. There must be some limit beyond which, if a binary component expands, its outermost layers cease to be identifiable with that star alone. Observation shows that one cross-section of real secondaries in Algol systems never significantly exceeds, and rarely (if ever) is much less than, the corresponding cross-section of the Roche lobe. To say that the Roche lobe is a good approximation to the real limiting surface does not imply that we believe any actual star to have the neat pointed egg-shape we so confidently draw in our diagrams. We may then reasonably assume that the circumstellar matter in Algol-type systems is coming from the (presently) secondary component and the question posed in the title of this section reduces to: is any substantial amount of this matter collected by the primary?

We should consider first how much matter is in the stream. Consider, again, U CrB, and grant, for the sake of argument, that the distortion of the velocity curve illustrated in Figure 2 is caused by a gas stream falling on to the primary star from the same direction as our line of sight near phase $0^{P}.83$ (that is, making an angle of about 60° with the line joining the two stars). In section 4 it was argued that the spectral characteristics of the stream indicate much the same state of excitation (and density) as the atmosphere of the B-type primary - about 10^{15} atoms cm^{-3} (Allen 1976). I have shown elsewhere (Batten 1973) that the particle density n of a stream, where it falls onto the primary star, is related to the rate of mass-transfer 10^{-x} $m_\odot$ yr^{-1} by

$$n = 0.25 \times 10^{21-x}$$

if the terminal velocity of the particles falling on the B-type star is 100 km s^{-1}, and if the area of the star's surface covered by the stream is that of a disk having the same radius as the Sun. For $n = 10^{15}$ we find, under these assumptions, a rate of mass-transfer of about 4 x 10^{-6} $m_\odot$ yr^{-1}. There is no direct evidence for the velocity of the stream in U CrB: the value of 100 km s^{-1} was selected as characteristic of Algol systems in general. We can estimate the cross-sectional area of the stream from the fact that its effects first become visible at phase $0^{P}.72$ and persist (as far as we can tell during primary eclipse) for an approximately equal interval after phase $0^{P}.83$. This means that over an interval of about 80° in phase enough of the surface of the star is covered by the stream for the measured velocity to be affected by it. One diameter of the stream, therefore, is approximately equal to a chord subtending an angle of about 80° at the centre of the B-type star. As the radius of this star is 3.3 $R_\odot$ (Batten and Tomkin 1981), this diameter itself is over 4 $R_\odot$. If the stream had a circular cross-section, its area would be four times that assumed above and the corresponding rate of mass-transfer over 10^{-5} $m_\odot$. Several assumptions are made in reaching this figure that can fairly be questioned, but we actually have one or even two orders of magnitude in hand. There is no detectable monotonic period increase in U CrB and no-one would suppose that it is in the rapid phase of mass-transfer. A stream capable of transferring 10^{-6} $m_\odot$ yr^{-1} or even only 10^{-7} $m_\odot$ yr^{-1} would be in acceptable agreement with the stage of mass-transfer that U CrB is believed to have reached.

We must also, however, consider the effect that the infall of the matter has on the primary star. If this made some detectable change to the star receiving the mass, it would be more persuasive evidence of the theory of mass-transfer. Here, it must be admitted, is a weak point in the theory. Although, as Kopal charged, the evolution of the mass-receiving component has been, relatively speaking, ignored, some progress has been made (Webbink 1980) and it suggests that it is difficult to produce a system by mass-transfer in which the receiving component remains on the main sequence. The often-repeated statement that Algol-type primaries are main-sequence stars, however, rests on the comparison of their mass and luminosity. I have encountered difficulties in reconciling the spectral types of both U Cep A and U CrB A, as derived from the line ratios, with the observed colours. The effects - about one sub-class - may seem small and not to be compared with the departures Webbink is predicting. Nevertheless, the temperature range from B0 to B9 is nearly three times as great as that from A0 to the end of the main sequence, and one subclass for a B-type star is not a small discrepancy. Very careful photometry and spectrophotometry of the primaries of Algol systems and some standard single B-type stars might provide some clue whether or not (and how) the former are adjusting to infalling matter.

We might return now to the question why we do not see any systems in the sky that we can clearly recognize as being in the rapid phase of mass-transfer. One answer might be to point to β Lyr. If our current interpretation of the system is correct it is the answer. In this system, the circumstellar matter is concealing the component we believe to be gaining mass: indeed, large quantities of optically thick material in a system may make it less conspicuous rather than more. Yet β Lyr is a conspicuous object and spectroscopically, at least, not without some similarities to a nova. The nova that is so conspicuous has a high rate of mass loss, for it divests itself of the small fraction of its outer layers in a few days. The rate of 10^{-3} m$\odot$ yr^{-1} corresponding to a nova would be attained only by massive systems undergoing mass-transfer. Moreover it is not only the rate of mass loss (or transfer) that must be considered, but also the manner. Novae are conspicuous because thermonuclear reactions are involved.

Perhaps an ordinary terrestrial waterfall provides a good analogy to this problem of mass-transfer. No one waterfall is like another, and any given one changes with time. Constantly changing eddies modify a fall's appearance as do boulders or temporary obstructions. Sometimes a boulder may deflect a jet of water upwards again, or even out of the watercourse altogether. Certainly much water will be sent high in the air as spray - especially in great falls like Niagara. Confronted with a dry watercourse interrupted by a precipice, none of us would dream of predicting the detailed appearance of the fall in flood, still less the trajectories of individual drops of water. Nevertheless, we would all fairly confidently predict that the water, when it appeared, would not turn round at the edge of the precipice and go back the way it had come. Our confidence, of course, would be based on the belief that, for all the complications, gravity is still the dominant force acting on the water. Kopal's critique of mass-transfer is essentially an assertion that gravity is not the dominant force on the stream, even after it has left the mass-losing star. If it is, then at some stage in the evolution of a binary mass must be transferred from one star to another. If it is not (and certainly gravity is not the only force), then Kopal is quite right in claiming that the necessary equations have not been solved and, quite possibly, are beyond our ability to solve. Nevertheless, such attempts as have been made to generalize the over-simplifications of particle dynamics (e.g. Prendergast and Taam 1974, Lubow and Shu 1975) suggest that the naive supposition that a massive star will attract matter in its vicinity is not very wide of the mark. In one of his famous stories, Conan Doyle made Sherlock Holmes claim that, given one drop of water, it should be possible to deduce the existence of Niagara Falls. I think that Sherlock Holmes overestimated his powers of deduction, but he did not have to argue with sceptics: he could take them to Niagara, or Iguacu, or the Victoria Falls. Similarly astronomy is not only a matter of solving equations: we can look at the stars, at Algol, U Cep or β Lyr. We may have overestimated our powers of deduction in thinking our theories explain completely the origin of these systems, but what other theory makes sense of as much of the observational data?

Sometimes, I reflect ruefully that the first interplanetary probes showed us to have been so mistaken about our close neighbours Mars and Venus, that we can scarcely believe anything we try to deduce about the stars. As for what happened in "the first 10^{-30} seconds", I entirely share Kopal's scepticism. Perhaps the more recent probes to Jupiter and Saturn provide more cause for hope. The moons of Jupiter and the rings of Saturn have been shown to be richer and more interesting objects than we imagined. Our ideas about them have been significantly changed, but at least these objects were not optical illusions created by our telescopes - as Galileo's critics alleged. I do not believe gas streams are optical illusions created by our spectrographs. If ever a probe is sent to Algol (or better U Cep), I am prepared to bet that it will find gas streams and confirm their role in mass-transfer. Our fault will have been not that our basic ideas were hopelessly wrong, but that we did not imagine the half of it.

The theory of mass-transfer and the interpretation of spectroscopic observations by gaseous streams seem to me a prime example of a paradigm as defined by Kuhn (1970). The old paradigm was that each component in a binary was a separate entity interacting with the other only gravitationally (and sometimes radiatively). Struve showed this paradigm to be insufficient, and, as I have indicated, encountered much resistance. Most investigators now accept his framework and are content to do "normal" science within it. Kopal's critique is valuable because it forces us to look again at that basic framework. Having looked at it again, I am not yet ready to abandon it for still another paradigm. There are inconsistencies which we cannot ignore for ever. Why, for example, do U Cep and U CrB show similarly distorted velocity curves but only U Cep shows an (essentially) monotonic period increase? Why are velocity-curves of some systems (U Cep) distorted all the time while emission lines appear only very spasmodically? Why have features in the spectra, now ascribed to gas streams, given nearly correct values of the mass-ratio when interpreted as secondary spectra (U CrB)? Can we be sure, from only a century or two of observation, that monotonic period increases such as those observed in β Lyr and U Cep really reflect the behaviour of these systems over tens of thousands of years? (The recent claim - Kviz and Murray 1980 - that the period decrease of SV Cen has stopped should give us all pause.) Why are there explosive outbursts as observed in U Cep if the underlying mechanism is a secular expansion? This perhaps is the easiest question to answer. Nature may not make leaps, but she frequently proceeds by fits and starts.

The questions just raised may seem trivial to most people now, but they are possible indicators that not everything will fit into the mass-transfer paradigm, which may in its turn be superseded. But there are different kinds of scientific revolutions: Newton's theory of gravitation was succeeded by Einstein's, but not in the same way that the phlogiston theory of combustion was succeeded by the oxygen theory. We may well have to modify and generalize our ideas about the evolution of binary systems and include effects that we have recently neglected (such as, as Kopal has pointed out, tidal effects) but I, for one, shall be very surprised if future generations conclude that mass-transfer plays no role at all, or even only a minor one. Even if they should, I do not see the work of the last quarter-century as wasted, for it has stimulated much research whose results will be useful whatever their ultimate interpretation may be. Much useful work was done in chemistry under the influence of the belief in phlogiston, and in the physics of heat when everyone accepted the reality of caloric.

At the base of all the discussion that Kopal has stirred up by his 1978 book is a difference of opinion as to the nature and purpose of a scientific theory. If we require of a theory that it must be the "true description" of whatever facet of the universe we are studying, then not only mass-transfer in binary systems, but every scientific theory we care to name is under a cloud of suspicion. I suppose that all of us were originally motivated to scientific research by some such hope of discovering a "true description", many of us before we were old enough to understand the philosophical difficulties inherent in the concept. Truth is not the only characteristic we require of a scientific theory, however. We also want it to integrate and make sense of a number of diverse facts and to suggest further lines of inquiry (to be fruitful). When a theory fulfils these other two requirements, we are inclined to accept it as true until it can be clearly demonstrated to be false. No theory, ultimately, will survive such a clear demonstration, but many are given the benefit of a very considerable amount of doubt. I believe this attitude can be justified: certainty of knowledge is not vouchsafed to human beings, especially of knowledge derived from sense data. If we can never be certain that our theories are true, a degree of pragmatism on the part of the working scientist is justified. What works and is not clearly false may be accepted as a useful hypothesis to guide further investigations. It almost certainly will be modified, for history has shown that whenever scientists divide sharply on an issue into two schools of thought, only rarely is either side completely wrong. To end on a note on which we can all agree, let me once more quote Plavec (1973): If we eventually fail to explain the observed systems, we will have to change the theory.

REFERENCES

Allen, C.W. 1976, "Astrophysical Quantities", 3rd Ed. (corrected) London, The Athlone Press, p. 213.
Batten, A.H. 1956, Mon. Not. R. Astr. Soc., **116**, p. 552.
_____. 1964, Quarterly J. R. Astr. Soc., **5**, p. 145.
_____. 1973, "Binary and Multiple Systems of Stars", Oxford, Pergamon Press, p. 210.
_____. 1974, Pub. Dominion Astrophys. Obs., **14**, p. 191.
Batten, A.H., Baldwin, B.W., Fisher, W.A., and Scarfe, C.D. 1975, Nature, **253**, p. 174.
Batten, A.H., and Tomkin, J. 1981, Pub. Dominion Astrophys. Obs., **15**, 419.
Catalano, S., Cristaldi, S., and Lacona, G. 1966, Mem. Soc. Astr. Ital. **37**, p. 213.
Crawford, J.A. 1955, Astrophys. J., **121**, p. 71.
Crawford, R.C., and Olson, E.C. 1979, Pub. Astr. Soc. Pacific, **91**, p. 413.
de Grève, J.P. 1980, Astrophys. Space Sci., **72**, p. 411.
Gorbatskij, V.G. 1969, Astrophys. Space Sci., **3**, p. 179.
Hall, D.S. 1973, in "Extended Atmospheres and Circumstellar Matter in Spectroscopic Binary Systems" (IAU Symposium 51) ed. A.H. Batten, Dordrecht, D. Reidel, p. 193.
Haslegrove, C.B., and Hoyle, F. 1956, Mon. Not. R. Astr. Soc., **116**, p. 515.
Hjellming, R.M., Wade, C.M., and Webster, E. 1972, Nature (Physical Science), **236**, p. 43.
Hjellming, R.M., Webster, E., and Balick, B. 1972, Astrophys. J., **17**, p. L139.
Horn, J. 1971, Bull. Astr. Inst. Csl., **22**, p. 37.
Huang, S.-S. 1956, Astr. J., **61**, p. 49.
Joy, A.H. 1942, Pub. Astr. Soc. Pacific, **54**, p. 35.
Kippenhahn, R., and Weigert, A. 1967, Zs. Astrophys., **65**, p. 251.
Kondo, Y.,McCluskey, G.E., and Stencel, R.E. 1980, Astrophys. J., **233**, p. 906.
Kopal, Z. 1955, Ann. Astrophys., **18**, p. 379.
_____. 1956, Ann. Astrophys., **19**, p. 298.
_____. 1957, in "Non-Stable Stars" (IAU Symposium No. 3) ed. G.H. Herbig, Cambridge, The University Press, p. 123.
_____. 1978, "Dynamics of Close Binary Systems" Dordrecht, D. Reidel, chap. 8.
_____. 1980a, Astrophys. Space Sci., **70**, p. 329.

Kopal, Z. 1980b, Astrophys. Space Sci., **71**, p. 65.
Kuhn, T.S. 1970, "The Structure of Scientific Revolutions" 2nd Edn., University of Chicago Press.
Kvíz, Z., and Murray, K.J. 1980, Comm. 27 IAU Inf. Bull. Var. Stars No. 1864.
Lubow, S.H., and Shu, F.H. 1975, Astrophys. J., **198**, p. 383.
Morton, D.C. 1960, Astrophys. J., **132**, p. 146.
Olson, E.C. 1976, Astrophys. J., **204**, p. 141.
_____. 1978, Astrophys. J., **220**, p. 251.
Olson, E.C. 1980a, Astrophys. J., **237**, p. 496.
_____. 1980b, Astrophys. J., **241**, p. 257.
Paczyński, B. 1967, in "On the Evolution of Double Stars" ed. J. Dommanget, Comm. Roy. Obs. Belgium, Serie B, No. 17, p. 111.
Pearce, J.A. 1936, Pub. Amer. Astr. Soc., **8**, p. 219.
Peters, G.J. 1980, in "Close Binary Stars: Observations and Interpretation" IAU Symposium 88, eds. M.J. Plavec, D.M. Popper, and R.K. Ulrich, Dordrecht, D. Reidel, p. 287.
Piotrowski, S. 1964, Acta Astr., **14**, p. 251.
Plaskett, J.S. 1920, Pub. Dominion Astrophys. Obs., **1**, p. 187.
Plavec, M. 1973, in "Extended Atmospheres and Circumstellar Matter in Spectroscopic Binary Systems" (IAU Symposium 51) ed. A.H. Batten, Dordrecht, D. Reidel, p. 251.
_____. 1980, in "Close Binary Stars: Observations and Interpretation" IAU Symposium 88, eds. M.J. Plavec, D.M. Popper and R.K. Ulrich, Dordrecht, D. Reidel, p. 251.
Plavec, M., and Polidan, R.S. 1975, Nature, **253**, p. 173.
_____. 1976, in "Structure and Evolution of Close Binary Systems" (IAU Symposium 73) eds. P. Eggleton, S. Mitton, and J. Whelan, Dordrecht, D. Reidel, p. 289.
Polidan, R.S., and Peters, G.J. 1980, in "Close Binary Stars: Observations and Interpretation" IAU Symposium 88, eds. M.J. Plavec, D.M. Popper, and R.K. Ulrich, Dordrecht, D. Reidel, p. 293.
Popper, D.M. 1973, Astrophys. J., **185**, p. 265.
_____ . 1980, Ann. Rev. Astr. Astrophys., **18**, p. 115.
Popper, K.R. 1959, "The Logic of Scientific Discovery" revised 1968 (English translation of "Logik der Forschung" Vienna 1934) Harper Torchbook Edition, New York.
Prendergast, K.H., and Taam, R.E. 1974, Astrophys. J., **189**, p. 125.
Refsdal, S., and Weigert, A. 1971, Astr. Astrophys., **13**, p. 367.
Refsdal, S., Roth, M.L., and Weigert, A. 1974, Astr. Astrophys., **36**, p. 113.
Sahade, J., and Struve, O. 1945, Astrophys. J., **102**, p. 480.
Schnopper, H.W., Delaville, J.P., Epstein, H., Helmken, H., and Murray, S.S. 1976, Astrophys. J., **210**, p.L75.
Smak, J. 1962, Acta Astr., **12**, p. 28.
_____. 1980, in "Close Binary Stars: Observations and Interpretation" (IAU Symposium 88) eds. M. Plavec, D.M. Popper and R.K. Ulrich, Dordrecht, D. Reidel, p. 577.
Struve, O. 1949, Mon. Not. Roy. Astr. Soc., **109**, p. 487.

Tomkin, J. 1981, Astrophys. J. (in press).
Tomkin, J., and Lambert, D.L. 1978, Astrophys. J., **222**, p. L119.
Wade, C.M., and Hjellming, R.M. 1972, Nature, **235**, p. 270.
Walter, K. 1931, Schrift. Konigsberger, Gelehrten Gesells. (Naturw. Klasse), 8, p. 55.
Webbink R.F. 1980, in "Close Binary Stars: Observations and Interpretation" (IAU Symposium 88) eds. M. Plavec, D.M. Popper and R.K. Ulrich, Doredrecht, D. Reidel, p. 127.
Wood, D.B. 1958, Astrophys. J., **127**, p. 351 and 128, p. 595.
Wood, F.B. 1950, Astrophys. J., **112**, p. 196.
Ziółkowski, J. 1969, Astrophys, Space Sci., **3**, p. 14.

AN ANALYTICAL APPROACH TO CONTACT BINARY STABILITY

John Hazlehurst
Hamburg Observatory, Hamburg-Bergedorf

ABSTRACT

It is argued that the Laplace transformation affords a convenient means of investigating the secular stability of contact binaries.

1. INTRODUCTION

Writing on the 'unsolved problem' of the nature of the W UMa stars, Kopal (1959) remarked that:-

"Notwithstanding a considerable amount of speculation in this field, an acceptable solution of this problem does not yet appear to be anywhere within sight."

One might be tempted to echo these sentiments even at the present time. Indeed, despite much effort and innovation in the intervening twenty years, the problem of understanding the existence and properties of the W UMa stars still appears to be a near-intractable one.

In view of this situation, it seems appropriate to concentrate on developing methods of analysis which at least promise to shed some light on certain aspects of the W UMa problem.

Contact binary stability theory was developed with this aim in mind. In particular, stability theory is one means - but of course not the only means - of testing various rival hypotheses concerning the nature of the W UMa stars. For this reason it seems worthwhile to go into stability theory in some detail.

F. D. Kahn (ed.), Investigating the Universe, 227–240.

2. AIMS OF PRESENT DISCUSSION

In this paper we shall be concerned exclusively with the problem of the secular stability of contact binaries. Hydrostatic equilibrium is always assumed to hold. The two component stars share a common envelope in which both mass and energy transfer can occur; the latter takes place by means of convective-type motions.

An instability would correspond to a situation in which a small perturbation in the transferred mass or energy would ultimately grow 'without limit'. This can of course only occur within the framework of a linear theory, but it should not be forgotten that it is the purely linear theory which determines whether or not the system is subject to the instability.

The problem of the secular stability of contact binaries has already been considered by Hazlehurst and Refsdal (1980), hereafter referred to as HR. The aim of the present paper is to complement this previous work, which was largely numerical, by introducing certain analytical methods. As in HR, we shall restrict our discussion to unevolved contact binaries.

It is hoped that this analytical approach will help in understanding why instability (or stability) prevails in any particular case.

3. THE BASIC EQUATIONS

Provided we restrict our considerations to small changes of the various physical quantities, Eqs. 1-4, 6, 8-12 and 20 of HR can be combined to give:-

$$\alpha \dot{E} + \beta \dot{M} = -\psi_{R_p} * \dot{E} - \phi_{R_p} * \dot{M} + \epsilon_1(t) \tag{1}$$

$$\gamma \dot{E} + \delta \dot{M} = \psi_{R_s} * \dot{E} + \phi_{R_s} * \dot{M} + \epsilon_2(t) \tag{2}$$

where the constants on the L.H.S. have the values:-

$$\alpha = \frac{f_1(q)\, d_0}{m D L_0 R_p} \qquad \beta = \frac{f_2(q)}{M_p} \tag{3, 4}$$

$$\gamma = \frac{f_3(q)\, d_0}{m D L_0 R_s} \qquad \delta = \frac{f_4(q)}{M_s} \tag{5, 6}$$

and the meaning of the symbols is as follows:-

a) $\dot{M}(t)$, $\dot{E}(t)$ are the unknowns to be determined; they are, respectively, the mass transfer rate and the variation in the energy transfer rate between the two components. The quantity

$$M(t) = \int_0^t \dot{M}(\tau)\, d\tau \tag{7}$$

is the total mass transferred up to the time t.

b) $\epsilon_1(t)$, $\epsilon_2(t)$ are arbitrarily assigned applied perturbations; the precise form of these functions is unimportant when we are simply concerned with the stability of the system represented by Eqs. 1 and 2.

c) The functions Ψ, ϕ are the so-called stellar response functions (see HR); they are assumed to be known or given. The suffix R_p, R_s denotes "stellar radius" where, as elsewhere, p = primary, s = secondary.

d) The symbol $A * B$ is introduced for convenience to denote the convolution:-

$$A * B = \int_0^t A(t-\tau)\, B(\tau)\, d\tau \tag{8}$$

e) Certain constants have the following meanings:-

d_o = initial depth of contact

DL_o = rate of energy exchange (transferred luminosity) in the unperturbed system

R_p, R_s, M_p, M_s = initial radii and masses of the components (p = primary, s = secondary)

f) The dimensionless constant m represents the sensitivity of energy exchange to changes in contact depth:-

$$m = \frac{\partial \log DL}{\partial \log d} \qquad\qquad DL = DL_o + \dot{E} \tag{9, 10}$$

In taking the partial derivative, the entropy difference ΔS between the components (see HR) is held constant.

g) The dimensionless quantities $f_1 - f_4$ are known, analytical functions of the initial mass-ratio:-

$$q = M_s/M_p \tag{11}$$

so that in Eqs. 3-6 the $f_1 - f_4$ are simply given constants.

The above equations form the starting-point of the present discussion. The problem, in its general form, is to determine $\dot{M}(t)$, $\dot{E}(t)$ given the ϵ's, the Ψ's, the ϕ's and the various constants.

Mathematically, the problem is essentially one of finding an appropriate analytical technique for achieving a deconvolution of the integrals on the R.H.S. of Eqs. 1 and 2.

4. THE FORMAL SOLUTION

We first note that Eqs. 1 and 2 can be simplified by using the identity:-

$$M = \int_0^t \dot{M}(\tau)\,d\tau = 1 * \dot{M} \tag{12}$$

Equations 1 and 2 then take the form:-

$$\alpha \dot{E} = -\Psi_1 * \dot{E} - \phi_1 * \dot{M} + \epsilon_1(t) \tag{13}$$

$$\gamma \dot{E} = \Psi_2 * \dot{E} + \phi_2 * \dot{M} + \epsilon_2(t) \tag{14}$$

where:-

$$\Psi_1 = \Psi_{R_p} \qquad \Psi_2 = \Psi_{R_S} \tag{15, 16}$$

$$\phi_1 = \phi_{R_p} + \beta \qquad \phi_2 = \phi_{R_s} - \delta \tag{17, 18}$$

We shall refer to ϕ_1, ϕ_2 as 'modified response functions'.

A formal solution of Eqs. 13 and 14 can be obtained with the help of the Laplace transformation:-

$$\bar{F}(p) = \mathcal{L}(F(t)) = \int_0^\infty e^{-pt} F(t)\,dt \tag{19}$$

where F(t) is an arbitrary function. Here and elsewhere we shall use bars to denote the transformed functions.

Taking the Laplace transforms of both sides of Eqs. 13 and 14 and using the well-known deconvolution property (see e.g. Doetsch 1955, p. 24) of these transforms:-

$$\mathcal{L}(A * B) = \mathcal{L}(A)\mathcal{L}(B) \equiv \bar{A}(p)\,\bar{B}(p) \tag{20}$$

we obtain:-

$$\alpha \mathcal{L}(\dot{E}) = -\bar{\Psi}_1(p)\mathcal{L}(\dot{E}) - \bar{\Phi}_1(p)\mathcal{L}(\dot{M}) + \bar{\epsilon}_1(p) \tag{21}$$

$$\gamma \mathcal{L}(\dot{E}) = \bar{\Psi}_2(p)\mathcal{L}(\dot{E}) + \bar{\Phi}_2(p)\mathcal{L}(\dot{M}) + \bar{\epsilon}_2(p) \tag{22}$$

which, on eliminating $\mathcal{L}(\dot{M})$ gives:-

$$\mathcal{L}(\dot{E}) = \frac{\bar{\Phi}_2(p)\,\bar{\epsilon}_1(p) + \bar{\Phi}_1(p)\,\bar{\epsilon}_2(p)}{\gamma\bar{\Phi}_1(p) + \alpha\bar{\Phi}_2(p) - \bar{\Phi}_1(p)\bar{\Psi}_2(p) + \bar{\Psi}_1(p)\bar{\Phi}_2(p)} \tag{23}$$

and the desired solution $\dot{E}(t)$ is then found by taking the inverse transform of the R.H.S. of Eq. 23. This therefore represents the formal solution of Eqs. 13 and 14.

5. THE STABILITY PROBLEM

If we are merely interested in the question as to whether the system is stable or not, the form of the applied perturbations $\epsilon_1(t)$, $\epsilon_2(t)$ is of course immaterial. It then suffices (see e.g. Doetsch 1955, p. 287) to investigate the roots of the equation

$$\bar{\Phi}_1(p)\bar{\Psi}_2(p) - \bar{\Psi}_1(p)\bar{\Phi}_2(p) = \gamma\bar{\Phi}_1(p) + \alpha\bar{\Phi}_2(p) \tag{24}$$

corresponding to the condition that the denominator of the R.H.S. of Eq. 23 vanishes.

In this connection it must however be noted that all roots of this equation (real and complex values of p) should in principle be taken into consideration. The stability of the system is then decided by the sign of the real part of the root (or roots) lying farthest to the right in the complex plane. A positive sign indicates instability, a negative one stability.

When discussing the solutions of Eq. 24, considerable simplifications are found to ensue when: -

a) The functions $\Psi(t)$, $\phi(t)$ vary only slowly with time

b) The constants α, γ have relatively small values.

With certain qualifications to be discussed later, both conditions a) and b) are satisfied in the systems in which we are interested.

Under the above conditions the complex roots of Eq. 24 either disappear or play a subordinate rôle (i.e. are not of decisive importance for the stability issue). It then suffices to investigate the solutions of Eq. 24 for real values of p only.

A further simplification arising when Ψ, ϕ vary only slowly with time is that the Laplace transforms $\bar{\Psi}$, $\bar{\Phi}$ can be readily calculated from a simple approximate formula. The procedure for doing this will be described in the following section.

6. LAPLACE TRANSFORMS OF SLOWLY-VARYING FUNCTIONS

Let us assume that F(t) is a slowly-varying function of time, and that we wish to determine its transform $\bar{F}(p)$. For this purpose it is convenient to introduce a change of variable :-

$$\tau = \log_e t \qquad\qquad f(\tau) = F(t) \tag{25, 26}$$

so that, with the help of Eq. 19, we find:-

$$\bar{F}(p) = \int_{-\infty}^{+\infty} e^{-pe^{\tau}+\tau} f(\tau)\, d\tau \tag{27}$$

The weighting function:-

$$e^{-pe^{\tau}+\tau}$$

has a fairly sharp maximum at

$$\tau = \tau_0 = \log_e 1/p \tag{28}$$

and is small elsewhere. Thus, in a first approximation we could take the slowly-varying function $f(\tau)$ outside the integral, evaluating this function at $\tau = \tau_o$ where the integrand is significantly different from zero. The evaluation of the remaining integral

$$\int_{-\infty}^{+\infty} e^{-pe^{\tau}+\tau} \, d\tau = \frac{1}{p} \tag{29}$$

is then relatively straightforward.

This suggests that, in order to improve the approximation, we assume that $f(\tau)$ can be developed in a Taylor series:-

$$f(\tau) = f(\tau_o) + (\tau-\tau_o)\,\phi'(\tau_o) + \frac{1}{2}(\tau-\tau_o)^2\phi''(\tau_o) + \ldots,$$

about the point:-

$$\tau_o = \log_e \frac{1}{p} \tag{31}$$

Substitution of the above series in Eq. 27, followed by a term-by-term integration, and finally a transformation back to the original time variable, then gives:-

$$\bar{F}(p) = \frac{1}{p} F\left(\frac{1}{p}\right) - \frac{C}{p}\left[\frac{dF(t)}{d\log_e t}\right]_{t=\frac{1}{p}} + \ldots \tag{32}$$

where C is Euler's constant:-

$$C = 0,5772 \ldots \tag{33}$$

and the series on the R. H. S. of Eq. 32 is a function of p only.

In principle, higher members of the series on the R. H. S. of Eq. 32 (corresponding to higher derivatives of F(t)) should be considered; however, if we are mainly interested in a rough estimate of the extent to which F(t) may be considered to be 'slowly varying', we can say:-

The function F(t) can be taken to be slowly-varying when:-

$$\left[\frac{dF}{d\log_e t}\right]_{t=\frac{1}{p}} \ll \frac{1}{C} F\left(\frac{1}{p}\right) \tag{34}$$

or, reverting to the original time variable, and converting from natural to common logarithms,

$$\left[\frac{dF}{d\log_{10} t}\right] \ll \frac{\log_e 10}{c} F(t) \tag{35}$$

which can be re-written:-

$$\frac{dF}{d\log_{10} t} \ll 4 F(t) \tag{36}$$

When this condition is satisfied, we may approximate the Laplace transform of F(t) by the expression:-

$$\bar{F}(p) \simeq \frac{1}{p} F\left(\frac{1}{p}\right) \tag{37}$$

corresponding to the first term of the series expansion on the R.H.S. of Eq. 32.

7. ARE THE RESPONSE FUNCTIONS SLOWLY-VARYING ?

The stability equation, Eq. 24, contains the Laplace transforms of the response functions $\Psi(t)$ and the modified response functions $\phi(t)$. Moreover, in section 6 we argued that the calculation of the Laplace transform becomes considerably simplified when the functions are slowly-varying. We now discuss whether, in practice, the functions $\Psi(t)$, $\phi(t)$ can be taken to be of the slowly-varying type.

Using criterion 36, and the response function data given in Hazlehurst, Refsdal and Stobbe (1977) we find that the functions $\phi(t)$ can indeed be considered to be reasonably slowly varying. The functions $\Psi(t)$ (the energy response functions) can also be taken to be slowly varying, except for very large times ($t \to \infty$). This is due to the fact that $\Psi(t)$, but not $\phi(t)$, approaches zero (in roughly exponential fashion) as $t \to \infty$. This qualification of the 'slowly-varying' nature of $\Psi(t)$ is important for systems near to marginal stability. We shall discuss in section 11 the way in which our analysis needs to be modified in such cases.

With the above qualification in mind, we shall now, as a crude approximation, assume the $\Psi(t)$, $\phi(t)$ to be slowly-varying, so that with the help of Eq. 37, we can now write:-

$$\phi_1(\tfrac{1}{p})\psi_2(\tfrac{1}{p}) - \psi_1(\tfrac{1}{p})\phi_2(\tfrac{1}{p}) = p\left\{\gamma\phi_1(\tfrac{1}{p}) + \alpha\phi_2(\tfrac{1}{p})\right\} \tag{38}$$

for the stability equation (Eq 24).

8. DISCUSSION IN TERMS OF THE TIME VARIABLE

For the convenience of those who may be unfamiliar with, or unattracted by, the use of Laplace transforms, it is worth noting that the stability equation (Eq. 38) may be re-expressed in terms of the more 'natural' time variable t, in place of the variable p, in the case when the solutions of Eq. 24 correspond to non-complex p.

Using the results of section 6, we can imagine a 'correspondence'

$$t \rightleftharpoons \frac{1}{p} \tag{39}$$

between the two coordinates. Defining now:-

$$F(t) \equiv \frac{\phi_1(t)\psi_2(t) - \psi_1(t)\phi_2(t)}{\phi_1(t) + \frac{\alpha}{\gamma}\phi_2(t)} \tag{40}$$

and using Eqs. 38 and 39 we find:-

$$F(t) = \frac{\gamma}{t} \tag{41}$$

as the new form of the stability equation.

9. THE CONSTANTS α AND γ

The further discussion of the stability equation (Eq. 41) requires some knowledge of the constants α and γ. These constants (see Eqs. 3 and 5) may be written:-

$$\alpha = \Gamma^{-1} f_1(q) \qquad \gamma = \Gamma^{-1} f_5(q) \qquad f_5 = \frac{R_p}{R_s} f_3 \tag{42,43,44}$$

where the constant Γ is defined as:-

$$\Gamma = \left\{ \frac{m \, DL_0}{d_0} \right\} \times R_p \tag{45}$$

The bracketed quantity on the R.H.S. of Eq. 45 is simply the change in transferred luminosity, per unit change in contact depth. It has the very high value:-

$$\frac{m \, DL_0}{d_0} \approx 300 L_\odot \text{ per stellar radius} \tag{46}$$

due to the relatively small value of the contact depth d_0 . This leads in turn to a very high value for the constant Γ :-

$$\Gamma \approx 300 \, L_\odot \tag{47}$$

Since the mass-ratio functions f_1 etc are all of order unity, the very high value of Γ leads to extremely small values of α and γ (see Eqs. 42 and 43). These values are, indeed, so small in physical terms that no essential error is committed if we consider the formal limits:-

$$\alpha \to 0 \qquad\qquad \gamma \to 0 \tag{48}$$

when discussing the properties of Eq. 41. At the same time, the smallness of α and γ helps to guarantee (see section 5) that the solutions of Eq. 24 will indeed correspond to real (positive or negative) values of p.

Finally, we note that since all quantities on the R.H.S. of Eqs. 3 and 5 are positive, it follows that both constants α and γ are positive. Moreover, the ratio

$$\frac{\alpha}{\gamma}$$

is of order unity (1) and unaffected by the limiting process $\alpha, \gamma \to 0$.

10. THE THREE TYPES OF STABILITY BEHAVIOUR

As already noted in section 9, the constant γ has a very small (positive) value. This means that the function on the R.H.S. of Eq. 41 can be regarded as a delta function when compared with the relatively slowly-varying function on the L.H.S. We can then distinguish three separate possibilities:-

i) $F(0) > 0$

This case clearly corresponds to instability, since the functions $F(t)$ and γ/t will intersect at some positive value $F(t_1)$ where:-

$$F(t_1) \simeq F(0) = \frac{\gamma}{t_1}$$

so that the root $p_1 = \frac{1}{t_1}$ is positive.

ii) $F(t) < 0$ for all $t > 0$.

This case is conditionally stable, since no intersection of $F(t)$ and γ/t is ever possible for $t > 0$; there is thus no root corresponding to $p > 0$.

iii) $F(0) < 0$ but $F(t) > 0$ at some time (or times) $t > 0$.

This case corresponds to what we might call a 'delayed instability'. The functions $F(t)$ and γ/t intersect at one or more points. The time-scale of the instability corresponds of course to the smaller value of t at which such an intersection occurs.

We have referred to case ii) as 'conditional stability' for the following reason: As already noted, our assumption that $\Psi(t)$ can be considered to be 'slowly varying' breaks down when very long time scales ($t \rightarrow \infty$) are involved. Since however a rigorous proof of stability has only been achieved when the possibility of even very long-time-scale instabilities ('very slow instabilities') has been eliminated, we are for the sake of completeness obliged to consider the form which our 'stability equation' takes when $t \rightarrow \infty$ or, equivalently, when $p \rightarrow 0$. We shall discuss this question in the following section.

11. VERY SLOW INSTABILITIES

As noted in section 7, the approximation so far used:-

$$\bar{\Psi}(p) \simeq \frac{1}{p}\Psi\left(\frac{1}{p}\right) \tag{50}$$

to represent the Laplace transforms of the energy response functions break down for $t \rightarrow \infty$ since these functions are then no longer slowly-varying in the sense of section 6.

This means that an alternative method must be found of calculating or approximating $\overline{\Psi}(p)$ when we are interested in either i) the problem of marginal stability or ii) eliminating the possibility of very slow instabilities when proving 'stability'.

Such an alternative approximation to $\overline{\Psi}(p)$ can readily be found from the definition of the Laplace transform (Eq. 19). We note that as $p \rightarrow 0$:-

$$\overline{\Psi}(p) \rightarrow \int_0^{\infty} \Psi(t)\, dt \tag{51}$$

so that, in this limit, the transforms can be determined directly from the areas under the energy response functions. Since, moreover, the approximation represented by Eq. 37 is still applicable for the ϕ-functions even as $t \rightarrow \infty$ $(p \rightarrow 0)$, the condition for marginal stability follows from Eq. 24 in a straightforward manner.

12. DISCUSSION

In order to investigate whether a given case is stable or unstable we can proceed as follows.

We first decide whether the system is close to marginal stability by investigating the long-time-scale behaviour in the manner described in the last section. Only in very exceptional cases will it turn out that marginal stability is indicated.

Having excluded this possibility, we know that the time-scale of the growing (unstable) modes or the decaying (stable) modes is sufficiently short to allow us to use the approximations applied in section 6. This means that the stability behaviour can be determined on the basis of the properties of F(t) alone, where

$$F(t) = \frac{\phi_1(t)\Psi_2(t) - \Psi_1(t)\phi_2(t)}{\phi_1(t) + \frac{\alpha}{\gamma}\phi_2(t)} \tag{52}$$

This result, that the stability question can be resolved simply on the basis of an inspection of the response functions, was discovered by Refsdal (unpublished) and confirmed by means of extensive numerical integrations of equations of the same type as Eqs. 1 and 2.

13. CONCLUDING REMARKS

Collecting the results of sections 10-12 together, we see that the secular stability behaviour can be one of three different types, namely

A) Initial instability

B) Delayed instability

C) Stability

thus confirming the expectations expressed in HR. Underlying our discussions is the requirement (see section 5) that p shall be real. One further requirement - not so far discussed - is that the systems under consideration shall always be dynamically stable; otherwise tests for secular stability have little meaning.

The question remains as to which of the three possible types of stability behaviour will actually occur in practice.

In this connection, we note that the forms of the response functions (and particularly of ϕ) show such considerable variations with stellar mass that it cannot a priori be excluded that all three possibilities will arise in practice, according to the particular mass combination considered.

Two of the three possibilities have already been found to occur - namely delayed instability (HR) and stability (Hazlehurst, Höppner, Refsdal, 1981) - both within the range of component masses appropriate to the W UMa systems.

It follows that those theories of W UMa stars which - in common with the present paper - are based upon a common envelope model must be able to accommodate the existence of both stable and unstable configurations. Taken together with another necessary condition - namely, the need to explain the observed light curves - this represents a very considerable 'double demand' which, in fact, no current theory comes anywhere near to satisfying.

ACKNOWLEDGEMENT

The author is grateful to S. Refsdal for valuable discussions.

Footnote to Section 9

1. It is in fact not difficult to show that the ratio α/γ is identical with the factor F introduced in HR to describe the geometry of the equipotential surfaces.

REFERENCES

Doetsch, G.: 1955, Handbuch der Laplace-Transformation, Vol. 2, Birkhäuser Verlag, Basle and Stuttgart.

Hazlehurst, J., Refsdal, S., Stobbe, C.: 1977, Astron. Astrophys. 58, 47

Hazlehurst, J., Refsdal, S.: 1980, Astron. Astrophys. 84, 200 (HR).

Hazlehurst, J., Höppner, W., Refsdal, S.: 1981, in preparation.

Kopal, Z.: 1959, Close Binary Systems, Chapman and Hall, London.

THE ROCHE DIMENSIONS OF ROTATING GASEOUS DISKS IN CLOSE BINARY SYSTEMS

Masatoshi Kitamura
Tokyo Astronomical Observatory
University of Tokyo
Osawa 2-21-1, Mitaka
Tokyo 181, Japan

ABSTRACT

The problem of rotating emission disks (or rings) in close binary systems is discussed from a simple hydrodynamic point of view for the inviscid gas flow with the aid of the Roche coordinates. Assuming, as a sufficient approximation, that the rotating gas flows around one component in a close binary system move along zero-velocity curves in the orbital plane, a certain simple relation can be deduced between the flow velocity and the outward density gradient. By regarding the observed rotational velocity of the emission disk as corresponding to the velocity of the stream of maximum density in the flows, we can easily evaluate, with the relation, the dimension of the rotating disk from the velocity. Application is made to emission disks in Algol-type close binaries and also to accretion disks in cataclysmic variables.

1. INTRODUCTION

Spectroscopic observations of a number of close binary systems have established that a rotating gaseous emission disk (or ring) exists around the equator of one of the components in such a system. Close binary stars with rotating gas streams are those of the semi-detached type and such gaseous disks rotate with velocities of several hundred km/sec, always in the same direction as that of the orbital motion (e.g., Kopal 1959; Sahade 1960).

There are two types of semidetached close binaries with rotating gaseous disks. One of them is Algol-type close binary systems with circulating emission disks (or rings), in which the components surrounded by such disk-like streams are main-sequence stars earlier in spectral type than their companions. The other

F. D. Kahn (ed.), Investigating the Universe, 241–270.

one is cataclysmic variables which are known to be very short-period close binaries of the semi-detached type. In the latter, the components with rotating accretion disks are degenerated compact stars (e.g., Walker 1954; Robinson 1976).

Owing to the mathematical difficulty, the formation and behaviour of such rotating gaseous disks in semi-detached close binary systems have been treated mainly with particle mechanics for the restricted three-body problem (e.g., Kopal 1959; Huang 1967; Warner and Peters 1972; Flannery 1975). However, the motion of such gas flows should be best treated on the bases of hydrodynamics. Numerical integrations of the hydrodynamic equations in rectangular coordinates have been attempted for the behaviour of the gas flow in a semi-detached close binary system by several workers (e.g., Prendergast and Taam 1974; Sorensen, Matsuda and Sakurai 1975). However, such numerical computations in the rectangular coordinates need a considerable amount of computer time and their treatments are rather complicated.

One of the important contributions by Kopal to the binary astronomy should be his introduction of the Roche coordinates in complete form (Kopal 1969). In the hydrodynamic treatment of gas motions in close binary systems, Prendergast (1960) first used the same coordinates but later, in more detail and in thorough form, Kopal showed that an introduction of a new system of orthogonal curvilinear coordinates (ξ , η , ζ) with the zero-velocity curves (ξ) as one of them can make their mathematical treatment much easier. The topological properties of the Roche coordinates were further discussed by Kitamura (1970).

In the present paper devoted to Professor Zdenek Kopal in honour of his retirement from the Victoria University of Manchester, I would like to show, as an example of application of Kopal's Roche coordinates, that their use in discussing the dimensions of rotating gaseous disks in close binary systems can present a beautiful analytical relation on the flow velocity circulating around the parent star which enables us to accurately estimate the dimension of the rotating gaseous disk

The behaviour of the Roche coordinates (ξ , η) in the orbital plane is illustrated for the cases of the mass ratio q = 1.0, 0.6 and 0.2 in Figures 1, 2 and 3, respectively. A schematic view of the other coordinate ζ = constant is also shown in three-dimensional perspective in Figure 4 (Kitamura 1970).

2. FLOW EQUATIONS

We start from the well-known Eulerian equation of motion

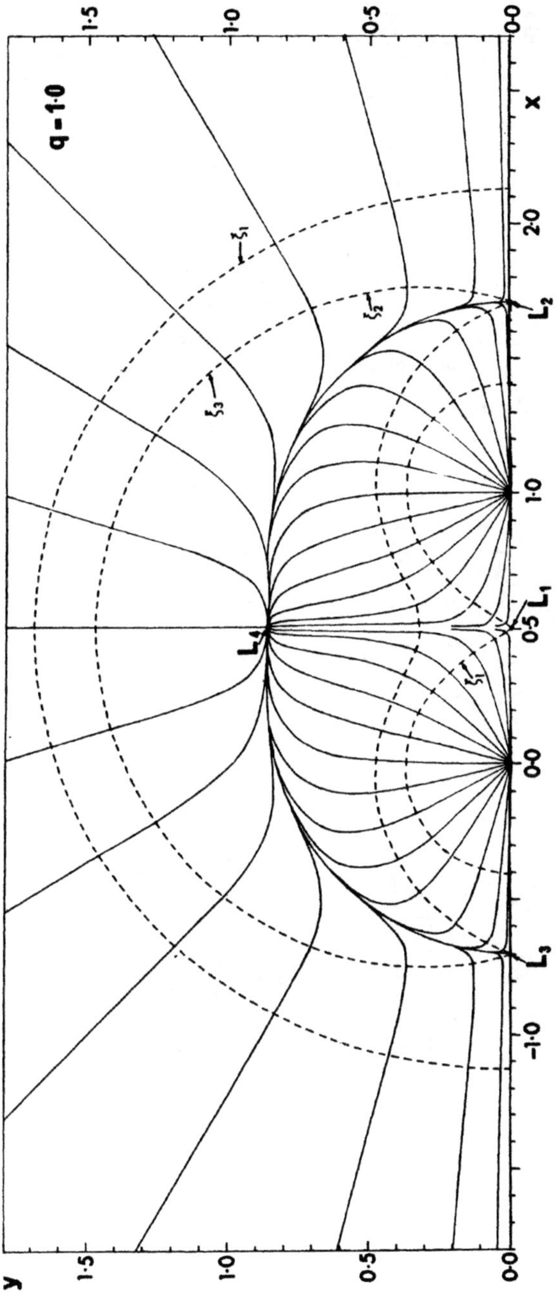

Fig. 1. The geometry of the curves η = constant in the xy-plane (the orbital plane) for $y \geq 0$, corresponding to the mass-ratio $q = 1.0$. The trajectories of $\eta(x, y, 0)$ are marked with full curves, while the dotted curves represent the equipotentials ξ = constant for the values which the potential assumes at the Lagrangian collinear points L_1, L_2, and L_3 (Kitamura 1970).

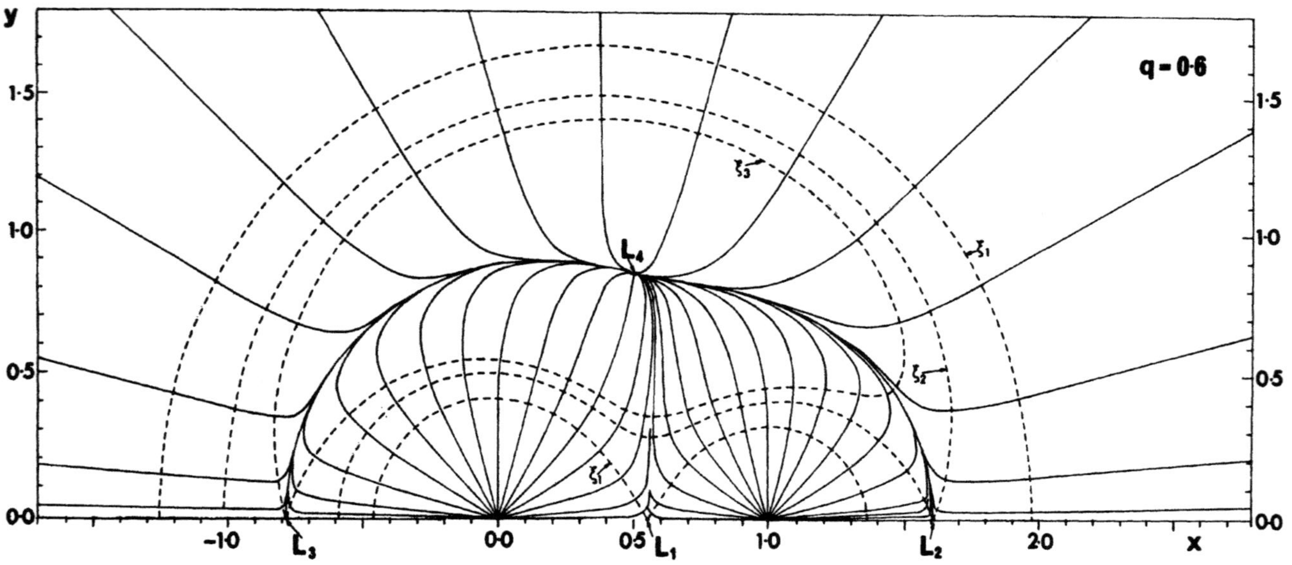

Fig. 2. The geometry of the curves η = constant in the xy - plane (the orbital plane) for $y \geq 0$, corresponding to the mass-ratio $q = 0.6$. The trajectories of $\eta\,(x, y, 0)$ are marked with full curves, while the dotted curves represent the equipotentials ξ = constant for the values which the potential assumes at the Lagrangian collinear points L_1, L_2 and L_3 (Kitamura 1970).

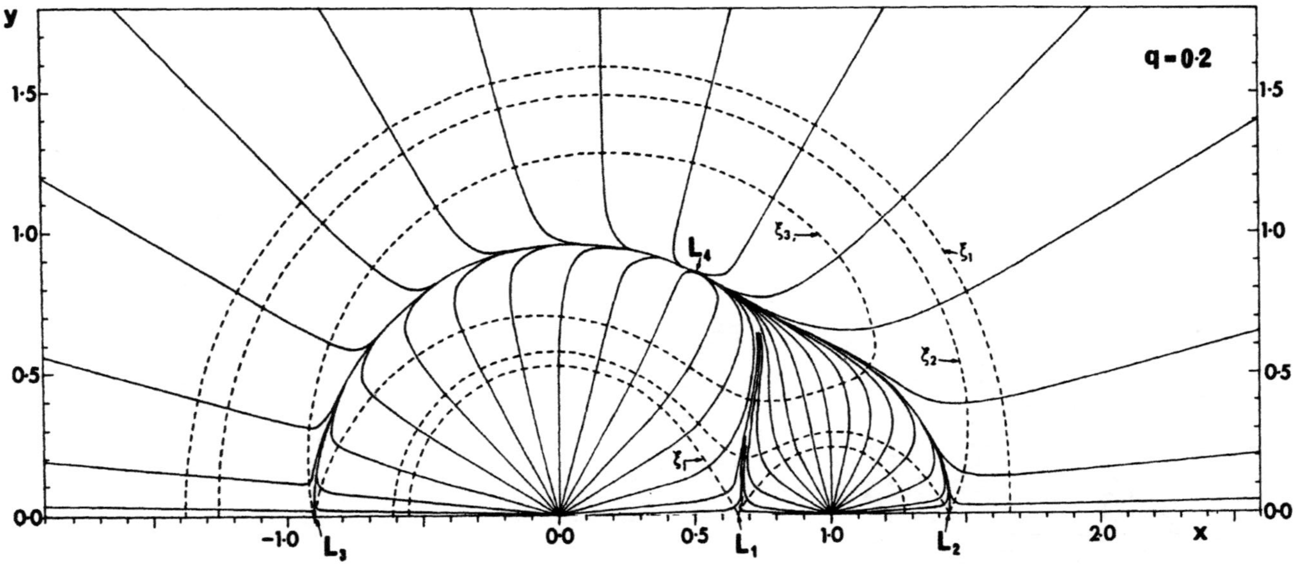

Fig. 3. The geometry of the curves η = constant in the xy-plane (the orbital plane) for $y \geq 0$, corresponding to the mass-ratio $q = 0.2$. The trajectories of η (x, y, 0) are marked with full curves, while the dotted curves represent the equipotentials ξ = constant for the values which the potential assumes at the Lagrangian collinear points L_1, L_2 and L_3 (Kitamura 1970).

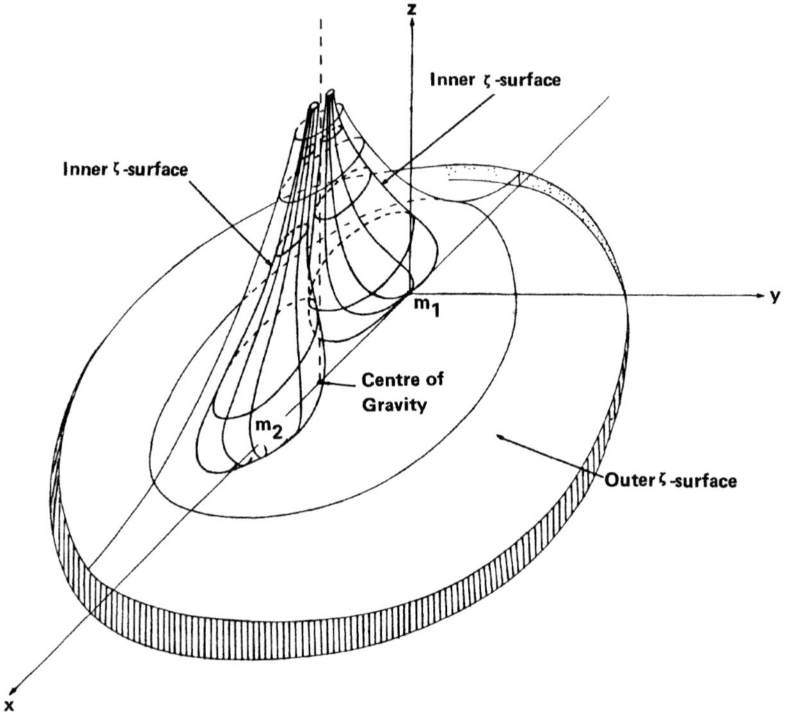

Fig. 4. Three-dimensional representation of the surfaces ζ (x,y,z) = constant (Kitamura 1970).

for the inviscid fluid flow:

$$\frac{Du}{Dt} + \frac{1}{\rho} \text{grad } P = K, \qquad (1)$$

where u denotes the velocity of the fluid at any point, P the pressure, ρ its density and K the external forces acting upon any element of the fluid.

Let us consider the case of circular orbits of the component stars and also confine our attention to steady flows in the orbital plane. Then the time derivative vanishes in Equation (1). We shall take the separation of the components $a_1 + a_2$ as unit of length, the total mass $m_1 + m_2$ as unit of mass, and select the unit of time to be Period/2π.

Using the Roche coordinates (ξ , η) in the orbital plane with the metric coefficients h_1 and h_2 in the transformation

$$(dx)^2 + (dy)^2 = h_1^2 (d\xi)^2 + h_2^2 (d\eta)^2, \qquad (2)$$

Equation of motion (1) reduces to two scalar equations on the orbital plane as follows (Kopal 1969);

$$u_1 \frac{\partial u_1}{\partial \xi} + \frac{h_1}{h_2} u_2 \frac{\partial u_1}{\partial \eta} + \frac{u_1 u_2}{h_2} \frac{\partial h_1}{\partial \eta} - \frac{u_2^2}{h_2} \frac{\partial h_2}{\partial \xi}$$

$$- 2u_2 h_1 + \frac{1}{\rho} \frac{\partial P}{\partial \xi} = K_\xi \qquad (3)$$

and

$$u_1 \frac{\partial u_2}{\partial \xi} + \frac{h_1}{h_2} u_2 \frac{\partial u_2}{\partial \eta} - \frac{u_1^2}{h_2} \frac{\partial h_1}{\partial \eta} + \frac{u_1 u_2}{h_2} \frac{\partial h_2}{\partial \xi}$$

$$+ 2u_1 h_2 + \frac{1}{\rho} \frac{\partial P}{\partial \eta} = K_\eta \qquad (4)$$

where subscripts 1 and 2 are used for ξ and η respectively such as

$$u_1 = \frac{\partial \xi}{\partial t} , \quad u_2 = \frac{\partial \eta}{\partial t} . \qquad (5)$$

It follows from the definition of the Roche coordinates that

$$\xi_x \eta_x + \xi_y \eta_y = 0 \qquad (6)$$

with

$$\xi = \frac{1}{r_1} + \frac{1}{2} r_1^2 + q \left(\frac{1}{r_2} + \frac{1}{2} r_2^2\right) - \frac{1}{2} q, \qquad (7)$$

where r_1 and r_2 denote the distances from the centres of the components and the mass ratio q is taken such that

$$q \equiv \frac{m_2}{m_1} \leq 1.0. \qquad (8)$$

The motion of fluid flows in rotating coordinates is known to be governed by an effective potential C given by the sum of the usual gravitational potential and a centrifugal potential per unit mass. Thus we may put

$$K_\xi = \frac{\partial C}{\partial \xi}, \quad K_\eta = \frac{\partial C}{\partial \eta}, \qquad (9)$$

where the effective potential C can be written as

$$C = \frac{G}{d}\left\{ m_1 \xi + \frac{m_2^2}{2(m_1 + m_2)} \right\}, \tag{10}$$

with G as the gravitational constant.

Differentiating Equation (10) with respect to ξ and η, we easily have $K_\xi = Gm_1/d$ and $K_\eta = 0$. As it follows from our definition of units that $Gm_1/d = 1/(1+q)$, the two equations of (9) reduce to

$$K_\xi = \frac{1}{1+q} \tag{11}$$

and

$$K_\eta = 0 \tag{12}$$

respectively.

The equation of continuity

$$\frac{D\rho}{Dt} + \rho \operatorname{div} u = 0 \tag{13}$$

can also be written in terms of Roche coordinates for a fluid flow in steady motion as

$$\frac{\partial}{\partial \xi}(h_2 \rho u_1) + \frac{\partial}{\partial \eta}(h_1 \rho u_2) = 0. \tag{14}$$

Let us assume from the thermodynamical point of view that the change of state in the gaseous disk is adiabatic and the motion is isentropic. Then the relation between pressure and density is given by

$$P = k\rho^\gamma \tag{15}$$

where γ denotes the ratio of specific heats and k is constant. We also assume the law of ideal gas

$$P = \frac{R}{m} T \tag{16}$$

where R is the gas constant and m the molecular weight. Differentiating Equation (15) with respect to ξ and η, we have

$$\frac{\partial P}{\partial \xi} = a^2 \frac{\partial \rho}{\partial \xi}, \quad \frac{\partial P}{\partial \eta} = a^2 \frac{\partial \rho}{\partial \eta} \tag{17}$$

with

$$a^2 = \gamma \frac{P}{\rho} \tag{18}$$

where a corresponds to the local sound velocity.

3. BOUNDARY CONDITIONS

The surface of the parent star surrounded by a gaseous disk may be best approximated by one of the zero-velocity surfaces given by

$$\xi = \xi_o \tag{19}$$

Therefore we should consider physical conditions over the surface given by Equation (19) as boundary ones.

First of all, it is evident that the velocity component u_1 should vanish on the stellar surface. Thus

$$(u_1)_o = \left(\frac{\partial \xi}{\partial t}\right)_o = 0, \tag{20}$$

where subscript $_o$ corresponds to the stellar surface. As the stellar surface is now equipotential, the pressure and density should be constant over the surface. Therefore we may put

$$\left(\frac{\partial P}{\partial \eta}\right)_o = 0, \quad \left(\frac{\partial \rho}{\partial \eta}\right)_o = 0. \tag{21}$$

As the velocity component of the flow, u_1, along the stellar surface should vanish as given by Equation (20), we may further put

$$\left(\frac{\partial u_1}{\partial \eta}\right)_o = 0. \tag{22}$$

The other boundary conditions may be imposed at the outer end of the gaseous disk such that

$$P_e = 0, \quad \rho_e = 0 \tag{23}$$

and also

$$\left(\frac{\partial P}{\partial \xi}\right)_e = 0, \quad \left(\frac{\partial \rho}{\partial \xi}\right)_e = 0 \tag{24}$$

where subscript e corresponds to the outer end of the disk.

4. NATURE OF THE FLOWS AT THE STELLAR SURFACE

In the previous section we have set up the boundary conditions of flows over the stellar surface as expressed by $\xi = \xi_o$. In this section we shall consider some additional properties of these flows by means of the equations of motion and continuity.

Substituting the surface boundary conditions (20), (21) and (22) in the equations of motion (3) and (4), we obtain

$$\left(\frac{1}{h_2}\frac{\partial h_2}{\partial \xi}\right)_o (u_2)_o^2 + 2(h_1)_o(u_2)_o = \left(\frac{1}{\rho}\frac{\partial P}{\partial \xi}\right)_o - \frac{1}{1+q} \tag{25}$$

and

$$\left(u_2\frac{\partial u_2}{\partial \eta}\right)_o = 0 \tag{26}$$

respectively, where K_ξ and K_η in the right-hand sides of Equations (3) and (4) are replaced by (11) and (12). Similarly the Equation of continuity (14) reduces, at the surface, to

$$\left(\frac{\partial h_1}{\partial \eta}\right)_o(u_2)_o + (h_2)_o\left(\frac{\partial u_1}{\partial \xi}\right)_o + (h_1)_o\left(\frac{\partial u_2}{\partial \eta}\right)_o = 0. \tag{27}$$

From a combination of the above three Equations (25), (26) and (27), we can have some relations for the surface flows. First of all, it follows from Equation (26) that

$$(u_2)_o = \text{constant}, \tag{28}$$

because Equation (26) may be rewritten as $\left(\frac{\partial u_2^2}{\partial \eta}\right)_o = 0$. Therefore, the derivative of u_2 with respect to η should vanish at the surface such that

$$\left(\frac{\partial u_2}{\partial \eta}\right)_o = 0. \tag{29}$$

Next, if we solve Equation (25) algebraically, we have

$$(u_2)_o = \frac{1}{\left(\frac{1}{h_2}\frac{\partial h_2}{\partial \xi}\right)_o}\left[-(h_1)_o + \sqrt{(h_1)_o^2 + \left(\frac{1}{h_2}\frac{\partial h_2}{\partial \xi}\right)_o\left\{\left(\frac{1}{\rho}\frac{\partial P}{\partial \xi}\right)_o - \frac{1}{1+q}\right\}}\right] \tag{30}$$

where the positive square root has been taken according to the following reason. As $(u_2)_o$ is constant from (26), the left-hand side of Equation (30) should be independent of η. Such independence generally occurs only when

$$\left(\frac{1}{\rho}\frac{\partial P}{\partial \xi}\right)_o = \frac{1}{1+q}, \tag{31}$$

because the quantities $(h_1)_o$ and $\left(h_2^{-1}\frac{\partial h_2}{\partial \xi}\right)_o$ appearing on the right-hand side of Equation (30) are functions of both of ξ and η. Obviously, Equation (31) represents the hydrostatic condition which governs the relation between pressure and density at the surface. Substituting the relation (31) in the right-hand side of Equation(30) it follows that

$$(u_2)_o = 0. \tag{32}$$

Thus, at the surface of the parent star the gravitational flows vanish in any direction in the orbital plane.

Substituting Equation (29) in the equation of continuity (27), we obtain another relation for $(u_2)_o$:

$$(u_2)_o = -\frac{1}{\left(\frac{1}{h_2}\frac{\partial h_1}{\partial \eta}\right)_o}\left(\frac{\partial u_1}{\partial \xi}\right)_o. \tag{33}$$

Combining Equation (33) with Equation (32), we eventually have

$$\left(\frac{\partial u_1}{\partial \xi}\right)_o = 0. \tag{34}$$

Thus, the gradient of the velocity component of the flow along the η -constant curves should vanish at the surface.

5. FLOWS ALONG EQUIPOTENTIALS

If the rotating gas flows exist steadily around the parent star, the velocity components along the curves η =constant should be small compared with the other components along the curves ξ = constant (equipotentials), because these disk-like flows do not seem to exist very far from the parent star and perhaps are within the Roche limit (this will be verified in Section 6). So, as a sufficient approximation, we may assume that the disk-like flows move along equipotentials within the Roche limit. In this connection, it may be recalled that Sobouti (1971) showed that the streamlines of the flows should be confined within appropriate equipotentials. Thus, we shall consider such equipotential flows as a good approximation to actual emission disks(or rings) in close binary systems.

Putting $u_1 = 0$ in Equations of motion (3) and (4) and in Equation of continuity (14), we have

$$\frac{1}{h_2}\frac{\partial h_2}{\partial \xi} u_2^2 + 2 h_1 u_2 = \frac{1}{\rho}\frac{\partial P}{\partial \xi} - \frac{1}{1+q}, \tag{35}$$

$$u_2 \frac{\partial u_2}{\partial \eta} + \frac{1}{\rho}\frac{\partial P}{\partial \eta} = 0 \tag{36}$$

and

$$\rho h_1 \frac{\partial u_2}{\partial \eta} + h_1 u_2 \frac{\partial \rho}{\partial \eta} + \rho u_2 \frac{\partial h_1}{\partial \eta} = 0 \tag{37}$$

respectively.

Solving Equation (35) for u_2, we obtain*

$$u_2 = \frac{1}{\frac{1}{h_2}\frac{\partial h_2}{\partial \xi}}\left\{- h_1 + \sqrt{h_1^2 + \frac{1}{h_2}\frac{\partial h_2}{\partial \xi}\left(\frac{1}{\rho}\frac{\partial P}{\partial \xi} - \frac{1}{1+q}\right)}\right\}, \tag{38}$$

where the positive square root has been taken according to the same consideration as done in Equation (30) in the previous section.

From Equation (36), we obtain for the pressure gradient along the equipotential

$$\frac{1}{\rho}\frac{\partial P}{\partial \eta} = - u_2 \frac{\partial u_2}{\partial \eta}. \tag{39}$$

Cancelling $\frac{\partial u_2}{\partial \eta}$ from Equations (37) and (38), we have

---

*) For comparison, we show the velocity component u_θ in the polar coordinates as

$$u_\theta = - r_1 + \sqrt{\frac{1}{1+q}\left\{\frac{1}{r_1} + q r_1\left(\frac{r_1 - \cos\theta}{r_2^3} + \cos\theta\right)\right\} + \frac{r_1}{\rho}\frac{\partial P}{\partial r_1}}$$

$$\frac{\partial P}{\partial \eta} = u_2^2 \left(\frac{\partial \rho}{\partial \eta} + \frac{\rho}{h_1} \frac{\partial h_1}{\partial \eta} \right). \qquad (40)$$

By combining the above relation (40) with Equation (37), we obtain

$$\frac{\partial \rho}{\partial \eta} = \frac{\rho\, u_2^2}{a^2 - u_2^2} \frac{1}{h_1} \frac{\partial h_1}{\partial \eta} \qquad (41)$$

and

$$\frac{\partial P}{\partial \eta} = \frac{\rho\, a^2 u_2^2}{a^2 - u_2^2} \frac{1}{h_1} \frac{\partial h_1}{\partial \eta} \qquad (42)$$

The quantity $\frac{1}{\rho}\frac{\partial P}{\partial \xi}$ appeared in Equation (38) can be rewritten with the aid of the adiabatic relation (15) as

$$\frac{1}{\rho} \frac{\partial P}{\partial \xi} = k \frac{\gamma}{\gamma - 1} \frac{\partial \rho^{\gamma-1}}{\partial \xi} \qquad (43)$$

I. the disk-like gas flows rotate stationarily around the parent star sustaining certain distances from the stellar surface, the gas density should take a maximum value along an inner η-constant curve Kitamura 1970) where it holds that

$$\left(\frac{\partial \rho}{\partial \xi} \right)_{max} = 0. \qquad (44)$$

As the value of γ is larger than the unity for the practical cases, we can write

$$\left(\frac{\partial \rho^{\gamma-1}}{\partial \xi} \right)_{max} = 0. \qquad (45)$$

Therefore, it follows from Equation (43) that

$$\frac{1}{\rho} \left(\frac{\partial P}{\partial \xi} \right)_{max} = 0. \qquad (46)$$

Substituting the above result (46) in Equation (38), the velocity of circulating streams corresponding to the maximum density in the flows is given by

$$(u_2)_{max} = \frac{1}{\frac{1}{h_2}\frac{\partial h_2}{\partial \xi}} \left\{ - h_1 + \sqrt{h_1^2 - \left(\frac{1}{h_2} \frac{\partial h_2}{\partial \xi} \right) \frac{1}{1+q}} \right\} \qquad (47)$$

Equation (47) reveals that the u_2-velocity at maximum density includes no pressure term. It may be noted that the form of Equation (47) is in agreement with Prendergast's previous result (1960) which was obtained by neglecting the pressure terms from the beginning in his treatment. However, Equation (47) obtained here should have a different meaning from that of Prendergast. It is important that Equation (47) should hold rigorously for the densest stream within the hydrodynamic flows along the equipotentials (ξ = constant curves) surrounding the parent star.

Referring to the surface condition (31), it follows from Equation (43) that at the surface of the parent star

$$\frac{\partial \rho^{\gamma-1}}{\partial(-\xi)} = \frac{\gamma-1}{k\gamma}\,\frac{1}{\rho}\,\frac{\partial P}{\partial(-\xi)} = \frac{1}{1+q} < 0 \tag{48}$$

On the other hand, for an outwards decreasing density we should have $\partial\rho^{\gamma-1}/\partial(-\xi) < 0$ while for an outwards increasing density $\partial\rho^{\gamma-1}/\partial(-\xi) > 0$. Therefore, if the gases constituting the disk extend down to the stellar surface, a minimum density with $\partial\rho^{\gamma-1}/\partial(-\xi) = 0$ may exist between the surface and the densest stream within the disk. However, this would not be plausible in actual emission disks in close binary systems. Considering the observational evidence of emission phenomena in such close binaries, the disk would detach from the surface of the parent star like rotating gaseous rings.

The metric coefficient h_1 in Equation (47) is known to be given by

$$h_1 = \frac{1}{\sqrt{\xi_x^2 + \xi_y^2}} \gtrless 0. \tag{49}$$

The metric quantity $\frac{1}{h_2}\frac{\partial h_2}{\partial \xi}$ can be written by the formula (e.g., Kopal 1969)

$$\frac{1}{h_2}\frac{\partial h_2}{\partial \xi} = \frac{\xi_y^2 \xi_{xx} - 2\xi_x \xi_y \xi_{xy} + \xi_x^2 \xi_{yy}}{(\xi_x^2 + \xi_y^2)^2} \tag{50}$$

where ξ_x, ξ_y, ξ_{xx}, ξ_{xy} and ξ_{yy} are explicitly given by differentiating Equation (7) with respect to ξ as follows.

$$\xi_x = -x\,r_1^{-3} + q\left\{(1-x)\,r_2^{-3} - 1\right\} + (1+q)\,x, \tag{51}$$

$$\xi_y = - y \left\{ r_1^{-3} + q\, r_2^{-3} - (1 + q) \right\}, \tag{52}$$

$$\xi_{xx} = (3x^2 - r_1^2)\, r_1^{-5} + q \left\{ 3(1-x)^2 - r_2^2 \right\} r_2^{-5} + (1+q), \tag{53}$$

$$\xi_{xy} \equiv 3y \left\{ x\, r_1^{-5} - q\,(1-x)\, r_2^{-5} \right\} \tag{54}$$

and

$$\xi_{yy} = (3y^2 - r_1^2)\, r_1^{-5} + q\,(3y^2 - r_2^2)\, r_2^{-5} + (1+q). \tag{55}$$

Numerical evaluation reveals that the values of the quantity $\frac{1}{h_2}\frac{\partial h_2}{\partial \xi}$ are always negative within the Roche limit; thus,

$$\left(\frac{1}{h_2} \frac{\partial h_2}{\partial \xi} \right)_{in} < 0, \tag{56}$$

where the subscript in represents for within the Roche limit.

Figures 5, 6 and 7 illustrate the result of numerical computation of $\frac{1}{h_2} \cdot \partial h_2 / \partial \xi$ for the cases of q= 1.0, 0.5 and 0.2 respectively, where the dotted curves represent the Roche limit and the full curves represent the boundary with $\frac{1}{h_2}\frac{\partial h_2}{\partial \xi} = 0$.

If in Equation (47) we take into account the relations (49), (50) and (56), the velocity of the flow circulating around the parent star should be negative; thus,

$$(u_2)_{in} \equiv \left(\frac{\partial \eta}{\partial t} \right)_{in} < 0. \tag{57}$$

On the other hand, the η - coordinate is taken so as to make a left-handed coordinate system in the orbital plane, for which Prendergast (1960) already showed that near the parent star it behaves like φ (usual aximuthal angle). Thus, if there exists a steady gas flow around the parent star, the circulating velocity should be faster than the Keplerian one in the same direction. This quite explains the observation (e.g., Struve 1950).

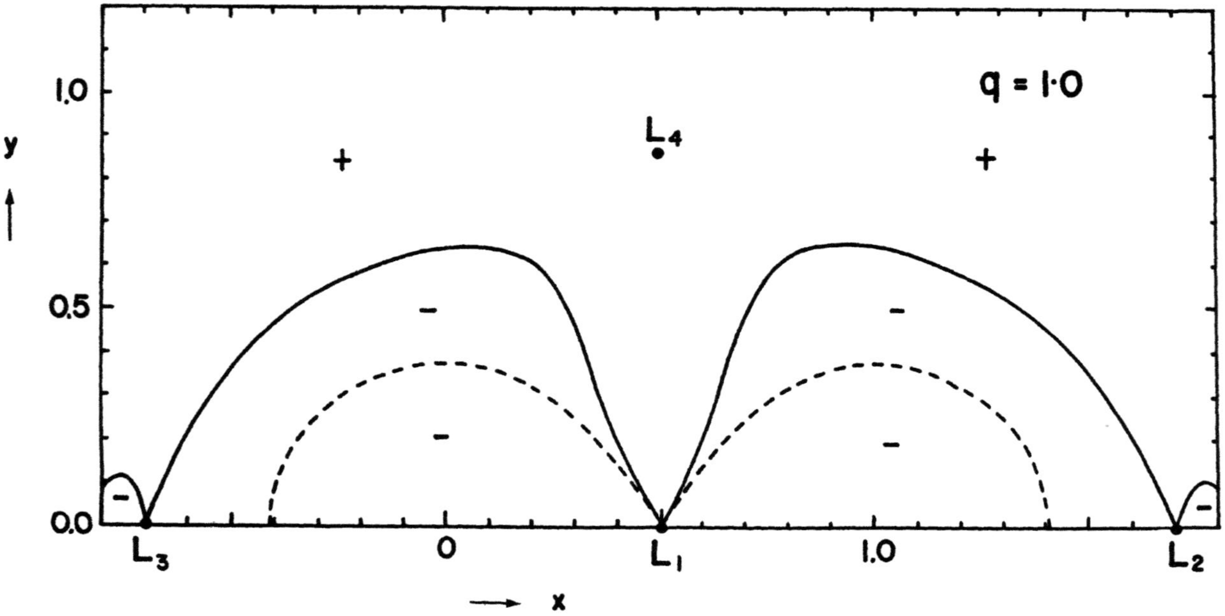

Fig. 5. Distribution of the sign of the metric quantity $1/h_2 \frac{\partial h_2}{\partial \xi}$ on the orbital plane for the case of the mass-ratio q = 1. The full curve represents the boundary with $1/h_2\, \partial h_2/\partial \xi = 0$ while the dashed curve represents the Roche limit.

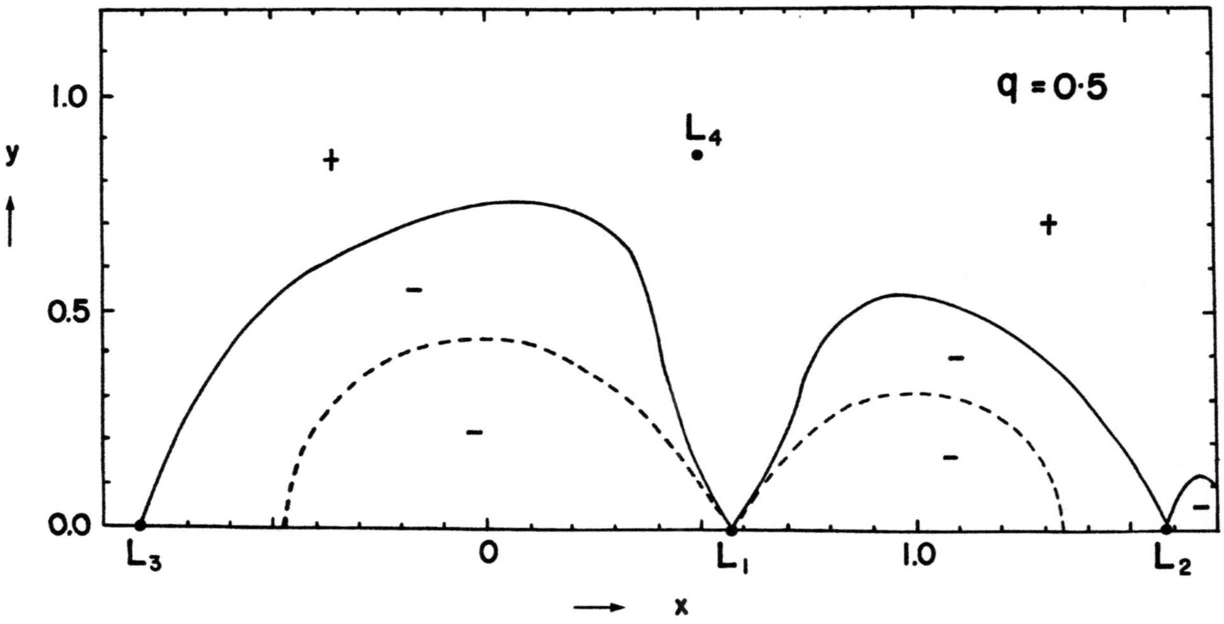

Fig. 6. Distribution of the sign of the metric quantity $1/h_2\, \partial h_2/\partial \xi$ on the orbital plane for the case of the mass-ratio $q = 0.5$. The full curve represents the boundary with $1/h_2\, \partial h_2/\partial \xi = 0$ while the dashed curve represents the Roche limit.

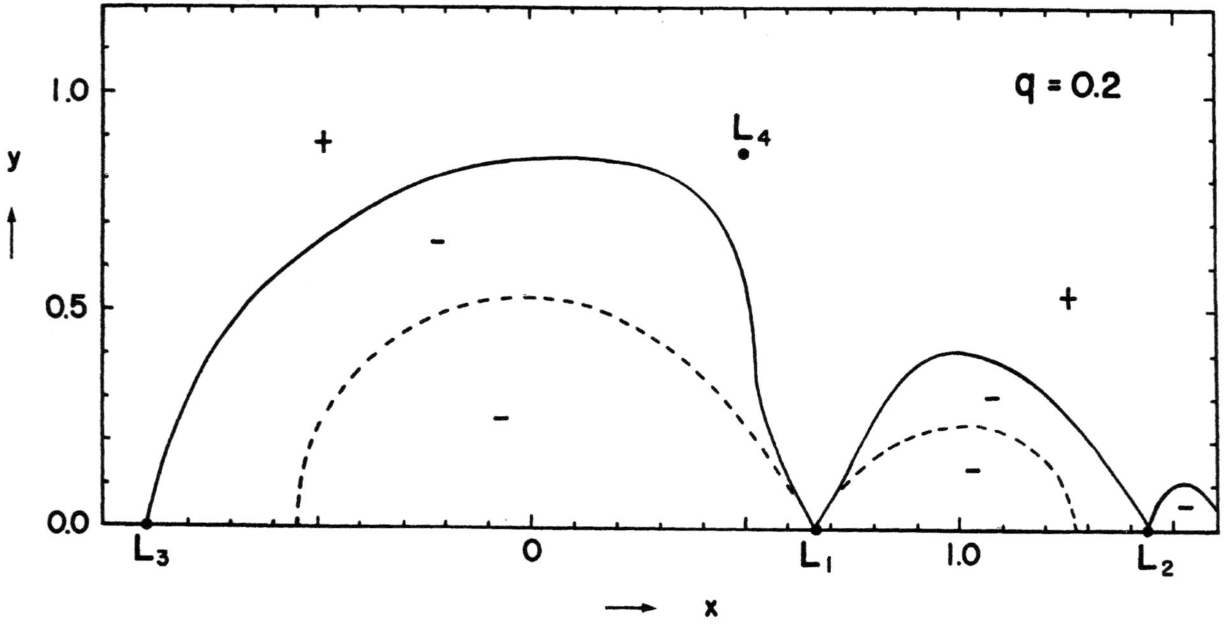

Fig. 7. Distribution of the sign of the metric quantity $1/h_2\ \partial h_2/\partial \xi$ on the orbital plane for the case of the mass-ratio $q = 0.2$. The full curve represents the boundary with $1/h_2\ \partial h_2/\partial \xi = 0$ while the dashed curve represents the Roche limit.

6. COMPARISON WITH SU-SHU HUANG'S RESULT FROM THE RESTRICTED THREE-BODY PROBLEM

The problem of gaseous rings in close binary systems has been discussed as the restricted three-body problem by a number of workers (e.g., Kopal 1959; Huang 1967; Flannery 1975; Paczynski 1977). In this section we shall take Huang's result and compare it with the result from our hydrodynamic consideration based on Roche coordinates.

Huang considered the rotating gaseous rings to be the stable periodic orbits of an infinitesimal particle in the restricted three-body problem. He sought a periodic solution by expanding the necessary physical parameters in terms of the reduced mass of the less massive component μ and gave numerical results of velocities of gaseous rings at time of eclipse from his periodic solution for the cases of μ = 0.1, 1/7, 1/6, 0.2, 0.25 and 0.3. We shall compare these results with ours as follows.

The velocities of the rotating flows for the same cases of μ have been calculated with the use of our Equation (47) for the same positions on the y-axis as used by Huang, which correspond to velocities of gaseous rings at time of eclipse in Table 2 of Huang's paper. All the results of calculations are shown in Table 1. The second and third columns show Huang's results from the restricted three-body problem and the two-body problem, respectively, while the results from the hydrodynamic consideration with Equation (47) are shown in the fourth column.

In Table 1, Huang's velocities given in the second and third columns correspond to the velocity component parallel to the x-axis, dx/dt, while the velocities in the fourth column correspond to the velocity component along the equipotential, $d\eta/dt$. A simple reduction by differentiating Equation (7) with respect to x and y gives for the difference between the directions of dx/dt and $d\eta/dt$ on the y-axis

$$\left(\frac{dy}{dx}\right)_{x=0} = -\left(\frac{\frac{\partial\xi}{\partial x}}{\frac{\partial\xi}{\partial y}}\right)_{x=0}$$

$$= \frac{q\left\{1 - \dfrac{1}{(1+y^2)\sqrt{1+y^2}}\right\}}{y - y^{-2} + qy\left\{1 - \dfrac{1}{(1+y^2)\sqrt{1+y^2}}\right\}} \qquad (58)$$

Numerical check with the above Equation (58) indicates that the differences between dx/dt and $d\eta/dt$ at the relevant positions on the y-axis are only of the order of 10^{-4} at most in the present unit.

TABLE 1. VELOCITIES OF GASEOUS RINGS AT TIME OF ECLIPSE

r=y	from 3-body	from 2-body	Present	r=y	from 3-body	from 2-body	Present
μ=0.1 (q=0.1111...)				μ =0.2(q=0.25)			
0.10006	2.898	2.999	2.8997	0.10022	2.724	2.825	2.7264
.15028	2.295	2.447	2.2985	.15063	2.151	2.305	2.1573
.20092	1.9122	2.116	1.9186	.2021	1.7799	1.9895	1.7940
.2524	1.6292	1.888	1.6408	.2559	1.4976	1.7681	1.5227
.3058	1.3979	1.7155	1.4167	.3152	1.2504	1.5932	1.2934
.3634	1.1897	1.5738	1.2196	.3901	0.9876	1.4321	1.0637
.4316	0.9745	1.444	1.0244	.4876	0.6951	1.2809	0.8225
.5329	0.6899	1.2996	0.7819	.5579	0.5121	1.1975	0.6734
μ=1/7 (q=0.16666....)				μ =0.25(q=0.33333...)			
0.10009	2.826	2.926	2.8272	0.10017	2.635	2.736	2.6378
.15042	2.234	2.387	2.2391	.15085	2.075	2.230	2.0833
.2014	1.8566	2.063	1.8661	.2029	1.7101	1.9227	1.7281
.2538	1.5744	1.8379	1.5911	.2584	1.4264	1.7042	1.4599
.3092	1.3378	1.6649	1.3661	.3226	1.1638	1.5248	1.2228
.3722	1.1125	1.5175	1.1594	.4139	0.8513	1.3462	0.9627
.4564	0.8501	1.3704	0.9327	.4903	0.6341	1.2369	0.7844
.5646	0.5582	1.2321	0.6911	.5350	0.5127	1.1841	0.6909
μ=1/6 (q=0.2)				μ =0.3(q=0.428571...)			
0.10010	2.784	2.885	2.7863	0.10022	2.541	2.643	2.5447
.15051	2.200	2.353	2.2053	.15109	1.9960	2.152	2.0068
.2017	1.8251	2.033	1.8362	.2038	1.6373	1.8533	1.6600
.2546	1.5429	1.8092	1.5631	.2613	1.3503	1.6368	1.3928
.3115	1.3026	1.6357	1.3365	.3340	1.0605	1.4478	1.1407
.3786	1.0642	1.4836	1.1221	.4313	0.7395	1.2740	0.8825
.4721	0.7765	1.3285	0.8795	.4777	0.6137	1.2105	0.7784
.5662	0.5282	1.2132	0.6750	.5069	0.5238	1.1751	0.7172

Numerical values given in Table 1 are shown in Figures 8, 9 and 10 for the cases of μ = 0.1, 0.2 and 0.3 respectively. In these figures, the upper dashed curves correspond to the two-body problem and the lower dashed curves to the restricted three-body problem, while the full curves correspond to the present result from the hydrodynamic consideration based upon Equation (47). It is easily seen that for the smaller y-values the result from the restricted three-body problem is quite in agreement with the present result from Equation (47) but the difference between the two becomes more and more appreciable with y.

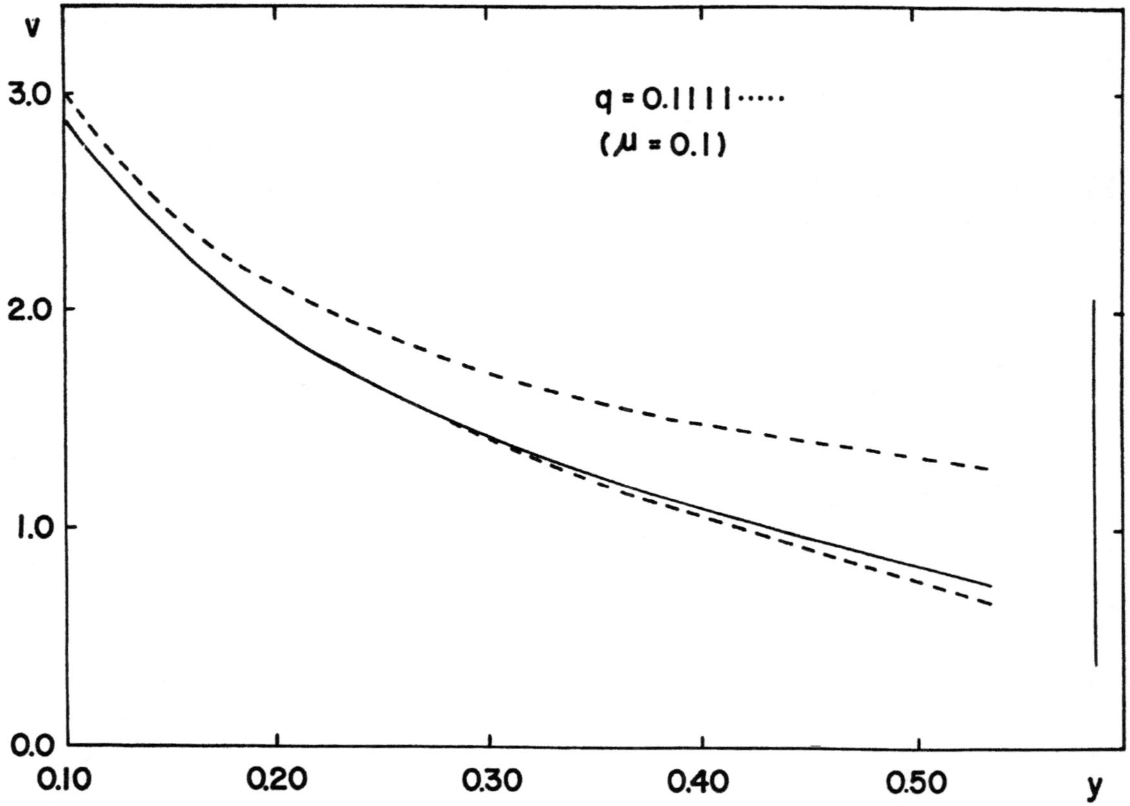

Fig. 8. Comparison of the velocity at points on the y-axis through the centre of the parent star predicted by different approximations for the case of $\mu = 0.1$ ($q=0.1111\ldots.$). The upper dashed curve corresponds to the two-body problem and the lower dashed curve to the restricted three-body problem, while the full curve corresponds to the present result. The bar corresponds to the Roche limit.

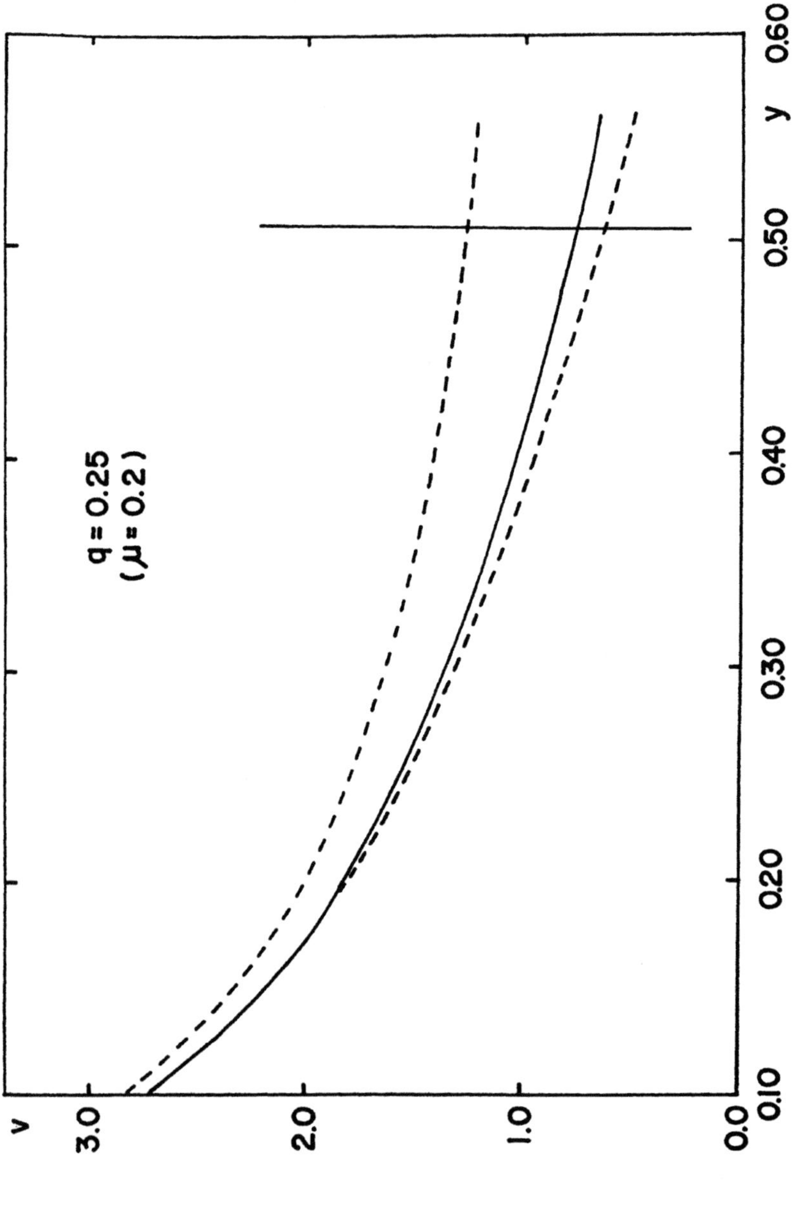

Fig. 9. Comparison of the velocity at points on the y-axis through the centre of the parent star predicted by different approximations for the case of $\mu = 0.2$ (q=0.25). The upper dashed curve corresponds to the two-body problem and the lower dashed curve to the restricted three-body problem, while the full curve corresponds to the present result. The bar corresponds to the Roche limit.

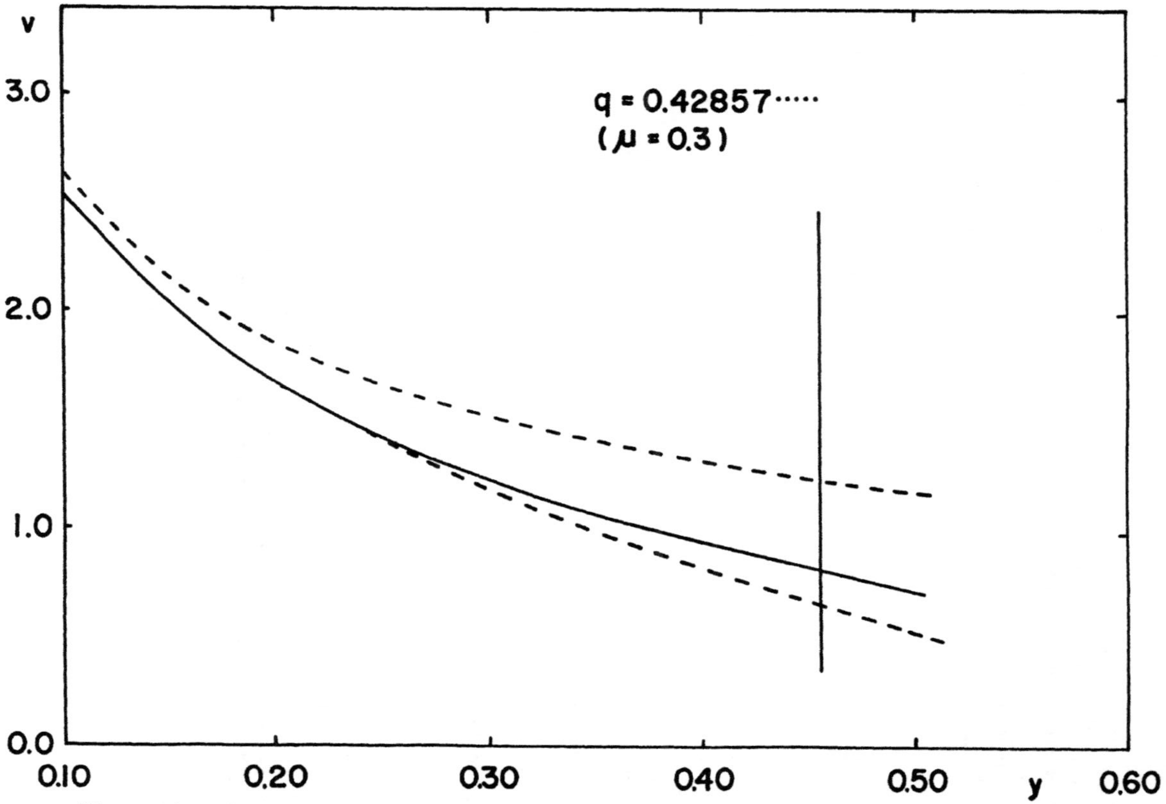

Fig. 10. Comparison of the velocity at points on the y-axis through the centre of the parent star predicted by different approximations for the case of $\mu = 0.3$ ($q=0.42857...$). The upper dashed curve corresponds to the two-body problem and the lower dashed curve to the restricted three-body problem, while the full curve corresponds to the present result. The bar corresponds to the Roche limit.

7. ROCHE DIMENSIONS OF GASEOUS DISKS IN ACTUAL CLOSE BINARY SYSTEMS

A simple relation (47) obtained in Section 5 enables us to accurately evaluate the velocity of the circulating flow around the parent star, which should correspond to the observed velocity of the emission disk (or ring), if its dimension is known. In the opposite way, if we substitute the observed radial velocity of the emission disk in $(u_2)_{max}$ in the left-hand side of Equation (47), the value of the dimension - Roche dimension - of the disk can be easily evaluated.

(a) ALGOL-TYPE CLOSE BINARY SYSTEMS WITH EMISSION DISKS

Table 2 lists Algol-type eclipsing binaries with observed emitting rings(or disks), for which details concerning the masses of the components and the orbital inclinations are available. Of the listed six systems, three are double-line binaries and the other three are single-line binaries with main-sequence primary components (parent stars). Their spectroscopic elements are taken from the Seventh Catalogue of the Orbital Elements of Spectroscopic Binary Systems by Batten, Fletcher and Mann(1978). For the three single-line binaries, the approximate masses of their primary components have been assumed from their spectral types as main-sequence stars (Allen 1973), and the masses of the secondary (fainter) components have been calculated from the respective mass-functions with their orbital inclinations obtained from analysis of light curves. These orbital inclinations are published in various papers, but in the present discussion we have used the latest data, mainly taken from A Catalogue of Graded Photometric Studies of Close Binaries by Koch, Plavec and Wood (1970) and A Catalogue of Modern Lightcurve Synthesis Solutions of Close Binary Systems by Cester, Fedel, Giuricin, Mardirossian, Mezzetti and Predolin (1979). The observed rotational velocities of the emission rings expressed by V_{em}Sin i are taken from the compilation of Kruszewski (1967).

The rotating velocity, V_{em}, of the emission disk shown in Table 2 has been rewritten in terms of the velocity unit

$$\sqrt{\frac{G\ (m_1 + m_2)}{a_1 + a_2}}\ ,$$

as shown in the fourth column of Table 3. The second and third columns represent the mass-ratio q and the velocity unit of each system.

TABLE 2. ALGOL-TYPE ECLIPSING BINARIES WITH EMISSION DISKS

(Double-lined)	P	V_{em} Sin i	i	m_1 (in ⊙	m_2 unit)	$a_1 + a_2$
RX Cas	32.3days	150km/sec	75°	0.66	0.62	3.22×10^7km
KU Cyg	38.4	184	90	4.7	0.83	5.89
AW Peg	10.6	210	87	2.02	0.32	1.88

(Single-lined)	P	V_{em} Sin i	i	Sp_1	m_1^*	f(m)	m_2	$a_1 + a_2$
RY Gem	9.30days	200km/sec	87°	A2Ve	2.6	0.0185	0.57	1.90×10^7km
U Sge	3.38	280	89	B9Ve	3.7	0.119	1.47	1.14
RW Tau	2.77	350	90	B8Ve	4.3	0.0382	1.03	1.00

*) The m_1-values of the primary components in single-line binaries have been assumed, corresponding to their main-sequence spectral types (Allen 1973).

Substituting the velocity of the rotating disk in standard unit, V_{disk} (as given in the fourth column of Table 3), in the left-hand side of Equation (47), the corresponding Roche dimension of the disk, r_{disk}, in standard unit can be deduced numerically as shown in the fifth column of Table 3. The r_{Roche}-value (dimension of the Roche limit) for each system is also shown in the last column, which has been taken from Plavec's work (1964) corresponding to the respective mass-ratio.

TABLE 3. ROCHE DIMENSIONS OF ROTATING DISKS IN ALGOL-TYPE ECLIPSING BINARY SYSTEMS

	q	Vel.Unit	V_{disk}	r_{disk}	r_{Roche}
RX Cas	0.94	7.26×10^6cm/sec	2.13	0.105	0.380
KU Cyg	0.17	11.14	1.65	0.240	0.546
AW Peg	0.16	12.86	1.63	0.245	0.552
RY Gem	0.22	14.87	1.34	0.308	0.521
U Sge	0.40	24.54	1.14	0.339	0.462
RW Tau	0.24	26.56	1.32	0.310	0.512

From the comparison of the values of r_{disk} and r_{Roche} in Table 3, it is evident that the rotating disk in an Algol-type close binary system is kept well inside of the Roche limit.

(b) CATACLYSMIC VARIABLES

The cataclysmic variables such as novae, recurrent novae, dwarf novae and nova-like variables are believed to be short-period (period < 16 hours) close binary systems and each system consists of a degenerated dwarf star and a non-degenerated companion. The non-degenerated component fills its Roche limit and the overflowing matter from the latter secondary component transfers onto the degenerated star, causing an accretion disk around it with emission features.

In order to evaluate the Roche dimensions of the accretion disks in cataclysmic variables, let us take the dwarf novae with available values of i, V_eSin i and q from the list of Warner (1976). They are shown in Table 4.

TABLE 4. DWARF NOVAE WITH AVAILABLE i, V_eSin i, m_1 and m_2

	P	V_eSin i	i	m_1	m_2	K_1	a_1+a_2
AE Aqr	$0^d.412$	550km/sec	47°	1.25	1.18	135km/sec	2.16×10^6km
RU Peg	0.371	500	44	0.97	1.12	137	1.88
Z Cam	0.290	700	59	1.17	0.86	144	1.59
SS Cyg	0.275	500	43	0.97	0.83	122	1.46
RX And	0.212	410:	29	1.02	0.65	77	1.18
DQ Her	0.194	680	64	0.83	0.60	150	1.06
SS Aur	0.194	450:	32	0.89	0.57	85	1.10
V603 Aql	0.139	240	15	0.87	0.40	37	0.865
TT Ari	0.133	350	23	0.79	0.38	55	0.794

The a_1+a_2 values given in the last column of Table 4 are calculated with

$$a_1+a_2 = a_1(1+1/q) = 0.01375\ 10^6 P^4 K_1 (1+1/q) . 1/\mathrm{Sin}\ i \qquad (59)$$

where P is given in days and K_1 in km/sec.

Similarly as in the case of Algol-type binaries in the previous sub-section (a), application of our Equation (47) has been done to each of the accretion disks of the nine cataclysmic variables in Table 4. The result is shown in Table 5.

TABLE 5. ROCHE DIMENSIONS OF ACCRETION DISKS IN CATACLYSMIC VARIABLES

	q	Vel.Unit	V_{disk}	r_{disk}	r_{Roche}	r'_{disk}
AE Aqr	0.94	3.87×10^7 cm/sec	1.94	0.122	0.380	0.124
RU Peg	1.15	3.83	1.87	0.119	0.361	0.140
Z Cam	0.73	4.11	1.98	0.130	0.403	0.100
SS Cyg	0.86	4.04	1.81	0.142	0.388	0.129
RX And	0.64	4.33	1.95	0.140	0.416	0.090
DQ Her	0.72	4.22	1.79	0.155	0.404	0.116
SS Aur	0.64	4.19	2.02	0.132	0.416	0.091
V603 Aql	0.46	4.41	2.10	0.138	0.448	0.075
TT Aur	0.48	4.42	2.02	0.145	0.444	0.076

In evaluating the dimension of an accretion disk in a cataclysmic variable, the following Equation has been frequently used (e.g., Huang and Struve 1956; Robinson 1976);

$$r'_{disk} = \frac{1+q}{q} \left(\frac{K_1}{V_e \text{ Sin } i} \right)^2 \qquad (60)$$

which was derived from the assumption of Keplerian rotation of the disk around the parent star (the primary component), neglecting the perturbation of the orbit of the disk by the secondary component. Such a Keplerian dimension of the disk is given as r'_{disk} in the last column of Table 5.

Comparison between the values of r'_{disk} and r_{disk} in Table 5 indicates that the deviation of the Keplerian dimension from the Roche dimension seems to be greater for smaller q-value. In the cases of V603 Aql and TT Aur with q-values less than 0.5, the respective Keplerian dimension amount to only one half of the corresponding Roche dimension .

(c) COMPARISON OF ROTATIONAL VELOCITIES OF THE ACCRETION DISK CALCULATED WITH VARIOUS MODELS

Next, we shall compare numerically the rotational velocities of the accretion disk of cataclysmic variable calculated with the present model (Equation (47)) and the theoretical models of Warner and Peters (1972), Flannery (1975) and Lubow and Shu (1975).

Warner and Peters (1972) published a series of models for the rotational velocity of the accretion disk in a cataclysmic variable in connection with the hot spot's size and location as a function of the mass ratio of the components based upon particle trajectory calculations in the restricted three-body approximation. In the following Table 6, we take the values of the disk dimension r and the rotation velocity V_{WP} calculated for a given mass-ratio q from their paper. For the same values of a pair of q and r, the rotational velocity calculated with Equation (47) is given in the fourth and eighth columns of Table 6 as V_{Roche}.

TABLE 6. COMPARISON OF THE ROTATIONAL VELOCITY OF THE ACCRETING DISK BETWEEN WARNER AND PETER'S MODEL AND THE PRESENT

q	r	V_{WP}	V_{Roche}	q	r	V_{WP}	V_{Roche}
0.01	0.3929	1.430	1.196	0.80	0.1321	2.350	1.925
0.025	0.3252	1.694	1.409	0.90	0.1279	2.328	1.907
0.05	0.2778	1.903	1.576	1.00	0.1244	2.304	1.887
0.10	0.2340	2.108	1.741	1.5	0.1118	2.180	1.787
0.20	0.1945	2.285	1.880	2.0	0.1039	2.065	1.694
0.30	0.1737	2.357	1.936	3.0	0.0940	1.877	1.544
0.40	0.1602	2.385	1.957	4.0	0.0877	1.734	1.430
0.50	0.1505	2.391	1.960	5.0	0.0831	1.622	1.340
0.60	0.1430	2.384	1.954	10.0	0.0704	1.290	1.073
0.70	0.1371	2.369	1.941				

Comparison between V_{WP} and V_{Roche} in Table 6 shows that the rotational velocity of the accretion disk calculated by Warner and Peter deviates rather seriously from that from the present model based upon the hydrodynamics. The main cause of the deviation of V_{WP} from V_{Roche} would not be due to the model adopted by Warner and Peters, but perhaps it may be attributed to some program error in their calculation as pointed out by Flannery (1975) and by Lubow and Shu (1975).

Correcting an error in Warner and Peter's program, Flannery (1975) calculated the rotational velocities of the disk V_F for various pair values of q and r, which are shown in Table 7. The rotational velocities V_{Roche} based upon the present model are also calculated as shown in the fourth and eighth columns of Table 7.

Comparison between the values of V_F and V_{Roche} in Table 7 indicates that the result of Flannery based upon his particle models agrees satisfactorily with the present hydrodynamic result of V_{Roche}.

Finally we shall take Lubow and Shu's numerical result based upon the gas dynamics (1975). Calculation of V_{Roche} with

TABLE 7. COMPARISON OF THE ROTATIONAL VELOCITY OF THE ACCRETING DISK BETWEEN FLANNERY'S MODEL AND THE PRESENT MODEL

q	r	V_F	V_{Roche}	q	r	V_F	V_{Roche}
19.000	0.0402	1.12	1.078	0.667	0.1028	2.42	2.316
9.000	0.0475	1.45	1.406	0.429	0.1237	2.39	2.258
4.000	0.0575	1.87	1.810	0.250	0.1567	2.26	2.107
2.333	0.0665	2.12	2.060	0.111	0.2227	2.01	1.791
1.500	0.0762	2.29	2.218	0.053	0.2984	1.78	1.489
1.000	0.0878	2.39	2.302				

Equation (47) in the present model has been done for the same pair values of 1/q and r as listed by Lubow and Shu in their paper, where 1/q is replaced by M_D/M_C, r by ϖ_d and V_{LS} by u_φ in their notations. The result of calculations is shown in Table 8. As is easily seen from Table 8, the agreement between the results of Lubow and Shu and the present is excellent.

TABLE 8. COMPARISON OF THE ROTATIONAL VELOCITY OF THE ACCRETION DISK BETWEEN LUBOW AND SHU'S MODEL AND THE PRESENT MODEL

1/q	r	V_{LS}	V_{Roche}	1/q	r	V_{LS}	V_{Roche}
0.0667	0.0404	1.203	1.206	3.0000	0.1363	2.209	2.213
0.1500	0.0492	1.578	1.581	4.5000	0.1631	2.076	2.080
0.3000	0.0589	1.921	1.923	5.5000	0.1781	2.001	2.005
0.5000	0.0683	2.141	2.144	7.0000	0.1978	1.905	1.909
0.7500	0.0780	2.266	2.269	8.5000	0.2147	1.826	1.830
1.0000	0.0865	2.317	2.321	10.0000	0.2297	1.759	1.763
1.6667	0.1058	2.324	2.328	15.0000	0.2698	1.594	1.598

8. CONCLUSION

As is evident from the results in the previous Section, the simple analytical relation (47)

$$V_{disk} \equiv V_{Roche} = \frac{1}{\frac{1}{h_2}\frac{\partial h_2}{\partial \xi}} \left\{ - h_1 + \sqrt{h_1^2 - \left(\frac{1}{h_2}\frac{\partial h_2}{\partial \xi}\right)\frac{1}{1+q}} \right\}$$

is very useful for accurate evaluation of the dimension of the rotating gaseous disk in an Algol-type close binary or of the accretion disk in a cataclysmic variable, if its rotational velocity is known from observation. The result of discussion with the above simple Equation should be equivalent to the result obtained from time-consuming numerical integrations of hydrodynamic equations of the gas flows, as the comparison between the present result and Lubow and Shu's one in Table 8 reveals.

REFERENCES

Allen,C.W.: 1973, Astrophysical Quantities, 3rd ed. Athlone Press.
Batten,A., Fletcher,J.M., and Mann,P.J.: 1978, Seventh Catalogue of Orbital Elements of Spectroscopic Binary Systems, Publ.Dom. Astrophys. Obs. XV, No.5.
Cester,B., Fedel,B., Giuricin,G., Mardirossian,F.,Mezzetti,M.,and Predolin,F.: 1979, Mem.Soc.Astr.Italiana 50, No.4.
Flannery B.P.: 1975, MNRAS 170, 325.
Huang, Su-Shu: 1957, Ap. J., 148, 793.
Huang, Su-Shu and Struve,O.: 1956, A. J., 61, 300.
Kitamura, M.: 1970, Astrophys. Sp. Sci., 7, 272.
Kitamura, M.: 1973, IAU Symp. No. 51, ed. A.Batten, pp.107-110.
Kopal, Z.: 1959, Close Binary Systems, Chapman-Hall and John Wiley.
Kopal, Z.: 1969, Astrophys. Sp. Sci., 5, 360.
Kopal, Z.: 1978, Dynamics of Close Binary Systems, D.Reidel.
Koch,R., Plavec, M., and Wood, F.B.: 1970, Catalogue of Graded Photometric Studies of Close Binaries, Publ.Univ.Pennsylvania,Vol.11.
Kruszewski, A.: 1967, Acta Astr., 17, 297.
Lubow, S.H. and Shu, F. H.: 1975, Ap. J., 198, 383.
Paczynski, B.: 1977, Ap..J., 216, 822.
Piotrowski, S.L.: 1975, Acta Astr. 25, 21.
Plavec, M.: 1964, BAC, 15, NO. 5.
Prendergast, K.H.: 1960, Ap. J., 132, 162.
Prendergast, K.H. and Taam, R.: 1974, Ap. J., 189, 125.
Robinson, E.L.: 1976, Ann.Rev.Astr.Astrophys. 14, 119.
Sahade,J.: 1960, Stellar Atmospheres ed.Greenstein, Univ.Chicago Press, Chap. 12.
Sobouti, Y.: 1971, Ap. Sp. Sci., 12, 408.
Sorensen,S., Matsuda,T., and Sakurai,T.: 1975, Ap.Sp.Sci.,33, 465.
Struve, O.: 1950, Stellar Evolution, Princeton Univ. Press.
Walker, M.: 1954, PASP 66, 230.
Warner, M. and Peters, W.L.: 1972, MNRAS 160, 15.
Warner, B.: 1973, MNRAS 162, 189.
Warner, B.: 1976, IAU Symp. No.73, Structure and Evolution of Close Binary Systems, ed. Eggleton,Mitton and Whelan, p.85.

CLOSE BINARY SYSTEMS OF SHORT PERIOD

E. Budding
Department of Astronomy, University of Manchester, Manchester, England.

ABSTRACT

Information on 406 close binary systems with periods less than 1.5 days for which the primary is of known spectral type, is assembled from recent sources. The organization of the presentation depends on a light curve descriptor which takes the values EA1, EA2, EA3, EB, EW and E 11. Within these groups the distribution in the spectral type (s): period (P) plane is examined. An additional distribution of this kind is presented for 311 EA2 candidate stars (evolved Algols), with periods of up to 15 days.

Recent empirical information on Main Sequence stars is presented for comparison, and the condition that a Main Sequence component of a binary system be in "contact" with its surrounding Roche lobe provides an interesting discriminant in relation to the (s:P) distributions of the different light curve groups. Certain anomalous cases come to light.

Concerning the EW (W UMa) group, detailed investigation of a particular case reveals the extent of parameter indeterminacy. A compelling case for "contact" can only be found among the accumulation which occurs at later spectral type.

A recommendation is made for a simple but more quantitative means of defining light curve type. Such a parameter could be of considerable significance in relation to statistical tests of binary evolution theory.

1. INTRODUCTION

There are practical reasons motivating attention to close binary systems of short period, not least of which is the central role played by the orbital period in relation to the organization for and throughput of a practical information acquiring system. Or to borrow Kopal's

F. D. Kahn (ed.), Investigating the Universe, 271–319.

(1979) analogy, such systems may be regarded as playing their "signature tunes" to us at a more brisk rate, and, since our time is finite, we may learn more from them than from their more allargando fellows. Alternatively, since shortness of periods generally implies a greater relative proximity, the extent to which stars will divulge their inner make-up, in response to neighbourly inducements (proximity or interactive effects), may be enhanced.

An upper limit of 1.5 days has been, somewhat arbitrarily, placed on the periods of most of the systems to which particular attention will be paid, which include all the main "classical" groups of close binary system. Highly evolved systems in which there may be collapsed objects, or cataclysmic binaries will lie outside the ambit of this paper, however. Further restrictions associated with the intended scope of the article and choice of period range mean that systems of very early spectral type or great mass are also excluded, though except for the occurrence of Wolf-Rayet stars in some such binaries, most would fall into the same groupings that have their counterparts among the more frequently encountered less massive stars.

The recent appearance of the extended and detailed "Finding List" of Wood et al. (1980) (hereafter referred to as WOFK) with information on over 3500 binary systems, and Popper's (1980) latest reappraisal of empirical absolute data on stars, together with the "General Catalogue" of Kukarkin et al. (1969) and its more recent supplements (1974, 1976) (hereafter GCVS 1, 2 and 3), have all made a quest, like the one considered here, a much more feasible possibility. These sources, especially, provided much of that valuable particular information from which generalities can be drawn (howbeit sometimes precariously), and the author is clear to acknowledge this.

2. BROAD CATEGORIZATIONS OF OBSERVATIONAL DATA

The photometric variations characteristic of close binary systems are customarily divided into three or four categories, which GCVS 1 denotes as EA, EB, EW or just E (i.e. not clearly one of the first three main groups). The distinction between the groups (which are alternatively described as Algol-type, β-Lyrae type, W UMa-type or just eclipsing type light curves) is somewhat qualitative and open to various criticisms. Kopal (1959), for instance, restated that the photometric appearance of an eclipsing binary light curve depends on the wavelength of observation - an Algol type light curve may look more like a β-Lyrae type light curve if observed at sufficiently long wavelengths (well into the IR). Also there are, no doubt, numerous boundary cases in which the assignation to one or other group must be somewhat arbitrary. However, these descriptions of different photometric appearances have the merits that they exist for almost all photometric binaries, i.e. sample sizes can be relatively large, and they do not depend much on preconceived ideas (other than the notion of eclipses). There may well be genuine and quite marked physical differences between systems which have some correlations with this descriptor, the exploration of which is one of the main purposes of this study. As to the wavelength of observation - in the great majority of cases included in the catalogues the original classification has been made from photographic data. In the few cases of a "visual" wavelength origin the shift from 0.45 to 0.55 μ is unlikely to significantly alter the gross characteristics of the light curve.

Apart from the light curve descriptor a catalogued system will generally be given (in addition to a name and position in the sky) maximum and minimum brightness levels, a period, some indication of the duration of an eclipse and maybe whether it is estimated to be partial or complete, and sometimes, usually for relatively brighter stars, a spectral type. The question of how long an eclipse lasts may be difficult to answer for the EB and, especially, the EW systems, however. There are also a few cases in which photometric variations, though often the most notable characteristic of close pairs, are not of the sort of scale that is associated with the informative eclipse effect (Ell type variables). No doubt with the increasing exploitation of interferometric and other techniques the importance of these numerous non-eclipsing close binary systems as information sources will increase.

Concerning spectral type, the advantages of a spectrographic designation are that, where reasonable dispersion spectrograms are available, the type can usually be assigned fairly easily and reliably to within one or two subclasses, and luminosity class can also often be inferred from the same data. Such information will be free from the relative complications of colour calibrations and accounting for reddening effects. On the other hand, spectral classification (as indeed the stellar magnitude system) is an inheritance from times when an entirely quantitative approach to the problem was unavailable. The subject has passed through a long history of subtle developments before reaching its present form (c.f. Keenan and McNeill, 1976, Yamashita et al. 1977), though some features of a semiqualitative origin still remain in the matter of type assignation. This can be expected to affect our procedure to the extent that since spectral type is one of the basic and useful descriptors it will be required to symbolise it by means of the variable s. It is well known that other stellar parameters (mass, radius etc.) have a somewhat irregular functional form in relation to s even within a given sequence. This is not an essential impediment to our envisaged presentation, however, in which spectral type, as a descriptor, stands intermediate between, say, magnitude and light curve class in terms of the degree of quantification.

We shall attempt to assess the implications of the information expressed by the foregoing general descriptors by making reference to fairly well established empirical generalities, such as Kepler's third law, the existence of a stellar Main Sequence or correlations between some of the major characteristics of certain kinds of star, and photometric consequences of various eclipse situations, which can usually be described, in essence, by suitable formulae. Such implications could then be related to various theoretical schemes, though there is not space to say much about these in the present paper. The actual descriptors concentrated upon will be period, spectral type and light curve type. It is of particular importance that there are a number of relatively bright well observed and studied representatives; the prototype names of the light curve classes give some indication of this. It will be presumed that certain attributes of these highly studied objects can apply also to less studied cases.

3. THE EMPIRICAL MAIN SEQUENCE

In order to proceed further with the categorization and analysis some yardstick criteria are required, and as a first task an empirical Main Sequence has been extracted from the data of Popper (op. cit.). Partly since data on the very hottest stars are somewhat limited, and partly because of the restrictions mentioned in the introduction, there will be little, if anything, to say about stars of spectral type earlier than B0. The adopted Main Sequence given in Table 1 has been constructed from running smooth curves through those binary data given by Popper which seem to correspond to an unevolved limit. Inevitably there must be some errors of personal judgement in such a procedure, but it is hoped that the scale of these can be indicated by the number of retained digits in Table 1, where it would be over optimistic to attach much importance to the third figure after the decimal point which is included only for smoothness.

Table 1: Empirical Main Sequence

Spectral Type	log (Mass) $M_\odot$	log (Radius) $R_\odot$	M_V
B0	1.18 5	0.76 0	-3.3
B5	0.72 0	0.50 5	-0.88
A0	0.32 4	0.26 2	0.96
A5	0.21 5	0.18 5	1.87
F0	0.12 6	0.11 7	2.72
F5	0.07 0	0.07 2	3.38
G0	0.00 6	0.01 8	4.25
G5	-0.04 4	-0.02 6	5.01
K0	-0.10 6	-0.08 3	5.93
K5	-0.25 9	-0.22 7	8.20
M0	-0.37 5	-0.33 9	9.90
M5	-0.58	-0.54	13.0
M8	-1.	-0.96	19.5

The adopted data comes mainly from an intensive study of about thirty "Main Sequence" eclipsing binaries, together with a few resolved spectroscopic binaries and several well documented visual binaries, the latter being particularly useful in the definition of the cooler end of the Sequence.

A comparison of the foregoing data with theoretical sequences such as those indicated in Fig. 6 of Iben (1967), using approximate temperatures and bolometric corrections interpolated from Allen's (1973) tables, shows that the empirical sequence lies sensibly between the ZAMS and TAMS-GAMS curves with some proclivity towards the GAMS curve at the lower end and a close approach to the ZAMS curve at the upper end.

Mostly the binaries used to define the sequence are relatively bright objects, and the data of Table 1 can be inferred to be representative of a normal Population I sample of the solar neighbourhood.

4. DISTRIBUTION OF OBSERVED TYPES FOR MAIN SEQUENCE STARS

Counts of star numbers, taken from a few sources, in certain ranges of spectral type and brighter than a given magnitude have been carried out. Fig. 1 shows results of these counts, and, for comparison, a theoretical distribution using a Salpeter function (Salpeter, 1955), $\phi(M) \propto M^{-\gamma}$, with some estimate of mean interstellar extinction, and the Main Sequence luminosity values of Table 1. The effect of variation of log (mass) interval with spectral type is indicated.

One difficulty of the Bright Star Catalog (hereafter BSC - Hoffleit (1964)) is that though many stars are assigned a spectral type in it (about 40% of the stars in the catalogue) only a certain proportion of these are also given luminosity classes. Moreover, even then the luminosity classification is of either Mt. Wilson (W) or Morgan-Keenan (MK) types. It has been supposed that the Mt. Wilson "d" classification, which applies mainly to intermediate and late type stars, is essentially the same as the "V" designation of Morgan-Keenan and that both of these refer to Main Sequence dwarfs. To arrive at some final representations allowances have to be made for this heterogeneity of make up, and this can be ascertained from the numbers in Table 2.

Numbers of classified stars taken from every tenth page of the catalogue are given in the first row, while the corresponding numbers of W and MK dwarfs among these are given on the next two rows. The columns correspond to a spectral type representative for a small surrounding range, for example A0 would include all stars from B8 to A2 inclusive. (This procedure will often be repeated). The numbers of stars among those in the first row which have actually been given a luminosity class of any kind are given in the fourth row (MK given first in those columns where both MK and W are included), from which the estimated numbers of dwarfs in each type range and corresponding frequency is given in the fifth and sixth rows.

Keeping in mind the inhomogeneities of the BSC, and also that this sample is likely to contain a representation of early stars influenced by some local regions of star formation, some fairly uniform samplings of Kapteyn areas due to Chuadze (1973) were also checked. These areas correspond to intermediate disk population stars. Luminosity classes as well as spectral types were given for a high percentage (> 80%) of the

Table 2: Main Sequence Stars of Various Types in the BSC

Type	O5	B0	B5	A0	A5	F0	F5	G0	G5	K0	K5	M
Totals	4	70	101	338	143	124	113	88	120	392	126	114
W(d)	0	0	0	0	5	19	44	21	5	3	0	0
MK(V)	0	27	40	79	24	8	12	16	10	7	2	0
Total L	0	65	67	119	39+7	28+27	31+57	36+33	30+49	152+99	52+57	20+66
Est.Total MS	0	30	60	224	90	61	72	47	23	16	2	0
Frequency	0.0	0.048	0.096	0.358	0.144	0.098	0.115	0.075	0.037	0.026	0.003	0.000

Table 3: Frequencies of Main Sequence Stars in Kapteyn Areas 25, 48 and 50 of Chuadze (1973)

Type	B0	B5	A0	A5	F0	F5	G0	G5	K0	K5	M
Frequency (Area 25)	0.0	0.012	0.368	0.152	0.047	0.168	0.199	0.009	0.031	0.012	0.0
Frequency (Area 48)	0.0	0.032	0.305	0.263	0.074	0.105	0.137	0.031	0.031	0.021	0.0
Frequency (Area 50)	0.0	0.018	0.283	0.226	0.065	0.114	0.218	0.007	0.040	0.025	0.002

stars in the samples, though the data become somewhat uneven at fainter than 13th magnitude, and such stars were not included in the counts. The results for three such areas are given in Table 3.

The theoretical distribution was transformed to the s axis by writing

$$\phi(s) = \text{const.} \times M^{\gamma}(s)\ \rho^{3}(s)\ \frac{dM}{ds}$$

where mass M(s) is obtained from Table I and the distance ρ(s) out to which a Main Sequence star of type s will be brighter than some given magnitude is obtained from the relation

$$5 \log \rho + 0.002\rho = \text{const.} - M_V(s)$$

where absolute magnitude $M_V(s)$ is also obtained from Table I.

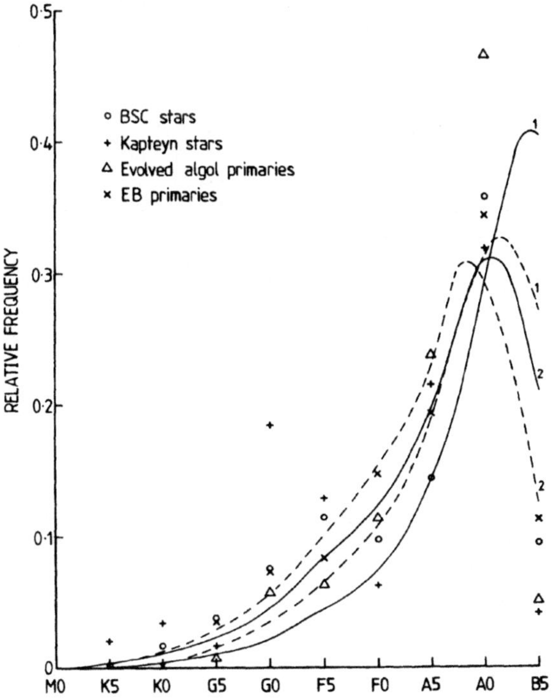

Figure 1: Star counts and theoretical comparisons. The index γ was taken to be 2.75 (Taff, 1974). Curves 1 correspond to limiting M_V of 10 while for 2, M_V = 15, i.e. the proportion of early stars is reduced due to reddening. The variation $(\frac{dM}{ds})$ is accounted for in the full curves but taken to be constant in the dashed ones.

The general behaviour of all these distributions is roughly similar though there are some peculiarities. The under-abundance of early types relative to the theoretical description is presumably due to age effects, though the position of the maximum may also be sensitive to the assumed value of some constants. The occurrence of a minimum at around F0 can be noted in the observational counts, though it is not clear how to relate this to a form such as (4.3)on the basis of the data of Table 1.

5. THE ALGOL-TYPE (EA) LIGHT CURVE SYSTEMS

Eclipsing binary light curves of this type are by far the most frequently observed kind ($\gtrsim 2/3$), though certain factors may favour their discovery. The main features of the light curve are that there should be at least one minimum of quite distinct duration during the cycle of light changes, and that outside the minima the light level is relatively constant ($\lesssim 0^m.1$), without the characteristic curved maxima associated with pronounced proximity effects. Generally speaking, the periods are longer than the 1.5 day limit which was adopted for this study, though even in the given range of periods a significant number of EA light curves may be counted.

With regard to circumstances favouring discovery, one fact which quickly comes to light and, in effect, can serve as the basis for a further distinction, is that some of the observed minima are remarkably deep (e.g. > 2 mag. in V light for X Tri), and this point alone allows us to deduce that such an eclipse cannot be due to a pair of Main Sequence stars. Aided by the knowledge of certain well studied examples, it is possible to distinguish between three possible categories of EA light curve as follows: EA1 those due to Main Sequence pairs, EA2 those due to "classical evolved Algols", (i.e. semi-detached or near semi-detached pairs) EA3 those others which cannot be clearly associated with either of these cases.

6. EA1 SYSTEMS - MAIN SEQUENCE PAIRS

The maximum depths of minima, λ_1, λ_2, in the case of Algol-type light curves, which can be associated with a pair of Main Sequence stars, can be approximately obtained as follows

$$\lambda_1 = -2.5 \log \left\{1 - \left(\frac{R(s_2)}{R(s_1)}\right)^2 L(s_1)\right\},\ \lambda_2 = -2.5 \log \left\{L(s_1)\right\}$$
$$\text{where } L(s_1) = 1/(1 + 10^{-.4(M_v(s_2) - M_v(s_1))}) \qquad (6.1)$$

where R(s) and $M_v(s)$ are available from Table 1. Curves showing the relationship of λ_1 and λ_2 for given combinations of s_1 and s_2 are presented in Figure 2. In the case of systems where the maximum obscuration is still only partial the foregoing depths would be reduced by approximately similar proportions. Hence where EA type light curves exhibit two minima

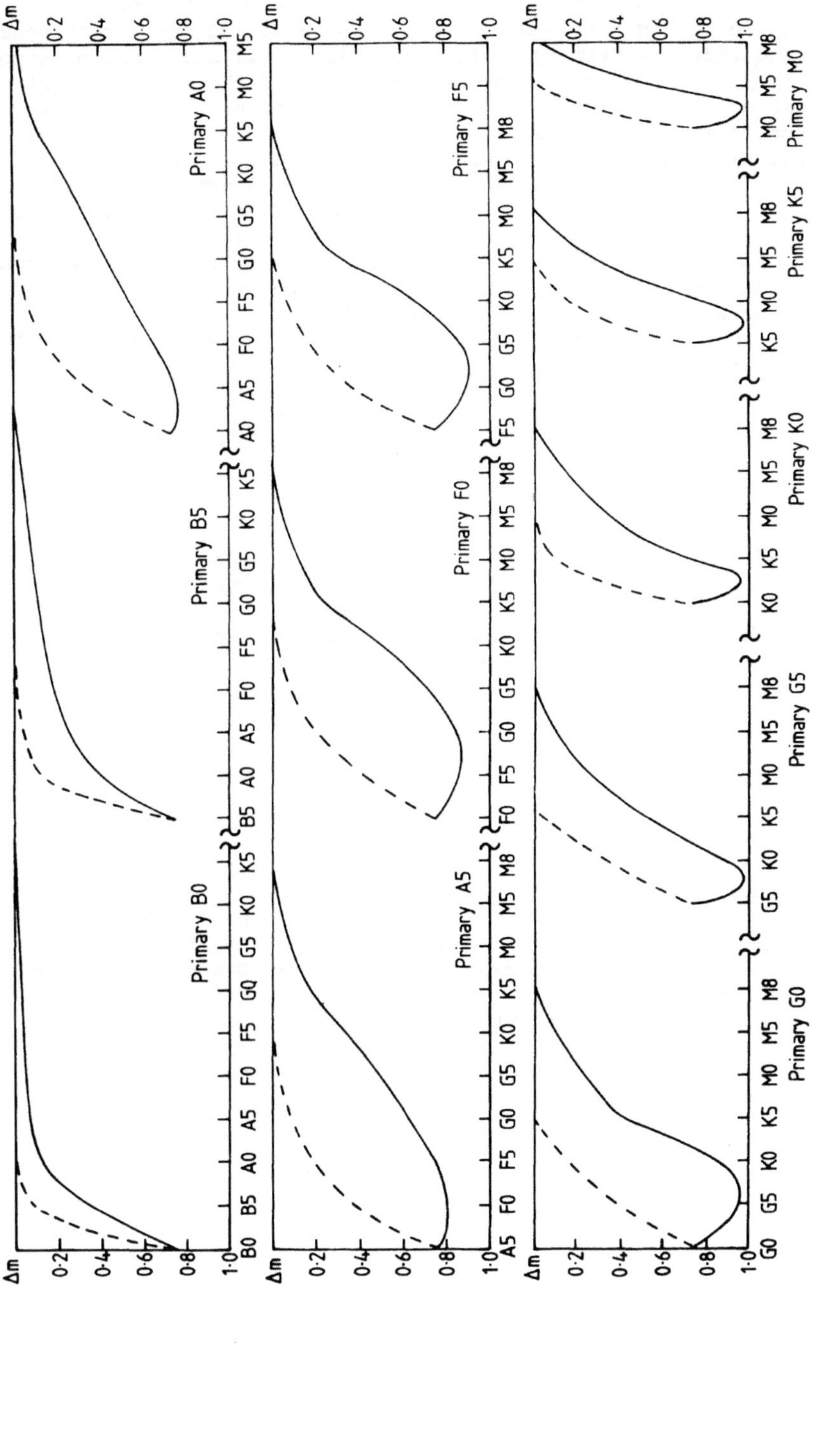

Figure 2: The values of maximum depths λ_1 (primary, full curve) and λ_2 (secondary, dashed curve) of eclipses, in magnitudes, for EA1 systems.

and the secondary minimum depth is in a similar ratio to the primary minimum depth, as would correspond, for a given spectral type, with some other Main Sequence spectral type, to within about 0.1 mag. (making some allowance for proximity effects on the definition of the maximum light level or limb darkening effects in the primary minimum) it will be taken as a good indication that the stars are probably both of Main Sequence type. In certain cases this has been attested by the identification of a secondary spectral type of the appropriate kind. Stars with light curves of this kind have been styled EA1. A sample of stars showing light curves of this type which have known spectral type primaries and are brighter than magnitude 12.5 is listed in Table 4 and Figure 3 illustrates the corresponding distribution in the spectral type-period (s:P) plane. The data of Table 1 can allow us to define a "contact domain" on such a diagram as follows: a point (s, P) is within the contact domain if, for given s, P satisfies

$$P(s) = \exp\left\{\frac{3}{2}\log_e 10\left(\log R(s) - \frac{1}{3}\log M(s) - \frac{1}{3}\log(1+q) - 0.624 - \log r_1^*(q)\right)\right\} \quad (2.5)$$

In the right hand expression the mass ratio $q (= M_2/M_1)$ acts as a parameter, and since for Main Sequence pairs $M_2 < M_1$, q must lie in the range $0 < q < 1$. $q = 0, 1$ define the lower and upper bands of the contact domain for the purposes of this paper. Equation (2.5) comes from nothing more than Kepler's third law slightly rearranged, so that the fractional size of the primary component in terms of the mean orbital separation of the two mass centres r_1 is volumetrically the equivalent to the Roche lobe quantity $r_1^*(q)$ (Kopal, 1959, Ch. 3), which thus expresses the "contact" condition. The values of log R(s) and log M(s) are obtained from Table 1.

It should be stressed that the presence of a star in the contact domain of the s:P plane does not necessarily mean that the primary is in contact with its surrounding Roche lobe, nor does the presence of a star above the contact domain preclude this possibility. The relationship of the position of the (s, P) point for a particular star to this domain will give some indication of the proximity of the primary to its Roche lobe on the assumption of a Main Sequence or near to Main Sequence character. If there is strong evidence of a "contact" condition for a star outside this domain it would similarly indicate an evolved condition. The usefulness of the demarcation can at least be justified by the sieve-like role it played in throwing up a number of curious cases which have possibly been misrepresented in catalogues, or otherwise show abnormal behaviour.

The following deductions can be made from Figure 3

1) All EA1 primaries are safely outside of the contact domain except the three systems EG Cep, RU UMi and GR Tau.

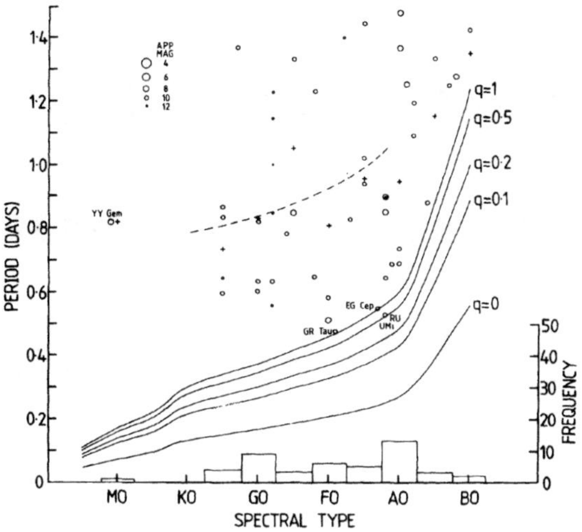

Figure 3: Distribution of EA1 stars in the s:P plane. Mean periods within each five subdivisions of spectral class are indicated by + signs. The dashed curve follows the trend of mean period values in the region where the samples may be more complete.

2) The mean periods tend to follow a path which is roughly parallel with the contact domain but such that $r_1/r_1{}^* = (P_1/P_1{}^*)^{-2/3} \sim 0.5$. This is in keeping with a selection effect to detect eclipses, which increases with greater r_1, subject to the constraint that $r_1, r_2 \lesssim 0.2$, in order that proximity effects do not start to produce too rounded maxima (corresponding to the EB light curves to be discussed later).

It may also be worth noting that six of the seven brightest stars of the sample with spectral type later than F5 have been reported to show CaII H and K lines in emission.

Concerning the exceptional systems we may make the following observations:

(i) RU UMi

There is apparently a difference in opinion over the spectral class of this star. The GCVS 1 follows Götz and Wenzel (1964) in publishing an A2 type. Macdonald (1964), however, found it to be F5, which would move it out of the contact domain as well as making the primary depth ($0^m.9$) more consistent with Fig.3. WOFK repeats the GCVS 1 data, on the basis of which the system could almost qualify as a classical (evolved) Algol candidate (see next section). Errors in type assignation could, of course, be present in a number of cases, some of which could cause other systems to move significantly in relation to the contact domain.

Table 4: 52 Close Binary Systems of Type EA1

Constellation	Name or No.	Period	Primary s Type	Mag.	Constellation	Name or No.	Period	Primary s Type	Mag.
And	RT	0.629	F8	9.34	Cyg	V387	0.641	G5	11.5
	AA	0.935	A5	10.03		V456	0.891	A2	11.0
Aqr	EE	0.509	FO	8.3		V466	1.392	A8	10.8
Aur	IM	1.247	B9	7.9	Del	DM	0.845	A2	8.6
Cam	SV	0.593	G5	9.8	Eri	WX	0.823	A7	9.5
Cnc	WX	1.225	F2	10.5	Gem	YY	0.814	M1	9.1
	WY	0.829	G5	9.6	Hya	AV	0.683	A1	10.7
CVn	VZ	0.842	F5	9.4		EU	0.778	F6	10.1
CMa	FZ	1.273	B2	8.07	Leo	UV	0.600	GO	9.5
CMi	XZ	0.579	FO	10.2	Lib	SS	1.438	A5	10.4
Cap	UW	1.399	A	10.9	Lyn	SW	0.644	F2	9.51
Car	DW	1.328	B5	9.6	Mon	BB	0.733	AO	9.9
	EZ	1.189	B8	9.6		V380	0.996	F8	12.3
Cas	ZZ	1.244	B3	10.8	Oph	V456	1.016	A5	10.2
	MM	1.158	F	11.3	Psc	RV	0.554	F8	11.3
	MS	1.173	F	12.0		UV	0.861	G5	9.4
	OR	1.246	F3	11.4	Sco	V616	0.844	F8	12.0
	PV	0.875	B6	10.0		V634	1.224	A	11.7
	V374	1.045	A	11.8	Sct	ER	1.361	AO	9.1
Cen	ST	1.223	F8	10.8	Tau	GR	0.474	A9	10.3
	V380	1.087	B8	9.7	Tel	HO	0.892	A2	7.9
	V775	1.327	F5	9.9	UMi	RU	0.525	A2	10.0
Cep	EG	0.545	A3	9.31	Vel	GG	1.475	AO	8.0
CrA	TZ	0.687	AO	9.6	Vir	BF	0.641	A2	10.5
Cru	AI	1.418	BO	9.6		BH	0.817	GO	9.6
Cyg	CG	0.631	GO	10.06	Vul	BT	1.141	F8	11.8

(Note: Stars for which the spectral subclass is not given are not plotted out).

(ii) EG Cep

The primary must be close to its Roche critical surface if the type assignation is correct. The ratio of depths of the minima suggests a mass ratio $q \sim 0.8$ for Main Sequence stars, while the primary actually lies on the curve representative of contact for a $q = 1$ condition. Period irregularities noted by Pohl and Kizilirmak (1977) may therefore be of special significance.

(iii) GR Tau

This star is in a similar situation to EG Cep. It is curious that in both these cases the apparent proximity effects are low, which could be suggestive of additional photometric effects tending to cover the normal proximity variations (which are known to operate in, for example, CG Cyg). Even if our interpretation was wrong and the systems are in reality more like evolved Algols, the deep secondary minima suggest such relatively bright secondaries that the large physical distortions then implied should be noticeable.

(iv) DM Del

This system is quoted in WOFK as having a period of close to 1/3 day but the remarks column shows that suspicion surrounds this value. The earlier GCVS 1 value of 0.845 certainly places the star in a more normal part of the diagram, and, in fact, Diethelm (1980), from whom the WOFK period originates, appears to have changed his original estimate.

7. EA2 SYSTEMS - CLASSICAL EVOLVED ALGOLS

Mention has already been made of the fact that certain Algol type light curves have such deep primary minima that they cannot be explained as resulting from eclipses involving Main Sequence stars, and indeed the shape of the minima in certain such cases suggests that the bright primary is totally eclipsed - while the absence of significant proximity induced "ellipticity" at once suggests that this larger occulting star is also of relatively low mass. Thus the familiar Algol paradox could almost be deduced (with the benefit of hindsight) from a judicious inspection of optical photometry done.

The great majority of such light curves which have been analysed in detail have been found to conform to the semi-detached configuration, though two relatively small anomalous subgroups have been identified as: (i) the "undersize subgiant" type, which has come to be closely identified with the RS CVn-like stars (see Morgan and Eggleton, 1979), having quite separate and recognizable properties (Hall, 1976) and (ii) the RCMa type systems which, though now usually regarded as being essentially similar to classical evolved Algols (see e.g. Sisteró, 1980, Okazaki, 1979) (i.e. the same semi-detached sort of configuration exemplified by Algol itself, see e.g. Söderhjelm, 1980), may nevertheless have certain distinguishing features associated with the combination of their very low mass ratio and usually relatively low periods, keeping in mind evolution schemes which have been proposed to account for the properties of evolved Algols (see e.g. Batten's review in this same publication).

There is a further small subgroup which could give rise to some confusion, associated with a few systems thought to be semi-detached but showing more of an ellipticity effect in their light curves such as V Pup. Such stars account for < 10% of the 70 or so semi-detached systems reveiwed by Svechnikov (1969). The essential characteristic here is that the secondary star is also of early type so that its distortion has a more noticeable effect, while the primary in such cases appears to be unusually early in comparison to the classical (evolved) Algols (earlier than B5). The difficulty with which reliable radial velocities can be determined for early type stars (Popper, 1967) should be kept in mind in considering these cases. They may be physically similar to classical Algols, but we have no easy way of picking them up from purely a priori general information on their light curves, and in the present classification scheme they would probably appear as EB type (Section 9).

The RS CVn (or AR Lac) stars occur at longer periods and among cooler systems. The difference in depths of minima tends to be much less than for the classical Algols, where the specific brightness contrast is large. Also there tend to be strong additional peculiarities like CaII emission and "wave like" photometric distortions. The list in Table 5 is expected, therefore, to be dominated by semi-detached Algols with the chance of the odd non-conforming binary being still present. The criteria used in forming this list were that if the system has not been previously generally recognized to be of semi-detached type it should show deep ($\gtrsim$ 1 mag) primary eclipses and very shallow secondaries as well as having an EA type GCVS designation. Figure 4 shows the distribution of classical evolved Algol systems with periods less than 1.5 days and brighter than $12^m.5$ in the s:P plane.

The main impressions which can be gathered from Figure 4 are:

1) The overall number of EA2 systems in the short period range is comparable to the number of Main Sequence pairs.

2) The present primaries in EA2 systems are with few exceptions clearly detached from their Roche lobes by significant distances.

3) Compared with the EA1 systems there is a considerable concentration in the distribution toward primaries with spectral type around A0.

4) There are, however, a number of cooler primary systems, and these are often relatively close semi-detached pairs, such as RS Sct, containing low mass primaries and entailing low overall mass for the systems.

5) The mean separation of the components at a given spectral type

Table 5: 55 Close Binary Systems of Type EA2

Constellation	Name or No.	Period	Primary s Type	Mag.	Constellation	Name or No.	Period	Primary s Type	Mag.
And	UU	1.486	F5	11.2	Her	SZ	0.818	A0	10.2
	XZ	1.357	A0	10.02		MT	0.488	F5	11.8
Aqr	CX	0.556	F2	10.55		V338	1.306	A9	10.07
	CZ	0.863	A5	10.98	Lac	VX	1.074	F0	10.9
Aql	V346	1.106	A0	9.0	Leo	VZ	1.090	A5	10.6
Aur	KU	1.320	F8	11.4		WZ	1.406	A1	11.3
Cnc	RY	1.093	F5	12.5	Lep	Z	0.994	F0	11.0
	WW	1.116	A8	10.5		RS	1.289	A0	9.3
CMa	R	1.136	A9	6.05	Lup	DT	1.453	A0	9.8
	RT	1.294	F8	11.4	Ori	ET	0.951	G3	11.2
Cap	TY	1.423	A5	10.5	Peg	AT	1.146	A0	9.5
Car	EX	1.396	G0	10.0	Per	RT	0.849	F2	10.46
Cas	RZ	1.195	A2	6.38		XZ	1.152	G1	10.6
	TW	1.428	B9	8.3		IU	0.857	A4	10.5
	AB	1.367	A0	10.3	Pic	X	0.862	A2	10.7
	IV	0.999	A2	11.2	Psc	SX	0.826	F1	11.0
Cet	RW	0.975	A5	10.43	Sgr	V505	1.183	A1	6.48
CrB	RW	0.726	F0	10.1	Sct	RS	0.664	F5	9.2
Cyg	ZZ	0.629	F6	10.61	Ser	AO	0.879	A2	10.7
	AE	0.969	A5	11.8	Tau	AQ	1.216	A5	11.8
	BR	1.333	B9	9.4	Tri	X	0.972	A3	8.9
Del	TY	1.191	B9	9.7		RV	0.754	A	11.4
	FZ	0.783	F	10.2	TrA	RR	0.713	B8	10.57
Dra	Z	1.357	A5	10.8	Tuc	V	0.871	A2	10.79
	AI	1.199	B9	7.2	UMa	XZ	1.222	A5	10.1
Gem	AF	1.244	A0	10.2	Vel	EQ	1.080	B7	11.0
	AL	1.391	F5	9.8	Vul	AB	1.461	A8	12.4
						AW	0.806	F0	10.8

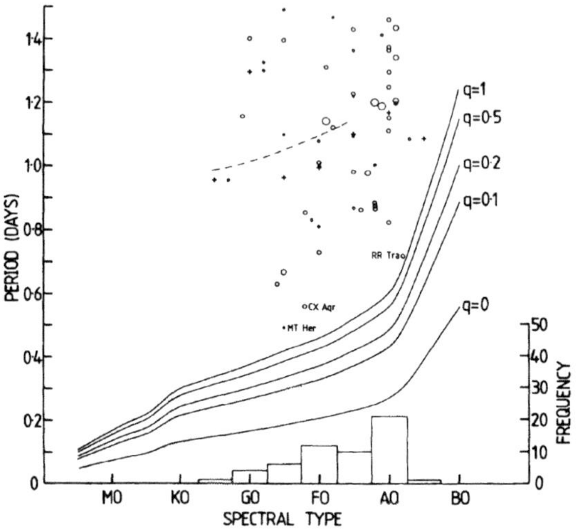

Figure 4: The format in this and Figures 7, 8 and 9 is the same as for Figure 3. In this case it is the distribution of EA2 stars which is presented.

appears to be greater than for Main Sequence systems. This seems to show up at the later types and it is also likely to hold for the earlier types, but the mean periods of earlier type systems are not clear due to the cutting off of numerous systems with period greater than 1.5 days. We shall return to this point subsequently.

6) There are a few systems for which the primary is located rather close to the contact domain. As mentioned before if the secondaries in these systems are of sufficiently low mass it is still quite possible that the primary is appreciably detached, however, such systems may be worthy of special note. They include the following:

RR Tr A

There may be some doubt about spectral type for this star. GCVS 1 gives B8, without reference, while WOFK does not give a spectral type.

CX Aqr

GCVS 1 reports unusually broad hydrogen lines. Period anomalies have been reported, though there has been no recent study.

MT Her

This may be the shortest period and one of the least massive of the evolved Algol systems. Although the primary minimum is quite deep ($1^m.1$)

it is still only a partial eclipse, while its duration is over a quarter of the entire light cycle.

ZZ Cyg

Undoubtedly a very interesting short period system which is one component of a visual binary. The primary exhibits short period low amplitude oscillations and the system may be similar to AB Cas, whose primary is probably a δ Sct variable. Hall and Cannon (1974) obtained a slightly later spectral type for the primary (F7), and considered either membership of EA1 or EA2 groups equally likely.

The system MM Ara (P = 1.278, s = A2) could also possibly qualify as an EA2 system, though the E designation in GCVS 1 suggests photometric irregularities which cause doubts to be raised.

Since the smallness of the window imposed by the period range concentrated upon seems to seriously impede our view of these systems, a list was compiled subject to similar criteria as for Table 5, but with the period restriction removed. The identifications of these stars are given in Table 6: further details on them may be found in the source catalogues. It has been attempted to summarize the main features of this source information as relevant to the present enquiry diagrammatically in Figures 5 and 6. Known RS CVn systems have been removed from this sample.

In these isometric orthogonal projections of the three-dimensional distribution, the upper parts of the diagrams indicate the actual counts at intervals corresponding to the given increments (5 spectral subclasses, and 0.5 days in period) centred on each spectral type and period value as marked along the co-ordinate axes. The corresponding values are listed in Table 7. The distribution is somewhat peaky, so, in order to make out its major features, it has been smoothed and interpolated by a least squares Chebyshev series surface fit, as shown in the lower parts of the diagrams.

The following features can be observed on these diagrams:

1) There is a definite maximum in the distribution at spectral type around A0 and period $\sim$ 2.5 days.

2) The maximum in the distribution with period occurs at somewhat shorter period for types later than A0, i.e. for type F5 it occurs at around 2 days; alternatively, at shorter period the maximum in the distribution with type occurs at later type - at the low period limit it may be as late as F0.

3) There is a very swift decline in the number of EA2 systems for types earlier than A0.

4) There is a long tail of systems with A-F type primaries stretching to long ($\gtrsim$ 15 days) periods.

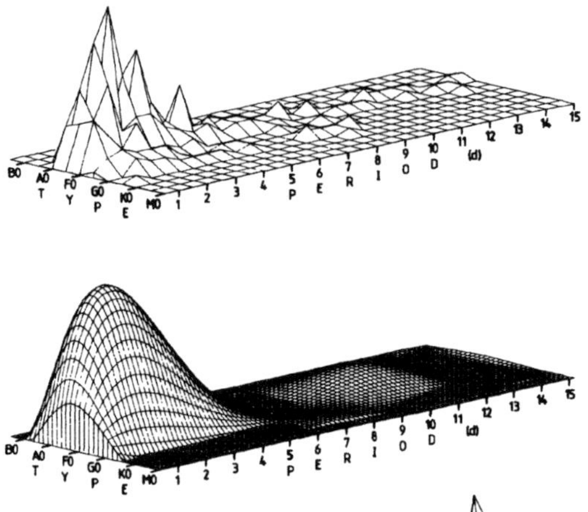

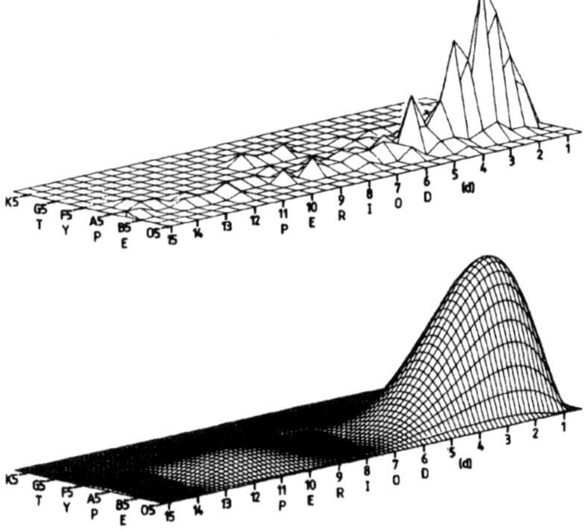

Figures 5 (upper pair) and 6 (lower pair): Distribution of 311 EA2 candidate stars in s:P plane. Figure 6 is a view of the same sample shifted by 180° from that of Figure 5. A surface fitter has been used to show up the main features, but this may have introduced some slight oscillatory behaviour of its own. Thus the suggestion of a second maximum at longer periods cannot be realistically taken as other than a noise effect. Certain borderline cases (such as OR Cas and BZ Eri, where the sources disagree) have been included in the counts. Also a few cases where spectral subclass was not given were distributed evenly between the adopted intervals.

Table 6: 323 EA2 Candidate Stars Brighter than $12^{m}.0$ With Known Primary Spectral Type

And	TT	Cas	IV	Del	FZ	Lac	VX	Peg	AT	Sco	V393
	TW		KR	Dra	Z		DG		AW		V496
	UU		MN		RR	Leo	Y		BG		V594
	WW	Cen	SS		SX		RT		BN	Sct	RS
	XZ		SU		TW		UU		DF		RZ
	CO		SW		TZ		VZ		DK		AC
	CP		SY		AI		WZ		DI		BS
Aps	DW		CT		BU	LMi	T		DO		CT
Aqr	RY		LP	Equ	S	Lep	Z		EH	Ser	AK
	CZ		V379	Eri	RY		RS		ER		AO
Aql	XZ		V402		TZ	Lib	δ	Per	β	Tau	λ
	YZ		V621		AS		EI		Z		RW
	FK		V646		BZ	Lup	DT		RT		SV
	KO	Cep	U	For	SU	Lyn	SX		RV		AC
	KP		RS	Gem	RW	Lyr	UZ		RW		AM
	QY		XX		RX		EW		RY		AQ
	V342		DW		RY	Mon	RW		ST		ET
	V343		EK		SV		TU		WY		GQ
	V409	Cet	RW		TX		AU		XZ	Tel	RV
	V415		SS		AE		AV		AB	Tri	X
Ara	RW		TU		AF		AW		AY		RV
	SY	Cir	T		AL		BO		DM	TrA	RR
	UW	CrA	RW		AY		FW		IU		EQ
Ari	RS		IS		BD		HO		IZ	Tuc	V
Aur	RY	CrB	U		BO		HY	Pic	X		AN
	AM		RW	Gru	U	Mus	UV		RV	UMa	TX
	HP	Cru	AE		X	Nor	RR	Psc	Y		XZ
	KU	Cyg	SW	Her	SZ		HH		SX		ZZ
Boo	UW		SY		TU	Oph	RV	Pup	RR		AC
Cam	Y		UW		UX		SW		SW	UMi	W
	XZ		VW		AD		SZ		XY	Vel	S
Cnc	S		WW		BC		UU		XZ		RX
	TU		ZZ		BO		V391		YY		TT
	WW		AE		CC		V441		ZZ		AS
	ZZ		BR		CT		V501		AA		CW
CMa	R		GM		DH		V535	Sge	U		DX
	RT		MY		FN		V735		UZ		EQ
CMi	AG		V346		HS		V806		DE		FU
Cap	RW		V370		MT		V913	Sgr	RS	Vir	UW
	TY		V463		MX	Ori	Z		SX		BD
Car	SW		V474		V338		CP		WX	Vul	Z
	CV		V512		V359		DN		WY		RR
	CX		V687	Hya	RX		EQ		XY		RS
	DE		V728		SX		ET		XZ		XZ
	EX		V748		SY		EY		ZZ		AB
	GU		V788		TT		FF		BN		AW
	KQ		V959		TY		FH		BQ		AX

Table 6 (continued)

Cas	RZ		V995		VW		FL		DV		AY
	TV	Del	W		VY		FM		EG		BE
	TW		RR		DE		FO		V354		BO
	AB		TT	Hyi	Y		OS		V505		BP
	BZ	Del	TY	Hyi	VX	Peg	TY	Sgr	V524	Vul	EY
	GG		AV	Lac	TW		UX	Sco	AL		
	IS		BI		UW		AQ		BS		

Table 7: Period-Type Distribution of EA2 (Evolved Algol) Systems

P (d) \ s	05	B0	B5	A0	A5	F0	F5	G0	G5	K0
0.0 - 0.5	0	0	0	0	0	0	1	0	0	0
0.5 - 1.0	0	0	0	7	8	9	3	0	1	0
1.0 - 1.5	0	0	1	13	8	6	4	3	0	0
1.5 - 2.0	0	0	1	21	14	3	3	2	0	0
2.0 - 2.5	0	0	3	27	5	7	3	2	0	0
2.5 - 3.0	0	0	2	14	7	2	3	1	0	0
3.0 - 3.5	0	0	1	18	7	2	1	1	0	0
3.5 - 4.0	0	1	2	8	4	1	0	1	0	0
4.0 - 4.5	0	0	2	4	3	2	0	1	0	0
4.5 - 5.0	0	0	0	10	2	0	1	0	0	0
5.0 - 5.5	0	0	2	2	3	0	1	1	0	0
5.5 - 6.0	0	0	0	3	1	0	0	0	0	0
6.0 - 6.5	0	0	0	2	2	0	0	0	0	0
6.5 - 7.0	0	0	1	2	2	0	0	0	0	0
7.0 - 7.5	0	0	0	1	1	1	0	2	0	0
7.5 - 8.0	0	0	0	1	1	0	0	1	0	0
8.0 - 8.5	0	0	0	3	0	0	0	1	1	0
8.5 - 9.0	0	0	0	0	0	1	0	2	0	0
9.0 - 9.5	0	0	0	2	0	0	0	0	0	0
9.5 - 10.0	0	0	0	0	1	0	0	0	0	0
10.0 - 10.5	0	0	0	0	1	0	0	0	0	0
10.5 - 11.0	0	0	0	0	1	0	0	0	0	0
11.0 - 11.5	0	0	1	2	0	0	0	0	0	0
11.5 - 12.0	0	0	0	1	0	0	0	0	0	0
12.0 - 12.5	0	0	0	1	1	0	0	0	0	0
12.5 - 13.0	0	0	0	1	0	0	0	0	0	0
13.0 - 13.5	0	0	0	0	1	0	0	0	0	0
13.5 - 14.0	0	0	0	0	1	1	0	0	0	0
14.0 - 14.5	0	0	0	1	0	0	0	0	0	0
14.5 - 15.0	0	0	0	1	0	0	0	0	0	0
15.0 - 15.5	0	0	0	0	0	0	0	0	0	0

8. EA3 SYSTEMS

The numbers of EA1 and EA2 systems in the sample being of the same order, it is reasonable to suppose that this group, which consists of tho eclipsing binaries with EA type light curves which cannot be prima facie

Table 8: 55 Close Binary Systems of Type EA3

Constellation	Name or No.	Period	Primary s Type	Mag.	Constellation	Name or No.	Period	Primary s Type	Mag.
Aqr	DD	0.721	A5	10.9	Hya	DI	0.615	A6	11.0
Aql	V843	1.498	B9	9.8	Leo	TY	1.185	G	10.5
Ari	RX	1.030	F5	9.4		UX	1.007	F4	10.3
Cam	AL	1.328	A2	10.5	Lyn	UU	0.468	A4	11.4
	AW	0.771	A0	8.0	Mon	EP	1.148	A3	10.5
	AY	1.367	G0	9.9		PR	1.124	A	11.0
CMi	AC	0.789	F4	11.0	Oph	V501	0.968	A5	10.6
Cap	δ	1.023	A7	2.88	Ori	ES	0.816	A2	10.1
Car	DV	0.841	B8	10.0		V392	0.659	A5	10.9
Cas	V361	1.229	A	12.3	Pav	KZ	0.950	F2	7.8
	V364	0.772	A7	11.2	Peg	BN	0.713	F8	10.5
	V380	1.357	A0	10.4		DI	0.712	K0	9.45
	V384	1.108	A0	11.8	Per	KR	0.996	B0	10.4
Cen	V685	1.191	A0	9.4		NZ	0.938	F2	11.7
Cet	TX	0.741	A2	10.0	Pyx	TX	1.125	A3	9.5
Col	U	1.246	A	10.4	Sge	CU	0.494	F5	10.9
Crt	V	0.702	A6	10.8	Sgr	V1177	0.963	B9	11.3
	RS	0.817	G0	9.9	Sco	V626	1.034	A7	11.4
Cyg	V442	1.193	F4	10.0	Ser	BI	1.205	A3	11.5
	V753	0.476	F8	11.2		CX	0.997	F8	11.0
	V909	1.403	A0	9.3	Tau	TY	1.077	K0	11.5
Del	YY	0.793	A3	11.1		EN	1.239	G0	11.6
Dra	TZ	0.866	A7	9.6	UMa	VV	0.687	A0	10.13
	AR	0.676	F8	12.2	Vel	DX	1.117	A5	10.2
Eri	BZ	0.664	F2	9.0	Vir	DL	1.315	A2	7.0
Gem	SX	1.367	A0	11.0	Vul	BU	0.569	G0	10.6
Her	LT	1.084	A1	10.1		FR	0.942	A2	10.3
	V450	0.913	A0	10.1					

assigned to either first or second groups is composed of stars drawn from these two groups in roughly equal numbers. In actuality there may be a majority of EA2 systems in the sample, since most EA1 systems with deep enough primary minima to be recognized would tend to show secondary minima, which are usually absent, or at least unreported, for the EA3 stars.

In any case, the character of the distribution shown in Figure 7 appears to be intermediate between that of Figures 3 and 4 showing some more concentration towards earlier types than Figure 3, but also some more of a tendency towards the contact domain than Figure 4.

Once again the diagram brings to light a few interesting exceptional cases including

i) KR Per

GCVS 1 gives the spectral type as OB, which is followed by WOFK. Nothing is known of a secondary minimum, which could mean, as in the apparently similar case of DN Cas, that the originally quoted period was actually half the real period. Alternatively, since the minimum is 0.7 m deep despite its very short duration the primary may be some kind of hot subdwarf.

ii) UU Lyn

Not so much has been reported about this rather faint system. It is possible that the quoted spectral type, at least, is in significant error. If the period is correct this system is the shortest period binary to show an EA type light curve.

iii) CU Sge

WOFK suggests some history of error in the classification of this star and that the period may, in fact, be double the quoted value.

9. EB SYSTEMS - "β LYR" TYPE LIGHT CURVES

The sample of systems of this kind is tabulated in Table 9 and their distribution in the s:P plane shown in Figure 8, where the following major features may be noticed.

1. There are somewhat fewer EB than EA systems altogether in the period range, but twice the expected total number of Main Sequence EA binaries (EA1 + ½ EA3).

2. The mean periods of EB binaries are located a little nearer to the contact domain than the EA1 systems (Figure 3), however, there is a considerable overlap in the positions occupied by EA1 and EB stars in the s:P plane.

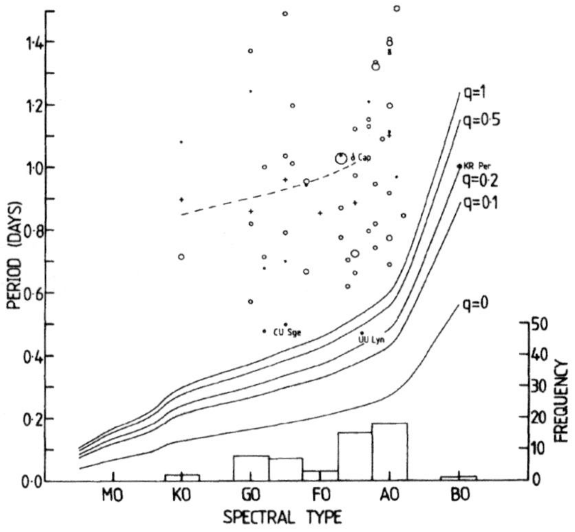

Figure 7: Distribution of EA3 stars in the s:P plane.

3) The distribution of these systems appears to be concentrated somewhat towards earlier spectral types than EA1 systems, though this distribution is not so markedly different from the distribution of single stars (especially BSC stars) as can be ascertained from Figure 1 - and more like such a distribution than is that of the EA1 stars in the given sample.

4) Some EB light curves occur at quite late types (even at KO), though the mean period diverges, to some extent, from the contact domain boundary at later type.

5) A few early type pairs can be found actually within the contact domain. Among these we may note one or two systems of exceptionally short period including

i) EM Cep
A bright binary located in the region of cluster NGC 7160 with very shallow minima. It is possible that some physical variability, perhaps of β Cep type, is at least partly responsible for the observed variation.

ii) DD Mon

The unequal depths of the two minima ($0.^m9$ and $0^m.6$) no doubt confirm an EB classification for this star which must be close to a contact configuration.

Table 9: 120 Close Binary Systems of Type EB

Constellation	Name or No.	Period	Primary s Type	Mag.	Constellation	Name or No.	Period	Primary s Type	Mag.
And	WZ	0.696	F5	11.6	Cyg	V1034	0.977	A0	9.6
	AD	0.986	F	10.9		V1425	1.252	B9	7.9
	BD	0.463	F8	11.3	Del	ET	1.011	A0	12.0
	DS	1.011	F3	10.8		GG	0.563	F7	12.4
Aqr	ST	0.781	A7	9.2	Dra	RZ	0.551	A5	10.0
	SU	1.045	A5	10.2		AX	0.568	F5	10.9
	DX	0.945	A2	6.2		BE	0.522	B	11.3
Aql	V609	0.797	F8	11.7	Eri	RU	0.632	F0	9.9
Ara	V349	1.138	A3	8.6		BQ	1.430	A8	10.5
Aur	SX	1.210	B3	8.2	Gem	WW	1.238	B6	9.9
	TT	1.333	B3	8.3		GW	0.659	A4	10.6
	ZZ	0.601	A7	10.8		GX	1.350	G5	10.9
	AP	0.569	A2	10.9	Her	TT	0.912	A2	9.61
	CL	1.244	A0	11.7		V342	0.852	F2	10.5
	EP	0.591	F8	10.8	Ind	RS	0.624	F4	9.9
	GI	1.208	F2	11.6		RY	0.712	A5	10.6
	HL	0.623	F4	10.8		SU	0.986	G5	9.3
Boo	AD	1.034	G0	9.8	Lac	VY	1.036	A2	10.2
Cam	AZ	1.319	F0	10.1		AW	1.143	B2	10.6
Cnc	XZ	1.114	A5	9.8		CN	0.637	G3	12.4
CMa	FF	1.213	B3	7.38		IZ	0.799	A3	12.5
CMi	YY	1.094	F5	8.5		LZ	1.251	A0	12.1
Car	X	1.083	A0	8.0	Leo	XX	0.971	A8	10.7
	ST	0.902	A0	9.1	Lep	V	1.070	A5	9.9
	GW	1.129	B5	9.8		RR	0.915	A2	10.2
	PX	0.795	A2	10.2	Lib	ES	0.883	A2	7.1
Cas	DO	0.685	A1	8.6	Lyn	RZ	1.147	A2	10.6
	DZ	0.785	A9	11.6	Lyr	TZ	0.529	K0	10.4
	IR	0.681	F4	10.8	Mon	UY	1.261	F8	9.6
	V375	1.473	B3	10.1		DD	0.568	B5	11.1

Table 9 (continued)

Constellation	Name or No.	Period	Primary s Type	Mag.	Constellation	Name or No.	Period	Primary s Type	Mag.
Cen	BH	0.792	B5	10.1		IM	1.190	B5	6.4
	LW	1.003	B1	9.3		MX	0.686	A5	11.6
Cen	V606	1.495	B1	10.4	Mon	V448	1.118	A2	9.4
	V701	0.738	A0	8.4	Mus	TU	1.387	B3	8.17
	V716	1.490	B9	6.2	Oph	V506	1.060	F0	11.2
Cep	SU	0.901	B8	10.0		V1010	0.661	A3	6.2
	WY	1.249	A7	10.7	Ori	UW	1.008	B	11.1
	EM	0.806	B0	7.02		VV	1.485	B1	5.14
	GK	0.936	A0	6.89		FR	0.883	A7	11.0
	GS	0.772	A0	10.2	Per	IK	0.676	A2	11.3
Crv	RV	0.747	F0	8.6		KW	0.931	A2	10.5
Cyg	WZ	0.584	F0	10.5	Pup	V	1.454	B1	4.74
	GO	0.718	B9	8.3		UZ	0.795	A5	9.6
	KR	0.845	A2	9.2		VY	0.817	F2	10.8
	V366	1.096	A5	10.0		AU	1.126	A0	8.6
	V388	0.859	A3	9.7		MQ	1.469	B9	8.0
	V505	0.668	F0	11.7		NO	1.257	B9	6.7
	V680	1.199	B9	10.2	Pyx	RZ	0.656	B7	8.85
	V796	1.481	A0	11.0	Sgr	V525	0.705	A1	7.9
	V836	0.653	A0	8.59		V2509	1.087	A0	7.6
	V841	0.761	A5	11.1	Sco	μ^1	1.440	B1	2.8
Sco	V701	0.762	B5	8.2	UMa	XY	0.479	G5	10.2
	V883	1.295	B2	7.34		AA	0.764	G0	11.3
Scl	RT	0.512	A5	10.54	Vel	BC	1.174	F8	11.8
Sct	U	0.955	F0	10.0		FT	1.060	A0	10.1
	EQ	1.322	B	11.7		FX	1.053	B	9.2
Ser	AS	0.466	F2	11.0	Vir	AX	0.703	A0	10.3
	CQ	0.760	A8	11.6		CG	0.935	F8	10.6
Tau	GW	0.641	A3	11.2		CX	0.746	F5	9.2
Tri	V	0.585	A3	10.9	Vul	CD	0.684	F0	11.5

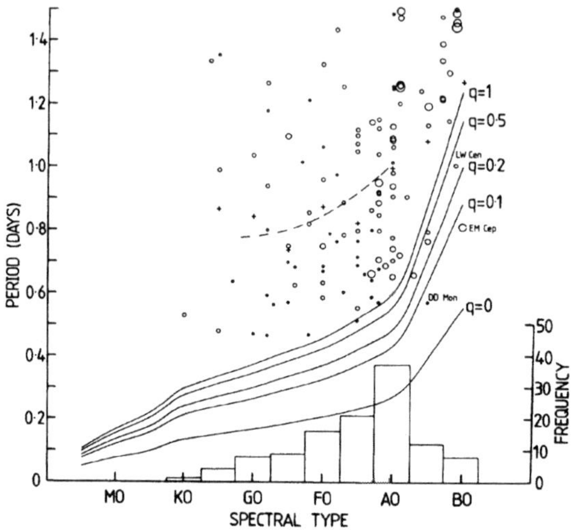

Figure 8: Distribution of EB stars in the s:P plane.

10. EW SYSTEMS - W UMa TYPE LIGHT CURVES

Systems showing this type of light curve have attracted a number of special studies since Kopal's (1955) proposal that they can be explained as due to stars which are so close together as to be actually in contact - a point of view which seems to have been sometimes taken for granted by subsequent authors. Later we shall find reason to question the general nature of this assertion, though it cannot be doubted that a good many such pairs must be relatively very close.

In Table 10 we list the EW systems for which the spectral type is known, and these are plotted in the s:P plane in Figure 9. Binnendijk (1970) subdivided these systems into two further subgroups - A and W - according as the more massive component approaches after primary minimum (A type) or recedes (W type). In Figure 9 the identification of certain members of these groups have been made according to the listings of Yamasaki (1975). Figure 9 shows up the following points:

1) There are slightly fewer EW than EB systems in the given period range and about half of them are located within the contact domain.

2) Though there is some area of overlap in the s:P plane, in fact with all the major light curve types, the EW systems are mostly closely concentrated towards the contact domain.

3) The EW light curves occur at all spectral subgroups from B0 to K5, but there is a concentration towards later spectral types with a pronounced peak around spectral type G0. The distribution contrasts sharply

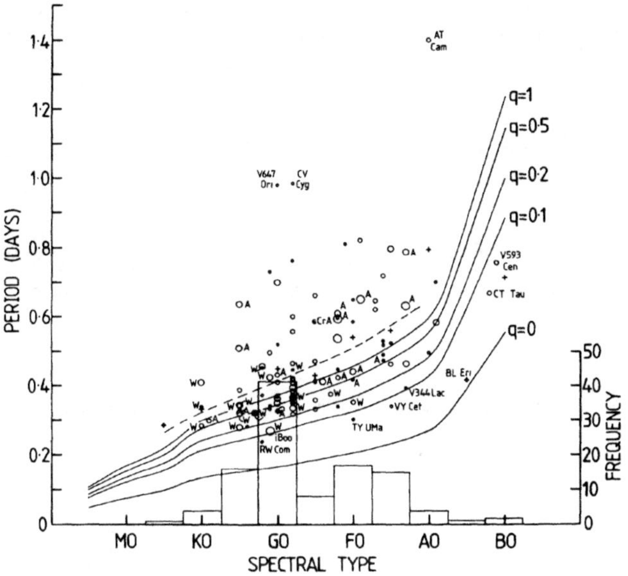

Figure 9: Distribution of EW stars in the s:P plane. A has been placed alongside A-type subgroup stars, W similarly with the other subgroup, as listed by Yamasaki (1975). A number of extreme cases have been marked, otherwise the diagram lay out is similar to previous ones.

with that of single Main Sequence stars where the ratio is, on average, about 2:1, for the group of stars centred on s = AO to that centred on s = GO in Tables 2 and 3. In the present sample it is 10:1 in favour of the GO group.

4) The W type subgroup concentrates more towards later types and the contact domain than does the A subgroup, though at least one A system (CW Cas) is found in a typical W position. Ruciński (1974) does not agree with Yamasaki (1975) about TZ Boo, while the "secondary" (total) minimum has become the deeper one on Hoffmann's (1980) light curve. U Peg, outside the contact domain, has a spectral type more typical of the A subgroup. Some systems may indeed change from A to W types (Hoffmann, 1979).

5) A number of exceptional systems can be noticed in this sample, with either anomalously long or short periods. They include the following:

i) AT Cam

This star has an exceptionally long period at 1.396 for its light curve to be of W UMa type. WOFK suggests period variability.

Table 10: 118 Close Binary Systems of Type EW

Constellation	Name or No.	Period	Primary s Type	Mag.	Constellation	Name or No.	Period	Primary s Type	Mag.
And	AB	0.332	G5	10.33	Dor	RW	0.285	K5	10.8
	BX	0.610	F2	8.9	Dra	BV	0.350	G0	8.4
	CN	0.463	F8	10.0	Eri	UX	0.445	F8	10.6
Ant	S	0.648	A9	6.7		YY	0.321	G5	8.8
Aqr	AO	0.698	F	11.5		BC	0.528	A6	10.83
Aql	OO	0.507	G5	9.2		BL	0.416	B5	11.5
	V417	0.370	G2	11.0		CU	0.634	G5	8.0
	V724	0.518	G0	11.1	Gem	AZ	1.006	A	11.8
Ara	V535	0.629	A3	7.4	Her	AK	0.422	F2	8.83
Ari	SS	0.406	F8	10.0		LS	0.462	A5	10.5
Aur	AH	0.494	G1	10.5	Hya	WY	0.716	A6	10.4
Boo	i	0.268	G1	6.5		DF	0.331	G0	11.0
	TU	0.324	G3	11.7		EZ	0.450	G2	11.0
	TY	0.317	G3	10.81		FG	0.328	G0	10.38
	TZ	0.297	G9	10.6	Lac	SW	0.321	G3	8.51
	VW	0.342	G5	10.5		V344	0.392	A3	12.2
	XY	0.371	F8	10.0	Leo	UZ	0.619	A7	9.8
	AC	0.352	F0	9.1		XY	0.284	K0	10.43
Cam	AT	1.396	A0	9.8		XZ	0.488	A6	10.6
Cnc	TX	0.383	G1	10.45		AM	0.366	F8	8.2
CMi	UZ	0.762	F8	11.6		AP	0.430	G0	9.7
Cas	CW	0.319	G5	11.8	LMi	RT	0.374	G0	11.4
	V366	0.729	G1	12.0	Lib	VZ	0.358	F5	10.1
Cen	RR	0.606	F5	7.46	Lup	FT	0.470	F5	9.2
	V593	0.755	B1	10.5	Lyn	UV	0.415	F8	9.41
	V752	0.370	G0	9.1	Lyr	PY	0.386	F	12.5
	V757	0.343	G5	8.3	Men	TY	0.462	A3	7.7
	V758	0.581	B9	8.8	Mus	TV	0.446	F2	11.0
	V759	0.394	F8	7.4	Oph	V502	0.453	G2	8.34

Table 10 (continued)

Constellation	Name or No.	Period	Primary s Type	Mag.	Constellation	Name or No.	Period	Primary s Type	Mag.
Cep	VW	0.278	G5	7.8		V508	0.345	G5	10.2
	WZ	0.417	F5	11.4		V566	0.410	F4	7.6
	GW	0.319	G3	11.4		V839	0.409	GO	9.4
Cet	TT	0.486	A	10.0	Ori	ER	0.423	G1	9.33
	TW	0.317	G5	10.43		V343	0.809	F1	10.7
	TY	0.647	FO	11.7		V647	0.978	GO	11.5
	VV	0.522	A5	10.3	Pav	MW	0.795	A5	8.49
	VY	0.341	A5	11.1	Peg	U	0.375	F3	9.23
	AA	0.536	F2	6.2		BB	0.362	F8	10.8
Cir	AQ	1.146 (?)*	B	11.0		BX	0.280	G4	10.9
Col	RS	0.672	G	9.54		DV	0.946	F	11.8
Com	RW	0.237	G2	11.0	Phe	AE	0.362	GO	7.7
	RZ	0.339	KO	10.96	Pup	TY	0.819	A9	9.2
	SS	0.413	F5	10.9		AV	0.556	F8	10.2
CrA	ε	0.591	F2	4.96		GY	0.412	G	11.9
Crv	Y	0.495	AO	11.0	Sge	CW	0.660	F5	11.0
	SX	0.317	F8	8.99		EL	0.333	F5	10.7
Cyg	CV	0.983	F8	10.8	Sgr	V779	0.445	F8	11.5
	DK	0.471	A6	10.3	Ser	RS	0.598	F8	10.8
	V401	0.583	FO	10.87		AU	0.387	G5	10.5
	V700	0.340	F2	11.7		CC	0.516	A6	11.14
	V1073	0.786	A3	8.0	Sex	Y	0.420	F8	9.9
Tau	RZ	0.416	FO	10.5	UMa	TY	0.301	FO	11.7
	WY	0.693	A	11.2		AW	0.439	FO	6.84
	AH	0.333	G1	11.25		BH	0.699	B9	11.5
	CT	0.667	B2	10.34	Vir	AG	0.643	A7	8.4
	CU	0.413	GO	12.1		AH	0.408	KO	9.2
	EQ	0.341	G1	10.5		AW	0.354	F8	11.4
Tuc	AQ	0.595	F2	10.5		AZ	0.350	F8	11.0
UMa	W	0.334	F8	7.9	Vul	ER	0.698	GO	7.27

* GCVS(1) gives half this period, but suggests it should be doubled if the star is EW type.

ii) CV Cyg

Another star with a long period ($0^d.983$) for W UMa type system, but, according to Eggen (1967) "it is not excluded that the star is of type RRc (i.e. RR Lyrae variable with symmetrical light curve) with $P = 0^d.49$". Some support for the view that it is a Population II physical variable comes from its inclusion by Perek in a list of high velocity stars.

iii) V647 Ori

Another long period case ($P = 0^d.978$). There seems to be some confusion over the spectral type - GCVS1 reports GO while WOFK gives A6.

Possibly similar long period EW stars are AZ Gem ($P = 1^d.006$) and DV Peg ($P = 0^d.946$). It is conceivable that some, if not all, of these stars are actually RRc type variables.

iv) V593 Cen

The spectral type Bl Vn is given in GCVS 2, though none is given in WOFK. GCVS 2 also reports that it may be a member of the cluster Stock 16.

v) CT Tau

Both GCVS and WOFK agree on a B2n spectrum but WOFK reports that the star is suspected of some kind of physical variability. The amplitude of the primary minimum (1.3 mag.) is certainly large for a W UMa like star.

vi) BL Eri

There is apparently no definite information on the secondary minimum of this rather faint star. Its position in the s:P plane may therefore be treated circumspectly.

There are a number of other cooler systems which are situated rather well in the contact domain. The mass ratios could be small, but then noticeable light variations become less likely. Also the primaries may be appreciably smaller than Main Sequence stars of the same type; though it could still be regarded possible, on purely empirical grounds, for the great majority of primaries to be not far from Main Sequence stars of the same spectral type. A similar result obtained by examining fewer systems but in more detail was expressed by Kopal as early as 1955 (op. cit.).

The possibility of primaries being below corresponding Main Sequence size may be intimately connected to the contact or over contact configuration. This was certainly suggested by Kuiper (1941) in the first theoretical investigation of such a configuration, though Kuiper's work stopped short of detailed modelling by integration of the structure equations. Though the emphasis in this paper is on empirical evidence, it should be noted that support for a possible close to Main Sequence character of the primaries in these systems is available from a number of more recent theoretical studies, as referred to, for example, by Moss and Whelan (1970), Whelan et al. (1973), Hazelhurst (1976), Robertson and Eggleton (1976), Lucy (1976) or Lubow and Shu (1977), though most of these authors have also considered situations in which contact components are more evolved, even to great age (see also Webbink, 1979).

The observed tendency of the EW systems to congregate towards the contact domain can be regarded as just another transformation of Eggen's (1967) period colour relation which, together with Kopal's (1955) identification of W UMa like light curves with those of contact systems, form the two main observational foundations on which Lucy (1968) built his well known theoretical account of contact binaries. The raison d'être of this theory was to produce systems which could remain in contact for nuclear timescales in order to account for the relatively frequent observational occurrence of these stars, and also to explain the case of TX Cnc (and presumably other similar examples) which was taken to be an unevolved member of the galactic cluster NGC 2632 (Praesepe). The original theory was eventually abandoned and, though there have been numerous subsequent attempts, it has proved to be very difficult, uncertainties in convection theory notwithstanding, to construct models of contact binaries which can be shown to remain stable for nuclear timescales and meet even the more salient properties which can be disentangled from the observations.

In order to appreciate this latter point more fully we compare, in Table 11, three sets of solutions for some of the brightest and best studied examples of this class of star. The elements labelled I come from Ruciński (1974), II come from Kopal (1968), while III come from Niarchos (1977). The extent of the variation between the different authors, all using different methods of analysis, is very apparent. To be fair to III, however, both I and II assumed the "contact" condition, which for a given mass ratio, which is determined by separate spectrographic means for all the foregoing systems, fixes both r_1/r_2 and almost $r_1 + r_2$. In III the requirement for "contact" was relaxed. The ratio of radii for given q in I and II comes essentially from Kopal's Table 3.3 of his well known (1959) book. To be sure, I does not give this ratio, but instead gives the mass ratio as the corresponding quantity. The difference between I and II about this element merely reflects their adoption of different spectrographic quotations for the value.

In the case of the quantity $r_1 + r_2$, this is also given in Table 3.3 in Kopal's book, just mentioned. In I an "over-contact" condition is actually what is assumed in the model, hence it is necessary to add to $r_1 + r_2$ a small quantity which depends on the value of the "fill-out" parameter . This has been obtained approximately from measurements on the diagrams given on pages 507 and 508 in Kopal's (1959) book for different mass ratios, noting the close correspondence between the polar

Table 11: Basic Geometric Parameters of Some EW Systems

	r_2/r_1			(r_1+r_2)			i		
System	I	II	III	I	II	III	I	II	III
AB And	.79	.79	.52	.76	.75	.73	88	86	73
i Boo	.71	.71	.53	.76	.75	.76	77	56	46
TX Cnc	.73	.73	-	.78	.75	-	60	44	-
CW Cas	-	.64	.44	-	.75	.79	-	73	64
RR Cen[1]	.43	.40	.60	.81	.78	.72	79	85	79
VW Cep	.58	.57	.42	.77	.76	.84	67	49	50
TW Cet	.73	.73	-	.76	.75	-	88	84	-
RZ Com	.69	.69	.53	.79	.75	.68	88	76	82
DK Cyg[1]	.58	.45	.54	.79	.77	.68	84	77	80
YY Eri	.77	.81	.49	.78	.75	.71	84	79	73
SW Lac	.94	.92	.45	.82	.75	.68	83	83	71
XY Leo[2]	.89	.89	-	.75	.74	-	73	68	-
V502 Oph	.59	.64	.38	.77	.76	.77	74	60	61
V566 Oph	.49	.49	.45	.79	.75	.69	76	73	81
ER Ori	.74	.78	.51	.75	.75	.71	86	77	69
U Peg	.90	.90	.43	.76	.75	.73	75	68	66
RZ Tau	.74	.74	.47	.80	.75	.63	79	76	81
W UMa	.81	.73	.53	.76	.75	.67	83	78	80
AH Vir[3]	.65	.65	.47	.77	.76	.69	79	61	78

1) Elements in I due to Lucy (1972)
2) Elements in I partly due to Mochnacki and Doughty (1972a)
3) Elements in I partly due to Mochnacki and Doughty (1972b)

height of the critical Roche lobe $y_{4,5}$ in Table 3.1 and the volumetric radius $(r^*)_{1,2}$ in Table 3.2, the latter quantity being the one usually specified as radius by Kopal and utilized, for comparison purposes, in foregoing Table 10.

There is an evident close correspondence, due to the contact assumption, between I and II for the first two of the three basic geometric parameters which usually characterize an eclipsing binary light curve. Moreover, there can be little scope for disagreement over fractional luminosities (L_1, L_2), since the temperatures of the two stars for this kind of light curve are, by the criterion of almost equal depths of minima generally taken to be almost the same. Hence the disagreement in the inclination values between I and II is all the more embarrassing since this is the only largely unconstrained quantity in the problem.

10.1 V566 Oph - Analysis of a "good" W UMa System

Only in a few cases in Table 10 are the inclination values approaching moderate agreement, and one of these is the system V566 Oph, which Rucinski (1973) describes as "well known for its regular behaviour and very good conformity to the contact model", citing papers by Mochnacki and Doughty 1972b and Lucy 1972 in support of this. In order to investigate the question of parameter determinacy, by utilizing some ideas from a recent discussion of this subject (Budding and Najim, 1980), the same good quality light curves used by these different authors, due to Bookmyer (1969), were reanalysed.

It is fully appreciated that the form of the fitting function used for this purpose, stopping short of terms $O(r^6)$, may be insufficiently precise to give an accurate representation of the light curve, or even if it does that, through intercorrelation effects between parameters introduce systematic errors into these parameters as a result of the true parameters not being able to quite match the observations due to the series truncation. The scale of such a departure (true light curve from calculated light curve with true parameters) can be estimated in advance to be of the order of a few percent. The problem is really this: if the estimation of parameter values from the observational conditions is essentially highly determinate, a similar small percentage error can be expected to pass unamplified through the inversion of the "normal matrix" (or its equivalent) in the determination of the unknowns. If, however, this "normal matrix" is essentially ill-conditioned as a result of inherent correlation effects which are independent of the series truncation or other numerical approximations, then parameter values cannot be expected to be estimated accurately no matter how accurate the form of the fitting function.

In a non-linear optimization problem the equivalent of the "normal matrix" is the curvature Hessian of second order derivatives of the χ^2 variate (assuming the "least squares" principle) at optimum. Indeterminacy is generally signalled by the presence of relatively large off-diagonal elements in this matrix, and shown, in fact, by the appearance of negative elements on the leading diagonal of its inverse. Truncation or quadrature inaccuracy may alter the absolute sizes of the matrix elements by a few percent but cannot be expected to change the relationship of off-diagonal to diagonal elements to appreciably alter the determinacy question, though it has to be admitted that this point has not been tested in great detail.

In a first optimization run (procedural details are given in Budding and Najim, 1980) compromise values of the fractional luminosities were taken from the different sources as 0.81 and 0.19, and the three basic geometric quantities only were sought. From an initial trial value of $k = r_2/r_1 = 0.5$ this quantity settled at 0.58, which is encouragingly close to the ratio (0.6) following from Kuiper's (1941) often quoted mass ratio radius ratio relation. $(r_1 + r_2)$ settled at 0.80 while the inclination dropped from an initial trial value of 80° to 74° - an inclination value not too distant from those in Table 10.

The adopted fractional luminosities were not in good agreement with the quoted spectral types F5 and F8, and a second solution was attempted, in which the luminosities were constrained by adopted temperatures of 6400 for the primary and 6100 for the secondary, but the quality of the fit deteriorated significantly, while k wandered far from its "contact" value. Subsequently, starting from luminosity values corresponding to these temperatures an optimization run, in which L_1 was allowed to vary while L_2 was subject to the constraint $L_1 + L_2 = 1$ and r_2/r_1 kept at the previous best (0.58), sent L_1 back from the corresponding value of 0.73 close to its previously assigned value of 0.81. It was deduced that the difference of surface mean temperatures between the components is likely to be greater than 300 rather than ∿ 100 as found by Rucinski - this despite the almost equal depths of minima. This point concerning depths of minima turns out to be very significant and we shall return to it later. A small correction of 2 - $0^{o}.16$ to the zero of the phase scale was also noted.

The theoretical light curve at this point was as shown in Figure 10, where, if the (standard deviation) error of a single observation point is taken as a reasonably good $0^{m}.007$, the solution produces a χ^2/ν value of 1.09 (ν is the number of degrees of freedom = 70 with the four unknowns unit of light, primary luminosity L_1, r_1 and i). However, a close inspection of this diagram shows that the departures of the theoretical from the observed light curve are not purely random, as might be expected if the foregoing observational accuracy estimate was right, but include some systematic effects. Neglecting the possibility that these systematic departures are due to some process related to observational reduction; our conclusion must be that the observations are more accurate than $0^{m}.007$ (i.e. probable error $0^{m}.005$ - Bookmyer actually estimated $0^{m}.003$ for this quantity). The systematic departures are most noticeable in the secondary minimum, which certainly seems to suggest a total eclipse.

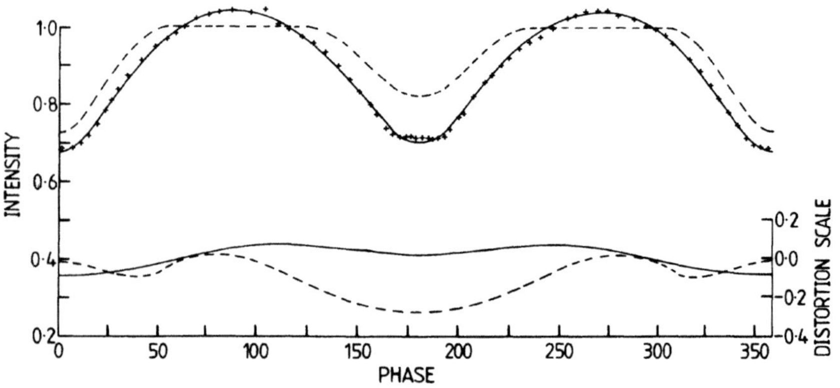

Figure 10: Optimal light curve fit to (Bookmyer's (1969) "good" light curve (V) of V566 Oph, with the lower inclination (74°) representative of earlier solutions. (The solution suggests over-contact, but the ratio of radii is somewhat greater than previously found values).

If this is the case then a strong constraint on the inclination is implied, since it must then satisfy

$$\cos i = \sqrt{\frac{r_1^{\ 2}(1 - k)^2 - \sin^2\theta_1}{1 - \sin^2\theta_1}} \qquad (10.1)$$

where θ_1 is the phase displacement from 180° at which totality begins. Reading θ_1 as $\sim 10^\circ$ from the light curve, and with k and r set by the "contact" premise as 0.59 and 0.48 respectively (Kopal's tables 3.3, 3.2 op. cit.) we obtain $i \simeq 85^\circ$! It was found impossible to match the observed curve with anything like the accuracy shown in Figure 10 with such high inclination values, which give rise to significantly deeper eclipses. One possibility which was tried was to allow the coefficient of gravity darkening of the primary component (which has considerably greater effect than that of the secondary) to enter as a free parameter. Instead of diminishing from its calculated value of 1.04, which some authors, preferring a lower value, suggest would be inappropriate for this type of star, net improvements were actually found by making it increase. The only solution which could be found to fit the light curve well enough to allow an improvement over Figure 10 was obtained when the condition $L_1 + L_2 = 1$ was dropped, i.e. that third light be admitted. This third light amounts to 13% of the mean uneclipsed light level from the system at $\lambda = 5500$. The same effect was obtained also for the blue light curve with a similar required proportion of third light. The possibility of this circumstance for V566 Oph has not been previously discussed as far as the author is aware. Its admission allows a dramatic improvement to the light curve fit as can be judged from comparing Figures 10 and 11.

Finally it was attempted to discern how many independent parameters could be estimated from the light curve from the condition that the curvature Hessian stay positive definite (see Budding and Najim, 1980). The answer appears to be about seven (including the two relatively uninformative fiducial quantities corresponding to mean uneclipsed light level, or "unit of light", and correction to the zero phase value). The three geometric elements and two separate luminosities can just be resolved for the V light curve, but the curvature Hessian comes perilously close to indeterminacy, as can be seen from the scale of error estimates in Table 12, and remarks following it. This point can also be judged from a subsequent experiment in which the "convective" value of gravity darkening coefficient (Lucy's (1968) $\beta = 0.08$) was substituted in a trial optimization sequence. The obtained minimum χ^2 value was some 10% greater than with the $\beta = 0.25$ von Zeipel value, but this could hardly be counted as a very significant variation. The primary radius was required to increase by some 6% over its preceding value, however, with obvious implications concerning the "degree of contact". There are actually 16 parameters requiring specification in the fitting function, (which is itself an oversimplification). Any uncertainty surrounding any one of the nine remaining quantities, whose values were adopted, must mean that the error estimates in Table 12 are actually underestimates.

Table 12: Parameters Characterizing the EW System V566 Oph

Parameter	Filter B	±	Filter V	±
L_1	0.75	0.06	0.70	0.12
L_2	0.14	0.01	0.17	0.01
L_3	0.11		0.13	
r_2/r_1	0.59		0.59	0.05
r_1	0.52	0.02	0.53	0.03
i	$83^\circ.0$	$0^\circ.5$	$82^\circ.3$	$0^\circ.7$
$\Delta\theta$	$-0^\circ.07$	$0^\circ.1$	$-0^\circ.17$	$0^\circ.07$
q	0.34		0.34	
u_1	0.5		0.5	
u_2	0.5		0.5	
τ_1	1.5		1.0	
τ_2	1.5		1.1	
E_1	1.6		1.1	
E_2	1.4		1.1	
$\Delta\ell$ (s.d.)	0.006		0.005	
χ^2/ν	1.33		1.16	

Explanation: A full discussion of the formalism is given by Budding and Najim (1980). The V light curve appears to be somewhat better defined than the B light curve, which exhibits a slight asymmetry and can only support a five element specification (i.e. the ratio of radii was adopted from the V light curve solution). The contact interpretation is certainly possible from these solutions ($r_1 + r_2 > 0.75$), though an effect of the truncation is to push up the mean fractional radii to increase the relative amount of "ellipticity" distortion at quadrature. Other quantities like the limb darkening (u_1, u_2), gravity darkening (τ_1, τ_2), and reflection coefficients (E_1, E_2) have simply been assigned "off-the-shelf" values (compromise "average", von Zeipel law and black body approximations respectively).

10.2 Gravity Darkening

In view of our finding that this quantity cannot be reliably determined simultaneously with the other principal geometric and fractional luminosity unknowns from photometric analysis of a single light curve - a point of view apparently shared by Anderson, Raff and Shu (1980); though its value is of evident significance in relation to the "degree of contact" (see also Kopal, 1968), the degree of observational support for the contact hypothesis appears to be influenced by an a priori theoretical position on this quantity. Anderson et al. (op. cit.) propose spectrographic analysis as a way out of this methodologically undesirable state - alternative suggestions will be made presently, but let us first pause to take note of the theoretical stances on this subject.

There is a well known problem (von Zeipel, 1924) in a distorted star where a barotropic (conservative) condition is taken to hold (i.e. P, ρ ≡

$P(\Phi)$, $\rho(\Phi)$, Φ being the potential), in that though the heating flux is different in the different directions of, say, equator and poles, the temperature is required to remain constant on equipotentials, $T \equiv T(\Phi)$, (through the equation of state, assuming uniformity of composition). It was shown by Eddington (1929) and Sweet (1950) that a good barotropic approximation can be maintained with the aid of an additional circulation of the stellar fluid. The circulation velocity may become rapid in regions where the anisotropic contribution to the local internal energy is large in comparison to the isotropic part, but this is thought to be in the outermost regions of the atmosphere, and the dominance of the anisotropic contribution there implies that the essential character of this emergent flux has already been defined in the lower layers where the barotropic approximation is not so disturbed. The case is well discussed in Anderson and Shu's (1977) paper in terms of their parameter δ.

The upshot is that in an envelope in which a diffusive character of heat transfer operates, which includes radiative transfer in the regions where the radiation field is largely isotropic (Mihalas, 1970, p. 28), and throughout the atmosphere in the Eddington approximation $\left\{H = 1/3 \dfrac{dJ(T)}{d\tau}\right\}$ the flux will vary with the local gravity, since $H = f_1(\Phi)\,\nabla(T) = f_2(\Phi)\,\nabla(\Phi)$, or $H \propto g$. This result is usually known as von Zeipel's law of gravity darkening (or, more recently, brightening).

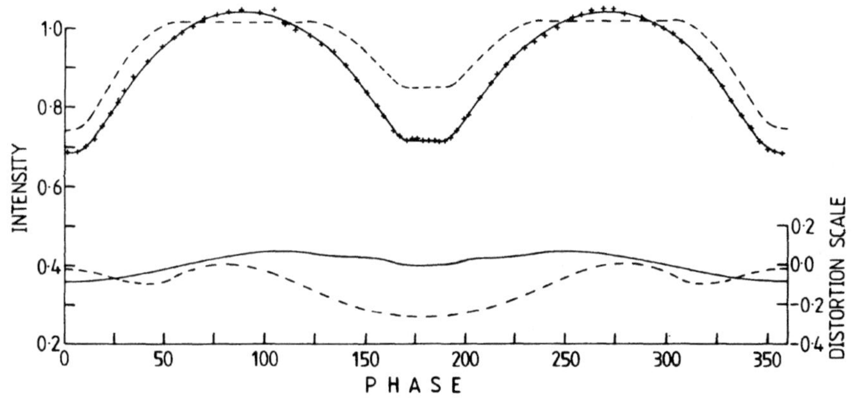

Figure 11: Optimal light curve fit to the same data as in Figure 10, but with the constraint $L_1 + L_2 = 1$ relaxed. The parameters are listed in Table 12.

Kuiper (1941) first deduced that the strict validity of the von Zeipel theorem $H \propto g$ should not apply, in a global sense, for stars in contact, on the basis that the outer boundary conditions on the mean radii of the two lobes could not be met by the "physical equilibrium" of the two stellar masses whose ratio to each other was "mechanically" imposed by the given ratio of radii and the Roche model prescription. Kuiper's interpretation was that in trying to find some compromise steady relations to each other the mean luminosity of the more massive star would be squeezed up while that of the less massive star would drop down on "physical" values, entailing that the effective temperature be somewhat higher over the larger star while the gravities would remain about the same for both surfaces.

It seems clear, however, that in this global sense the proportionality of mean flux to mean gravity need not be far wrong if the near equality of depths of minima required for a EW type light curve allow us to deduce almost equality of the two effective temperatures and we presuppose contact. In this case the luminosities L_1, L_2 must satisfy

$$\frac{L_1}{L_2} \simeq \frac{R_1^{\,2}}{R_2^{\,2}} \qquad (10.2)$$

while the mass: radius relation for contact systems requires

$$\frac{R_1}{R_2} \simeq \left(\frac{M_1}{M_2}\right)^{0.46} \qquad (10.3)$$

so that,

$$\frac{\left(\dfrac{L_1}{R_1^{\,2}}\dfrac{R_1^{\,2}}{M_1}\right)}{\left(\dfrac{L_2}{R_2^{\,2}}\dfrac{R_2^{\,2}}{M_2}\right)} = \frac{\left(\dfrac{\bar{H}_1}{\bar{g}_1}\right)}{\left(\dfrac{\bar{H}_2}{\bar{g}_2}\right)} \simeq \left(\frac{M_2}{M_1}\right)^{0.08} \qquad (10.4)$$

i.e. a very weak dependence on mass ratio. For a typical mass ratio 0.5 we find $\bar{H}_1/\bar{g}_1 = 0.95\ \bar{H}_2/\bar{g}_2$.

Though this relationship of mean surface flux to mean gravity may apply globally, Lucy (1967), argued that for stars with convective outer regions it did not apply locally. By examining a grid of convective envelope models characterized by parameters log g and log T_e. Lucy showed that the value of the adiabatic constant $K = PT^{-\gamma/(\gamma-1)}$ was fairly insensitive to the value of log g for fixed T_e. He arrived at the formula $H \propto g^{0.32}$ from the requirement that K be constant at the level where the ionization of hydrogen and helium is complete.

Anderson and Shu (1977) have argued that this procedure is not strictly legitimate, however, since it is the value of the convective flux which is required, rather than the adiabatic constant itself, and this flux is very sensitive to slight departures from the adiabatic state; reducing, for example, to zero if the numerically imposed strict adiabaticity were to hold exactly. Instead they go directly to the expression for the mixing length theory of the convective flux, which is of the form

$$H = (\ell\ g)^2\ f(\Phi) \qquad (10.5)$$

where ℓ is the mixing length. It is generally taken that this quantity is numerically of the order of the pressure scale height i.e. $\ell = \propto P/\nabla P$ where α is a coefficient of order unity. Since pressure gradient varies directly with gravity in the barotropic regime Anderson and Shu allow this dependence to cancel out in (10.5), hence making the flux quite independent of temperature gradient. Though α may be a quantity of order unity, so is the ratio of local to mean gravity, and any functional relationship between these quantities - normally irrelevant in a global problem - is not clearly known, as recognized by Anderson and Shu. The imposition of $\ell \propto 1/g$ will make the local convective velocity to be independent of surface position, however, while if the density is to remain constant on equipotentials, continuity of mass (i.e. ρvA = const.) seems to require that the proportion of available local surface area occupied by a convective cell at the equator would be less than at the pole. Even the mixing length theory may therefore not be entirely adequate to describe net flux situation.

By way of alternative empirical procedures two suggestions are offered. Firstly, in Budding and Najim (1980) it was argued that the geometrical parameters obtained at different observation wavelengths should be sensibly the same. Any systematic variation with wavelength may reflect model inadequacies, such as an imposed incorrect gravity darkening parameter. Secondly, though the inclination values of I and II in Table 11 are strikingly different individually, there is a reasonably good general correlation. The coinclination values (90 - i) of I are about 0.6 those of II on average. This factor can be traced to the assumed gravity darkening coefficients - 0.32 for I and 1 for II. Other things being equal, I would require higher inclination values to produce the same "ellipticity" effect, since the gravity darkening, which contributes to this effect, is scaled down. A few EW stars are sufficiently bright that the promising new technique of differential polarimetry could allow some possibility of an independent check on the inclination value.

10.3 EW Systems as Contact Systems

There are good reasons to believe that many binary systems such as those listed as EA1 will, at some time in the future, evolve into a condition in which the primary and almost as certainly the secondary too, for sufficiently short period systems, will be in contact with their Roche lobes. This strong expectation coupled with the fact that theoretical light curves of stars which are in contact or over-contact produce light curves which look like those of W UMa type give force to the contact hypothesis.

Difficulties have been met, however, in relation to the question of whether the observed incidence of such systems, which is generally taken to imply their rather frequent spatial occurrence, is of the same order as or greater than should be expected theoretically. In this section we shall consider what the actual incidence of contact systems could be.

Let us note first that the probability of observing eclipses in a binary composed of two spherical stars of fractional radii r_1, r_2 (using the customary notation) is $1 - \sqrt{1 - (r_1 + r_2)^2}$ (= 1 - sin i min). We observed previously that $r_1 + r_2$ for contact stars can be taken to be about 0.75.

Next suppose that the morphological difference between different eclipsing light curve types can be quantified in some way, and symbolized by ℓ. We shall assume that the value of ℓ depends essentially on the relative stellar radii expressed as a fraction of their two mass centres a, and their relative luminosities, i.e. $\ell \equiv \ell\,(a, R_1, R_2, L_1, L_2)$. Now a satisfies $a = \text{const.} \times \{M_1(1 + q)\}^{1/3} P^{2/3}$ (note the weak dependence on q), i.e.

$$\ell \equiv \ell\,(P, M_1, R_1, R_2, L_1, L_2, q), \tag{10.6}$$

but also within some stellar "sequence" we can write

$$M_1 \equiv M_1(s), R_1 \equiv R_1(s),\ L_1 \equiv L_1(s) \text{ and } L_2/L_1 \sim q^m \text{ and } R_2/R_1 \sim q^n.$$

This implies

$$\ell \equiv \ell(s, P, q). \tag{10.7}$$

The condition that eclipses be visible could be taken to restrict the range of q, moreover, the condition that the binary components belong to a particular sequence may also mean, through whatever manner of their formation actually applies (c.f. Lucy, 1980), that q tends to concentrate around a particular value. In any case, the relative dependence of ℓ on q comes mainly through the secondary luminosity and we shall presume that the effect of q for a given s, P binary is not high, so that

$$\ell \simeq \ell(s, P) \tag{10.8}$$

The fact that binaries of different ℓ are found in the same part of the s:P plane is thus an indication that such stars do not all belong to the same "sequence", or that evolution as well as mere proximity gives rise to the different light curve types.

Nevertheless, the mean periods of the different ℓ types are in somewhat different locations, and as a first line of enquiry one might seek to test the hypothesis that the types EA1, EB and maybe some EW are actually Main Sequence like stars at different degrees of relative proximity. Utilizing the separation in period ΔP of these groups we can deduce typical relative radii of order

$$\frac{(r_1 + r_2)}{0.75} = \left(\frac{P^*}{P^* + \Delta P}\right)^{2/3} \tag{10.9}$$

and hence the probability of eclipse can be expressed as,

$$\mathcal{P}(\text{ecl}) \simeq 1 - \{1 - 0.56\ (1 + \frac{\Delta P}{P^*})^{-4/3}\}^{\frac{1}{2}} \tag{10.10}$$

Comparing actual incidences we find

Spectral Type(s)	A0 + A5			F0 + F5			G0 + G5		
Light Curve Type(ℓ)	N	ΔP	$\mathcal{P}$	N	ΔP	$\mathcal{P}$	N	ΔP	$\mathcal{P}$
EA1 + ½EA3	34	0.42	0.14	15	0.47	0.11	17	0.45	0.10
EB	49	0.55	0.12	37	0.12	0.23	12	0.51	0.09
EW	19	0.06	0.28	24	0.07	0.27	58	0.06	0.26
P^*		0.55			0.44			0.36	

An inspection of the, admittedly rough, figures shows that among A and F type stars (down to F7) the EW stars are not overfrequent but somewhat underfrequent, if we suppose that all these systems derive from the same parent distribution with roughly constant expectation of incidence at low periods. Only among G type stars is the incidence of EW light curves conspicuously high, which echoes the anomaly found previously when EW incidence was compared with that of single Main Sequence stars at late relative to early types.

It has been remarked before that there is a significant overlap in the s:P plane between the EB and EW systems. Suspicions concerning a more inherent connection between these two groups are aroused by a system like AG Vir which GCVS 1 and 2 classify as EB but GCVS 3 classifies as EW. The main point of difference between these two groupings, it will be recalled, apart from vague indications that eclipses are somewhat more recognizable on EB light curves, are that while the minima are of comparable depth for EW systems they are of significantly different depth for EB systems.

But that there should be significant secondary minima at all in EB systems may seem, at first sight, to be curious. In the well studied case of VV Ori the mass ratio is ∿0.45 (Duerbeck, 1975) implying fractional overall luminosities in the ratio 94% to 6%, which are confirmed from the photometric solutions (Budding and Najim, 1980) yet the secondary minimum depth is ∿50% that of the primary (at λ = 4250Å). The explanation may be seen from Figure 11, where at secondary minimum the large minimum due to the primary ellipticity effect adds to the light loss, while such an effect is eclipsed out at the primary minimum. Could a pair of stars which were actually highly distorted Main Sequence stars give rise to a W UMa type light curve?

If we return to the formula (6.1) for relative primary and secondary depths and add in the second harmonic tidal distortion (which, of course, <u>underestimates</u> the scale of this distortion) we obtain (in intensity units)

$$\left.\begin{aligned} \lambda_1 &\simeq q^{2n} L_1 + 2q^{m-1} L_1 (0.38 + 0.2 \log q)^3 \\ &\text{and} \\ \lambda_2 &\simeq q^m L_1 + 2q L_1 (0.38 - 0.2 \log q)^3 \end{aligned}\right\} \quad (10.11)$$

where, as before, $L_2/L_1 = q^m$; $r_2/r_1 = q^n$

Use has been made of the Roche lobe approximate formula (Plavec, 1964) $r_i = 0.38 + 0.2 \log \frac{M_i}{M_{3-i}}$, while the effect of the geometrical distortion alone has been doubled due to the action of gravity darkening (Kopal, 1959, p 181)

For the upper part of the Main Sequence (BO - AO) the slope of the log L:Log M curve (at V wavelengths) falls to about 2, while n ≃ 0.6. Cancelling L_1 in the above expressions the depth ratios turn out to be

q	λ_2/λ_1
0.8	0.88
0.6	0.76
0.4	0.69

If the ratio of secondary to primary depth was as much as 0.88 the system would normally be classified as EW type. Moreover, there are two additional facts favouring equality of the minima. The foregoing formulae apply to complete eclipses in which the inclination ∿ 90°. For smaller inclination values (partial eclipses), the eclipse contribution, corresponding to the first right hand members in (10.11), diminish with respect to the second.

Secondly the value n = 0.6 corresponds to the Table I values, which were estimated from the assumed position of the unevolved limit in Popper's (1980) data. Most of the individual pairs in this data, however, particularly among those of early type (E and earlier) show a relationship between themselves for which n is significantly greater than 0.6, while the corresponding variation of luminosity is relatively little affected.

It is relatively simple to show that a pair of stars need not be in contact to produce a light curve which looks like that of a W UMa star (an example is shown in Figure 12); the problem has usually been that if the stars were detached and of the observed low mass ratios how could the minima be of approximately equal depth. The foregoing discussion shows that this could be possible at least for early type stars, especially if the primary is somewhat evolved from the Main Sequence and the eclipses are partial. It is feasible that at least some of the A-type W UMa systems could be regarded in this way, i.e. as particular cases of a more general and essentially uncomplicated, detached kind which give rise to EB type light curves.

A more compelling case for contact comes with the W UMa type systems of later type and shorter period. For stars in the range F5 - G5 the index m in (10.11) grows to be $\gtrsim$ 5 according to Table 1, while the individual pairs in Popper's data adhere much more closely to the mass radius relation, though n does rise to $\sim$ 0.85. If we form the corresponding depth ratios from Equation 10.6 we have;

q	λ_2/λ_1
0.8	0.60
0.6	0.40
0.4	0.43

i.e. it seems very difficult to reconcile the observed almost equality of depths with highly distorted stars of something like Main Sequence character. Let us note, in passing, two other features of the stars in question: Firstly they occupy a region of the s:P plane in which the overlap with systems of different light curve type is small. Secondly they are predominantly made up of the W subgroup of W UMa systems which exhibit the decidedly non Main Sequence feature of an inverse ratio of surface brightness to mass, as well as period and other irregularities suggestive of interactive effects on a thermal timescale (Yamasaki, 1975).

11. Ell SYSTEMS - "ELLIPSOIDAL" VARIABLES

Only a few such variables are known but they are included in Table 13 for the sake of completeness. A large number of variables actually exist which give rise to small amplitude quasi-sinusoidal fluctuations but mostly an explanation in terms of physical variability is given. The systems must be fairly bright and well studied in order that binarity be established, since as we have seen, physical variables can sometimes be confused with eclipsing variables despite the usually clear character of the eclipse effect.

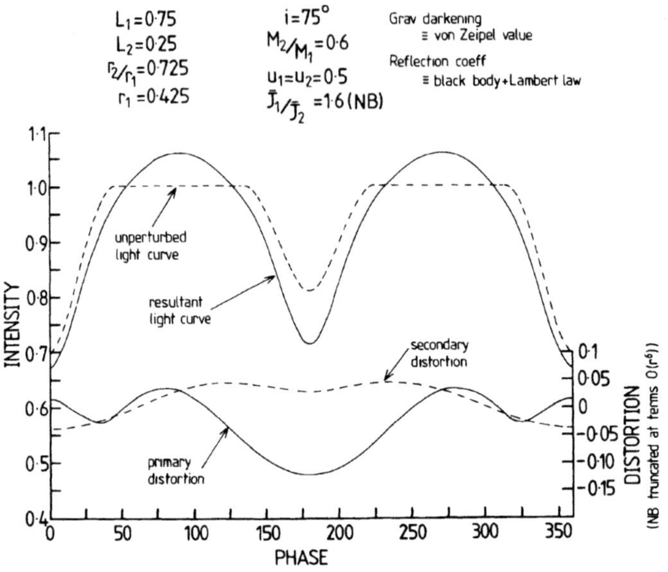

Figure 12: An EW type light curve from a non-contact system with standard modelling parameters. It is clear that the form of the light curve alone is not a strong argument concerning contact.

On the other hand, it can be argued that many low amplitude binary variables must be still awaiting recognition, particularly among the EB and EW systems. If the contact interpretation applies, Kopal has shown (op. cit. Table 3.4) that eclipses should be possible down to an inclination of $\sim 34^{\circ}$, though 44° is the lowest value obtained for any such system, even with Kopal's more generous co-inclination values. About 70% of such systems have inclination values less than 44°.

Table 13: Close Binary Systems of Type Ell

Constellation	Name or No.	Period	Primary s Type	Mag.
Aqr	Y	1.302	B8	5.03
Per	IW	0.917	A2	5.8
	IX	1.326	F2	6.6
Psc	UU	0.841	F0	6.0

12. CONCLUSIONS

12.1 What Has Been Attempted

Though the use of statistics makes for an inherent reduction of the total information content of an ensemble of particular cases, nevertheless, the foregoing presentation of a reduced form of information recently made available through the quoted sources can serve two basic purposes. Firstly it can be used to clarify certain terms in the vocabulary on close binary systems - Algol systems, contact binaries, or some others. Secondly, it may provide some means of testing general statements made about such terms.

With regard to the second point a few simple empirical tests have been included in this presentation. This is not meant as a criticism of detailed discussions, but only to educe more on the factual side, against which the merits of more intensive explicative efforts can be assessed.

12.2 Results

Results appear as the various tables and diagrams which have been presented. A few summary statements are also possible, at this stage, from intercomparison of these data.

The character of the different distributions does suggest an interpretative role for the light curve descriptor if this can be systematized, and though it is expectable that physical proximity of the two mass centres is the most strongly correlated variable, there are indications of a relationship with the evolutionary state of the binary thus described. This is most clear with the distinction between what has been referred to as EA1 and EA2 type light curves, but, on the opposite side of the same coin, the relationship of the EW descriptor to contact systems has been shown to be far from clear.

In connection with EW systems two specific points emerge: (i) even with high quality photometric data, only six or seven positive eigenvalues of the optimal curvature Hessian could be found. In this sense parameter determinacy for this type of light curve appears to be somewhat less than, for example, the EB light curves of VV Ori (in which proximity effects are less pronounced) analysed recently by Budding and Najim (1980); (ii) as a consequence of this, independent confirmation of the appropriate gravity darkening parameter for this type of light curve, whose adopted value definitely correlates with the "degree of contact" required to allow theoretical models to fit observed light curves,is probably not available from the analysis of individual light curves.

Data presented for comparison on the likely statistical properties of single dwarfs may be used to show the extent to which such statistical properties of stars are affected by the circumstance of binarity. EB primaries, for example, show a distribution with spectral type which is not unlike that of dwarf stars in the Bright Star Catalogue. The same may also be true for EA1 binaries, but in this case, the cut-off effect associated with the 1.5 day imposed period limit should more seriously affect the distribution at earlier type. Certainly the application of an empirical Main Sequence and contact criterion to such distributions brings to light a number of interesting anomalous cases.

12.3 Further Remarks

The interesting role played by the light curve descriptor in relation to the preceding discussion calls to mind a possible simile with the early qualitative days of spectral classification. The enormous application which further quantitative refinements of that scheme have had in relation to the understanding of stars in general, promotes the idea of some clearer empirical quantification of the definition of light curve types. Fourier analysis seems to have much to recommend itself in this context - not just as a possible means of procuring "elements" in which a "solution" is seen, but as a set of conceptual filters from which the colour or colours (in this filtered sense) of particular light curve could be measured. In order to be practicable on a large scale such a scheme should, of course, be simple. For example, with just the three components $a_0 + a_1 \cos\theta + a_2 \cos 2\theta$ by considering the relation of a_2/a_0 to a_1/a_0 it may be possible to define the difference between EA, EB and EW light curves. Just as U-B versus B-V provided a second dimension of colour representation in which effects indicative of evolutionary status could be distinguished from mere temperature effects, further refinements of the single dimension scheme (e.g. $a_2/a_0 - a_1/a_0$) could perhaps show up the binary evolution implying difference between types EA1 and EA2.

There has not been left much space to say something about what such binary evolution implying effects are referring to. Let us briefly consider a couple of points which emerge from the foregoing compilation in this connection. Firstly it has been found that there are comparable if not more numbers of evolved (EA2) to unevolved (EA1) Algol systems within the sample. The trend of opinion among specialists in detailed calculation of binary evolution appears to be that Case B rather than Case A type mass transfer provides the better explanation of how most of the EA2 systems got into their present state (Plavec, 1973; Ziołkowski, 1976). Since Case B allows less than nuclear timescales (e.g. Kippenhahn et al. 1967) for the semi-detached configuration, however, the EA2 systems appear to be in a similar situation of overabundance as the cool EW binaries, though the problem is clearly more acute for the latter in the present sample. The Algol problem could be investigated further by producing information like that of Table 6 and Figures 5 and 6 also for the EA1 systems, i.e. by dropping the period restriction.

The role of non-conservative processes in mass transfer in relation to the EW systems has been emphasized by van t'Veer (1980), and Vilhu (1981), while similar remarks have been made in relation to evolved Algols by Svechnikov (1969), and more recently by de Grève and

Vanbeveren (1980), and also massive close binaries (Vanbeveren and de Loore, 1980). Though more likely than not "too early to hazard" the speculation pushes itself forward that the observed overabundance of evolved close binary systems could be due to angular momentum loss cramming the aftermath of episodes of large scale mass variation into a smaller range of separations than existed initially.

Returning to the data presented after (10.10), it could be inferred that there are ∿ 200 G type EW-like systems brighter than ∿ 12th magnitude (i.e. including those that do not eclipse and assuming sample completeness). Is it possible to find (Case A) "parents" for these stars from some other group in the sample? It could indeed be estimated that there are correspondingly ∿ 300 F-type unevolved (EA and EB-like) binaries in all - however the volume of space occupied by these systems is ∿ 8 times larger than that of the G-type systems (Table 1). More detailed comparisons of the expected lifetimes in the respective evolutionary stages could perhaps resolve this discrepancy - for the present the hypothesis that certain stars originate in a contact configuration (Lucy, 1968, 1976) has not been shown to be unnecessary.

I would like to conclude on a more personal note. My interest in and opportunity to study and observe close binary systems over the last eleven years has been in no small way due to Professor Z.Kopal. I have found in this occupation a considerable source of mental enlivenment and memorable experience. Throughout there has been the possibility of the precious acquisition of some measure of real understanding. If, as a result of this particular involvement, a more general diffusion of learning were to occur, so much the better. From my side, as the more immediate recipient, I express my sincere thanks.

REFERENCES

Allen, C. W.: 1973, *Astrophysical Quantities*, Athlone Press, London.
Anderson, L., Raff, M. and Shu, F. H.: 1980, in *Close Binary Stars: Observation and Interpretation*, IAU Sump. No. 88 (Ed. M. J. Plavec, D. M. Popper and R. K. Ulrich) p. 485.
Anderson, L. and Shu, F. H.: 1977 Astrophys. J. 214, p. 798.
Binnendijk, L.: 1970, Vistas in Astronomy 12, p. 217.
Bookmyer, B. B.: 1969, Astron. J. 74, p. 1197.
Budding, E. and Najim, N. N.: 1980, Astrophys. and Space Sci.72, p. 369.
Chuadze, A. D.: 1973, Bull. Abastumani Astrophys. Obs. 44, p. 107.
Diethelm, R.: 1980, IBVS No. 1822 (Konkoly Obs. Budapest).
Duerbeck, H. W.: 1975, Astron. Astrophys. Suppl. 22, p. 19.
Eddington, A. S.: 1929, Mon. Not. Roy. Astron. Soc. 90, p. 54.
Eggen, O. J.: 1967, Mem. Roy. Astron. Soc. 70, p. 111.
Götz, W. and Wenzel, W.: 1964, Mitt. Ver. Sterne, Berlin,2, H4, p. 85.
de Grève, J. P. and Vanbeveren, D.: 1980, Astrophys. and Space Sci. 68, p. 433.
Hall, D. S.: 1976, IAU Colloq. No. 29 (Budapest), p. 287.
Hall, D. S. and Cannon, R. O.: 1974, Acta Astron. 24, p. 79.
Hazlehurst, J.: 1976, in *Structure and Evolution of Close Binary Systems*, IAU Symp. No. 73, (Ed. P. Eggleton, S. Mitton, J. Whelan), p. 321.
Hoffleit, D.: 1964, *Catalogue of Bright Stars*, Yale Univ. Obs., New Haven, Conn.
Hoffmann, M.: 1979, Astron. and Astrophys. Suppl. Ser. 40, p. 263.
Hoffmann, M.: 1980, IBVS No. 1877 (Konkoly Obs., Budapest).
Iben, I.: 1967, Ann. Rev. Astron. Astrophys. 5, p. 571.
Keenan, P. C. and McNeill, R. C.: 1976, *An Atlas of Spectra of the Cooler Stars, Part 1. Introduction and Tables*, Ohio State Univ. Press.
Kippenhahn, R., Kohl, K. and Weigert, A.: 1967, Zeit. für Astrophys., 65, p. 251.
Kopal, Z.: 1955, Ann. d' Astrophys. 18, p. 379.
Kopal, Z.: 1959, *Close Binary Systems*, Chapman and Hall Ltd., London.
Kopal, Z.: 1968, Astrophys. and Space Sci., 2, p. 23.
Kopal, Z.: 1979, *Language of the Stars*, Reidel.
Kuiper, G. P.: 1941, Astrophys. J. 93, p. 133.
Kukarkin, R. V., Kholopov, P. N., Efremov, Yu. N., Kukarkina, N. O., Kurochkin, N. E., Medvedeva, G. I., Perova, N. B., Fedorovich, V. P. and Frolov, M. S.: 1969, *General Catalogue of Variable stars*, also Vol. 2 (1970), Second Suppl. (1974), Third Suppl. (1976) Publishing House "Nauka", Moscow.
Lubow, S. H. and Shu, F. H.: 1977, Astrophys. J. 216, p. 517.
Lucy, L. B.: 1967, Zeit. für Astrophys. 65, p. 89.
Lucy, L. B.: 1968, Astrophys. J. 151, p. 1123.
Lucy, L. B.: 1972, (cited by Ruciński, 1973), manuscript presented at IAU Coll. 16, *Analytical Methods for the Solution of Light Curves*, (Philadelphia).
Lucy, L. B.: 1973, Astrophys. and Space Sci. 22, p. 381.
Lucy, L. B.: 1976, Astrophys. J. 205, p. 208.
Lucy, L. B.: 1980, in *Close Binary Stars: Observations and Interpret-*

ation IAU Symp. No. 88 (Ed. M. J. Plavec, D. M. Popper and R. K. Ulrich, p. 7.

MacDonald, D. D.: 1964, McCormick Publ. 12, Part 5.

Mihalas, D.: 1970, *Stellar Atmospheres*, Freeman, San Francisco.

Mochnacki, S. W. and Doughty, N. A.: 1972a, Mon. Not. Roy. Astron. Soc. 156, p. 51.

Mochnacki, S. W. and Doughty, N. A.: 1972b, Mon. Not. Roy. Astron. Soc. 156, p. 243.

Morgan, J. G. and Eggleton, P. P.: 1979, Mon. Not. Roy. Astron. Soc. 187, p. 661.

Moss, D. L. and Whelan, J. A. J.: 1970, Mon. Not. Roy. Astron. Soc. 149,p. 147.

Niarchos, P.: 1977, Ph.D. Thesis, Univ. of Manchester (unpublished).

Okazaki, A.: 1977, Publ. Astron. Soc. Japan 29, p. 289.

Plavec, M.: 1964, Bull. Astron. Inst. Czech. 15, p. 156.

Plavec, M.: 1973 in *Extended Atmospheres and Circumstellar Matter in Spectroscopic Binary Systems,* (Ed. A Batten), Reidel IAU Symp. No. 51.

Pohl, E. and Kizilirmak, A.: 1977, IBVS No. 1358, (Konkoly Obs., Budapest).

Popper,D. M.: 1967, Ann. Rev. Astron. Astrophys. 5, p. 85.

Popper, D. M.: 1980, Ann. Rev. Astron. Astrophys. 18, p. 115.

Robertson, J. A. and Eggleton, P. P.: 1977, Mon. Not. Roy. Astron. Soc. 179, p. 359.

Ruciński, S. M.: 1973, Acta Astron. 23, p. 79.

Ruciński, S. M.: 1974, Acta Astron. 24, p. 119.

Salpeter, E. E.: 1955, Astrophys. J., 121, p. 161.

Sisteró, R. F.: 1980, The Observatory 100, p. 121.

Söderhjelm, S.: 1980, *Close Binary Stars: Observations and Interpretation* IAU Symp. No. 88 (Ed. M. J. Plavec, D. M. Popper and R. K. Ulrich), Reidel.

Svechnikov, M. A.: 1969, *Katalog, Orbitalnyich Elementov, Mass i Svetimostei Tesnyich Dvoinyich Zvezd,* A. M. Gorky University of the Urals, Sverdlovsk.

Sweet,P. A.: 1950, Mon. Not. Roy. Astron. Soc., 110, p. 548.

Taff, L. G.: 1974, Astron. J., 79, p. 1280.

Vant' Veer, F.: 1980, in *Close Binary Stars: Observations and Interpretation* IAU Symp. No. 88 (Ed. M. J. Plavec, D. M. Popper and R. K. Ulrich), Reidel. p. 517.

Vanbeveren, D. and de Loore, C.: 1980, in *Close Binary Stars: Observations and Interpretation,* IAU Symp. No. 88 (Ed. M. J. Plavec, D. M. Popper and R. K. Ulrich), Reidel. p. 115.

Yamasaki, A.: 1975, Astrophys. and Space Sci. 34, p. 413.

Yamashita, Y., Nariai, K. and Norimoto, Y.: 1977, *An Atlas of Representative Stellar Spectra*, Univ. Tokyo Press.

von Zeipel, H.: 1924, Seeliger Festschrift, p. 144.

Ziolkowski, J.: 1976, in *Structure and Evolution of Close Binary Systems* IAU Symp. No. 73 (Ed. P. Eggleton, S. Mitton and J. Whelan) p. 321.

BIFURCATIONS OF PLANAR TO THREE-DIMENSIONAL PERIODIC ORBITS IN THE RESTRICTED THREE-BODY PROBLEM

V.V.Markellos, Dept. of Applied Mathematics, Faculty of Engineering, University of Patras, Greece.

C.L.Goudas and G.A.Katsiaris, Dept. of Mechanics, Faculty of Science, University of Patras, Greece.

ABSTRACT

We study the mechanism of bifurcation of three-dimensional periodic orbits from the plane of motion of the primaries. The branchings of simple and double periodic three-dimensional orbits occur at the plane periodic orbits of indifferent vertical stability. These vertical-critical plane orbits are doubly important. Not only are they necessary in order to specify the vertically stable segments of the families of plane periodic orbits, but they are also the starting points of the families of the simplest possible three-dimensional periodic orbits, namely, the simple and double periodic. We describe here suitable numerical techniques for their accurate determination and for their continuation through the range of the mass parameter μ of the problem. We use these techniques to determine the first few vertical-critical orbits of the basic families of periodic orbits of the problem, namely, families ℓ, m, i, h, a, b and c, for the entire range of μ. The initial conditions of over three hundred vertical-critical orbits are given, classified in twenty-one series. Their horizontal stability is also examined and this leads to predictions of where in the range of μ stable three-dimensional periodic orbits may not branch from these basic families.

1. INTRODUCTION

The importance of the periodic orbits for their occurrence in nature and for their significance in the structure of the set of all dynamically possible orbits, has provided strong motivation for extensive investigations of the periodic solutions of the restricted three-body problem, this being considered an appropriate model for the study of the motions especially of the smaller members of the Solar system. The near flatness of the orbits in the Solar system and the simplifications achieved in considering plane and symmetric periodic solutions of the problem of three bodies are good reasons for computing and studying the properties of large number of such solutions. But, it is evident that with a view

F. D. Kahn (ed.), Investigating the Universe, 321–351.

to applications in the Solar system a more relevant type of periodic orbits would be the three-dimensional, low inclination, ones. However, while a great number of numerical results are available on the plane periodic orbits of the restricted problem relatively few results exist on the three-dimensional ones. It is known that the plane periodic orbits of indifferent vertical stability are the bifurcation points of families of three-dimensional periodic orbits. In this paper we aim at the large-scale detection of three-dimensional periodic orbits. The occurrence of such orbits is characterized by the existence of their bifurcations from the plane. We aim therefore to determine such bifurcations, i.e., to determine many of the known bifurcations, given in Hénon (1973) for $\mu=1/2$, for the entire range of the mass parameter. We shall restrict our attention to simple or double vertical bifurcation orbits leading to families of simple or double three-dimensional periodic orbits. Also, we consider here only symmetric periodic orbits and refer to Markellos (1977) for a similar study of non-symmetric orbits.

2. EQUATIONS OF MOTION AND VARIATION

The equations of motion of the restricted problem of three bodies in three dimensions can be written as follows:

$$\dot{x}_i = f_i(x_1, \ldots, x_6), \quad i=1,\ldots,6, \tag{1}$$

with

$$f_1 = x_4, \quad f_2 = x_5, \quad f_3 = x_6,$$

$$\begin{aligned} f_4 &= 2x_5 + x_1 - \frac{(1-\mu)(x_1+\mu)}{r_1^3} - \frac{\mu(x_1+\mu-1)}{r_2^3}, \\ f_5 &= -2x_4 + x_2 - \frac{(1-\mu)x_2}{r_1^3} - \frac{\mu x_2}{r_2^3}, \\ f_6 &= -\frac{(1-\mu)x_3}{r_1^3} - \frac{\mu x_3}{r_2^3}, \end{aligned} \tag{2}$$

where

$$x_1 = x, \; x_2 = y, \; x_3 = z, \; x_4 = \dot{x}, \; x_5 = \dot{y}, \; x_6 = \dot{z}$$

and

$$r_1 = [(x_1+\mu)^2 + x_2^2 + x_3^2]^{1/2}, \qquad r_2 = [(x_1+\mu-1)^2 + x_2^2 + x_3^2]^{1/2}.$$

The coordinates of the infinitesimal particle in phase space $x_1,\dots,x_6$ depend uniquely, along any solution, on the initial conditions and the time

$$x_i=x_i(x_{01},\dots,x_{06};t),\quad i=1,\dots 6,$$

their partial derivatives with respect to the initial conditions satisfying the equations of variation

$$\frac{d}{dt}\left(\frac{\partial x_i}{\partial x_{0j}}\right)=\sum_{k=1}^{6}\frac{\partial f_i}{\partial x_k}\frac{\partial x_k}{\partial x_{0j}},\quad i,j=1,\dots,6, \tag{3}$$

If we denote by υ_{ij} the quantities $\partial x_i/\partial x_{0j}$ we can write these last equations more explicitly as follows

$$\begin{aligned}&\dot\upsilon_{ij}=\upsilon_{(i+3)j},\quad i=1,2,3,\ j=1,\dots,6,\\ &\dot\upsilon_{ij}=f_{i1}\upsilon_{1j}+f_{i2}+\upsilon_{2j}+f_{i3}\upsilon_{3j}+f_{i4}\upsilon_{4j}+f_{i5}\upsilon_{5j},\quad i=4,5,6,\ j=1,\dots,6,\end{aligned} \tag{4}$$

where

$$f_{ij}=\partial f_i/\partial x_{0j},\quad (f_{i6}=0,\ i=4,5,6)$$

and

$$\begin{aligned}&f_{41}=-Q+1+R_1(x_1+\mu)^2+R_2(x_1+\mu-1)^2,\\ &f_{42}=-Q'x_2,\ f_{43}=-Q'x_3,\ f_{44}=0,\ f_{45}=2,\\ &f_{51}=f_{42},\\ &f_{52}=-Q+1+(R_1+R_2)x_2^2,\\ &f_{53}=-Q''x_3,f_{54}=-2,\ f_{55}=0,\\ &f_{61}=f_{43},\ f_{62}=f_{53},\\ &f_{63}=-Q+(R_1+R_2)x_3^2,\\ &f_{64}=f_{65}=0\ .\end{aligned} \tag{5}$$

We have abbreviated

$$R_1=\frac{3(1-\mu)}{r_1^5},\quad R_2=\frac{3\mu}{r_2^5},$$

$$Q=\frac{1-\mu}{r_1^3}+\frac{\mu}{r_2^3}, \tag{6}$$

$$Q'=-R_1(x_1+\mu)-R_2(x_1+\mu-1), \quad Q''=-(R_1+R_2)x_2 \quad . \tag{7}$$

3. HORIZONTAL AND VERTICAL STABILITY PARAMETERS

When the initial position and velocity of the infinitesimal particle are in the plane of motion of the primaries the particle will remain in this plane forever. In this case we continue to use the same notations as above but now the indices do not take the values 3 and 6 and the quantities indicated by one (but not both) of these values become redundant. However, even in the case of plane motion the quantities

$$a_\upsilon=\upsilon_{33}, \quad b_\upsilon=\upsilon_{36}, \quad c_\upsilon=\upsilon_{63}, \quad d_\upsilon=\upsilon_{66} \tag{8}$$

are important because they describe the stability properties of the orbit under perturbations opposing the planar nature of the orbit, i.e., perturbations in the x_3, x_6 components of the solution vector in phase-space. The quantities in (8) are called "vertical stability parameters" of the orbit.

On the other hand, the stability properties of the orbit under perturbations which are compatible with its planar nature are described by the "horizontal stability parameters" a, b, c, d. For the accurate determination of these parameters it is useful to integrate numerically some of the equations of variation, namely, those of the quantities υ_{ij} with $i=1,2,4$, $j=1,4,5$. The pertinent formulae (from Markellos, 1976) are

$$\begin{aligned} a&=(\upsilon_{11}+\upsilon_{15}D_1)+D_2(\upsilon_{21}+\upsilon_{25}D_1) \quad ,\\ b&=(\upsilon_{14}+\upsilon_{15}D_2)+D_2(\upsilon_{24}+\upsilon_{25}D_2) \quad ,\\ c&=(\upsilon_{41}+\upsilon_{45}D_1)-(D_1+2)(\upsilon_{21}+\upsilon_{25}D_1) \quad ,\\ d&=(\upsilon_{44}+\upsilon_{45}D_2)-(D_1+2)(\upsilon_{24}+\upsilon_{25}D_2) \quad , \end{aligned} \tag{9}$$

where

$$D_1=\frac{f_4}{x_5}-2 \ , \quad D_2=-\frac{x_4}{x_5} \ . \tag{10}$$

In the case of symmetric periodic orbits we have $D_2=0$ and expressions (9) become

$$\begin{aligned} &a=\upsilon_{11}+\upsilon_{15}D_1 \ , \quad b=\upsilon_{14} \ , \quad c=(\upsilon_{41}+\upsilon_{45})D_1-(D_1+2)(\upsilon_{21}+\upsilon_{25}D_1),\\ &d=\upsilon_{44}-(D_1+2)\upsilon_{24}. \end{aligned} \tag{11}$$

In the above expressions (8) and (11) the quantities involved in the right hand sides are evaluated by numerical integration at t=T, where T is the period of the orbit. For symmetric periodic orbits, however, there exist simple formulae for the calculation of the horizontal and vertical stability parameters using the values at t=T/2 of the quantities involved in the above expressions, i.e., requiring numerical integration of the equations of motion and variation for only half the period of the orbit (see, e.g., Hénon, 1973).

Here, for the numerical determination of plane periodic orbits and their horizontal and vertical stability parameters we integrate numerically 20 of the full set (42 equations) of (1) and (4). The reduction is illustrated by the following Table.

x_1	x_2	0	x_4	x_5	0	(motion)
υ_{11}	—	0	υ_{14}	υ_{15}	0	
υ_{21}	—	0	υ_{24}	υ_{25}	0	
0	—	υ_{33}	0	0	υ_{36}	(variation)
υ_{41}	—	0	υ_{44}	υ_{45}	0	(12)
υ_{51}	—	0	υ_{54}	υ_{55}	0	
0	—	υ_{63}	0	0	υ_{66}	

where - indicates qunatities that need not be computed and 0 indicates quantities that remain equal to zero during planar motion.

From equation (4) it transpires that for planar motion ($x_3=x_6=0$) the differential equations for the planar and the vertical variations decouple, while the latter reduce to the following simple form:

$$\dot{\upsilon}_{33} = \upsilon_{63} \quad , \qquad \dot{\upsilon}_{36} = \upsilon_{66} \quad ,$$

$$\dot{\upsilon}_{63} = -Q\upsilon_{33} \quad , \qquad \dot{\upsilon}_{66} = -Q\upsilon_{36} \quad .$$

Effectively we just have one equation of the form

$$\ddot{\eta}(t)+Q(t)\eta(t)=0 \tag{13}$$

and two independent solutions satisfying the initial conditions $\eta_1(0)=1$, $\dot{\eta}_1(0)=0$, $\eta_2(0)=0$, $\dot{\eta}_2(0)=1$.
When motion is actually periodic Q(t), the quantity abbreviated in (6), is a periodic function of time and (13) is a Hill type differential equation.

Second Variations

In the present work we also require some of the second order variations of the solution with respect to the initial conditions, namely, those referring to the vertical stability parameters. If we denote by υ_{ijk} the quantities $\partial^2 x_i/\partial x_{0j}\partial x_{0k}$, then the pertinent values of the indices are i,j=3,6 and k=1,4,5 and the corresponding differential equations can be written as follows

$$\dot{\upsilon}_{3jk}=\upsilon_{6jk}\ , \tag{14}$$

$$\dot{\upsilon}_{6jk}=-Q\upsilon_{3jk}-Q'(\upsilon_{1j}\upsilon_{3k}+\upsilon_{3j}\upsilon_{1k})-Q''(\upsilon_{2j}\upsilon_{3k}+\upsilon_{3j}\upsilon_{2k}), \tag{15}$$

where Q, Q', Q'' are the quantities abbreviated in (6),(7).

4. THE BRANCHING OF THREE-DIMENSIONAL PERIODIC ORBITS FROM THE PLANE

The three-dimensional periodic orbits of the restricted three-body problem are classified into symmetric and non-symmetric and there are three types of symmetry possible (see, e.g., Zagouras and Markellos, 1977 and Markellos, 1976). We shall now consider the bifurcation from the plane of a particular type of symmetric three-dimensional orbits, the plane-symmetric, i.e., those which are symmetric with respect to the plane xOz, but similar considerations can easily be seen to be valid for the other types of symmetric three-dimensional periodic orbits.

Periodicity of a plane-symmetric three-dimensional orbit is achieved by two perpendicular crossings of the plane of symmetry xOz, and the periodicity conditions are

$$\begin{aligned} &x_2(x_{01},\ x_{03},\ x_{05};\tau)=0,\\ &x_4(x_{01},\ x_{03},\ x_{05};\tau)=0,\\ &x_6(x_{01},\ x_{03},\ x_{05};\tau)=0. \end{aligned} \tag{16}$$

where τ is half the period. Let us assumme that these conditions are satisfied for a given orbit and suppose that we want to continue the family of which this orbit is a member, i.e., to find the "next" periodic orbit of the same family. We seek to satisfy now the conditions

$$\begin{aligned} &x_2(x_{01}+\delta x_{01},\ x_{03}+\delta x_{03},\ x_{05}+\delta x_{05};\tau+\delta\tau)=0\ ,\\ &x_4(x_{01}+\delta x_{01},\ x_{03}+\delta x_{03},\ x_{05}+\delta x_{05};\tau+\delta\tau)=0\ ,\\ &x_6(x_{01}+\delta x_{01},\ x_{03}+\delta x_{03},\ x_{05}+\delta x_{05};\tau+\delta\tau)=0\ . \end{aligned}$$

We linearize these conditions, by expanding the functions on the left into Taylor series and neglecting terms of order two or higher, to obtain

$$\upsilon_{21}\delta_1+\upsilon_{23}\delta_3+\upsilon_{25}\delta_5=-\upsilon_{2\tau}\delta\tau\ ,$$
$$\upsilon_{41}\delta_1+\upsilon_{43}\delta_3+\upsilon_{45}\delta_5=-\upsilon_{4\tau}\delta\tau\ , \qquad (17)$$
$$\upsilon_{61}\delta_1+\upsilon_{63}\delta_3+\upsilon_{65}\delta_5=-\upsilon_{6\tau}\delta\tau\ ,$$

where $\delta_i=\delta x_{0i}$, $i=1,3,5$. This system of equations for the unknown quantities δ_1, δ_3, δ_5, $\delta\tau$ describes a predictor scheme for the determination of the initial conditions and half-period

$$x^*_{01}=x_{01}+\delta_1\ ,\ x^*_{03}=x_{03}+\delta_3\ ,\ x^*_{05}=x_{05}+\delta_5\ ,\ \tau^*=\tau+\delta\tau\ ,$$

of the next periodic orbit along the family. The change in half-period $\delta\tau$ can be chosen arbitrarily (so long as it is small enough for the above linearization to be valid) and can serve as the parameter of the family (the fact that the family is a monoparametric set is apparent from the fact that the system (17) consists of three equations in four unknowns).

Let us now write the system in matrix form

$$S\cdot\underline{\delta}=-\delta\tau\cdot\underline{h}\ , \qquad (18)$$

with

$$S=\begin{bmatrix}\upsilon_{21} & \upsilon_{23} & \upsilon_{25}\\ \upsilon_{41} & \upsilon_{43} & \upsilon_{45}\\ \upsilon_{61} & \upsilon_{63} & \upsilon_{65}\end{bmatrix}\ ,$$

$$\underline{\delta}=\begin{bmatrix}\delta_1\\ \delta_3\\ \delta_5\end{bmatrix}\ ,\quad \underline{h}=\begin{bmatrix}\dot{y}(\tau)\\ \ddot{x}(\tau)\\ \ddot{z}(\tau)\end{bmatrix}\ ,$$

where we have substituted $\upsilon_{2\tau}=\dot{y}(\tau)$, $\upsilon_{4\tau}=\ddot{x}(\tau)$, $\upsilon_{6\tau}=\ddot{z}(\tau)$.

We wish to examine this linear predictor (18) in the special case when the periodic orbit from which we start (and at which the quantities S and $\underline{h}$ are determined) is actually planar. In this case we have $\upsilon_{23}=\upsilon_{43}=\upsilon_{61}=\upsilon_{65}=\ddot{z}=0$, for all time, and the predictor system becomes

$$\begin{bmatrix}\upsilon_{21} & 0 & \upsilon_{25}\\ \upsilon_{41} & 0 & \upsilon_{45}\\ 0 & \upsilon_{63} & 0\end{bmatrix}\begin{bmatrix}\delta_1\\ \delta_3\\ \delta_5\end{bmatrix}=-\delta\tau\begin{bmatrix}\dot{y}(\tau)\\ \ddot{x}(\tau)\\ 0\end{bmatrix} \qquad (19)$$

with

$$\det S = -\upsilon_{63} \det P, \tag{20}$$

where

$$P = \begin{bmatrix} \upsilon_{21} & \upsilon_{25} \\ \upsilon_{41} & \upsilon_{45} \end{bmatrix}.$$

Assuming that the orbit at hand is a "regular" plane periodic orbit in the sense that it can be continued into a family of plane periodic orbits, we can write

$$\det P \neq 0,$$

i.e., P is a regular matrix.
We shall consider separately the homogeneous case of the predictor (19), $\delta\tau=0$, and the non-homogeneous case $\delta\tau\neq 0$.

(a) $\delta\tau=0$. This means that we seek to continue the present orbit in such a way that the half-period τ of the orbit remains unchanged to first order terms. The predictor becomes

$$\begin{bmatrix} \upsilon_{21} & 0 & \upsilon_{25} \\ \upsilon_{41} & 0 & \upsilon_{45} \\ 0 & \upsilon_{63} & 0 \end{bmatrix} \begin{bmatrix} \delta_1 \\ \delta_3 \\ \delta_5 \end{bmatrix} = \begin{bmatrix} 0 \\ 0 \\ 0 \end{bmatrix} \tag{21}$$

and, since $\det P\neq 0$, it gives

$$\delta_1 = \delta_5 = 0 \quad ,$$

as the only possible values for the changes to the initial conditions in the "planar" components $x_{01}=x_0$, $x_{05}=\dot{y}_0$ of the initial state vector. The change δ_3 in the "vertical" component $x_{03}=z_0$ must satisfy the condition

$$\upsilon_{63}\delta_3 = 0 . \tag{22}$$

In general we have

$$\upsilon_{63} \neq 0 \ ,$$

(see (12)) and (22) gives

$$\delta_3 = 0 \quad ,$$

as a necessary consequence.
Therefore, in the general case when the present orbit is regular and $\upsilon_{63}\neq 0$, the only possible changes that would preserve periodicity are the zero changes, i.e., there is no periodicity preserving direction which would not imply a change in period (to first order terms).

If, however, our orbit is exceptional in that it satisfies the equality

$$\upsilon_{63} = 0 \ ,$$

then condition (22) will be satisfied for an arbitrary $\delta_3=\varepsilon\neq 0$ (provided it is small enough for the present linear analysis to be valid). Therefore, in this exceptional case the solution vector of the predictor system is

$$\underline{\delta}=(0,0,\varepsilon)^T , \qquad (23)$$

and this provides a bifurcation of the family of plane orbits with a family of three-dimensional orbits.

(b) $\delta\tau\neq 0$. This means that we seek to continue the present orbit in such a way that the half-period τ changes to first order terms. The predictor is given by (19) and provides unique values (det$P\neq 0$) for the changes δ_1, δ_5 in the "planar" components of the initial state vector, which are functions of the arbitrarily chosen (small) quantity $\delta\tau=\varepsilon'$, as follows:

$$\delta_1(\varepsilon')=\frac{1}{\det P}\begin{bmatrix} -\varepsilon'\dot{y} & \upsilon_{25} \\ -\varepsilon'\ddot{x} & \upsilon_{45} \end{bmatrix} ,$$

$$\delta_5(\varepsilon')=\frac{1}{\det P}\begin{bmatrix} \upsilon_{21} & -\varepsilon'\dot{y} \\ \upsilon_{41} & -\varepsilon'\ddot{x} \end{bmatrix} . \qquad (24)$$

As mentioned earlier we have in general $\upsilon_{63}\neq 0$ and the solution vector of the predictor system is

$$\underline{\delta}=(\delta_1(\varepsilon'),\ \delta_5(\varepsilon'),\ 0)^T \qquad (25)$$

This means that for a regular orbit with $\upsilon_{63}\neq 0$ (i.e. in the general case) periodicity is preserved only along the family of planar orbits and bifurcation with a family of three-dimensional periodic orbits is not possible if a change in period to first order terms is to occur.
In the exceptional case when $\upsilon_{63}=0$ a solution of the predictor system in the form

$$\underline{\delta}=(\delta_1(\varepsilon'),\ \delta_5(\varepsilon'),\ \varepsilon)^T \ , \qquad (26)$$

appears to be possible. This two-parameter solution would lead to an infinity of periodicity preserving directions, since both ε and ε' can be arbitrarily chosen. Such solutions leading to an infinite number of families of periodic orbits branching from the same bifurcation orbit have not been found in practice and appear to be spurious as a higher-order analysis would possibly show.

Our conclusion, then, from the above linear analysis of the bifurcation is that for a regular planar periodic orbit: either $\upsilon_{63}(\tau)\neq 0$ (general case) and periodicity is maintained only along the family of planar orbits, or $\upsilon_{63}(\tau)=0$ (bifurcation condition) in which case periodicity can be maintained not only along the family of planar orbits but also along a bifurcating family of three-dimensional orbits.

It also follows from the above analysis that the bifurcating family of three-dimensional periodic orbits branches from the plane "perpendicularly" in the sense that, to first-order terms, the changes δ_1, δ_5 to the "planar" components of the initial state vector are zero and the transfer from the planar orbit to the three-dimensional orbit at the bifurcation is achieved by a non-zero change δ_3 in the "vertical" component $x_{03}=z_0$.

A similar analysis for the other cases of symmetry (of the three-dimensional orbit) leads to similar conclusions. There are four types of "vertical bifurcation" depending on the type of three-dimensional orbit involved, and the corresponding bifurcation conditions are as follows:

Type of 3D orbit	Quantity that is zero at the bifurcation (type of bifurcation)
Doubly symmetric	$\upsilon_{33}(\tau)$
Axisymmetric	$\upsilon_{36}(\tau)$
Plane-symmetric	$\upsilon_{63}(\tau)$
Doubly symmetric	$\upsilon_{66}(\tau)$

There is no essential difference between the first and fourth case. If for a bifurcation orbit $\upsilon_{33}(\tau)=0$ (respectively $\upsilon_{66}(\tau)=0$) and we transfer the origin of time along this orbit from t=0 to t=τ then the "new" orbit satisfies $\upsilon_{66}(\tau)=0$ (respectively $\upsilon_{33}(\tau)=0$). For more details we refer to Hénon (1973). We note that the quantities $\upsilon_{33}(\tau)$, $\upsilon_{36}(\tau)$, $\upsilon_{63}(\tau)$, $\upsilon_{66}(\tau)$ are denoted in Hénon's paper by A_υ, B_υ, C_υ, D_υ, respectively.

As mentioned in section 3 the quantities

$$a_\upsilon = \upsilon_{33}(2\tau), \qquad b_\upsilon = \upsilon_{36}(2\tau),$$

$$c_\upsilon = \upsilon_{63}(2\tau), \qquad d_\upsilon = \upsilon_{66}(2\tau),$$

are called vertical stability parameters and describe the vertical stability properties of the orbit. For symmetric periodic orbits it can be shown that

$$a_\upsilon = d_\upsilon$$

and the condition for vertical stability is

$$-1 < a_\upsilon < 1.$$

It can also be shown that

$$\begin{aligned} \upsilon_{33}(2T) &= \upsilon_{66}(2T) = \upsilon_{33}(T)\,\upsilon_{66}(T) + \upsilon_{36}(T)\,\upsilon_{63}(T) \\ &= 1 + 2\,\upsilon_{36}(T)\,\upsilon_{63}(T) = -1 + 2\,\upsilon_{33}(T)\,\upsilon_{66}(T), \end{aligned} \tag{27}$$

$$\upsilon_{36}(2T) = 2\,\upsilon_{36}(T)\upsilon_{66}(T) \;,$$

$$\upsilon_{63}(2T) = 2\,\upsilon_{33}(T)\upsilon_{63}(T) \;.$$

It follows that the vertical bifurcation conditions derived in this section give

$$\begin{aligned} \upsilon_{33}(T) &= 0: & \upsilon_{33}(2T) &= -1\,, & \upsilon_{63}(2T) &= 0\,, \\ \upsilon_{36}(T) &= 0: & \upsilon_{33}(2T) &= 1\,, & \upsilon_{36}(2T) &= 0\,, \\ \upsilon_{63}(T) &= 0: & \upsilon_{33}(2T) &= 1\,, & \upsilon_{63}(2T) &= 0\,, \\ \upsilon_{66}(T) &= 0: & \upsilon_{33}(2T) &= -1\,, & \upsilon_{36}(2T) &= 0\,. \end{aligned} \tag{28}$$

This means that the vertical bifurcation orbits involved in the above discussion are also vertical-critical orbits ($|a_\upsilon|=1$), i.e., orbits at which we have a change of vertical stability along the family of planar periodic orbits. We shall therefore refer to these bifurcation orbits as <u>vertical-critical</u> orbits. The bifurcating family of three-dimensional orbits consists of simple periodic (one revolution) orbits if $a_\upsilon=1$, and of double-periodic (two revolutions) orbits if $a_\upsilon=-1$.

We note that three-dimensional periodic orbits also branch from the plane when one of the following conditions of "vertical self-resonance" is satisfied

$$\upsilon_{33}(nT)=0, \quad \upsilon_{36}(nT)=0, \quad \upsilon_{63}(nT)=0, \quad \upsilon_{66}(nT)=0,$$

with $n>1$. These conditions correspond to the bifurcation from the plane of multiple-periodic (many revolutions) three-dimensional orbits. For a discussion of this type of bifurcation we refer to Robin and Markellos (1980).

In the present paper we are not concerned with vertical self-resonant orbits but focus our attention to the vertical-critical orbits.

5. NUMERICAL DETERMINATION OF VERTICAL-CRITICAL ORBITS

Let us now consider the problem of the numerical determination of the vertical-critical periodic orbits. Suppose that for a given value of the mass parameter μ of the restricted problem we have a family of periodic orbits and that as we "move" along the family from orbit to orbit the existence of a vertical-critical orbit becomes apparent by a change of sign of the quantity

$$\upsilon_{ij}(T), \qquad i,j=3,6 \;.$$

We thus have approximate values of the initial conditions (x_{01},x_{05}) of the vertical-critical orbit and we wish to find the exact values of these initial conditions (to some desired accuracy, e.g., to 5 decimal places). The periodicity conditions that have to be satisfied by the initial conditions of the orbit are

$$x_2(x_{01}, x_{05}, \tau ; \mu) = 0 ,$$
$$x_4(x_{01}, x_{05}, \tau ; \mu) = 0 , \tag{29}$$

where the zero initial conditions $x_{02} = y_0 = 0$, $x_{03} = z_0 = 0$, $x_{04} = \dot{x}_0 = 0$, $x_{06} = \dot{z}_0 = 0$, have been omitted for clarity. These periodicity conditions arise from the fact that if an orbit "starts" on the x-axis with an initial velocity perpendicular to that axis, then this orbit will be periodic with period 2τ if at time τ it crosses the x-axis with a velocity which is perpendicular to that axis. This fact is due to the symmetry property of the equations of motion of the restricted problem (see, e.g., Szebehely, 1967).

Since we seek to establish a vertical-critical periodic orbit, to the periodicity conditions (29) we append the appropriate vertical-criticality condition and obtain the system

$$\begin{aligned} x_2(x_{01}, x_{05}, \tau ; \mu) &= 0 , \\ x_4(x_{01}, x_{05}, \tau ; \mu) &= 0 , \\ \upsilon_{ij}(x_{01}, x_{05}, \tau ; \mu) &= 0 , \end{aligned} \tag{30}$$

where the indices i, j take the values 3 or 6, the combination depending on the type of vertical bifurcation involved. Since the values x_{01}, x_{05}, τ satisfy the conditions (30) only approximately we seek "corrections" $\Delta x_{01}, \Delta x_{05}, \Delta\tau$ such that the values $x_{01} + \Delta x_{01}$, $x_{05} + \Delta x_{05}$, $\tau + \Delta\tau$ satisfy the conditions (30) exactly, i.e.,

$$\begin{aligned} x_2(x_{01} + \Delta x_{01}, x_{05} + \Delta x_{05}, \tau + \Delta\tau ; \mu) &= 0 , \\ x_4(x_{01} + \Delta x_{01}, x_{05} + \Delta x_{05}, \tau + \Delta\tau ; \mu) &= 0 , \\ \upsilon_{ij}(x_{01} + \Delta x_{01}, x_{05} + \Delta x_{05}, \tau + \Delta\tau ; \mu) &= 0 . \end{aligned}$$

Expanding the functions on the left into Taylor series "around" (x_{01}, x_{05}, τ) and neglecting terms of order higher than the first, we obtain

$$\begin{aligned} \upsilon_{21}\Delta_1 + \upsilon_{25}\Delta_5 + \upsilon_{2\tau}\Delta\tau &= -x_2(\tau) , \\ \upsilon_{41}\Delta_1 + \upsilon_{45}\Delta_5 + \upsilon_{4\tau}\Delta\tau &= -x_4(\tau) , \\ \upsilon_{ij1}\Delta_1 + \upsilon_{ij5}\Delta_5 + \upsilon_{ij\tau}\Delta\tau &= -\upsilon_{ij}(\tau) , \end{aligned}$$

where we have abbreviated $\Delta x_{01} = \Delta_1$ and $\Delta x_{05} = \Delta_5$. Substituting $\upsilon_{2\tau} = \dot{y}(\tau) = x_5(\tau)$, $\upsilon_{4\tau} = \ddot{x}(\tau) = \dot{x}_4(\tau)$, $\upsilon_{ij\tau}(\tau) = \dot{\upsilon}_{ij}(\tau)$ we can write this system of equations in matrix form as follows

$$\begin{bmatrix} \upsilon_{21} & \upsilon_{25} & x_5(\tau) \\ \upsilon_{41} & \upsilon_{45} & \dot{x}_4(\tau) \\ \upsilon_{ij1} & \upsilon_{ij5} & \dot{\upsilon}_{ij}(\tau) \end{bmatrix} \begin{bmatrix} \Delta_1 \\ \Delta_5 \\ \Delta\tau \end{bmatrix} = - \begin{bmatrix} x_2(\tau) \\ x_4(\tau) \\ \upsilon_{ij}(\tau) \end{bmatrix} \tag{31}$$

The elements of the coefficient matrix are determined from the numerical integration of the equations of motion and variation (see section 3). Solving (31) for the "corrections" $\Delta_1, \Delta_5, \Delta\tau$ we repeat the procedure for the new initial conditions $x_{01}+\Delta_1$, $x_{05}+\Delta_5$ and half-period $\tau+\Delta\tau$. Iterative application of this procedure produces the correct initial conditions x_{01}, x_{05} and half-period τ of the vertical - critical orbit with any desired accuracy (within the limits of the computer).

When one vertical-critical orbit has been established accurately the value of μ can be increased by an arbitrary (but small) amount, the above corrector procedure can be applied again to provide eventually another vertical-critical orbit, and the whole process can, in principle, be repeated as necessary. In the present paper we have computed monoparametric sets of vertical-critical orbits with μ as the parameter. Each such set we call a series of vertical-critical orbits. When μ changes the family characteristic "moves" in a continuous fashion and the point representing the vertical-critical orbit "moves" similarly, tracing a curve representing the series. This is illustrated schematically in Figure 1.

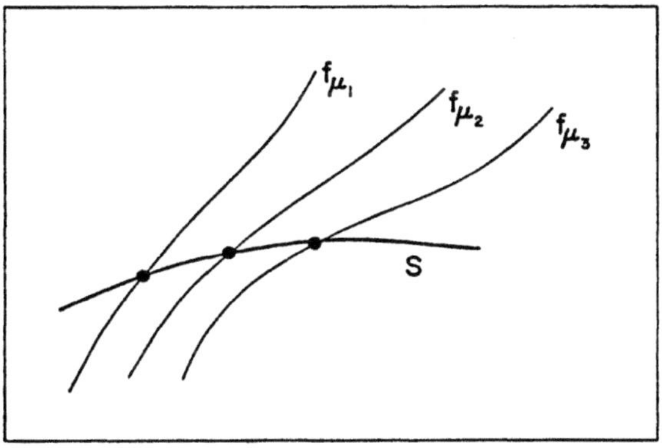

Figure 1. Schematic representation of the formation of a series of vertical-critical periodic orbits.

In order to compute a series efficiently the above process was not in practice repeated indefinitely except in so far as the corrector was concerned; the "next" vertical-critical orbit of the series was not sought by merely using the initial conditions of the last known member of the series, as a rough approximation. Rather, a predictor was used allowing a very good approximation of the initial conditions of the next member of the series to be derived from the known "previous" members. This was done by extrapolation based on the classical divided differences scheme. A linear formula was used to obtain the prediction of the third member of a series when the first two were known, and higher-order formulae were used when more members were known. This was found to be very effective. Formulae of up to the third order were used for the first 5 members of a series, and a fourth order formula was constantly employed thereafter to predict the next member from the previous 5. When only one member of a series of vertical-critical orbits is known no extrapolation can be performed and it is necessary to predict the initial conditions of the next (second) member differently (or simply use the corrector procedure directly as discussed above).

The described predictor-corrector process allows the economical determination of entire series of vertical-critical orbits. Most of the series given in this paper were obtained from single runs of the computer program.

6. RESULTS

Using the above process we have determined numerically the basic series of vertical-critical periodic orbits of the restricted problem. Specifically, we have continued in the entire range of the mass parameter μ the following vertical-critical orbits of the basic families ℓ, m, i, h, b, c given in Hénon (1973):

$\ell 1\upsilon$, $\ell' 1\upsilon$, $\ell 2\upsilon$, $\ell 3\upsilon$, $\ell 4\upsilon$, $\ell' 4\upsilon$, $\ell 5\upsilon$, $\ell 6\upsilon$,

m1υ, m'1υ, m2υ,

i1υ, i2υ, i11υ, i12υ,

h1υ, h2υ,

b1υ, b2υ,

c1υ, c2υ.

Characteristic curves representing, in the (x_0, μ) plane, the generated series are presented in Figures 2 to 6. Numerical data for the orbits of these series are given in the Appendix.

For each orbit given in the Tables I to IV in the Appendix, $x_{11}=x_1(\tau)$ denotes the position of the particle on the x-axis at the half period $\tau = T/2$, and C denotes the value of the Jacobi constant (see, Szebehely, 1967). In presenting the data in the Appendix and in Figures 2 to 6 we have adopted the same format of presentation as in the relevant papers of Hénon (1973) and Hénon and Guyot (1970). We, therefore, refer to these papers for further explanations of the presentation of our results.

We note that in Figures 2 and 3 only x_{01} has been plotted against μ while in Figures 4, 5 and 6 both x_{01} and x_{11} have been plotted. As can be seen in Figures 4 and 5 the series i1υ and i11υ actually connect with each other and are represented by a single curve (for each intersection) and this is also true of the pair i2υ and i12υ and the pair h1υ and h2υ. Further continuation of the series h2υ was prevented by a collision.

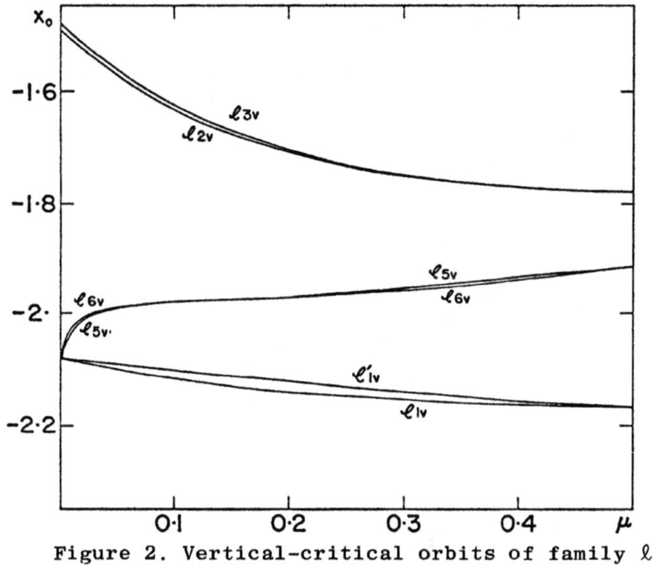

Figure 2. Vertical-critical orbits of family ℓ (direct around both primaries).

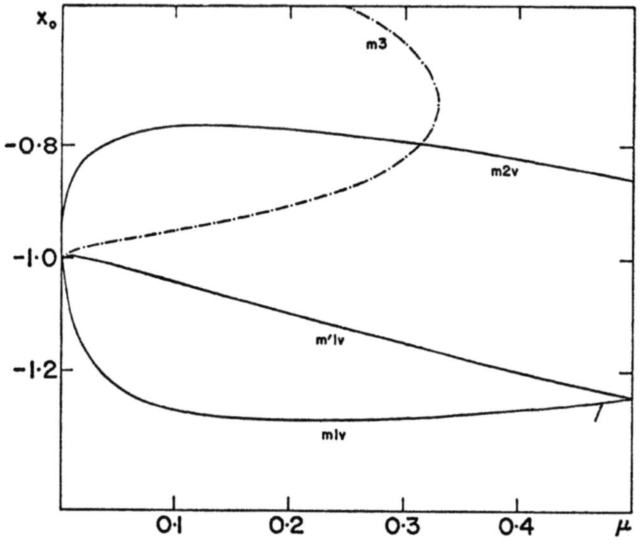

Figure 3. Vertical-critical orbits of family m (retrograde, in the rotating frame, orbits around both primaries).

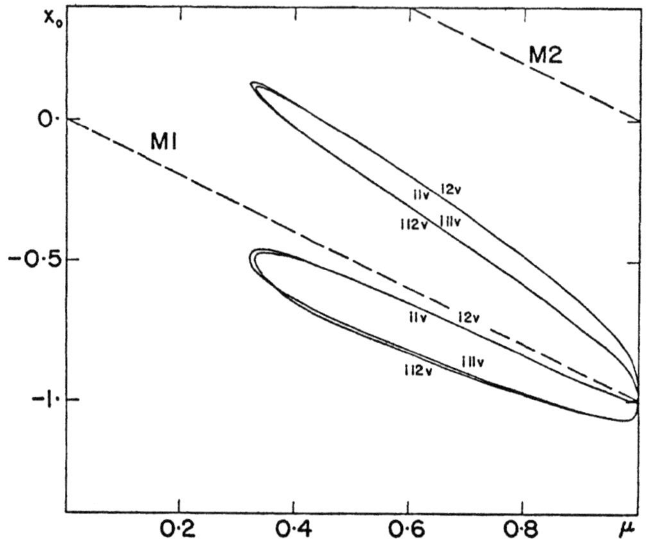

Figure 4. Vertical-critical orbits of family i (direct around one primary).

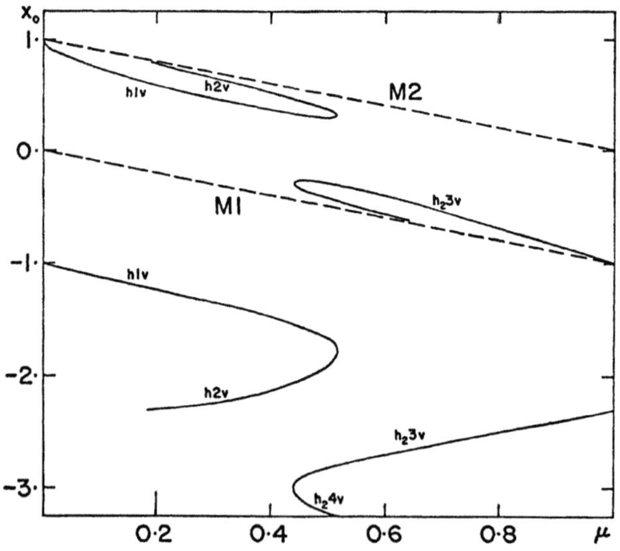

Figure 5. Vertical-critical orbits of family h (retrograde around one primary).

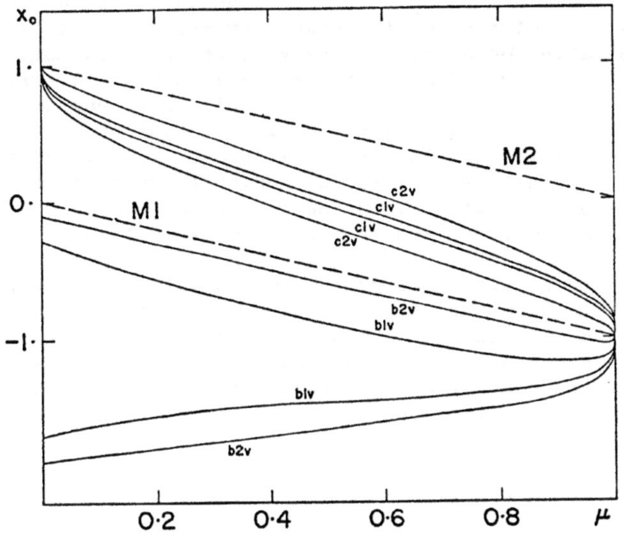

Figure 6. Vertical-critical orbits of families b and c (orbits around a collinear equilibrium point).

Horizontal stability of the vertical critical orbits.

The importance of the vertical-critical orbits as starting points for the determination of the basic families of three-dimensional periodic orbits is evident. Their usefulness in this respect is sufficient motivation for their direct computation by procedures such as the one described above. But an additional factor enhancing their importance is that by examination of their "horizontal" stability we can make predictions about the stability of the bifurcating three-dimensional orbits. Ultimately one aims at the three-dimensional problem and one of the most important qualitative aspects of the problem, from the practical point of view, is the stability of various types of three-dimensional motion. The existing theoretical and numerical evidence tends to associate each region of stability, or quasi-periodic motion, in phase space with the existence of a stable low-order periodic orbit as a central feature of each stability region. Therefore, in seeking to establish such stability regions with the method of periodic orbits, the first objective is to determine the location in phase-space of the stable simple - or double - periodic orbits of the three-dimensional problem, and these are precisely the periodic orbits generated from the vertical-critical orbits.

By determining the stability of the vertical-critical orbits we can predict the possibility of stable three-dimensional periodic orbits, at least for low inclination. More precisely we can predict that if a vertical-critical orbit is unstable then for low inclination the generated three-dimensional periodic orbits will-certainly be unstable too, because of the continuous evolution of the stability properties from the earlier orbit to the latter ones; but if the vertical-critical orbit is stable then the generated three-dimensional periodic orbits may well be stable. The stability of the vertical-critical orbit is a necessary condition for the stability of the generated low inclination orbits and in the search for stable low inclination orbits we can exclude, in the first instance, all the families of three-dimensional orbits which bifurcate from unstable plane orbits. This has the obvious advantage of considerable economy in computer time and in the organization of the search.

In the present paper we have computed the horizontal stability parameters a, b, c, d of all the orbits given. The parameter a which determines stability is given in the last column of each Table in the Appendix. In the cases where the value of a indicates a change of stability ($a = \pm 1$) one of the quantities b, c passes through the value zero. This quantity is indicated in the appropriate place in the Tables.

The calculation of a, throughout each series of vertical-critical orbits, shows that among the 21 series given in this paper only the series $\ell 1\upsilon$, $\ell' 1\upsilon$, $\ell 5\upsilon$, $\ell 6\upsilon$, $m1\upsilon$, $m'1\upsilon$, $m2\upsilon$, $i1\upsilon$, $i2\upsilon$ and $hi\upsilon$ contain horizontally stable orbits. These define "intervals of horizontal stability" on the μ-axis and the end points of these intervals are given below.

Intervals of horizontal stability of vertical-critical periodic orbits on the μ-axis. For orbits of the families ℓ and m the stability intervals are only given in the range (0,1/2).

Series	Type	μ_{min}	μ_{max}
$\ell 1\upsilon$	$\upsilon_{66}(\tau)$	0.	0.5
$\ell' 1\upsilon$	$\upsilon_{33}(\tau)$	0.231	0.5
$\ell 5\upsilon$	$\upsilon_{36}(\tau)$	0.116 0.134	0.121 0.137
$\ell 6\upsilon$	$\upsilon_{63}(\tau)$	0.114 0.130	0.118 0.134
$m1\upsilon$	$\upsilon_{66}(\tau)$	0.	0.5
$m'1\upsilon$	$\upsilon_{33}(\tau)$	0.	0.5
$m2\upsilon$	$\upsilon_{63}(\tau)$	0.313	0.5
$i1\upsilon$	$\upsilon_{66}(\tau)$	0.624	1.
$i2\upsilon$	$\upsilon_{33}(\tau)$	0.575	1.
$h1\upsilon$	$\upsilon_{33}(\tau)$	0.	0.320

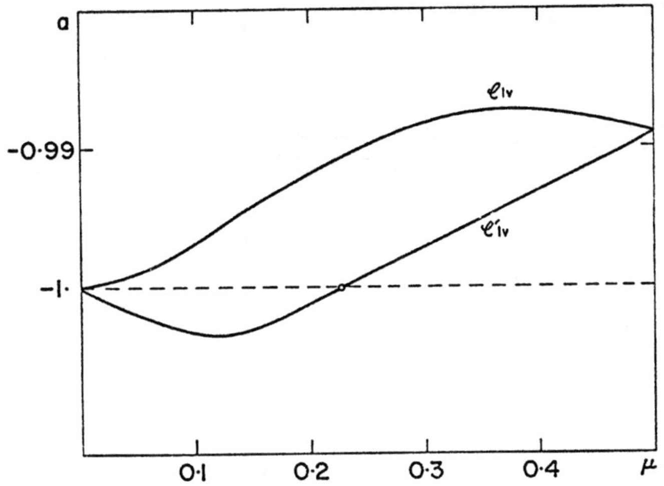

Figure 7. Horizontal-stability plots for the series ℓ1υ and ℓ'1υ.

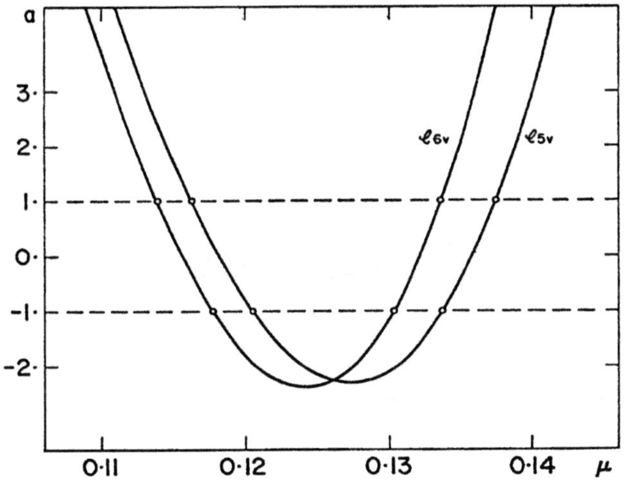

Figure 8. Horizontal-stability plots for the series ℓ5υ and ℓ6υ.

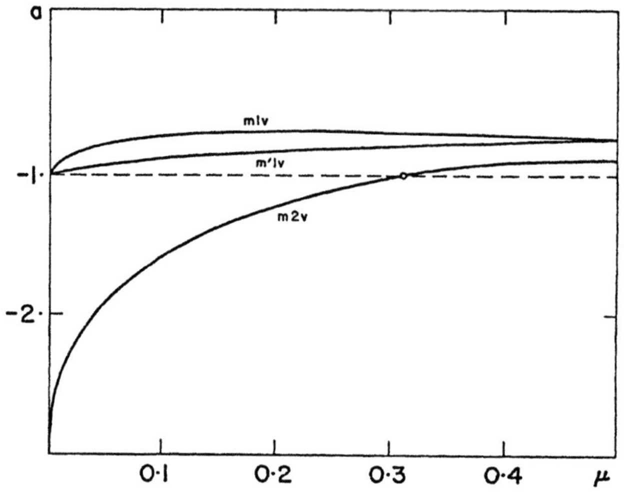

Figure 9. Horizontal-stability plots for the series m1υ, m'1υ and m2υ.

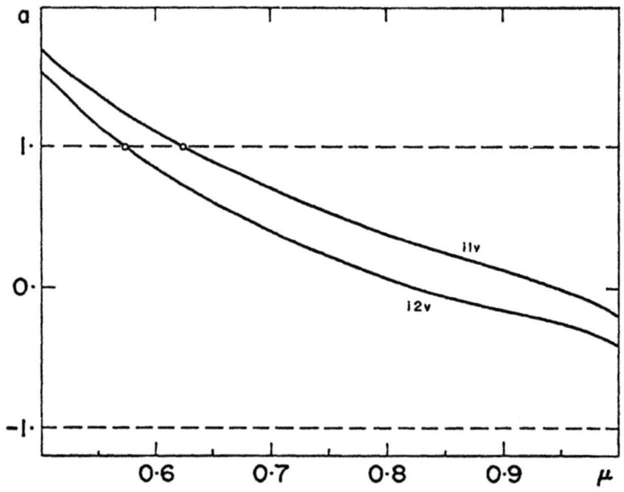

Figure 10. Horizontal-stability plots for the series i1υ and i2υ.

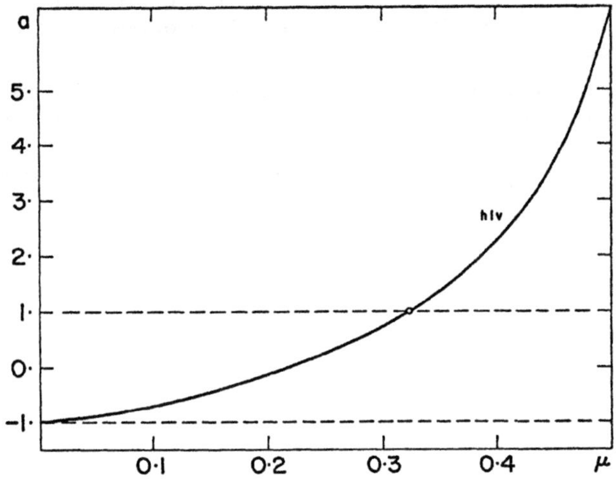

Figure 11. Horizontal-stability plot for the series h1υ.

Horizontal-stability plots for the series which contain stable vertical-critical orbits are presented in Figures 7 to 11.

Comment

As mentioned in the beginning, the present paper deals only with symmetric periodic orbits of the restricted problem. For a similar study of symmetric vertical-critical periodic orbits of the general three-body problem we refer to Markellos (1981). Also, for a study of symmetric vertical-critical orbits of the elliptic restricted problem we refer to Robin (1981).

Acknowledgment

Most of the computations in this paper were carried out while the first author was at the University of Glasgow under an S.R.C. research grant.

REFERENCES

Hénon, M.: 1973, *Astron. Astrophys.* 28, 415.

Hénon, M. and Guyot, M.: 1970, in G.E.O. Giacaglia (ed.), *Periodic Orbits, Stability and Resonances*, D. Reidel Publ. Co., Dordrecht-Holland, p. 349.

Markellos, V.V.: 1976, *Astrophys. Space Sci.* 43, 449.

Markellos, V.V.: 1977, *Monthly Notices Roy. Astron. Soc.* 180, 103.

Markellos, V.V.: 1981, "Bifurcations of Planar to Three-Dimensional Periodic Orbits in the General Three-Body Problem", *Celes. Mech.* in press.

Robin, I.A.: 1981, "Bifurcations of Plane with Three-Dimensional Periodic Orbits in the Elliptic Restricted Problem", *Celes. Mech.* 23, 97

Robin, I.A. and Markellos, V.V.: 1980, *Celes. Mech.* 21, 395.

Szebehely, V.: 1967, *Theory of Orbits*, Academic Press, New York.

Zagouras, C.G. and Markellos, V.V.: 1977, *Astron. Astrophys.* 59, 79.

APPENDIX

Numerical data for the vertical - critical periodic orbits of the restricted three-body problem.

Table I . Vertical - critical orbits of family ℓ

μ	T/2	x_{01}	x_{11}	C	a
		$\ell 1\upsilon(\upsilon_{66})$			
0.5	4.6496	-2.16672	2.16672	3.44480	-0.989
0.45	4.6464	-2.16550	2.17094	3.44525	-0.988
0.4	4.6442	-2.16311	2.17373	3.44448	-0.988
0.35	4.6430	-2.15949	2.17485	3.44239	-0.988
0.3	4.6432	-2.15456	2.17400	3.43886	-0.988
0.25	4.6451	-2.14816	2.17074	3.43367	-0.990
0.2	4.6492	-2.14005	2.16447	3.42653	-0.992
0.15	4.6562	-2.12989	2.15431	3.41703	-0.994
0.1	4.6674	-2.11716	2.13888	3.40453	-0.997
0.05	4.6847	-2.10104	2.11576	3.38797	-0.999
0.012	4.7044	-2.08567	2.09025	3.37145	-1.000
0.001	4.7117	-2.08056	2.08098	3.36579	-1.000
		$\ell' 1\upsilon(\upsilon_{33})$			
0.5	4.6496	-2.16672	2.16672	3.44480	-0.989
0.45	4.6535	-2.16127	2.16677	3.44321	-0.991
0.4	4.6581	-2.15474	2.16562	3.44048	-0.993
0.35	4.6632	-2.14726	2.16318	3.43662	-0.995
0.3	4.6689	-2.13894	2.15930	3.43157	-0.997
0.25	4.6749	-2.12990	2.15376	3.42525	-0.999 c
0.2	4.6814	-2.12025	2.14623	3.41752	-1.001
0.15	4.6883	-2.11016	2.13620	3.40814	-1.003
0.1	4.6957	-2.09983	2.12288	3.39676	-1.003
0.05	4.7036	-2.08960	2.10497	3.38282	-1.002
0.012	4.7102	-2.08225	2.08689	3.36989	-1.000
0.001	4.7122	-2.08026	2.08068	3.36565	-1.000
		$\ell 2\upsilon(\upsilon_{63})$			
0.5	6.0796	-1.77996	1.77996	3.14525	54.01
0.45	6.0799	-1.77598	1.78175	3.14644	52.34
0.4	6.0807	-1.76940	1.78163	3.14987	47.57
0.35	6.0823	-1.75971	1.77978	3.15521	40.36
0.3	6.0855	-1.74626	1.77628	3.16187	31.74
0.25	6.0911	-1.72817	1.77100	3.16899	22.84
0.2	6.1007	-1.70431	1.76361	3.17537	14.77
0.15	6.1168	-1.67303	1.75342	3.17938	8.382
0.1	6.1441	-1.63171	1.73904	3.17872	4.092
0.05	6.1916	-1.57557	1.71758	3.16975	1.796
0.012	6.2545	-1.51655	1.69242	3.15331	1.075
0.001	6.2806	-1.49519	1.68265	3.14609	1.003

Table I (continued)

		$\ell 3\upsilon(\upsilon_{36})$			
0.5	6.1279	-1.77687	1.77687	3.13399	64.23
0.45	6.1291	-1.77323	1.77813	3.13508	62.45
0.4	6.1327	-1.76685	1.77729	3.13828	57.29
0.35	6.1388	-1.75721	1.77450	3.14336	49.33
0.3	6.1476	-1.74361	1.76982	3.14994	39.44
0.25	6.1593	-1.72509	1.76324	3.15736	28.81
0.2	6.1744	-1.70031	1.75461	3.16465	18.73
0.15	6.1934	-1.66737	1.74368	3.17030	10.43
0.1	6.2171	-1.62344	1.73003	3.17191	4.776
0.05	6.2465	-1.56416	1.71312	3.16553	1.891
0.012	6.2736	-1.50465	1.69769	3.15128	1.084
0.001	6.2824	-1.48440	1.69278	3.14493	1.004
		$\ell 4\upsilon(\upsilon_{66})$			
0.5	7.6768	-1.84118	1.84118	2.71702	874.2
0.45	7.6628	-1.85163	1.82584	2.72301	875.5
0.4	7.6493	-1.85896	1.80802	2.73657	855.9
0.35	7.6370	-1.86255	1.78761	2.75791	808.9
0.3	7.6265	-1.86135	1.76424	2.78735	728.2
0.25	7.6188	-1.85364	1.73726	2.82530	610.9
0.2	7.6154	-1.83656	1.70551	2.87211	460.1
0.15	7.6186	-1.80496	1.66702	2.92782	290.2
0.1	7.6339	-1.74850	1.61809	2.99116	130.1
0.05	7.6784	-1.63906	1.55173	3.05589	22.41
0.032	7.7214	-1.54178	1.53596	3.07750	3.170
		$\ell' 4\upsilon(\upsilon_{33})$			
0.5	7.6768	-1.84118	1.84118	2.71702	874.2
0.45	7.6909	-1.85396	1.82793	2.71849	856.6
0.4	7.7047	-1.86389	1.81196	2.72748	824.7
0.35	7.7177	-1.87039	1.79310	2.74422	776.8
0.3	7.7295	-1.87247	1.77087	2.76917	707.9
0.25	7.7396	-1.86843	1.74442	2.80307	611.2
0.2	7.7478	-1.85537	1.71231	2.84700	481.4
0.15	7.7543	-1.82785	1.67210	2.90219	321.8
0.1	7.7615	-1.77480	1.61937	2.96907	154.7
0.05	7.7791	-1.66863	1.54465	3.04281	30.04
0.03	7.7967	-1.59065	1.50547	3.06941	5.874

Table I (continued)

$\ell 5\upsilon(\upsilon_{36})$

0.5	9.1475	-1.91249	1.91249	2.68209	4070	
0.45	9.1493	-1.92487	1.89908	2.68423	3862	
0.4	9.1535	-1.93405	1.88756	2.68900	3423	
0.35	9.1643	-1.94655	1.86876	2.70124	2453	
0.3	9.1779	-1.95574	1.85137	2.71600	1552	
0.25	9.1959	-1.96372	1.83184	2.73482	765.5	
0.2	9.2190	-1.97045	1.80920	2.75759	238.1	
0.15	9.2477	-1.97606	1.78148	2.78427	15.99	
0.14	9.2543	-1.97709	1.77500	2.79008	3.022	
0.138	9.2556	-1.97729	1.77365	2.79127	1.418	c
0.137	9.2563	-1.97739	1.77297	2.79186	0.730	
0.134	9.2583	-1.97770	1.77090	2.79366	-0.892	b
0.133	9.2590	-1.97780	1.77020	2.79426	-1.290	
0.13	9.2610	-1.97811	1.76808	2.79608	-2.081	c
0.12	9.2681	-1.97914	1.76064	2.80226	-0.802	b
0.116	9.2709	-1.97955	1.75749	2.80479	1.141	
0.1	9.2829	-1.98132	1.74369	2.81528	14.41	
0.05	9.3226	-1.99072	1.67368	2.85540	52.77	
0.012	9.3043	-2.03101	1.42706	2.93970	73.29	
0.005	9.2343	-2.05994	1.27520	2.97751	248.8	
0.001	9.0574	-2.09898	1.12845	2.99874	2400	

$\ell 6\upsilon(\upsilon_{63})$

0.5	9.2948	-1.91477	1.91477	2.68930	6540	
0.45	9.2941	-1.92690	1.90164	2.69126	6183	
0.4	9.2921	-1.93807	1.88742	2.69712	5197	
0.35	9.2892	-1.94826	1.87190	2.70686	3815	
0.3	9.2858	-1.95738	1.85472	2.72044	2356	
0.25	9.2827	-1.96535	1.83527	2.73786	1131	
0.2	9.2814	-1.97206	1.81246	2.75914	346.2	
0.15	9.2838	-1.97748	1.78413	2.78453	27.28	
0.134	9.2859	-1.97897	1.77320	2.79364	1.323	c
0.132	9.2862	-1.97915	1.77175	2.79481	-0.085	c
0.130	9.2865	-1.97934	1.77028	2.79600	-1.141	
0.124	9.2876	-1.97987	1.76571	2.79961	-2.382	
0.118	9.2888	-1.98040	1.76092	2.80331	-1.104	b
0.114	9.2897	-1.98076	1.75759	2.80583	0.909	b
0.1	9.2934	-1.98203	1.74488	2.81501	13.04	
0.05	9.3139	-1.98954	1.67183	2.85624	53.20	
0.012	9.2630	-2.02451	1.41703	2.94442	73.08	
0.005	9.0785	-2.04679	1.25666	2.98550	251.4	
0.0015	8.5831	-2.07154	1.13032	3.00720	1668	
0.001	8.3638	-2.07838	1.10383	3.01077	2933	

Table II . Vertical - critical orbits of family m

μ	T/2	x_{01}	x_{11}	C	a
			$m1\upsilon(\upsilon_{66})$		
0.5	1.7657	-1.24787	1.24787	-1.59055	-0.725
0.45	1.7821	-1.25830	1.26829	-1.61031	-0.712
0.4	1.7973	-1.26783	1.28669	-1.62667	-0.700
0.35	1.8110	-1.27613	1.30268	-1.63901	-0.690
0.3	1.8228	-1.28272	1.31566	-1.64640	-0.682
0.25	1.8324	-1.28686	1.32477	-1.64745	-0.678
0.2	1.8387	-1.28744	1.32864	-1.63993	-0.679
0.15	1.8402	-1.28249	1.32492	-1.61993	-0.689
0.1	1.8334	-1.26805	1.30879	-1.57938	-0.714
0.05	1.8084	-1.23324	1.26713	-1.49592	-0.772
0.012	1.7423	-1.15869	1.17745	-1.33183	-0.882
0.001	1.6540	-1.07294	1.07791	-1.14994	-0.972
			$m'1\upsilon(\upsilon_{33})$		
0.5	1.7657	-1.24787	1.24787	-1.59055	-0.725
0.45	1.7483	-1.22577	1.23678	-1.56778	-0.738
0.4	1.7300	-1.20221	1.22517	-1.54222	-0.753
0.35	1.7110	-1.17740	1.21311	-1.51394	-0.768
0.3	1.6914	-1.15151	1.20055	-1.48278	-0.785
0.25	1.6712	-1.12471	1.18732	-1.44829	-0.803
0.2	1.6507	-1.09718	1.17301	-1.40958	-0.824
0.15	1.6299	-1.06916	1.15674	-1.36478	-0.850
0.1	1.6092	-1.04109	1.13654	-1.30985	-0.884
0.05	1.5888	-1.01402	1.10675	-1.23330	-0.929
0.012	1.5743	-0.99722	1.06042	-1.12590	-0.979
0.001	1.5709	-0.99719	1.01951	-1.03946	-0.998
			$m2\upsilon(\upsilon_{63})$		
0.5	1.2338	-0.85939	0.85939	-0.99043	-0.882
0.45	1.2346	-0.84095	0.87853	-0.99097	-0.890
0.4	1.2368	-0.82342	0.89815	-0.99259	-0.915
0.35	1.2407	-0.80710	0.91805	-0.99524	-0.957
0.3	1.2466	-0.79236	0.93802	-0.99889	-1.018 c
0.25	1.2547	-0.77977	0.95777	-1.00344	-1.104
0.2	1.2660	-0.77021	0.97692	-1.00871	-1.218
0.15	1.2817	-0.76523	0.99484	-1.01437	-1.372
0.1	1.3048	-0.76810	1.01036	-1.01966	-1.586
0.05	1.3442	-0.78814	1.02048	-1.02238	-1.913
0.012	1.4152	-0.84629	1.01663	-1.01604	-2.388
0.001	1.4965	-0.92533	1.00488	-1.00473	-2.772
0.0003	1.5199	-0.94888	1.00239	-1.00234	-2.856

Table III. Vertical - critical orbits of family i

μ	T/2	x_{01}	x_{11}	C	a
		$i1\upsilon(\upsilon_{66})$			
0.999	1.0182	-0.94257	-1.00690	3.03945	-0.232
0.995	0.9949	-0.89934	-1.00769	3.10828	-0.187
0.988	0.9769	-0.86086	-1.00417	3.18371	-0.146
0.95	0.9398	-0.74879	-0.97357	3.40996	-0.016
0.9	0.9227	-0.64885	-0.92844	3.57742	0.115
0.85	0.9167	-0.56384	-0.88219	3.68776	0.243
0.8	0.9168	-0.48572	-0.83564	3.76528	0.379
0.75	0.9215	-0.41162	-0.78912	3.81982	0.528
0.7	0.9301	-0.34017	-0.74284	3.85639	0.696
0.65	0.9426	-0.27058	-0.69697	3.87768	0.889
0.6	0.9595	-0.20238	-0.65174	3.88502	1.113 c
0.55	0.9815	-0.13530	-0.60746	3.87868	1.378
0.5	1.0101	-0.06924	-0.56462	3.85790	1.696
0.45	1.0484	-0.00434	-0.52413	3.82034	2.082
0.4	1.1022	0.05870	-0.48808	3.76049	2.554
0.35	1.1907	0.11668	-0.46358	3.66104	3.122
0.325	1.2909	0.13474	-0.47382	3.55684	3.443
		$i11\upsilon(\upsilon_{66})$			
0.325	1.3566	0.11987	-0.50282	3.49456	3.526
0.35	1.4255	0.06334	-0.56588	3.43329	3.811
0.4	1.4596	-0.02137	-0.63637	3.39981	4.956
0.45	1.4700	-0.09630	-0.68828	3.38407	6.778
0.5	1.4737	-0.16746	-0.73327	3.37345	9.233
0.56	1.4750	-0.25079	-0.78327	3.36394	13.03
0.6	1.4752	-0.30588	-0.81533	3.35877	16.12
0.66	1.4751	-0.38853	-0.86222	3.35182	21.65
0.7	1.4750	-0.44398	-0.89279	3.34704	25.99
0.76	1.4750	-0.52829	-0.93753	3.33797	33.58
0.8	1.4752	-0.58570	-0.96638	3.32909	39.42
0.86	1.4757	-0.67491	-1.00730	3.30689	49.47
0.9	1.4763	-0.73790	-1.03188	3.28089	57.10
0.96	1.4774	-0.84505	-1.05853	3.20086	69.96
0.988	1.4774	-0.91325	-1.05541	3.11224	76.27
0.999	1.4756	-0.96726	-1.02935	3.02652	78.05
0.9998	1.4746	-0.98140	-1.01773	3.00965	77.83

Table III (continued)

			$i2\upsilon(\upsilon_{33})$		
0.999	1.0484	-0.94200	-1.00649	3.03941	-0.419
0.995	1.0240	-0.89842	-1.00699	3.10817	-0.391
0.988	1.0050	-0.85968	-1.00326	3.18351	-0.365
0.95	0.9653	-0.74710	-0.97215	3.40954	-0.273
0.9	0.9461	-0.64690	-0.92668	3.57687	-0.167
0.85	0.9384	-0.56175	-0.88018	3.68715	-0.054
0.8	0.9369	-0.48356	-0.83343	3.76468	0.073
0.75	0.9400	-0.40944	-0.78674	3.81929	0.220
0.7	0.9469	-0.33801	-0.74031	3.85598	0.395
0.65	0.9575	-0.26848	-0.69434	3.87745	0.603
0.6	0.9722	-0.20040	-0.64908	3.88501	0.856
0.55	0.9917	-0.13353	-0.60489	3.87894	1.165 c
0.5	1.0172	-0.06781	-0.56238	3.85839	1.547
0.45	1.0516	-0.00352	-0.52275	3.82085	2.016
0.4	1.1009	0.05821	-0.48897	3.75997	2.576
0.35	1.1894	0.11135	-0.47420	3.65057	3.127
0.34	1.2265	0.11675	-0.48208	3.60760	3.171
0.33625	1.2548	0.11455	-0.49353	3.57638	3.140
			$i12\upsilon(\upsilon_{33})$		
0.34	1.3104	0.09159	-0.53188	3.51916	2.974
0.35	1.3384	0.06663	-0.56208	3.49237	2.896
0.4	1.3827	-0.02471	-0.64654	3.44979	3.302
0.45	1.3946	-0.10034	-0.70087	3.43361	4.430
0.5	1.3994	-0.17088	-0.74571	3.42236	6.110
0.55	1.4025	-0.23951	-0.78670	3.41266	8.342
0.6	1.4054	-0.30753	-0.82582	3.40345	11.18
0.65	1.4088	-0.37561	-0.86380	3.39405	14.71
0.7	1.4128	-0.44424	-0.90088	3.38367	19.06
0.75	1.4176	-0.51388	-0.93702	3.37117	24.35
0.8	1.4234	-0.58512	-0.97194	3.35469	30.77
0.85	1.4303	-0.65892	-1.00501	3.33101	38.54
0.9	1.4385	-0.73708	-1.03485	3.29372	47.96
0.95	1.4489	-0.82458	-1.05736	3.22660	59.55
0.988	1.4608	-0.91293	-1.05595	3.11370	70.80
0.999	1.4685	-0.96719	-1.02944	3.02664	75.62

Table IV . Vertical - critical orbits of family h

μ	T/2	x_{01}	x_{11}	C	a
		$h1\upsilon(\upsilon_{33})$			
0.001	1.5710	-1.00453	0.97876	-0.95698	-0.998
0.012	1.5749	-1.02290	0.92193	-0.83754	-0.972
0.05	1.5942	-1.06850	0.82701	-0.62766	-0.861
0.1	1.6259	-1.12266	0.73873	-0.42226	-0.671
0.15	1.6635	-1.17542	0.66379	-0.23945	-0.427
0.2	1.7076	-1.22828	0.59576	-0.06551	-0.120
0.25	1.7592	-1.28217	0.53239	0.10504	0.266
0.3	1.8206	-1.33814	0.47278	0.27529	0.758
0.35	1.8954	-1.39775	0.41683	0.44772	1.404
0.4	1.9903	-1.46393	0.36521	0.62532	2.296
0.45	2.1209	-1.54369	0.32025	0.81429	3.669
0.5	2.3495	-1.66673	0.29211	1.04285	6.550
0.515	2.5328	-1.75933	0.30212	1.16992	9.413
		$h2\upsilon(\upsilon_{33})$			
0.515	2.6574	-1.82144	0.32126	1.24117	11.70
0.5	2.8542	-1.91938	0.36785	1.34488	16.01
0.45	3.1289	-2.05336	0.45989	1.50240	23.92
0.4	3.3081	-2.13489	0.53340	1.62940	30.92
0.35	3.4555	-2.19484	0.59958	1.75246	38.43
0.3	3.5880	-2.24089	0.66163	1.87617	47.19
0.25	3.7143	-2.27626	0.72112	2.00180	58.08

Table V. Vertical - critical orbits of family b

μ	T/2	x_{01}	x_{11}	C	a
		$b1\upsilon(\upsilon_{63})$			
0.001	3.1398	-1.70635	-0.29417	2.41475	1.030
0.012	3.1198	-1.69688	-0.30937	2.42341	1.378
0.05	3.0519	-1.66510	-0.36165	2.45801	2.858
0.1	2.9652	-1.62617	-0.42945	2.51318	5.592
0.15	2.8823	-1.59115	-0.49552	2.57684	9.414
0.2	2.8033	-1.56035	-0.55941	2.64645	14.52
0.25	2.7281	-1.53371	-0.62083	2.71979	21.08
0.3	2.6566	-1.51093	-0.67967	2.79506	29.28
0.35	2.5883	-1.49159	-0.73591	2.87079	39.28
0.4	2.5228	-1.47523	-0.78955	2.94579	51.28
0.45	2.4597	-1.46132	-0.84063	3.01902	65.46
0.5	2.3985	-1.44934	-0.88918	3.08951	82.07
0.56	2.3271	-1.43676	-0.94408	3.16911	105.6
0.6	2.2802	-1.42898	-0.97860	3.21828	123.8
0.66	2.2105	-1.41740	-1.02706	3.28451	155.3
0.7	1.1640	-1.40914	-1.05696	3.32240	179.7
0.76	2.0931	-1.39458	-1.09759	3.36686	222.6
0.8	2.0443	-1.38228	-1.12127	3.38571	256.4
0.86	1.9661	-1.35664	-1.14968	3.39068	318.7
0.9	1.9079	-1.33090	-1.16145	3.37072	372.1
0.96	1.7963	-1.26008	-1.15548	3.27196	492.5
0.988	1.7070	-1.18018	-1.11999	3.15111	607.1
0.999	1.6130	-1.07919	-1.05701	3.03515	745.8
		$b2\upsilon(\upsilon_{36})$			
0.001	3.1407	-1.90249	-0.09795	1.85905	1.004
0.012	3.1309	-1.89635	-0.10897	1.85909	1.049
0.05	3.0971	-1.87521	-0.14703	1.86169	1.249
0.1	3.0525	-1.84773	-0.19705	1.87091	1.618
0.15	3.0076	-1.82071	-0.24700	1.88674	2.124
0.2	2.9625	-1.79421	-0.29688	1.90917	2.793
0.25	2.9171	-1.76831	-0.34669	1.93820	3.659
0.3	2.8715	-1.74303	-0.39639	1.97382	4.769
0.35	2.8257	-1.71839	-0.44599	2.01600	6.177
0.4	2.7796	-1.69435	-0.49547	2.06471	7.955
0.45	2.7333	-1.67085	-0.54480	2.11987	10.18
0.5	2.6868	-1.64777	-0.59396	2.18140	12.96
0.56	2.6306	-1.62036	-0.65267	2.26348	17.19
0.6	2.5929	-1.60203	-0.69160	2.32306	20.65
0.66	2.5359	-1.57391	-0.74962	2.41943	27.05
0.7	2.4975	-1.55427	-0.78796	2.48809	32.27
0.76	2.4388	-1.52224	-0.84480	2.59708	41.90
0.8	2.3987	-1.49800	-0.88208	2.67317	49.81
0.86	2.3360	-1.45372	-0.93654	2.79077	64.73
0.9	2.2910	-1.41458	-0.97123	2.86955	77.56
0.96	2.2113	-1.31886	-1.01726	2.97929	105.2
0.988	2.1548	-1.21910	-1.02950	3.01386	128.1
0.999	2.1018	-1.09637	-1.01884	3.00800	150.8

Table VI . Vertical - critical orbits of family c

μ	T/2	x_{01}	x_{11}	C	a
		$c1\upsilon(\upsilon_{63})$			
0.5	1.1139	-0.02377	0.02377	3.92307	1744
0.45	1.1151	0.04715	0.09466	3.91874	1741
0.4	1.1186	0.11842	0.16588	3.90556	1734
0.35	1.1248	0.19042	0.23777	3.88294	1719
0.3	1.1339	0.26362	0.31078	3.84979	1698
0.25	1.1465	0.33865	0.38553	3.80431	1669
0.2	1.1638	0.41651	0.46291	3.74359	1629
0.15	1.1876	0.49887	0.54444	3.66268	1575
0.1	1.2220	0.58921	0.63324	3.55213	1498
0.05	1.2783	0.69736	0.73787	3.38905	1375
0.012	1.3722	0.82417	0.85550	3.17306	1179
0.001	1.4666	0.92421	0.94091	3.03698	996.7
0.0001	1.5065	0.96452	0.97292	3.00832	924.4
		$c2\upsilon(\upsilon_{36})$			
0.5	1.8555	-0.16969	0.16969	2.68622	182.6
0.45	1.8560	-0.09557	0.24298	2.69041	183.0
0.4	1.8575	-0.02026	0.31574	2.70297	184.3
0.35	1.8601	0.05666	0.38828	2.72384	186.4
0.3	1.8640	0.13577	0.46088	2.75291	189.4
0.25	1.8694	0.21789	0.53394	2.78994	193.2
0.2	1.8768	0.30429	0.60793	2.83449	197.6
0.15	1.8872	0.39716	0.68365	2.88566	202.4
0.1	1.9027	0.50101	0.76259	2.94144	206.8
0.05	1.9290	0.62848	0.84870	2.99619	208.7
0.012	1.9753	0.78254	0.93103	3.02138	200.2
0.001	2.0242	0.90646	0.97622	3.00861	183.1
0.0001	2.0453	0.95636	0.98971	3.00228	174.6

THE ROLE OF ELECTROSTATIC CHARGING OF SMALL AND INTERMEDIATE SIZED BODIES IN THE SOLAR SYSTEM

D. A. Mendis
Department of Electrical Engineering and Computer Science and The Center for Astrophysics and Space Science
University of California, San Diego
La Jolla, CA 92093, USA

ABSTRACT

The role of electrostatic charging of small and intermediate sized bodies in the solar system is reviewed. These bodies include planetary, interplanetary and cometary dust as well as cometary nuclei (at large heliocentric distances), asteroids and the larger bodies in the Saturnian ring system. While this charging has both physical and dynamical consequences for the small dust grains, it has only physical consequences for the larger bodies. The main physical consequences for the small grains are electrostatic erosion ("chipping") and disruption, whereas for the larger bodies they include electrostatic levitation and blow-off of fine loose dust from their surfaces.

A large variety of solar system phenomena, recently observed by the Pioneer and Voyager deep space probes as well as the HEOS-2 earth satellite, are explained in terms of these processes. Certain peculiar features observed in the dust tails of comets as well as the spatial orientation of the zodiacal dust cloud may also be explained along these lines.

The possible electrostatic erosion of the dust mantles of new comets as well as the electrostatic "polishing" of the smaller aseteroids are also discussed.

1. INTRODUCTION

Several bodies within the solar system, ranging in size all the way from interplanetary dust to many planetary satellites and even a couple of planets are not surrounded by significant atmospheres nor do they possess substantial intrinsic magnetic fields. Consequently solar ultraviolet radiation as well as solar wind or magnetospheric plasma (as the case may be) flows directly onto their surfaces. The

F. D. Kahn (ed.), Investigating the Universe, 353–384.

photoelectron flux from the surface produced by the ultraviolet photons, as well as the electron and ion fluxes onto the surface from the surrounding plasma, constitute primary electric currents which charge the body to some appropriate potential. The sign of the surface potential depends largely on the relative magnitudes of the photoelectron, plasma electron fluxes.

This electrostatic charging of the surfaces has important physical effects in the case of all bodies, large or small, but it will affect also the dynamics of the small dust grains, due to the presence of either the planetary or the interplanetary magnetic field.

In this review we will confine ourselves to the lower end of the mass spectrum (interplanetary, planetary and cometary dust, comets at large heliocentric distances, the larger Saturnian ring particles, and the smaller asteroids). This selection is partly influenced by the author's own interests in this area. But it is also influenced by the fact that there has been extensive work on the charging of our moon (which is a prototype of the larger bodies) over the past two decades and this has been adequately reviewed (e.g. Grard, 1973; Whipple, 1981). On the other hand, studies of the physics and dynamics of charged dust within the solar system is comparatively new and is presently in a state of vigorous development. Furthermore, the very recent Voyager 1 close-up observations of the Saturnian ring system showed up several phenomena which underscore the importance of these processes. The study of the charging of comets at large heliocentric distances has only just begun, but it has already shown a totally new way of dust release from comets as well as a large source of dust input into the outer solar system.

2. ELECTROSTATIC CHARGING

Central to the following discussion is the electrical charging of a body in a given environment. If Q is the electrical charge on a grain at a given time, then

$$\frac{dQ}{dt} = I \tag{1}$$

where I, the net current to the surface, depends on the surface potential, the surface characteristics (both physical and geometrical), the environmental conditions (both plasma and radiation) and the relative motion between the body and the plasma environment. Contributions to I may come from (a) thermal and nonthermal electron fluxes, (b) thermal and nonthermal ion fluxes, (c) photoemission, (d) field emission,

(e) thermionic emission, (f) secondary emission of electrons due to energetic electron and ion impact, (g) radioactivity, and (h) release of electrons during interparticle collisions.

The relative importance of these various currents will, of course, depend on the particular environment in question and the nature of the bodies.

It has been shown (Whipple, 1965) that in interplanetary space as well as within the terrestrial magnetosphere and upper ionosphere, (a), (b) and (c) make the only important contributions to I, while (e) may be important for meteoritic dust entering the lower ionosphere. The same is true for the Jovian and Saturnian magnetospheres too although (d) is of importance for the smallest sub-micron sized grains (Mendis and Axford, 1974; Mendis, 1979).

The thermal electron and ion currents to a body in relative motion with respect to the plasma may be obtained by integrating the differential flux over a drifting Maxwellian velocity distribution. If the body is small compared to the Debye length, L, it is effectively unshielded for many radii away from its center and the electric field is essentially Coulomb. On the other hand, bodies that are large compared with L have their charge shielded within a small fraction of their linear scale away from the surface. Consequently, from the point of view of plasma collection, the body may be approximated by a neutral body with a slightly larger surface area.

For the small micrometeoroids, the radius $R_g \ll L$, both in interplanetary space as well as in planetary magnetospheres. In this case, the omnidirectional thermal ion current (when the surface potential, $\phi_s < 0$) is given by (Whipple, 1965):

$$I_i(\phi_s < 0) = \frac{n_i e \alpha_i A}{4\sqrt{\pi}} \left[\frac{\sqrt{\pi}\,\alpha_i}{V} \operatorname{erf}\left(\frac{V}{\alpha_i}\right) \left(\frac{-e\phi_s}{kT_i} + \frac{V^2}{\alpha_i^2} + \frac{1}{2} \right) - e^{-(V/\alpha_i)^2} \right] \tag{2}$$

where

$$\left.\begin{aligned} \operatorname{erf}(x) &= \frac{2}{\sqrt{\pi}} \int_0^x e^{-t^2}\, dt \\ \alpha_i &= \left(\frac{2kT_i}{m_i} \right)^{1/2} \\ \text{and} \quad A &= 4\pi R_g^2 \end{aligned}\right\} \tag{3}$$

In the above n_i, T_i and m_i are respectively the ion density, temperature and mass, V is the relative speed between the grain and the plasma, e is the electronic charge (numerical) and k is the Boltzmann constant.

On the other hand, if $\phi_s > 0$ we have (Whipple, 1965),

$$I_i(\phi_s > 0) = \frac{n_i e \alpha_i A}{8} \left\{ \left[1 + \frac{\alpha_i^2}{2V^2} - \frac{U_i^2}{V^2}\right] \left[\operatorname{erf}\left(\frac{V + U_i}{\alpha_i}\right) + \operatorname{erf}\left(\frac{V - U_i}{\alpha_i}\right)\right] + \frac{\alpha_i}{\sqrt{\pi} V} \left[\left(\frac{U_i}{V} + 1\right) e^{-\left(\frac{V - U_i}{\alpha_i}\right)^2} - \left(\frac{U_i}{V} - 1\right) e^{-\left(\frac{V + U_i}{\alpha_i}\right)^2}\right] \right\} \tag{4}$$

where

$$U_i = \left(\frac{2e\phi_s}{m_i}\right)^{1/2} \tag{5}$$

It is to be noted that (2) applies generally to an attractive potential while (4) applies to a repulsive potential. Consequently, the thermal electron currents I_e can be obtained simply by noting that I_e ($\phi_s < 0$) will be given by the right hand side of (4) with the subscript i replaced by the subscript e everywhere. Likewise, I_e ($\phi_s > 0$) will be given by the right hand side of (2) with the same substitution.

Finally, the photoelectron current, I_p, at a heliocentric distance d (AU) is given by

$$I_p (\phi_s < 0) = \frac{1}{4} A e f_1 d^{-2} \tag{6}$$

and

$$I_p (\phi_s > 0) = \frac{1}{4} A e f_1 \exp\left(-\frac{e\phi_s}{kT_p}\right) d^2 \tag{7}$$

where kT_p is the average energy of the photo-electrons, and $f_1 \approx 2.5 \times 10^{10} \chi$, with the efficiency factor χ (≈ 1 for most metals and ≈ 0.1 for most dielectrics; e.g. Hill and Mendis, 1979). Also typically $kT_p \approx 2$ eV for most dielectrics.

Hill and Mendis (1979) considered the charging of micrometeoroids entering the Jovian magnetosphere which is dominated by the co-rotating thermal plasma with $kT_e \approx 400$ eV, up to the plasmapause. The co-rotation speed at the plasmapause ($\approx 35\ R_J$) ≈ 440 km s^{-1}, while the grain speeds are typically ≈ 10 km s^{-1}. On the other hand, the electron thermal speed $\approx 1.2 \times 10^4$ km s^{-1}. Consequently, $V \ll \alpha_e$ and the thermal electron current, reduces to

$$I_e(\phi_s < 0) = \frac{n_e e \alpha_e A}{2\sqrt{\pi}} e^{e\phi_s/kT_e} \tag{8}$$

and

$$I_e(\phi_s > 0) = \frac{n_e e \alpha_e A}{2\sqrt{\pi}} \left(1 + \frac{e\phi_s}{kT_e}\right) \tag{9}$$

On the other hand, the ion thermal speed ≈ 340 km s^{-1}. Consequently, V is comparable to α_i and it is necessary to use the full equations (2) or (4) for the ion currents.

In this case, also, $I_e \gg I_p$ and consequently the grains will be charged to negative potential, which is given by

$$\frac{d|Q|}{dt} = R_g \frac{d|\phi_s|}{dt} = I_e(\phi_s < 0) - I_i(\phi_s < 0) - I_p(\phi_s < 0) \tag{10}$$

The solution of (10) for several grain sizes both on the dayside as well as the nightside (where $I_p = 0$) are shown in Fig. 1, assuming that $\chi = 0.1$. It is seen that the grains attain quasi-equilibrium potential of about ~ -670 V on the dayside and -830 V on the nightside. It is also seen that a grain of radius 10 μ will charge up to its quasi-equilibrium potential in about one hour on either side of Jupiter. As seen from equation (10) the charging time scales inversely as the grain radius. Consequently, a grain of radius 1 μ will take about 10 h to attain its quasi-equilibrium potential.

In the terrestrial case the thermal plasma in the inner plasmasphere is much cooler ($kT \approx 1$ eV), although the tail plasma sheet, the tail lobes and the auroral regions are dominated by intermediate energy plasma ($kT \approx 1$ keV). Even more energetic plasma ($kT \gtrsim 1$ MeV) are found in the radiation belts. Also the uv radiation field at the earth is about 27 times greater than at Jupiter. Consequently, solid bodies immersed in the terrestrial magnetosphere would typically reach

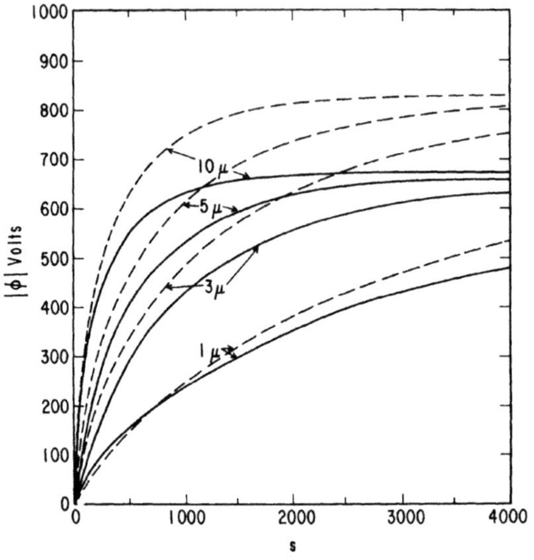

Figure 1. Charging rates of different sized grains in the Jovian magnetosphere. Solid lines correspond to the dayside while broken lines correspond to the nightside (from Hill and Mendis, 1979).

small positive potentials of a few volts on the dayside, but could attain numerically large negative potentials (-1 - 10 kV) while being in the geomagnetic tail or in the auroral regions as has been observed via the charging of spacecraft (DeForest, 1972).

3. ELECTROSTATIC DISRUPTION OF CHARGED DUST

As was first pointed out by Öpik (1956), perhaps the most important physical consequence of electrostatic charging is the possible disruption of the body. The mutual electrostatic repulsion of surface charges produces an electrostatic tension in the body. If this electrostatic tension in the body across any section exceeds the tensile strength there, the body will break up across that section. Öpik considered a sphere of uniform strength F_t (see also, Rhee, 1976). In that case, it is easy to show that the body disrupts if its radius R_g is less than a critical value R_c, given by

$$R_c = 6.65\, F_t^{-1/2}\, |\phi_s| \tag{11}$$

where R_c is measured in microns, ϕ_s is measured in volts and F_t is measured in dynes cm^{-2}.

There is good reason to believe that small dust grains in many environments, and particularly in comets, are highly non spherical, perhaps needle shaped, disc shaped or chain-like (e.g. Sekanina and Farrell, 1979, 1980). Consequently, Hill and Mendis (1981a) have studied the disruption of such grains idealizing them as prolate and oblate spheroids.

The electrostatic tension parallel to the polar (or rotation) axis (i.e. the polar tension P_x at any given distance x along this axis from the center may be evaluated by imagining the spheroid to be divided into two segments by a plane at that distance normal to the polar axis. Each element of charge dq located at (r, θ, ϕ) (see Fig. 2) experiences a force in the polar direction due to the distribution of elementary charges on the entire spheroid. Using the Maxwell stress tensor this is given by

$$P_x(x) = \frac{1}{\pi \rho^2} \int_0^{\rho} \frac{E^2}{8\pi} \cdot 2\pi\rho \, d\rho \qquad (12)$$

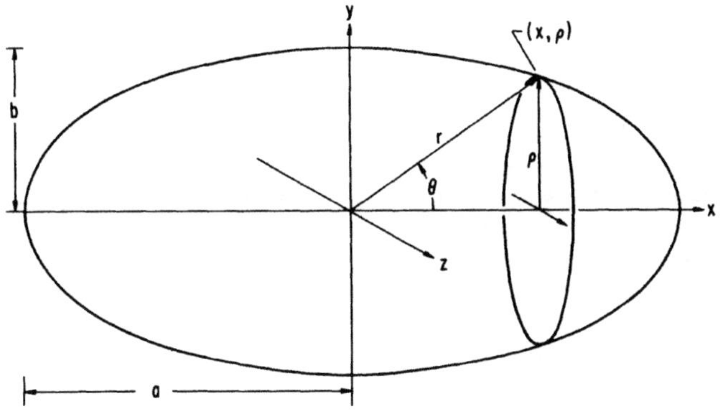

Figure 2. Specification of the coordinates used in evaluating the electrostatic stress in a charged prolate spheroid (from Hill and Mendis, 1981a).

where

$$E = 4\pi\sigma \tag{13}$$

$$\sigma = \frac{pQ}{4\pi\lambda b^3} \tag{14}$$

and

$$Q = \frac{b\phi_s\sqrt{\lambda^2 - 1}}{\tanh^{-1}\left(\sqrt{\lambda^2 - 1}/\lambda\right)} \tag{15}$$

for prolate spheroid, while

$$Q = \frac{b\phi_s\sqrt{1 - \lambda^2}}{\tan^{-1}\left(\sqrt{1 - \lambda^2}/\lambda\right)} \tag{16}$$

for an oblate spheroid (see Jeans, 1927). In the above, E is the electric field, σ is the charge density, b is the equatorial radius, a is the polar radius, $\lambda = a/b$, and p is the length of the normal from the center of the spheroid to the tangent plane at the point in question. Using (12), (13), (14), (15), one obtains after some manipulation that for a prolate spheroid ($\lambda > 1$)

$$P_x(x) = \frac{Q^2 \ln[1 + (\lambda^2 - 1)(1 - x^2/a^2)]}{8\pi b^4(\lambda^2 - 1)(1 - x^2/a^2)} \tag{17}$$

The variation of $P_x(x)$ with x between 0 and a for various values of λ (> 1) as well as for a sphere ($\lambda = 1$) normalized to unity at $x = 0$ is shown in Fig. 3. It is seen that $P_x(x)$ increases monotonically with x, for all prolate spheroids, reaching a maximum at the ends. Only in the case of a sphere is $P_x(x)$ a constant.

The consequences of charging up such a prolate spheroidal grain of uniform tensile strength toward its equilibrium potential are clear from this figure. The tensile strength of the grain will be exceeded by the electrostatic tension (if at all) first at the ends. As a consequence, the ends will be chipped off. As the potential increases this chipping process will continue with the grain becoming more and more spherical. The final outcome could be one of two things: If the tensile strength of

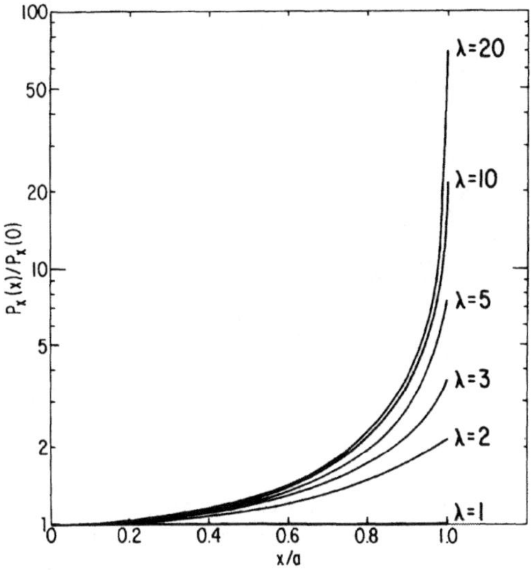

Figure 3. Variation of the ratio of the electrostatic pressure at the end to that at the center of a prolate spheroid with axis ratio λ (from Hill and Mendis, 1981a).

the grain is sufficiently large this end chipping process will eventually cease while the grain still remains prolate. Otherwise, it will continue chipping off until it becomes more or less spherical, at which time it will explode in toto.

If we consider grains (perhaps chain-like ones) having a sufficient central weakness to split at the center, we need only the electrostatic tension at the center, $P_x(0)$. The variation of this with λ at constant ϕ_s (normalized to unity at $\lambda = 1$) is shown by the curve marked P (for polar tension) in Fig. 4. The central equatorial tension P_ρ is also shown here by the curve marked E.

Sekanina and Farrell (1979) have argued that certain peculiar features called pseudo-synchronic bands, or striae, observed in the dust tails of several comets are a consequence of the sudden disruption of elongated or chain-like parent grains in the tail. Hill and Mendis (1980a) have argued that the specific mechanism for this disruption is the electrostatic one discussed above. This occurs when the grains are suddenly charged to high electrostatic potentials by energetic electron beams (1 - 10 keV). The energy source for these electrons is the

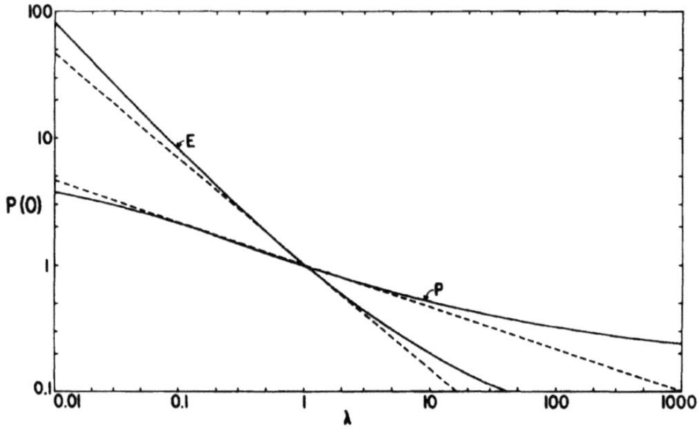

Figure 4. Variation of the electrostatic pressure at the center of a charged conducting spheroid with λ. The polar pressure is shown by the solid line marked P, while the equatorial pressure is shown by the solid line marked E. The dotted straight lines correspond to power laws which closely approximate the solutions for λ near unity (from Hill and Mendis, 1981a).

interaction of the solar wind with the cometary ionosphere producing currents and associated double layers. Noting that $P_x(0)$, in Fig. 4 varies approximately as $\lambda^{-1/3}$ it is easy to show that there is a critical value of b (= b_c) below which these elongated grains disrupt electrostatically. This is given by

$$b_c(\mu) \approx 6.65 \, |\phi_s| \, \lambda^{-1/6} \, F_t^{-1/2} \, . \tag{18}$$

The variation of b_c with ϕ, for various values of F_t and λ are shown in Fig. 5. The limiting value for grain discharge by field emission is also indicated. It is seen that for any reasonable value of the dust cross-section, sudden disruption due to this process occurs only if the tensile strength of the grains $\gtrsim 10^4$ dynes cm^{-2}. If the tensile strength were smaller the dust would disrupt soon after leaving the nucleus, due to attaining a potential of about +6 V in the solar radiation field attenuated by the cometary thermal plasma. On the other hand, if the tensile strength of the grains $\gtrsim 10^6$ dynes cm^{-2}, electrostatic disruption will not occur even with the enhanced potentials

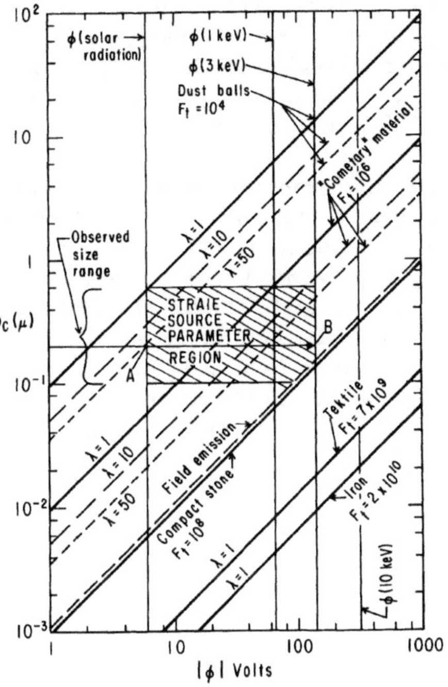

Figure 5. Critical semiminor axis for spheroidal grains of different tensile strengths and varying electrostatic potentials against electrostatic disruption are shown by the diagonal lines. The limiting values for grain discharge by field emission are also indicated. The stable regions are above the respective lines. Three lines corresponding to axis ratios of 1, 10, and 50 are shown for dust balls ($F_t = 10^4$ dyne cm^{-2}) and for "cometary" material ($F_t = 10^6$ dyne cm^{-2}). Vertical lines represent equilibrium potentials in (a) thermal plasma and solar radiation (+6 V), (b) with the addition of 1 keV energetic electrons (-63 V), (c) with the addition of 3 keV energetic electrons (-132 V), and (d) with the addition of 10 keV energetic electrons (-300 V). The charging of a parent grain of striae material (b = 0. 2 µm) to beyond ϕ (solar radiation) without disruption (point A) requires approximately $F_t = 8 \times 10^4$ dyne cm^{-3}, $\lambda = 1$; or $F_t = 2 \times 10^4$ dyne cm^{-2}, $\lambda = 50$. For the grain to reach the equilibrium potential of -132 V without disruption requires $F_t > 4 \times 10^7$ dyne cm^{-2}, $\lambda = 1$; or 10^7 dyne cm^{-2}, $\lambda = 50$. The shaded region allows determination of disruption parameters for the range of observed sizes (from Hill and Mendis, 1980a).

reached when subject to the energetic beams. These grains must remain in the normal dust streamers. Consequently, if electrostatic disruption due to sporadic charging is the process responsible for the striae, it implies that the tensile strength of the parent grains must lie in the range $\sim 10^4$-10^6 dynes cm^{-2}.

A sudden increase, by almost 2 orders of magnitude, of the micrometeoroid dust flux was measured by the dust penetration experiment on Pioneer 10 when it encountered Jupiter at about 35 R_J (Humes et al., 1975). This cannot be explained by gravitational focusing alone. Consequently, Hill and Mendis (1980b) suggested the electrostatic disruption of fragile Brownlee-type interplanetary dust grains entering the Jovian plasmapause (which is at about 35 R_J), followed by the subsequent magneto-gravitational trapping of the disruption fragments is responsible for this observation. The process of charging and disruption is illustrated in Fig. 6, where the grains are assumed

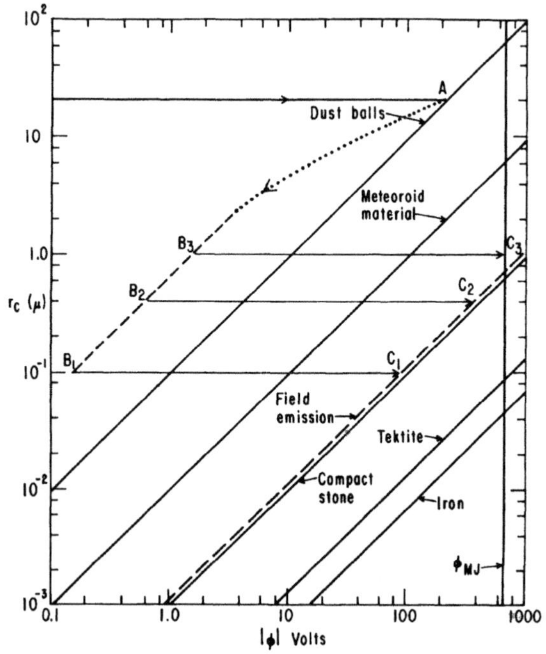

Figure 6. Critical grain radii for the electrostatic disruption of different materials; the stable regions being above the respective lines. The field emission limit is shown by the broken line, while the charging paths of the parent grain and its disruption fragments are shown by the horizontal lines.

to be roughly spherical. The tensile strengths of these grain aggregates are unknown, but it is assumed that they may be comparable to that of loose dust balls ($F_t \approx 10^4$ dynes cm^{-2}). Such a grain will charge up to a potential of about -220 V (flux point A) at which time it will be electrostatically disrupted to its constituent crystalline grains, presumably of much larger tensile strength (e.g. that of compact stone $\approx 10^8$ dynes cm^{-2}). Immediately after break up these fragments, they are shown to start off with negligible (negative) potentials. Fragments of radii 0.1 μ and 0.4 μ will then recharge to potentials -92 V and -360 V, respectively (points C_1 and C_2), these limits being set by field emission from these grains. This is because neither conductors nor insulators can support very large surface electric fields, and this leads to another critical grain radius R_{gf} given by

$$R_{gf} \approx 1.1 \times 10^{-3} \, |\phi_s| \tag{19}$$

where R_{gf} is in microns and ϕ_s is in volts (Gomer, 1961; see also, Mendis and Axford, 1974). A larger fragment of radius 1 μ will charge up to a potential determined by the plasma and radiative environment in the magnetosphere, which is about -670 V on the dayside (point C_3).

Fechtig et al. (1979) have also proposed that such an electrostatic disruption mechanism of interplanetary dust may be taking place within the terrestrial magnetosphere. They observed a 2-3 orders of magnitude increase in the micrometeoroid dust flux (10^{-15} g $\leqslant m_g \leqslant$ 10^{-12} g) within 10 earth radii with their dust experiment on the HEOS-2 earth satellite. They have argued that these small grains (which occur mostly in swarms as opposed to the larger grains) are sudden electrostatic disruption products of larger parent bodies traversing the earth's auroral zones, which are populated by energetic plasma ($kT \approx 1$-10 keV). They believe that the parent bodies are type III fireballs in the mass range (~ 10-10^6 g). Reference to equation (18), (with $\lambda = 1$, corresponding to spheres) shows that these masses for the parent bodies are too large. Even if $|\phi_s|$ is as large as 100 kV (which is a generous upper limit to the surface potential in the auroral regions), when $m_p \approx 10$ g (which corresponds to $r_p \approx 1.36$ cm with $\rho_p \approx 1$ g cm^{-3}), electrostatic disruption requires that $F_t \approx 2.4 \times 10^3$ dynes cm^{-2}. This is barely possible for very loose dust balls. On the other hand, when $m_p = 10^6$ g (corresponding to $r_p \approx 63$ cm, with $\rho_p \approx 1$ gm cm^{-3}) electrostatic disruption requires that $F_t \leq 2.5 \times 10^{-2}$ dynes cm^{-2}, which is incredibly low. Consequently, such large bodies are not likely to be electrostatically disrupted in toto on entering the auroral regions. On the other hand, such bodies are likely to be highly irregular in shape. Since the electrostatic tension at any point on such a body is inversely proportional to the radius of curvature there, provided the surface potential is constant, sharp edges and projections will be chipped away, as discussed earlier, when the body gets charged to a high enough potential. We think that this is the only way that parent bodies at the higher end of the mass range can contribute to the observed micrometeoroid swarms, whereas those at the lower end could presumably do so by electrostatic bursting if their tensile strengths correspond to those of very loose dust balls ($F_t \approx 10^3$ dynes cm^{-2}).

4. DYNAMICS OF CHARGED DUST IN PLANETARY MAGNETOSPHERES

Once dust grains are electrically charged, their dynamics, both in planetary magnetospheres as well as in the heliosphere, are effected by the presence of magnetic fields there. The effect is stronger the larger the charge to mass ratio.

The basic nature of these orbits in planetary magnetospheres has been discussed by Mendis and Axford (1974). More recently, Hill and Mendis (1979, 1980b) considered the detailed dynamics and orbital evolution of dust grains entering the Jovian magnetosphere in order to explain the distribution of micrometeoroid dust observed there by the Pioneer and Voyager spacecrafts. Within the region of the magnetosphere, where the plasma is co-rotating, the equation of motion of a charged grain in a planet-centered inertial frame, is given by

$$m\ddot{\underset{\sim}{r}} = \frac{q(t)}{c}\left[-(\underset{\sim}{\Omega} \times \underset{\sim}{r}) + \dot{\underset{\sim}{r}}\right] \times \underset{\sim}{B}(\underset{\sim}{r}) - \frac{GM_J m}{r^3}\underset{\sim}{r} - \underset{\sim}{F}(\underset{\sim}{r}) \tag{20}$$

where m and q(t) are, respectively, the mass and the varying charge on the grain; M_J, $\underset{\sim}{\Omega}$ and $\underset{\sim}{B}(\underset{\sim}{r})$ are, respectively, the mass, the angular velocity and the magnetic field of the planet and $\underset{\sim}{F}(\underset{\sim}{r})$ is the Coulomb drag on the grain, which is shown to be negligible. Here the term $-(\underset{\sim}{\Omega} \times \underset{\sim}{r}) \times \underset{\sim}{B}(\underset{\sim}{r})$ represents the plasma polarization electric field induced by the co-rotation.

While the co-rotation of the Jovian magnetosphere holds strictly only within $r \leqslant 10\ R_J$, it shows only about a 30% decrease in the outer regions (McNutt et al., 1979), which may be associated with radial plasma convection and the associated spiralling of the magnetic field. Neglecting this effect, assuming that the spin axis and the magnetic dipole axis are centered at the planet and exactly parallel (although in reality the dipole axis is displaced by about 0.2 R_J from the center and tilted at an angle of about 10° to the spin axis), and confining themselves to the equatorial plane, Hill and Mendis (1980b) computed the orbits of charged dust fragments that are injected to the magnetosphere due to the electrostatic disruption of their parents at the outer edge of the plasmasphere ($\approx$ 35 R_J). Typical orbits of different sized fragments of parent grains entering the Jovian plasmasphere at two different angles, one prograde and the other retrograde, are shown in Figs. 7 and 8. In the prograde case none of the grain fragments get to within the orbit of Io. In the retrograde case the grain fragments are pulled in closer to Jupiter by the co-rotational electric field and all except the largest ones are seen to cross the orbit of Io. The 1 μ fragment gets very close to Jupiter as it makes a tight, high-speed loop and ends up with a net retrograde orbit precession. The smaller fragments , however, end up with a net prograde precession. Fragments of parent grains entering the magnetosphere directed straight at the planet follow orbits intermediate to these two cases.

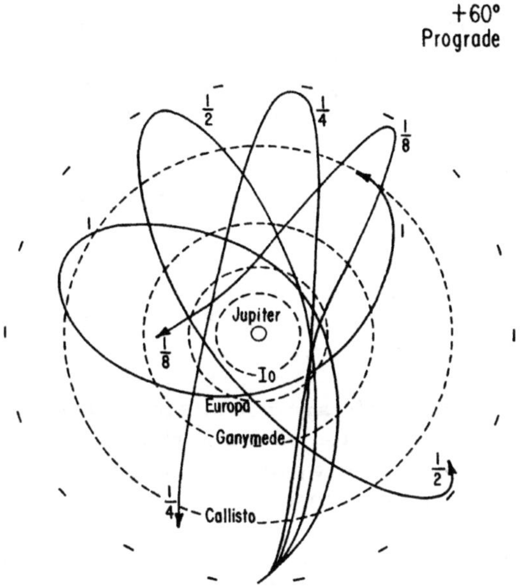

Figure 7. Orbits of fragments of a parent grain which enters the Jovian magnetosphere at 35 R_J at an angle 60° away from Jupiter in the prograde direction (from Hill and Mendis, 1980b).

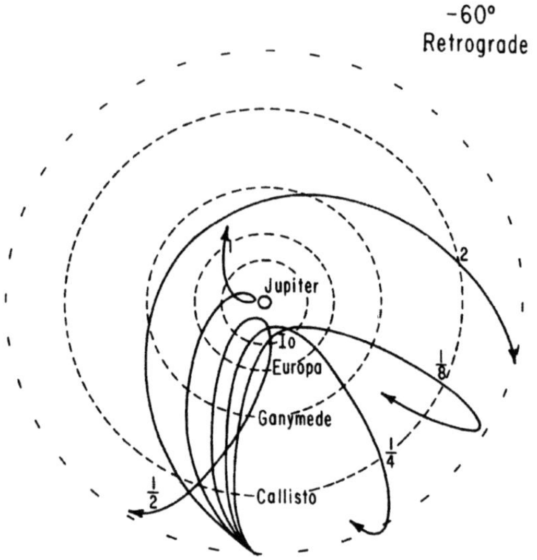

Figure 8. Orbits of the fragments of a parent grain that enters the Jovian magnetosphere at 35 R_J at an angle of 60° away from Jupiter in the retrograde direction (from Hill and Mendis, 1980b).

While the smallest fragments have their potentials set constant by field emission, fragments with radius $\geqslant 0.6\ \mu$ have their potentials set by the plasma and radiative environment. These potentials continuously change due to varying relative speed with respect to the co-rotating plasma. This is shown in Fig. 9. The grain is slightly discharged at larger Jovicentric distances. The finite times for charging and discharging also introduce a phase lag of about 6 h, so that the smallest (numerical) potentials are not attained exactly at apijove. This non-symmetric acceleration of the particles by the electric field causes them to lose orbital energy and get magneto-gravitationally captured by the Jovian magnetosphere.

Hill and Mendis (1980b) also considered the evolution of the spatial distribution of the charged grains in the equatorial plane of the Jovian magnetosphere. In their computer simulation, a grain is injected every hour at 35 R_J and hourly samples are then taken of the number of grains in each 1 R_J thick ring from 0 to 35 R_J. The distribution after 1 week,of grains initially projected in a retrograde direction (-88° to the radial direction) is shown in Fig. 10. The bimodal nature

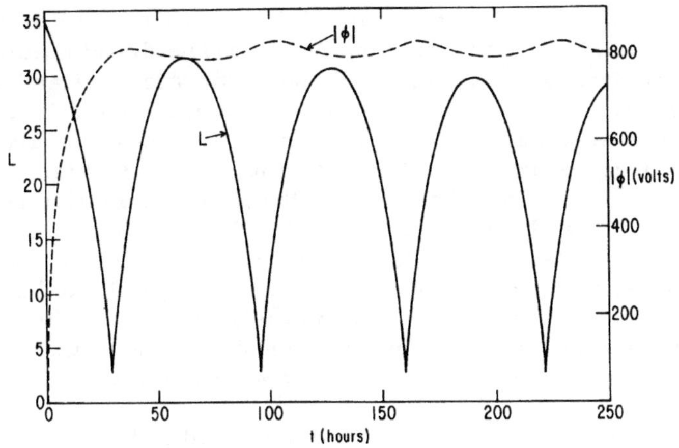

Figure 9. Time variation of L and $|\phi|$ of a 1 μ grain fragment. The grain is slightly discharged at larger Jovicentric distances, with a phase lag of about 6 h. This results in the non-symmetric acceleration of the particle by the co-rotational electric field and a consequent loss of energy.

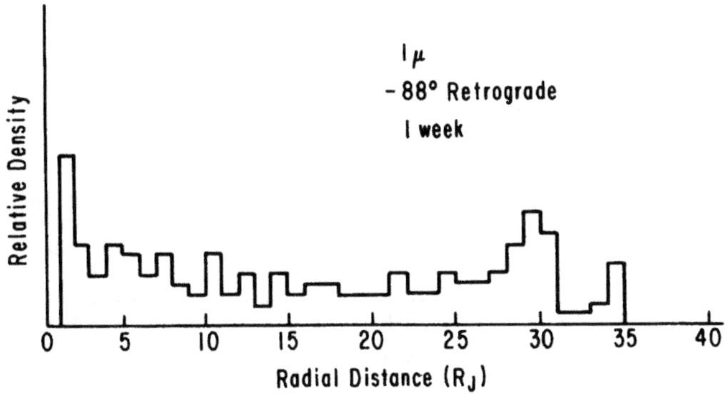

Figure 10. Spatial distribution of 1 μ grain fragments entering the Jovian magnetosphere at 35 R_J pointed 88° away from Jupiter in the retrograde direction, after one week (from Hill and Mendis, 1980b).

of the distribution is clearly a result of the longer times spent by the grains in the vicinity of their turning points. The distributions of these particles after 8 weeks and 1 year are shown in Figs. 11 and 12, respectively. Energy is being continuously sapped from the orbiting grains, as discussed earlier in this section, and consequently their apijoves are progressively moving in as they get more highly bound to Jupiter. It is seen, as a consequence, that the distribution is peaking more and more toward the planet. Similar distributions are obtained for other launch angles also. Due to the gradually slowing rate of evolution with time, the numerical calculations are terminated after one year. It is, however, expected that the distribution will eventually peak around the synchronous orbit at about $2R_J$, where the co-rotational electric field goes through a zero and changes sign.

The cameras aboard Voyager 1 discovered a thin dust ring around 1.8 R_J (Smith et al., 1979). The dust in this ring is also expected to be very small ($r_g \approx 1\ \mu$) based on forward scatter data. Consequently, it has been suggested by Hill and Mendis (1980b) that this dust ring corresponds to the calculated peak in the magneto-gravitationally captured charged dust distribution. The reason why the dust ring is centered around 1.8 R_J rather than near the synchronous orbit

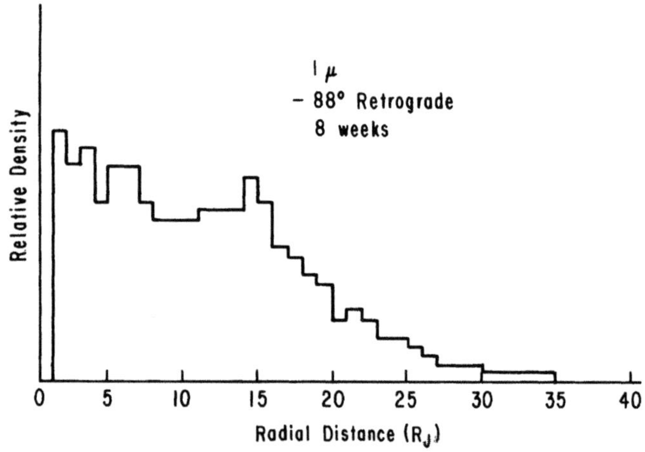

Figure 11. Spatial distribution of 1 μ grain fragments entering the Jovian magnetosphere at 35 R_J pointed 88° away from Jupiter in the retrograde direction, after eight weeks (from Hill and Mendis, 1980b).

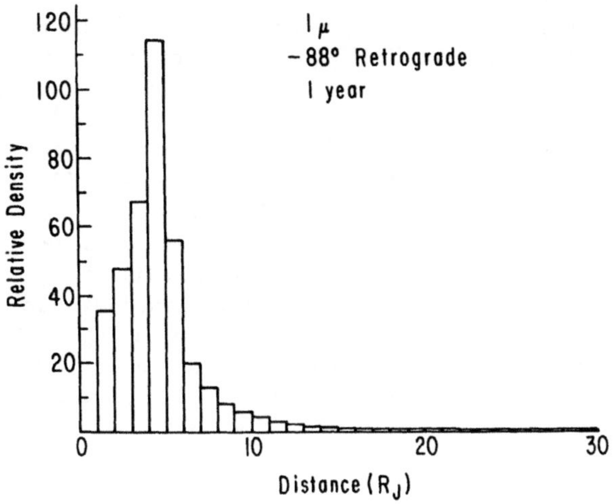

Figure 12. Spatial distribution of 1 μ grain fragments entering the Jovian magnetosphere at 35 R_J pointed 88° away from Jupiter in the retrograde direction, after one year (from Hill and Mendis, 1980b).

at about 2 R_J may very well have to do with the presence of the newly discovered small satellite at 1.81 R_J (Danielson and Jewitt, 1979). The impact of these orbiting grains may produce a secondary source of small dust grains which is concentrated around the satellite orbit.

The calculations of Hill and Mendis (1980b) described above are limited by the fact that they are restricted to the equatorial plane. They also do not take into account the highly compressible nature of the Jovian magnetosphere as well as the wobbling of the plasma disc as a result of the non-alignment of the magnetic axis and the spin axis. This can cause grains to leak out of the magneto-disc. Some of the grains will also be lost by impact with the satellites. Comparison of the time scales to cross these satellite barriers, with the orbital period of the satellites, indicate that significant depressions will be produced in the dust distribution, particularly near the inner satellites. Otherwise, the overall distribution is unaltered.

Despite all these limitations, the observed micrometeoroid dust distribution within the Jovian magnetosphere is in good agreement with the predictions of the above model, whereas it is inconsistent with a purely gravitational model.

The Galilean satellites are known to be in synchronous rotation with the planet. They also show a peculiar leading side-trailing side brightness asymmetry, in that Callisto (the outermost Galilean satellite) is darker on the leading face whereas the three inner Galilean satellites are darker on the trailing side. Hill and Mendis (1979) explain these brightness asymmetries in terms of the impact geometries of these charged micrometeoroids on their surfaces. Also the similarity in these brightness asymmetries with those of the larger inner and outer Saturnian satellites (which are also known to be in synchronous rotation with the planet) and the expectation that they had a similar cause, led these authors to predict that the magnetic moment and spin of Saturn are parallel as in the case of Jupiter (and unlike the case of the earth, where they are anti-parallel) with the limit of plasma corotation lying between the satellites of Rhea and Iapetus. Both these predictions were subsequently confirmed by the Pioneer 11 and Voyager 1 observations.

Several other authors have also considered the effects of dust charging within the Jovian magnetosphere. In particular, Johnson et al. (1980) show that 0.1 μ sized "smoke" particles in Io's volcanic plumes can be charged and ejected into the Jovian magnetosphere by the Lorentz force overcoming Io's gravity. They suggest that these grains may be an important source of erosive impacts on large Jovian ring particles and destructive collisions with small particles. Subsequently, Morfill et al. (1980a) have further discussed this problem. They conclude that these very small charged particles behave adiabatically, in the sense that the guiding center approximation to their motion applies. They also argue that random fluctuations of the surface charge are the major cause of cross-field transport. Finally, in a related study of the interaction of the Jovian energetic radiation belt electrons with the magnetospheric dust (Morfill et al., 1980b), they conclude that this constitutes a significant sink for these electrons and that the depressions observed by Voyager, in the energetic electron profile, are consistent with such a process.

Most recently, the Voyager 1 encounter with Saturn produced two very exciting observations of the Saturnian ring system, which strongly indicated the importance of electromagnetic forces in controlling the dynamics of the small dust particles there. One was the wavy braided nature of the thin outer F-ring. The second was the radial spokes which were observed in rotation across the inner part of the bright B-ring.

Observations of the F-ring and the spokes, in forward and back scattered sunlight, indicated that they were both composed of small (micron sized) grains. Hill and Mendis (1981b) pointed out that the electromagnetic forces on such grains (which are charged to high surface potentials as in the Jovian case) are responsible for both these effects.

With regard to the braided F-ring they pointed out that small charged dust grains which constitute the F-ring would eventually settle down to circular orbits moving under the combined effect of the planet's gravity and the electric field due to the relative motion between the grain and the co-rotating plasma. They will in fact have speeds intermediate to the Kepler speed and the local co-rotation speed. What is interesting is that the F-ring is flanked on either side by two small satellites. A dust grain close to either of these satellites will suffer a gravitational perturbation. If the grain is not charged, it will follow an oribt as shown by the dashed line in Fig. 13. This orbit, which is displaced

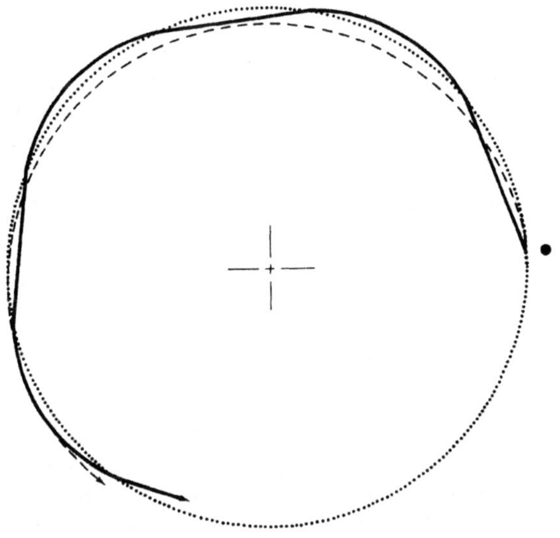

Figure 13. The undulating orbit of an electrically charged dust grain in Saturn's F-ring, once it has been perturbed by one of the newly discovered satellites, is shown by the solid curvy line. Its original circular orbit is shown by the dotted line. If the grain was not charged, the resulting orbit after the satellite perturbation is shown by the dashed line. The perturbing satellite is shown by the solid dot, while the center of Saturn is indicated by the cross (from Hill and Mendis, 1981b).

from the initial circular one, will intersect the latter diametrically opposite to the point of perturbation and return to that point. It will not show any wavy pattern. The situation is quite different if the grain is charged. As soon as it leaves the circular orbit it will be deflected back toward it by the magnetic field. As a result, the disturbed grain will oscillate back and forth about the circular orbit with a period equal to its gyro-period as it moves around the planet in a new wavy orbit, as shown by the solid line in Fig. 13. These authors used this fact to calculate the wavelength of the observed wave and found it to be in good agreement with the observations. The braids are explained in terms of a juxtaposition of different waves corresponding to different grain sizes and therefore different charge to mass ratios.

With regard to the rotating spokes, they seemed to be straight and well defined as they emerged from the shadow. They also appeared to remain more or less straight for some time, as they rotated across the sunlit side although they gradually lost definition and eventually dissipated. However, these features seemed to be ubiquitous, being continuously regenerated on the shadowed side of the ring. The fact that they remained straight and well-defined even as long as they did was considered surprising, because the ring particles were believed to rotate with Kepler speeds, and the resulting shear from the differential motion between the ring particles constituting the outer and inner parts of the spokes was expected to bend and break them up within tens of minutes.

Hill and Mendis (1981b) explain these rotating spokes the following way: Electrons accelerated to keV energies in the auroral regions of the planet are beamed along the field lines to the equatorial plane on the nightside, as is observed in the earth. These charge up the large bodies in the rings to kilovolt potentials. As a result, any loose micron sized dust that may be present on their surfaces, and which are also charged, are levitated and quickly brought into co-rotation by the co-rotational electric field which overwhelms the planetary gravitation on these small charged grains. Once these grains emerge from the shadow they are partially discharged by the solar radiation and brought to a potential corresponding to the cooler thermal plasma in the neighborhood of the synchronous orbit. Consequently, these particles will once again become more strongly affected by the gravitational field vis a vis the co-rotational electric field and begin to move in essentially Keplerian orbits, when the spokes gradually dissipate as observed.

5. DYNAMICS OF CHARGED DUST IN THE HELIOSPHERE

The dynamics of charged dust grains in the heliosphere was first investigated by Parker (1964), with the interplanetary magnetic field idealized by the dipole component of the solar magnetic field but stretched out by the radially flowing solar wind. He estimated that the grains were charged to positive potentials of about 10 V and showed that the polarization electric field of the solar wind normal to the equatorial plane produces large perturbations on the Kepler orbits of submicron sized particles moving under the combined effects of solar gravitation and solar radiation pressure. He also discussed the effects of sporadic field reversals in the equatorial plane every few days due to the "sector structure" of the interplanetary field. He adopted a stochastic treatment of the dynamics of the grains and showed that they had a mean lifetime of about 10^7 s. Taking into consideration the post-Pioneer 10 view of the interplanetary magnetic field, wherein the sector structure is treated as a consequence of a corrugated current sheet in the equatorial plane, Parthasarathy (1978) has calculated the orbits of such charged submicron grains. He found that during epochs when the solar rotation and dipole axes are parallel, the grains spiral away from the equatorial plane causing a maximum concentration in conical surfaces $\pm 25°$ off the equatorial plane. Although not discussed by Parthasarathy, it is clear from his treatment that the only stable orbits possible for positively charged grains during epochs when the solar rotation and the dipole axes are antiparallel are, in fact, in the equatorial plane because the polarization electric field on such a grain, when off the equatorial plane, is directed back toward this plane. Also the grain potential estimated by this author (≈ 1.1 V) appears to be too low (see following section).

Recently, Morfill and Grün (1979a) have re-examined the whole problem of electromagnetic perturbation of charged dust particle orbits in interplanetary space in the light of the present understanding of the large scale spatial and temporal interplanetary plasma and field topology. They find that stochastic variations induced by electromagnetic forces are less important than suggested by Parker. They also find that while systematic variations in the orbital parameters induced by electromagnetic forces, inside 1 AU, tend to shift the plane of symmetry of the zodiacal dust cloud toward the solar magnetic equatorial plane, outside 10 AU there is no such systematic effect. This is due to the solar cycle variation in the magnetic polarity, which tends to cancel out systematic effects in the case of grains with periods larger than the period of the solar cycle itself. Inside 0.3 AU they show that dust particles may enter a region of "magnetically resonant" orbits when the orbital period becomes comparable to twice the solar rotation period. In this case the particle "sees" an outward directed magnetic field during half its orbit and an inward directed field for the other half for a 4-sectored field structure. Rapid changes in the orbital parameters are then possible. Variation of the orbital parameters in an initially near circular orbit of a dust grain of radius 1 μ is shown in Fig. 14. These authors conclude that the observed similarity of the plane of symmetry of zodiacal light with the solar equatorial plane may be an effect of this interaction of the charged interplanetary dust particles with the interplanetary magnetic field.

Morfill and Grün (1979b) have also considered the effects of electromagnetic forces on charged interstellar grains entering the heliosphere. They conclude that such dust particles with radii ⩾ 0.1 μ can penetrate deeply into the heliosphere if their incidence direction at

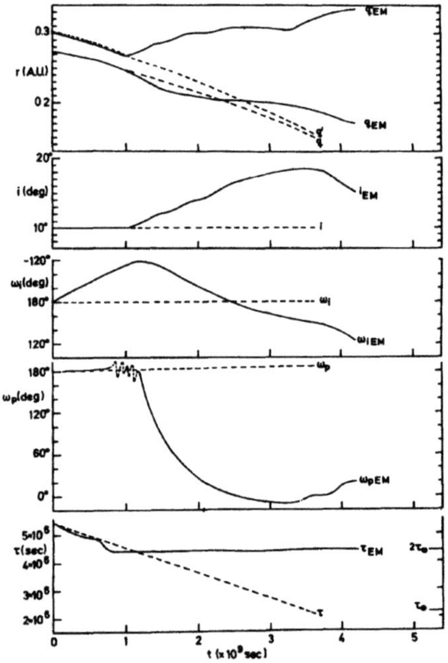

Figure 14. Numerical solution of the equation of motion of a dust grain (radius = 10^{-4} cm) in a near circular orbit about the sun with electromagnetic forces acting (solid lines) and without electromagnetic forces (dashed lines). From top to bottom are plotted the aphelion and perihelion distance, inclination, longitude of ascending node, longitude of perihelion and orbital period (from Morfill and Grün, 1979a).

the heliopause is almost radially inward and close to the solar magnetic equatorial plane, whereas those with radii $\leqslant 0.1\ \mu$ are prevented from reaching the inner heliosphere.

6. PHYSICAL EFFECTS OF CHARGING INTERMEDIATE SIZED BODIES IN THE SOLAR SYSTEM

From the smallest end of the mass spectrum of the condensed phase in the solar system, namely micrometeoroid dust, we now proceed to the class of intermediate sized bodies. These include comets, the smaller asteroids and the larger particles in the Saturnian ring system. These are characterized by a size in the range 0.1-10 km. These bodies, like the micrometeoroids, can be charged to large potentials in various environments. However, they are much too large for their orbital dynamics to be effected by the electric forces induced by their motion across the magnetic fields in their respective environments. They are also much too large to be electrostatically disrupted for any reasonable value of their tensile strength. On the other hand, there are other interesting physical effects that are produced by the surface charging. These include the electrostatic levitation and blow-off of any finely ground loose dust on their surfaces, since their surface gravity is sufficiently small that it can be overcome by the electric force of repulsion between grains and the body.

The effect of charging the cometary nucleus at large heliocentric distances has been discussed very recently (Mendis et al., 1981). It is shown that cometary nuclei, which for the most part are believed to be composed of H_2O ice, are not shielded from the solar uv radiation and the solar wind by protective atmospheres when they are beyond about 5 AU from the sun. The unimpeded flow of the solar wind electrons and ions constitute two important charging currents onto the cometary surface. The other important current is the flux of photoelectrons from the surface produced by the solar uv radiation. In considering the charging of the surface, it is recognized that the surface conductivity is unlikely to be sufficiently large to make the surface of these relatively large bodies equipotentials. Consequently, a point-wise current balance has to be solved in order to obtain the local potential. All these electric currents now depend, besides all the parameters discussed in section 2, also on the polar angle from the sub-solar point. The expressions for the currents are essentially the same as those given for the case of the Moon by Manka (1973) and have been adopted with only a slight modification by Mendis et al. (1981). It is noted here, as in the case of the Moon, that the dimensions of the body are larger than the electrostatic shielding distance.

In order to calculate the variation of the surface potential distribution with heliocentric distance, one needs to know how the plasma density and velocity, as well as the electron and ion temperatures, vary with heliocentric distance. These authors have adopted two different solar wind models: One a simple polytropic model with constant solar wind velocity and a second one due to Holzer (1972) which takes into account the gradual heating and deceleration of the solar wind beyond about 5 AU due to contamination with the interstellar gas that penetrates the heliosphere.

The variation of the surface potential ϕ_s with θ on the sunlit hemisphere for different heliocentric distances from 5 AU to the outer shock distance (estimated as about 70 AU) is shown, for the polytropic solar wind model, in Fig. 15. It is seen that while the potential is small and positive for most of the surface, it becomes numerically large and negative close to the limb. Qualitatively similar results are obtained with the Holzer model too, except that in this case the

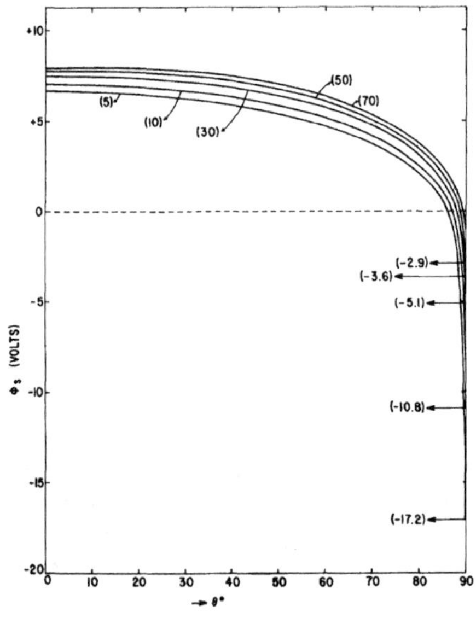

Figure 15. The variation of the surface potential with angular distance from the sub-solar point (on sunlit side) at various heliocentric distances of the comet (indicated by the numbers within parentheses) in the case of the polytropic solar wind model (from Mendis et al., 1981).

positive potentials decrease and the negative potentials numerically increase with increasing heliocentric distance, in contrast to the polytropic case, where the opposite is true.

The electric potential on the dark side is less easy to calculate exactly due to wake effects, but it is easy to see that it must be numerically large and negative. While the solar wind is highly supersonic with respect to the ions, it is subsonic with respect to the electrons. Consequently, there will initially be a large depletion of ions behind the comet, whereas the solar wind electrons can readily impinge on the back side, due to their large thermal speed vis-à-vis the bulk speed. Without going into detail, however, one can estimate the (negative) potential ϕ_s on the back side by noting that the potential will keep on increasing (numerically) until it is large enough to deflect the solar wind protons moving past the comet with bulk energy = $1/2\ m_p V_{sw}^2$. Consequently, $\phi_s \approx -m_p V_{sw}^2/2e$. With $V_{sw} = 350$ km s^{-1} (corresponding to the average value for a slow-speed stream), and $V_{sw} = 700$ km s^{-1} (corresponding to the average value for a high-speed stream), one gets ϕ_s to be respectively ≈ -630 V and -2540 V. Although secondary electron emission can reduce these (numerical) values somewhat, one does not expect any substantial reduction in the values, because the secondary yield at these electron temperatures is well below unity.

In order to calculate the surface electric field, one needs to calculate also the Debye shielding distance. This is given by

$$L = \left(\frac{kT}{4\pi ne^2} \right)^{1/2} \tag{21}$$

and its variation with θ over the sunlit surface and with heliocentric distance is shown in Fig. 16 for the polytropic model. For much of the sunlit hemisphere the shielding is provided not by the solar wind electrons but rather by the photoelectrons (which have a larger density than the solar wind electrons). This reduces the Debye shielding length substantially from that in the solar wind, and consequently, increases the surface electric field.

Just as in the case of the electrostatic potential, it is difficult to calculate the electrostatic shielding distance on the dark side since the value of n_i (the density of the attracted species) in the wake is unknown. It can, however, be argued that the electric field on the night side must extend far enough to attract ions streaming past the comet, and hence the screening length should approach the cometary radius.

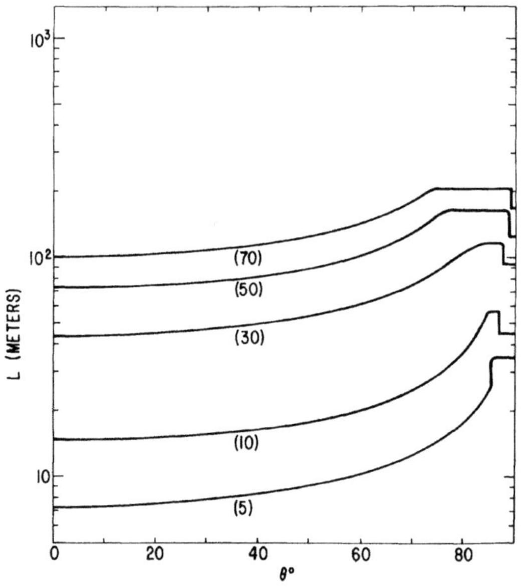

Figure 16. Variation of the electrostatic shielding distance (L) with angular distance from the sub-solar point, for various heliocentric distances (indicated by the numbers within parentheses), in the case of the polytropic solar wind model (from Mendis et al., 1981).

These large electric fields and potentials can have important physical consequence, particularly for loose dust on the surface. It has been suggested by Whipple (1977), that perhaps due to cosmic ray damage, even new comets could have an outer "frosting" a few meters thick. Such a layer might also exist due to the lack of consolidation of the last layers of dust that accumulated on the surface during the formation of the comet. Whatever may be the cause, it is supposed that a layer of thin, loose dust exists on the surface. As the surface gets charged, these dust grains will also get charged to the same potential. But they acquire only a small fraction of the charge they would possess if they were moving freely in space (see Singer and Walker, 1962). This charge is proportional to their projected surface area, and as a result, only a small fraction of the grains, at any given time, will have at least a unit charge.

Mendis et al. consider a comet of radius 1 km, whose rotation period is sufficiently long ($\geq$ 5 hrs) not to affect the surface gravity substantially. A grain of radius $a(\mu)$ having unit charge will levitate on the surface if the repulsive electric force is larger than the attractive gravitational force. This will be true for those grains where,

$$a(\mu) \leq 0.51 \left\{ \frac{|\phi_s(V)|}{L(m)} \right\}^{1/3} . \tag{22}$$

Also, all grains with radii such that

$$a(\mu) \leq 5.1 \times 10^{-2} \, |\phi_s(V)|^{1/3} \tag{23}$$

will escape from the surface. For instance, at 5 AU, grains will levitate on much of the sunlit hemisphere if $a \leq 0.5\ \mu$ and will escape from the surface entirely if $a \leq 0.1\ \mu$. On the dark hemisphere, during quiet times of the solar wind (V_{sw} = 350 km s^{-1}), grains will levitate if $a \leq 0.6\ \mu$ and will escape if $a \leq 0.4\ \mu$. During disturbed times (V_{sw} = 700 km s^{-1}) the corresponding values are, respectively, $a = 0.9\ \mu$ and $a = 0.7\ \mu$.

These authors also estimate upper limits to the mass of dust that can be blown off from such a cometary surface as it traverses the supersonic region of the solar wind. They estimate, due to blow-off from the sunlit face alone, that this amounts to about 5.2×10^{12} g with the polytropic model, while it amounts to about 1.25×10^{13} g with the Holzer model of the solar wind. The corresponding mantle erosion thicknesses are 0.4 - 1.0 m. It is pointed out that if masses close to these upper limits can indeed be blown off, they are comparable to the mass of dust that would be blown off a typical medium bright comet by the viscous drag of the sublimating gases during perihelion passage between about 3 AU and 0.5 AU. Also this process could provide a substantial new source of fine dust in the outer solar system. The authors also speculate that the highly solar wind modulated blow-off of dust from the dark side of the comet, predicted by their model, may explain the observed periodic dust outbursts from comet P/ Schwassman Wachman 1, which moves in an almost circular orbit between those of Jupiter and Saturn.

That a similar effect takes place on the larger bodies of the Saturnian ring system has already been discussed in section 4, with regards to the recently observed rotating spokes in the inner part of the Saturnian B-ring. When such large bodies are on the shadowed side of Saturn, they can get charged up to negative potentials around -1 kV. In that case, micron sized grains residing on their surfaces may be levitated and blown-off the surface. The only difference from the case of the comets is that these small grains (which acquire their full complement of charge corresponding to their potential in free space soon after leaving the surface) will be caught up by the strong magnetic fields there and swing around with the co-rotation speed as they bounce up and down along the field lines.

With regard to the asteroids, so far there is no existing discussion regarding the effects of surface charging, although work is now in progress (Hill and Mendis, 1981c). The distribution of surface potentials is similar to that of the case of distant comets, as discussed earlier. Also, with the polytropic model of the solar wind, appropriate to this case (2 AU $\leqslant$ d $\leqslant$ 4 AU), the positive potentials on the sunlit surface are somewhat smaller ($\approx$ 5-6 V) whereas the numerical values of the negative potentials at the terminator are significantly larger (-20 $\sim$ -30 V). Also numerically large negative potentials ($\sim$ -1 kV) can once again be attained on the dark side. A consequence of this is that the smaller asteroids (R $\leqslant$ 1 km) will be electrostatically "cleaned" of any small loose dust on their surface, whereas the larger ones will not be. This not only points to an important difference in the surface structure between small and large asteroids, it will also have important consequences to the dynamics of inter-asteroidal collisions (both now, and presumably, also in the past), particularly due to the importance of such "regolith layers" in cushioning the impacts and helping agglomeration (e.g. see Hartmann, 1978).

ACKNOWLEDGMENTS

The author acknowledges support from the following grants: NASA-NSG-7102 of the NASA Geophysics and Geochemistry program, NASA-NGR-05-009-110 of the NASA Planetary Atmospheres program, and NSF-AST-78-19008 of the Solar System Astronomy Program.

REFERENCES

Danielson, G.E. and Jewitt, D., 1979, California Institute of Technology, Pasadena, Press Release, Oct. 16, 1979.

DeForest, S. E., 1972, J. Geophys. Res., 77, 651.

Fechtig, H., Grün, E. and Morfill, G., 1979, Planet. Space Sci., 27, 511.

Gomer, R., 1961, "Field Emission and Field Ionization," Harvard Univ. Press.

Grard, R. J. L. (Ed.)., 1973, "Photon and Particle Interactions with Surfaces in Space, D. Reidel Pub. Co., Holland.
Hartmann, W. K., 1978, Icarus, 33, 50.
Hill, J. R. and Mendis, D. A., 1979, The Moon and Planets, 21, 3.
Hill, J. R. and Mendis, D. A., 1980a, Astrophys. J., 242, 395.
Hill, J. R. and Mendis, D. A., 1980b, The Moon and the Planets, 23, 53.
Hill, J. R. and Mendis, D. A., 1981a, Can. J. Phys. (in press).
Hill, J. R. and Mendis, D. A., 1981b, The Moon and the Planets (in press).
Hill, J. R. and Mendis, D. A., 1981c, in preparation.
Holzer, T. E., 1972, J. Geophys. Res., 77, 5407.
Humes, D. H., Alverez, J. M., Kinard, W. H. and O'Neal, R. L., 1975, Science, 188, 473.
Jeans, J. H., 1927, "The Mathematical Theory of Electricity and Magnetism," Cambridge Univ. Press, p. 246.
Johnson, T. V., Morfill, G. E. and Grün, E., 1980, Geophys. Res. Lett., 7, 305.
Manka, R. H., 1973, "Photon and Particle Interaction with Surfaces in Space (Ed. R. J. L. Grard), D. Reidel Pub. Co., p. 347.
McNutt, R. L., Jr., Belcher, J. W., Sullivan, J. D., Bagneal, F. and Bridge, H. S., 1979, Nature, 803, 280.
Mendis, D. A. and Axford, W. I., 1974, Rev. Earth and Planet. Sci., 2, 419.
Mendis, D. A., 1979, COSPAR Space Res., XIX (Ed. M. J. Rycroft), Pergamon Press, NY, p. 423.
Mendis, D. A., Hill, J. R., Houpis, H. L. F. and Mendis, D. A., 1981, Astrophys. J. (submitted).
Morfill, G. E. and Grün, E., 1979a, Planet. Space Sci., 27, 1269.
Morfill, G. E. and Grün, E., 1979b, Planet. Space Sci., 27, 1283.
Morfill, G. E., Grün, G. and Johnson, T. V., 1980a, Planet. Space Sci. (in press).
Morfill, G. E., Grün, G. and Johnson, T. V., 1980b, Planet. Space Sci. (in press).
Öpik, E. J., 1956, Irish Astron. J., 4, 84.
Parker, E. W., 1964, Astrophys. J., 139, 951.
Parthasarathy, R., 1978, Astrophys. J., 83, 1235.
Rhee, J. W., 1976, "Interplanetary Dust and Zodiacal Light (Ed. H. Elsässer and H. Fechtig), Springer-Verlag, p. 238.
Sekanina, Z. and Farrel, J. A., 1980, in Proceedings of IAU Symposium No. 90: Solid Particles in the Solar System (Ed. I. Halliday and B. A. McIntoch), p. 267.
Sekanina, Z. and Farrell, J. A., 1981, Astron. J., (in press).
Singer, S. F. and Walker, E. H., 1962, Icarus, 1, 112.

Smith, E. J., Davis, L., Jr., Jones, D. E., Colburn, D. S., Coleman, P. J., Jr., Dyal, P. and Sonett, C. P., 1974, Science 183, 305.
Whipple, E. C., Jr., 1965, NASA-GSFC publication X-615-62-296.
Whipple, E. C., Jr., 1981, Reports in Progress on Physics (in press).
Whipple, F. L., 1977, "Comets, Asteroids and Meteoroids," (Ed. A. H. Delsemme), Univ. of Toledo Press, p. 25.

INTERACTING GALAXIES

P. Stewart
Mathematics Department, Manchester University.

Summary

This paper is divided into two parts: (a) a short review of the literature on interacting galaxies, (b) a detailed account of several 20,000 body computer simulations of rotating galaxies in direct collision.

Review

Early observations on double galaxies can be found by Pease (1917), Curtis (1921), Lundmark (1926), and Hubble (1936). Of the 8000 NGC galaxies investigated by Lundmark (1926-8) 200 were double; Holmberg's findings (1937) on over 800 double and multiple galaxies were:
1. If a double galaxy is defined as having an apparent separation less than twice the sum of the largest apparent diameters, with a few exceptions, this definition implies a physical association;
2. Ellipticals, Spirals and Irregular types are found in multiple association;
3. Formation is probably by capture.

Jeans (1928) pointed out that the pairs of galaxies NGC 5278/9 and 4395/4401 are suggestive of tidal action. Zwicky (1956) summarised the situation 28 years later as follows: (a) Interconnected galaxies may have formed from one cloud or as a result of encounters with already fully formed galaxies. (b) Encounters will feed the intergalactic medium which may then be swept up by other galaxies. (c) The "internal viscosity" of galaxies is large as judged by filamentary strings holding together well away from the galaxies - see Toomre & Toomre (1972). (d) Collision will lead to internal oscillations with the filaments having radial motions towards and from the centre.

Morphology

Vorontsov-Velyaminov (1959, 1977) has given a detailed description of the morphology of interacting galaxies. Among the types he lists are the following:

F. D. Kahn (ed.), Investigating the Universe, 385–422.

(i) interacting galaxies with pronounced HII regions of varying sizes;
(ii) nests of 3 or more nearly touching galaxies;
(iii) dumb-bell systems;
(iv) those having the appearance of ejected twins or single satellites the latter including "bottle" forms;
(v) M51 types being spirals with a satellite;
(vi) spirals with elliptical-like associations the latter usually somewhat larger than in (v);
(vii) chains of galaxies;
(viii) coalescent pairs;
(ix) parallel flat systems and systems at angles;
(x) "mice" objects with, for example, long straight tails;
(xi) "antennae" : double systems with symmetries in the tail structures "V" shaped with the two galaxies at the vertex;
(xii) ring galaxies with a central core and/or companion;
(xiii) "comet" structures;
(xiv) galaxies with long jet-like tails with no visible companion : mice-like;
(xv) systems with "tidal streams" beside spiral arms.

Of these categories it is unclear whether (xiii) and (xiv) are interacting galaxies and whether (xv) are not just special cases of (x) or (xi).

Restricted Three Body Calculations and Vorontsov-Velyaminov's Assertions

Vorontsov-Velyaminov (1961) drew attention to the fact that connecting tails and filaments between galaxies consist mainly of stars with some gas. He further argued that these appendages are not tidal in origin because:
1. Tails occur more frequently than bridges; Wright's calculations (1972) confirm this.
2. Tails are often thin and sometimes are long, as long as 100kpc. The calculations of Wright (1972) and Toomre & Toomre (1972) do not agree with this.
3. There are parallel pairs of galaxies.
4. The percentage of interacting pairs is the same as or lower than for field galaxies.
5. Many galaxies have close neighbours but few show evidence of tidal interaction. Calculations e.g. Wright (1972) showed that distortion is critically dependent on the distance of closest approach.

No statistical evidence was listed for the above but many of these points were later taken up by Wright (1972) in his restricted 3-body simulations of interacting galaxies in which the masses of the two galaxies were concentrated at two points and massless particles placed in purely circular orbits around each mass with their Keplerian velocity. As Wright points out until this time our intuitive understanding of tidal effects was based on the solar system where one body is distant and small and, to this one might add, largely non-deformable. Wright's work on interaction between disk galaxies led him to conclude:

1. Tails normally form when the orbital and spin angular momentum are parallel and the interaction is with a galaxy of similar mass passing almost parabolically and close.
2. Tails form more easily than bridges, and live longer and related to this when both form, the tail mass is greater and volume less than that of the bridge.
3. Low velocity hyperbolic encounters lead to weak short-lived tails.
4. Retrograde interaction where the system angular momentum is opposite to the spin of the individual galaxies does not produce tails or bridges; inclinations which are not co-planar do however.
5. Close interaction between galaxies of comparable mass can lead to disruption and escape of disk material. Tails formed by non-coplanar interaction tend to lie in the plane of the parent galaxy regardless of the direction of the other galaxy's orbit.
6. The mass escaping a binary interaction is 10 to 20% of the total.

There are several important limitations in the approach of Wright: the calculations are not self-consistent (the stars are massless), the initial velocities of the stars have no dispersion, the model galaxies did not have haloes and the potentials are of point rather than distributed masses. Subject to these reservations Wright's work has demonstrated contrary to Vorontsov-Velyaminov's assertions that long tails and filaments can be tidal.

The work of Toomre & Toomre (1972) is particularly interesting in that they attempt to model specific observed pairs of galaxies. Here the approach is essentially that of Wright (1972) namely restricted 3-body with massless stars in disks. The four systems they looked at were: Arp 295,M51-NGC5195, the 'Mice': NGC4676 and the 'Antennae': NGC4038.

Their simulation of Arp 295 is one of a disk galaxy whose companion has a quarter the mass and is in direct motion with the disk; the inclination of the orbital plane is 15°, with closest approach 45° after crossing the disk plane. This produces a tilted bridge in the direction of the companion and a counterarm in the original disk plane. One cannot say looking at their simulations that the bridge has anything like the collimation of Arp 295. Furthermore, the fact that the companion was not itself given test particle stars leaves one unclear about the effect of the main galaxy on the companion. Another difficulty is with the neglected velocity dispersion: if this is 10% of the rotational velocity of 200km/sec say then in the 3.9×10^8y of the Toomre model and without self gravity, within the bridge the structure would by now have extended to a total thickness of about 8kpc.

The second case Toomre and Toomre examined is the classical M51 - a spiral with satellite. Their conclusion is that the spiral is close to face-on and that the satellite's orbit is highly eccentric and highly inclined. The time since pericentre is 5×10^6 years and the orbital period is 2.5×10^9 years. The spins of the two galaxies must be almost anti-parallel. They believe the satellite to have not more than one-tenth the mass of the spiral although self-gravity could be important in deciding this question. Extensive kinematical and dynamical work has since been carried out by Tulley (1974, a,b,c) who has found that most of the observations are consistent with the tidal hypothesis of Toomre & Toomre.

The third case simulated by the Toomres was Arp 242 : the 'Mice'. The very long straight tail of one is interpreted as a counterarm of one galaxy seen edge-on with the counterarm of the other galaxy seen closer to face-on as more obviously arched. Their model does however disagree with the somewhat uncertain measurement of the Burbidges (1961) of one galaxy's velocity relative to the other.

The last and most impressive of their simulations is of the 'Antenna' system, NGC4038/9, in which identical disk galaxies with 60° inclinations to the orbital plane are now seen having gone well past the pericentre with well developed counterarms thrown by disk rotation and tidal effects both above the orbital plane. If this is viewed looking in the orbital plane it is seen to resemble the observed system. Huchtmeier and Bohnenstengel (1975) showed that there is about $1.2 \times 10^{9} M_{\odot}$ of HI concentrated along the antennae i.e. about one third of the total for the system of galaxies.

Burbidge and Burbidge (1959), Theys and Spiegel (1977), Toomre (1978) and Eneev et al. (cited Toomre, 1977) have attempted to explain ring galaxies in terms of the penetration of a disk galaxy by a parabolic intruder causing a 'splash' ring in the disk; penetration of the disk is essential for the ring to develop. The development of the spokes is not explained by the 'splash' model being viewed instead as a gas dynamical response rather than stellar. Coherence of the rings will be reduced by incorporating random motion, but if self-gravity is also included the effect may be to maintain coherence.

Dostal and Metlov (1977), on the basis of the data of Vorontsov-Velyaminov and Krasnogorskaya (1962), estimate that the probability of occurrence of ring galaxies does not seem to depend on cluster membership. They argue that the probability of a central collision is 100 times less than that required to produce the observed frequency and they advocate looking for internally produced rings believing that the presence of a companion may then be relevant in a different way.

The work of the Toomres (1972) and Wright (1972) (see also Keenan and Innan, 1975) implies a rather less exotic interpretation of bridges and tails than that of Gold and Hoyle (1952) and Pilkelner (1968) involving magnetic fields, although one is caused to reflect on the observations noted in Arp (1966) that one of the examples Arp 295 chosen by the Toomres has a polarized bridge.

Of some 300 cases in Arp's atlas of peculiar galaxies (1966), the Toomres (1972) were able to simulate about 10%. Toomre (1974) points out that the self-consistent work of Clutton-Brock (1972) suggests longer lived bridges than are obtained in the restricted 3-body approach.

N-Body Simulations ($N > 2$)

The inadequacies of the restricted three-body approach have already been alluded to.

White (1978, 1979 and 1979a) has simulated colliding galaxies self-consistently using 250 particles for each galaxy. He found for parabolic collisions that rapid merging took place and that the products resembled elliptical galaxies. The inclusion of spin gave merged products whose structures depended only weakly on those of the progenitors. The rotation of the merged products was however greater than that observed in ellipticals, the best agreement being merging following head-on collision.

Miller (1980) has simulated the collision of two initially spherical non-rotating galaxies each of 50,000 stars with relative velocities of the order of 500km/sec. The galaxies had initial density profiles appropriate to $n = 3$ polytropes. He finds (a) galaxies contract to about half size shortly after closest approach implying (b) that the galaxies have ample time to respond, (c) the energy of bulk motion is transferred into ordered flow within each galaxy, (d) the impulsive approximation is invalid at velocities of approach of less than 1000km/sec.

From what has been said above, it is clear that the case for tidal interactions between galaxies is a strong one and that detailed calculations, particularly self-consistent simulations, are necessary to assess the effects of large scale deformations.

Experiments with Colliding Galaxies

Let us turn now to the results of some recent experiments performed by the author.

We shall consider the following cases involving direct collisions between equal mass galaxies each comprising 10^4 particles, initially uniform in density, spheroidal in shape and having angular momentum consistent with an oblate Maclaurin spheroid of axial ratio 0.6:
(1) A fast direct collision between 2 galaxies with spins parallel to the line of collision;
(2) A slower, direct penetration than (1) leading to merging, with the initial rotation of the galaxies parallel and anti-parallel to each other but with each galaxy's spin lined up with the line joining the centroids.
(3) A moderately fast collision between two galaxies with parallel spin vectors which are perpendicular to the line joining the initial positions of the centroids of the two galaxies.

Various physical parameters are quoted in normalized units such that the gravitational constant $G = (4\pi)^{-1}$, each of the 10^4 particles used to simulate a galaxy is taken to have unit mass and the calculations were performed in a cubic box divided into 32^3 mesh intervals each of unit length.

The integration procedure used to follow the evolution was as follows:

(1) Prescribe the particle positions, therefore the mass distribution, and the velocities.

(2) Using the 'cloud in cell' method, determine the density variation throughout the cubic space.

(3) Invert Poisson's equation to determine the gravitational potential, by taking fast Fourier transforms of the three-dimensional central difference operator - the approximation to the Laplacian operator of Poisson's equation.

(4) Obtain the gradient of potential and hence the force on each particle.

(5) Determine the change in velocity and position of each particle then return to step 1.

This method is described in detail elsewhere by Stewart (in preparation). Given below are the factors by which the normalized quantities referred to in the work below should be multiplied, for galaxies of specified total mass and scale length in order to obtain the real values of the physical variables. If M is the mass of a galaxy simulated by N particles each of mass $M_* = M/N$ and if the real length scale of one of the 32 intervals in each direction along the mesh is L, then if the normalized variables corresponding to time, distance, velocity, energy, moment of inertia and angular momentum are τ, x_n, v_n, E_n, I_n and h_n, respectively, the real variables are:

Time : $t = 4.2 \times 10^7 M_1^{-\frac{1}{2}} L_1^{3/2} \tau$ years

Distance : $x = L_1 x_n$

Velocity : $v = 23.6 M_1^{\frac{1}{2}} L_1^{-\frac{1}{2}} v_n$ km/sec

Energy: : $E = 1.1 \times 10^{53} M_1^2 L_1^{-1} E_n$ ergs

Moment of Inertia : $I = 1.8 \times 10^{83} M_1^2 L_1^{\frac{1}{2}} I_n$ gm cm^2

Angular Momentum : $h = 1.4 \times 10^{67} M_1^{3/2} L_1^{\frac{1}{2}} h_n$ gm cm^2 sec^{-1}

where M_1 is the mass M in units of $10^{11} M_\odot$ and L_1 is the length scale L in kpc.

In these normalized units each galaxy was given an initial geometric mean diameter of 6.0 corresponding to a minor axis of 4.3 and major axis of 7.1.

(1) Fast Collision of Parallel Spinning Galaxies

Figs 1 & 2 show the effect of colliding spheroids which are initially co-rotating along the line of collision. The galaxies are given a relative velocity of approach of $16.5\sqrt{3}$ at a separation of $8.6\sqrt{3}$. Fig 1.1 and 1.2 show the initial configuration $\underline{\ell} \perp \underline{\Delta r}$ and $\underline{\ell} \parallel \underline{\Delta r}$ where $\underline{\ell}$ is the line of sight direction and $\underline{\Delta r}$ is the difference between the initial position vectors of the centroids of the two galaxies. Close to collision at $\tau = .44$, the combined system ($\underline{\ell} \perp \underline{\Delta r}$) is shown in Fig 1.3. Galaxy 1, Fig 1.4, is moving to the left and a somewhat asymmetrical shape has developed (Gal 2 Fig 1.5). For $\underline{\ell} \parallel \underline{\Delta r}$ Fig. 1.6 shows some heating; relaxation of the outer region has occurred. At $\tau = .84$, the galaxies have passed through each other and are now at about the same separation on opposite sides as initially (see Fig 2.1 for $\underline{\ell} \perp \underline{\Delta r}$ and Fig 2.2 for $\underline{\ell} \parallel \underline{\Delta r}$). By $\tau = 1.24$, $\underline{\ell} \perp \underline{\Delta r}$, the effects of heating are evident: asymmetric forms with the leading head being the flat side (see Fig 2.3). Fig 2.4 shows both galaxies, $\underline{\ell} \parallel \underline{\Delta r}$, and compared with Fig 2.2 at $\tau = .84$ a general expansion is evident. By $\tau = 1.64$ the asymmetry is less pronounced for $\underline{\ell} \perp \underline{\Delta r}$ (see Fig 2.5) but the haloes in Fig 2.6 have developed tenuous edges.

(2) Slower Direct Penetration leading to Merging

In this set of experiments, the galaxies were released from rest with their spin axes parallel each passing through the other's centroid. The initial separation was $5\sqrt{3}$ units. The evolution was followed until $\tau = 6.56$. An initial increase in the internal energy by a factor 2.5 (Fig 15) over the initial value occurred with the maximum at the minima in the principal moments of inertia - full contraction; the increased internal kinetic energy leads to expansion with maxima in the inertia moments at $\tau = 1.5$ occurring at the minimum in internal kinetic energy (Fig 15). At this point it is I_{diag}, the moment of inertia of each galaxy about an axis parallel with $\underline{\Delta r}$, which is greatest i.e. the greatest expansion occurs at right angles to the line of spin and collision. As the system continues to oscillate a halo develops and I_{diag} decreases while $I_{\perp}$, the moment of inertia perpendicular to $\underline{\Delta r}$, increases. This is reflected in the Figs 3-8 where head and tail sectors develop along the line of collision. Beyond $\tau = 4.0$, I_{diag} follows, in anti-phase, the internal kinetic energy. Variation in $I_{\perp}$ seems less pronounced suggesting that it is the dimensions perpendicular to the line of collision that undergo the greatest variation. The terminal mean internal kinetic energy is 1.0×10^6 being only 25% greater than the original.

Fig 3.3 shows galaxies 1 and 2 at $\tau = .84$ and Figs 3.4 and 3.5,1 and 2 individually. The centroids of galaxies 1 and 2 are almost coincident at this first collision. Each galaxy has a conical shape. At $\tau = 1.68$, Fig 4.1 gives the combined appearance of reflected wings. Fig 4.2 shows that a central condensation and a conical halo have developed. Fig 4.3 gives galaxy 2 for $\underline{\ell} \perp \underline{\Delta r}$; Fig 4.4 shows core-halo structure of 1 and 2 (compare Fig 3.2 and .6 at $\tau = 0$ and .84, respectively). Fig 4.5 shows, at $\tau = 2.48$ how the conical structures have partially consolidated into disks. Fig 4.6 shows that the central core of galaxy 1 has become more elongated along $\Delta\underline{r}$ (e.g. Fig 4.3 at $\tau = 1.68$). Fig 5.1 ($\underline{\ell} \parallel \Delta\underline{r}$) is little different from Fig 4.4. Figs 5.2-.5 are taken at $\tau = 3.28$. Particularly noticeable are the more diffuse halo and more condensed core of the combined system. Fig. 5.3 and .4 show galaxy 1 and 2 respectively with slightly curved disks to the left and right respectively. This is more pronounced at $\tau = 3.28$ where the combined system looks like core, inner halo and disk (see Fig 5.6) and each galaxy, e.g. galaxy 1, see Fig 6.1, has a concentrated arc to the left of its core with an outer disk (see Fig 6.2 for galaxy 2). Fig 6.3 shows 1 and 2, the combined system ($\underline{\ell} \parallel \Delta\underline{r}$), indicating less even filling of the halo whose edge is now less clearly delineated. Fig 6.4 shows both galaxies at $\tau = 4.12$ (c.f. Fig 5.6 at $\tau = 3.28$) and indicates that many particles have gone from the disk, and (Fig 6.5) what was a fairly compact arc on the left side of galaxy 1 is now slightly more diffuse with a slight filling on the right hand side of the core having taken place. Fig 7.1 shows both galaxies, $\underline{\ell} \perp \Delta\underline{r}$, at $\tau = 4.96$ indicating a faint suggestion of a central "x" shaped halo which for galaxies 1 and 2 seen separately is almost separate from the left hand side of the core of "1" in Fig 7.2 and right hand side of core "2" in Fig 7.2. Fig 7.4, at the same time, shows, for $\underline{\ell} \parallel \Delta\underline{r}$, that both galaxies (c.f. Fig 6.3 at $\tau = 3.28$) have more condensed haloes than at $\tau = 3.28$. Fig 7.5, for $\underline{\ell} \perp \Delta\underline{r}$, shows both galaxies at $\tau = 5.76$ (c.f. Fig 7.1 at $\tau = 4.96$): the "x" shape has gone; for galaxy 1 only, see Fig 7.6 (c.f. Fig 7.2 at $\tau = 4.96$); is now slightly more diffuse in the outer sectors. Fig 8.2 shows the combined system for $\underline{\ell} \parallel \Delta\underline{r}$ (c.f. Fig 7.4 at $\tau = 4.96$) with a slightly less diffuse halo. At $\tau = 6.52$, Fig 8.4 shows $\underline{\ell} \perp \Delta\underline{r}$ for galaxy 1 and Fig 8.5 for galaxy 2. The core is now well formed with a faint halo. Fig 8.3 at $\tau = 6.52$ shows that the merged system is somewhat prolate caused by the equal and opposite pairs of sectors which are symmetrical about $\Delta\underline{r}$. Fig 8.6 shows a tighter halo formation than at $\tau = 4.96$ (see Fig 7.4) as is reflected in the difference between I_{diag} at those two times.

Another experiment was run, in which the two galaxies instead of spinning in parallel were taken anti-parallel. Although it is reasonable to suppose the results might be the same it is not clear whether say a collective instability might not occur. In the anti-parallel case, bulk potential and kinetic energy, moments of inertia, internal kinetic energy and overall configurations (for the latter, see Fig 7a compared frame for frame with Fig 7 for the same time) were identical within the statistical fluctuations. There is however an important difference in the parallel case: the net angular momentum of the merged system is the sum of the intial system angular momenta i.e. non-zero, whereas in the antiparallel case the net spin is zero. Figs 9, 10 & 11 and 12, 13 & 14 give the spectral distributions for $\underline{\ell} \perp \Delta\underline{r}$ for galaxies 1, 2 and 1 + 2 for the parallel and anti-parallel cases, respectively. Whereas in the parallel case, significant mean rotational velocities are observed in the merged systems (see Fig 11), the counter-streaming system, Fig 14 gives a negligible difference from zero at least in the inner boxes: (x,y) = (2,1), (2,2), (2,3), (2,4). Because of the anti-symmetry that develops in the outer regions of each galaxy, a rather delicate measurement of the stars mean velocity $\bar{v}$ here might reveal whether galaxies are products of oppositely, or nearly so, rotating systems. For example with reference to Fig 12, for galaxy 1, boxes (1, 1) and (4, 4) have $\bar{v} = -3.61 \pm 1.31$ and $+2.97 \pm 1.13$ containing .6% and 1.1% of each galaxy's stars, respectively, whereas Fig 13, for galaxy 2 have $\pm\ 3.65 \pm .73$ and -4.74 ± 1.76 with 1.2% and .4% respectively. The overall system as seen by an observer, $\underline{\ell} \perp \Delta\underline{r}$, gives, for boxes (1, 1) and (4, 4), $\bar{v} = +1.16 \pm .88$ and $.84 \pm 1.15$, i.e. zero, see Fig 14. Fig 11, for the combined system gives $\bar{v} = -3.96 \pm .88$ and $+4.98 \pm .76$ for the parallel case. Thus we see that by looking at the wings of a merged galaxy there may be a hint as to its origin i.e. whether the galaxies were originally counter-rotating.

Let us now turn to other properties: The principal moments of inertia of the individual galaxies were

initially	$\tau = 0$	$I_{diag} = 1.00 \times 10^5$
		$I_{\perp} = 9.11 \times 10^6$
and at	$\tau = 6.56$	$I_{diag} = 0.93 \times 10^6$
		$I_{\perp} = 1.90 \times 10^6$

This can be interpreted in the following way in terms of the shape of each galaxy at the end of the run compared with the start: if $\langle z^2 \rangle$ is the mean square dimension along the rotation axis and $\langle r^2 \rangle$ that for the radius perpendicular to z, then $\langle z^2 \rangle^{\frac{1}{2}}/\langle r^2 \rangle^{\frac{1}{2}} = .4$ at $\tau = 0$ for each galaxy and 1.2 at $\tau = 6.56$ i.e. each galaxy has become prolate; because of the negligible separation between centroids at $\tau = 6.56$, 1.2 is also the value for the merged system so that we see the system is marginally prolate at the end. Turning now to mass loss, ε, at $\tau = 6.56$: beyond r = 10 units $\varepsilon = 3.4\%$ and beyond r = 3 (the initial geometric mean radius) $\varepsilon = 38\%$ - here r is measured relative to the centre of the merged system.

The density run $\rho(r) \propto r^{-n}$ for the merged system, within r = 5 at $\tau = 6.56$, is given in Table I.

Table I

r	n
2	.2
3	1.3
4	3.9
5	7.4

The half density point is at about r = 2.5.

Finally, it is of interest to note that the internal kinetic energy of each galaxy stabilizes at a level about 25% above the initial value.

The total energy at $\tau = 6.56$ differed from that at $\tau = 0$ by <.3% and angular momentum <2.5%. The initial and final angular momenta were within 1°.1 of each other. The total number of stars lost from the mesh was less than 1%.

(3) Collision between parallel spheroids with spin perpendicular to the line of collision

Consider now the case where the galaxies initially have parallel angular momenta perpendicular to the line of collision. The galaxies having been released with a relative velocity of approach of $8\sqrt{3}$ with a separation of $8.6\sqrt{3}$, collide at $\tau = .74$, see Fig 16. Of particular interest is the way in which the motion of the centroids of the galaxies changes from rectilinear before $\tau = .74$ to curved thereafter. Fig 17 shows the paths of the centroids of the two galaxies with "1" starting from the right at $\tau = 0$; this view is taken looking in the direction of spin; the centroid of the whole system remains fixed for all τ by the conservation of momentum. The angular momentum $\underline{h}_1$ of individual galaxies about their respective centroids decreases monotonically (see Fig 16) so that while the contribution to the total system angular momentum comes initially from the spin of each galaxy, after collision this total angular momentum arises from the rotation of the centroids of each galaxy about the system centroid. The initially identical and smaller moments of inertia split and one moves in value closer to the initially larger inertia associated with the spin vector. Fig 16 shows I_1, I_2 & I_3 relative to an isolated Maclaurin spheriod. At $\tau = 1.5$, the spin and principal axis of the largest moment of inertia are within 3° of each other.

Fig 18.5 shows the configuration of the combined system looking along the spin axis at $\tau = .80$ i.e. almost at the coincidence of the two centroids. The configuration is completely symmetrical. Fig 18.6 shows galaxy "1" only; the centroid of "1" is slightly to the left of "2" at this time; the line of collision is perpendicular to the line of sight. Fig 18.3 shows the view at the same time with the line of sight perpendicular to both spin and $\underline{\Delta r}$. There is slight asymmetry in the individual galaxies, see Fig 18.4, with the leading (left) edge slightly denser. At $\tau = 1.00$, the galaxies have emerged from the encounter in contracted states (see Fig 19.5 looking down the spin axes). The leading edge, which is smaller than the rear end at $\tau = .8$, has expanded and is now larger than the rear end, which in turn has contracted. By $\tau = 1.32$ (see Fig 20.5) a symmetrical pattern has emerged looking along the spin axis; Galaxy "1" is now on the left. A fan (c.f. Fig 20.3, where the line of sight is perpendicular to the spin and $\underline{\Delta r}$) has deveoped, which has expanded further by $\tau = 1.64$ (see Fig 21.5) showing a density gradient. Views along the line of collision, Figs. 18.1, 19.1, 20.1 and 21.1 and perpendicular to both the spin and $\underline{\Delta r}$, Figs 18.4, 19.4, 20.4 and 21.4 at $\tau = .8$, 1.00, 1.32 and 1.64, respectively, show that a bar develops with the minor axis parallel with the spin axis. The bar is tapered at both ends from the latter view-point and at the end corresponding to the curved part of the fan in the former case.

Figs 22a and b show for galaxy "1" a sample of orbits of a number of stars. "1" and "2" mark the initial positions of the centroids of galaxies "1" and "2", respectively and the initially horizontal line emanating from just above "1" is the locus of the centroid of "1" and the other lines are the paths of stars with marks placed at regular intervals of time, $\Delta\tau = .08$. What is evident is that the stars with initially pronounced downward motion move down across the line of collision and under the restoring force of galaxy "1" and the increasing force of "2" are pulled back in a wide arc to the other side of the line again; such particles have, in the process, picked up kinetic energy and will in general move further out from "1". Stars which are further away from the line of collision when "1" and "2" overlap, will experience the greatest force from "2"; their energy increases will be greater, so that after collision these particles will move further out. This motion has the net effect of producing a rotating pattern in that the principal axes associated with the smallest two moments of inertia rotate about an axis which is almost exactly parallel with the angular momentum axis of the individual galaxies. Fig 23 shows the change θ in the direction of these principal axes relative to the initial value; it remains roughly constant until the galaxies overlap and then the axes start rotating rapidly. This can be seen by comparing of Figs 18.6 with 19.6. The pattern shown by galaxy "1" has clearly rotated in a clockwise sense between $\tau = .8$ and 1.0. The maximum rate of change of θ - the pattern speed, Ω_p - is reached at $\tau = 1.0$ close to the time for maximum internal kinetic energy; thereafter, θ starts swinging back. It is interesting to compare the pattern speed, Ω_p, with the material rotation rate, ω_m, which can be defined by the relation-

$$\omega_m \underset{\sim}{I} = \underset{\sim}{h}_1$$

where $\underset{\sim}{I}$ is a principal moment of inertia whose dierection is parallel with the angular momentum of galaxy "1" about its centroid. ω_m does in fact follow a similar but numerically different variation from that of Ω_p; the maximum value of $\Omega_p = 4.8$ and of $\omega_m = 2.6$. The maximum in ω_m^p occurs because of the severe contraction at the point of overlap of the two galaxies and the decrease in ω_m to zero is a consequence both of the rapid expansion of the galaxies and angular momentum decrease after collision. The rotation of the principal axis is about an axis which is within 2° of the angular momentum axes of the individual galaxies.

Discussion

Very little analytical work has been done on self-consistent models of interacting galaxies. Bisnovatyi-Kogan (1974) has however shown that it is possible to construct from two integrals of motion a self-consistent steady-state system of cold ellipsoids rotating such that the principal axes of the two galaxies are always aligned with each other. This is probably descriptive of a minority of interacting galaxies however and certainly not of galaxies with separations comparable with their sizes. Another approach is the impulsive approximation which enables one to estimate the total change in internal kinetic energy of an interacting system on the assumption that the relative velocity of the colliding galaxies is much greater than the internal velocities. In this approximation, the internal kinetic energy changes instantaneously. There is nc discussion in this approximation of the subsequent motion of the stars in the galaxy, which will obviously be one of implosion, at least in the first instance. The impulsive calculation has been carried out by Spitzer (1958) in the case of distant encounters and by Toomre (1977) for the case of direct collision of two identical $n = 5$ polytropes. Spitzer refined his impulsive calculation for distant encounters to allow for a slow passage by an intruder whose effect was considered to be a perturbation to the galaxy whose internal energy change was being sought. His result is of some interest in that, as one would expect, if the typical time for the intruder to pass the galaxy is of the order of the crossing time for a star in the galaxy, then a resonance occurs and the total change in the internal energy can be higher than that obtained in the impulsive approximation by as much as 15%. His calculation depends essentially on the smallness of the ratio $\lambda = (M_N/M_c)(r_c/p)^3$, where M_N is the mass of the intruder, M_c the mass of the galaxy, r_c the radius of the galaxy and p the distance of closest approach of the intruder. If however the masses of the galaxies are comparable and the distance of closest approach is comparable with the size of the galaxy then the ratio λ is no longer small. Indeed, the findings of Wright (1972) and the Toomres (1972) indicate that in order to obtain significant tidal effects this ratio should be of order unity. There are three possible regimes for the initial orbits prior to the collision of galaxies: bound, parabolic and hyperbolic; the latter is essentially the impulsive case. An interesting question is just how closely the impulsive calculation predicts the energy change for orbits which are bound or parabolic. Using Toomre's calculation (1977) for identical colliding $n = 5$ polytropes, the total change, ΔT, in the internal energy of a galaxy is

$$\Delta T = G^2 M^3/3U^2 a^2$$

where G is the gravitational constant, M is the mass of each colliding galaxy, U is the relative velocity and a is a characteristic radius. For a polytrope of this order, the radius within which half the mass of the galaxy is contained is 1.3a. A uniform sphere of radius a_u will have a half-mass radius of $.8a_u$. If the equivalent value of a is obtained by equating 1.3a with $.8a_u$, then in the dimensionless units $G = 1/4\pi$, $M = 10^4$, an estimate of ΔT may be obtained for these experiments by specifying U which we shall take to be the average of the speed at the start of the experiment and that at the point when the galaxies centroids first coincide. If this is done for experiments (2) and (3) the values of ΔT are 4.3×10^6 and 1.3×10^6 respectively compared with the increases obtained in numerical experiments of 1.4×10^6 and $.9 \times 10^6$. It is worth noting that in an experiment on colliding non-rotating spheres performed by the author (in preparation) for initial bulk motion and centroidal separation identical with that described in experiment (3) above, the value of ΔT obtained from the simulations was 1.2×10^6 much closer to the above-mentioned estimate than that obtained for the rotating case. In that paper the effect of spin orientation of the colliding galaxies on energy changes is described in detail. The ratio of the initial r.m.s. speed of the particles in each galaxy to the average relative speed of the two galaxies was .6 and 1.0 for experiments (2) and (3), respectively. From this we see that there is reasonable agreement between the energy estimates of internal energy change using the impulsive calculation when the ratio of dispersion to relative bulk velocity is of order of unity, but for lower values of about .6 the initial change in energy is well outside this estimate, as one would expect. The immediate consequence of this increase in kinetic energy is an expansion of the galaxies which in turn leads to a loss in internal kinetic energy. Experiment (2) was followed for a time sufficient to ensure merging during which period there were several expansions and contractions of the constituent galaxies with the internal energy finally settling down at a level which was about 25% higher than the initial value, a value considerably lower than the initial impulsive change.

Finally, to provide a numerical appreciation of the implications of these experiments, if the galaxies have each a mass of $5 \times 10^{11} M_\odot$ and a radius of 15kpc then the initial internal r.m.s. speed is 280km/sec.

Conclusions

The simulation of isolated and interacting galaxies using large numbers of particles promises to reveal much about the properties of deformable bodies which the majority of work described in the review is unable to do. Calculations which are self-consistent of the kind shown in the second part of this paper use so many particles that binary heating, so much a problem in the direct methods of simulation, occurs on time scales well outside any of interest in the majority of astronomical problems (times greater than 250 crossing times). The very large number will, in addition, help to define the morphology of interacting galaxies more clearly at least when taken in conjunction with modern visual display methods. One of the drawbacks of the indirect method used by the author involving fast Fourier transform techniques (see James, 1977) is that the space is restricted and, with present procedures, inadequate for interactions occurring over more extended regions of space than 2 or 3 times the galactic radii. There is, nevertheless, a large parameter space of initial motion and mass distribution to be investigated before this method is likely to be worked out.

With the fully self-consistent calculations we are at last able to see how tidally induced deformations in their turn lead to further modes of behaviour of galaxies. The rôle and limitations of the violent relaxation theory and the validity of the assumption of an achievable static virial equilibrium in both the scalar and tensor forms can now be tested.

It is to be hoped that the judicious choice of computer simulations will provide insights and finally a solid analytical basis for phenomena that are essentially non-linear and, as yet, outside of our experience in the dynamical sense.

Acknowledgement

The author wishes to thank Dr. R.A. James for discussions on a variety of computational problems and in particular for providing him with some very finely tuned fast Fourier transform routines. The difference scheme differs however from that currently used by Dr. James. I wish also to thank Mr. J. Murphy for his assistance with the spectral display programme and the S.R.C. for his support and finally Prof. F.D. Kahn for suggesting improvements to the manuscript.

References

Arp, H., 1966, *Atlas of Interacting Galaxies*, California Institute of Technology.
Burbidge, E.M., & Burbidge, G.R., 1959, Ap.J. 130, pp.12 & 23.
Clutton-Brock, M., 1972, Ap.Sp.Sci., 17, p.292.
Curtis, H.D., 1921 Bull. Nat. Res. Council Washington, 2, Pt.3, p.171.
Dostal, V.A. Metlov, V.G., 1977, Sov.A.J., 23, p.1.
Gold, T., Hoyle, F., 1959, *Paris Symposium on Radio Astronomy*, I.A.U. Symp. No.9 ed. Bracewell, R.N.
Holmberg, E., 1937, Ann. Obs. Lund., No.6.
Hubble, E.P., 1936, *The Realm of the Nebula*, Yale University Press.
Huchtmeier, W.K., Bochnenstengel, H.D., 1975, Astron. & Astrophysics., 41, p.477.
James, R.A., 1977, J. Comp. Phys., 25, p.71.
Jeans, J.H., 1928, *Astronomy & Cosmogony*, p.397, C.U.P.
Kennan, D.W., Innan, K.A., 1975, A.J., 80, p.290.
Lundmark, K., 1926, cited Zwicky, 1956.
Miller, R.H., 1980, Ap.J., 235, p.421.
Pease, F.G., 1917, Ap.J., 46, p.24.
Pikelner, S.B., 1968, Ann. Rev. Astron. & Ap., 6, p.165.
Spitzer, L., 1958, Ap.J., 127, p.17.
Theys, J.C., Spiegel, E.A., 1977, Ap.J., 212, p.616.
Toomre, A., 1974, *The Formation & Dynamics of Galaxies*, I.A.U. Symp. No. 58, p.347, ed: Shakeshaft, J., Reidel.
Toomre, A., 1977, *The Evolution of Galaxies & Stellar Populations*, p.401, ed: Tinsley, B.H., Larson, R.B., Yale University Observatory.
Toomre, A., 1978, *The Large Scale Structure of the Universe*, I.A.U. Symp. No. 79, eds: Longair, M.S., Einsato, J., Reidel.
Toomre, A., Toomre, J., 1972, Ap.J., 178, p.623.
Tulley, R.B., 1974a, Ap.J. Suppl. Ser., 27, p.415.
Tulley, R.B., 1974b, Ap.J. Suppl. Ser., 27, p.437.
Tulley, R.B., 1974c, Ap.J. Suppl. Ser., 27, p.449.
Vorontsov-Velyaminov, B.A., 1959, *Atlas & Catalogue of Interacting Galaxies*, PtI Moscow University.
Vorontsov-Velyaminov, B.A., 1961, *Problems of Extra-galactic Research*, I.A.U. Symp. No.15, ed: McVittie, G.C.
Vorontsov-Velyaminov, B.A., Krasnogorskaya, A.A., 1962, *Morphological Catalogue of Galaxies*, 1, Moscow University Press.
Vorontsov-Velyaminov, B.A., 1977, Astron. & Astrophys. Suppl. Ser. 28, p.1, *Atlas of Interacting Galaxies*, Pt2.
White, S.D.M., 1978, M.N.R.A.S., 184, p.185.
White, S.D.M., 1979, M.N.R.A.S., 189, p.831.
White, S.D.M., 1979a, Ap.J., 229, p.L9.
Wright, A.E., 1972, M.N.R.A.S., 157, p.309.
Zwicky, F., 1956, *Ergenbnisse der Exakten Naturwissenschaften*, 29, p.344. Springer Verlag.

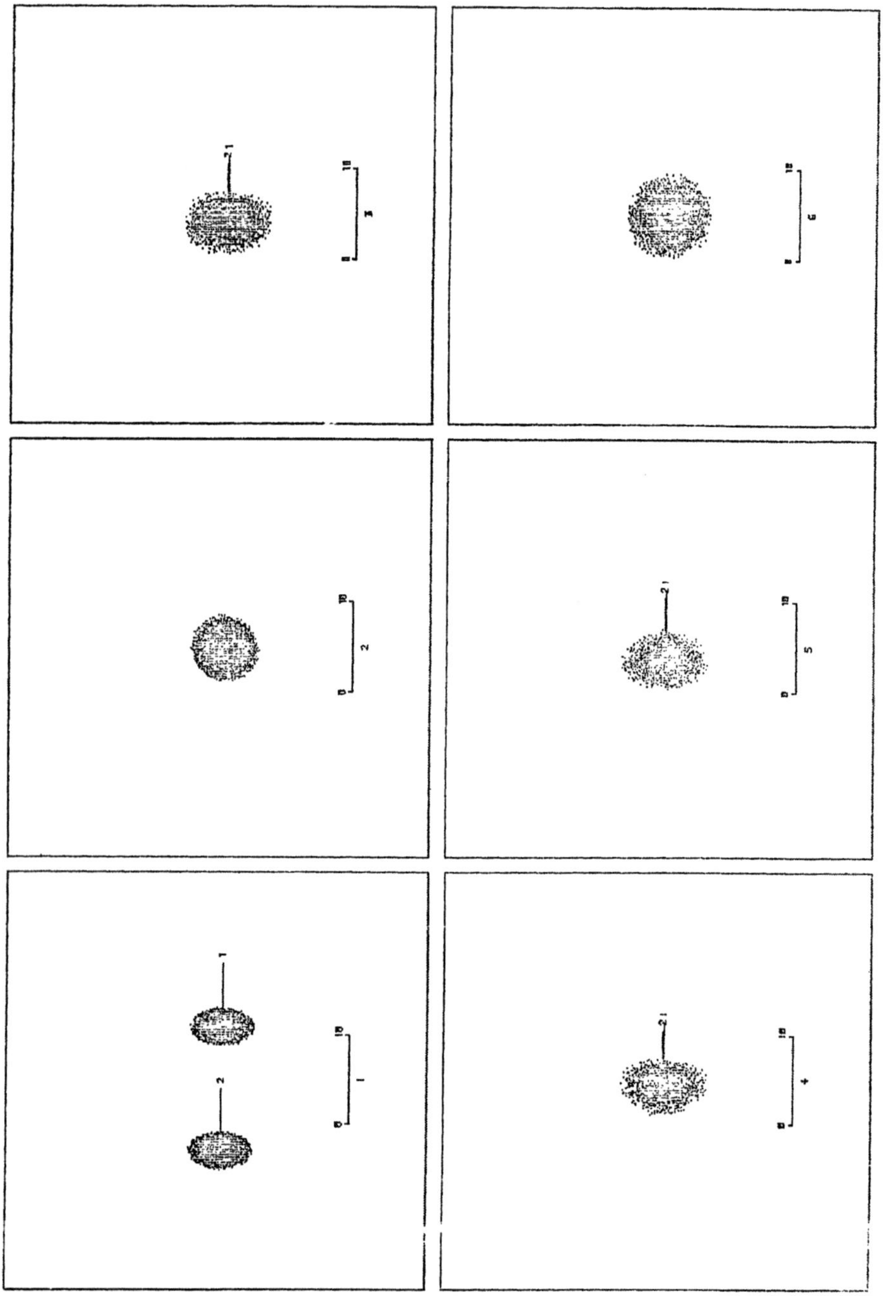

Figure 1. Fast Direct Collision

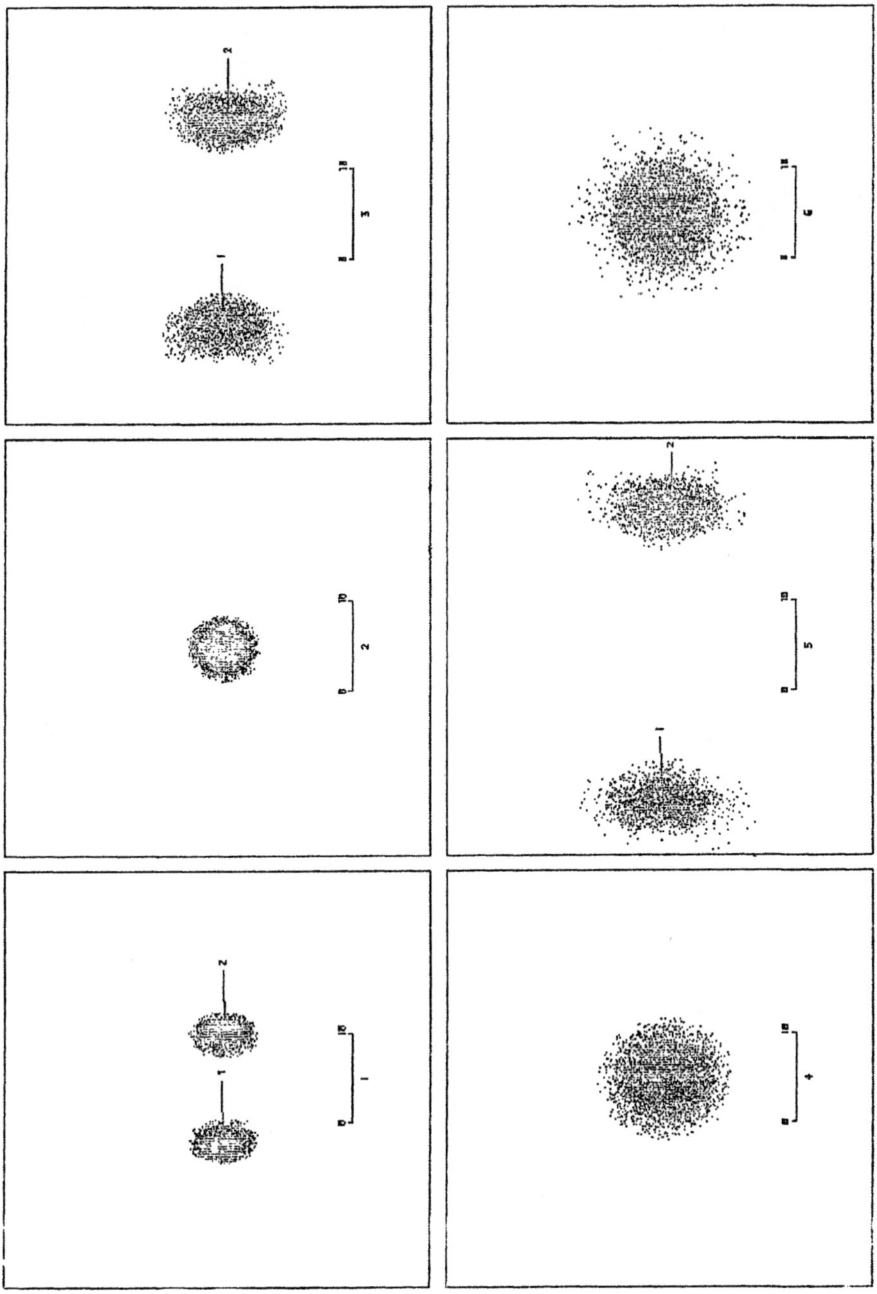

Figure 2. Fast Direct Collision (cont.)

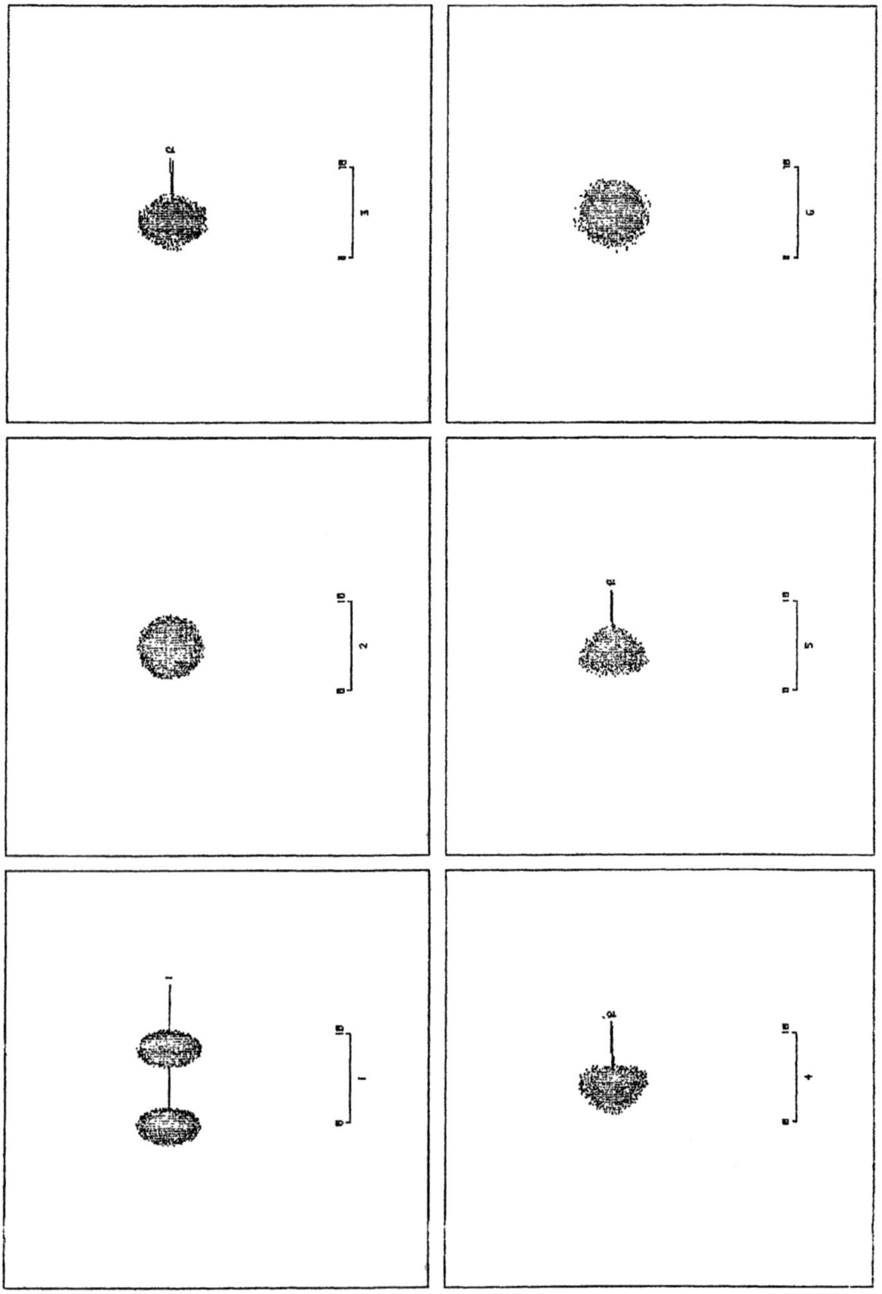

Figure 3. Slow Direct Collision & Merging

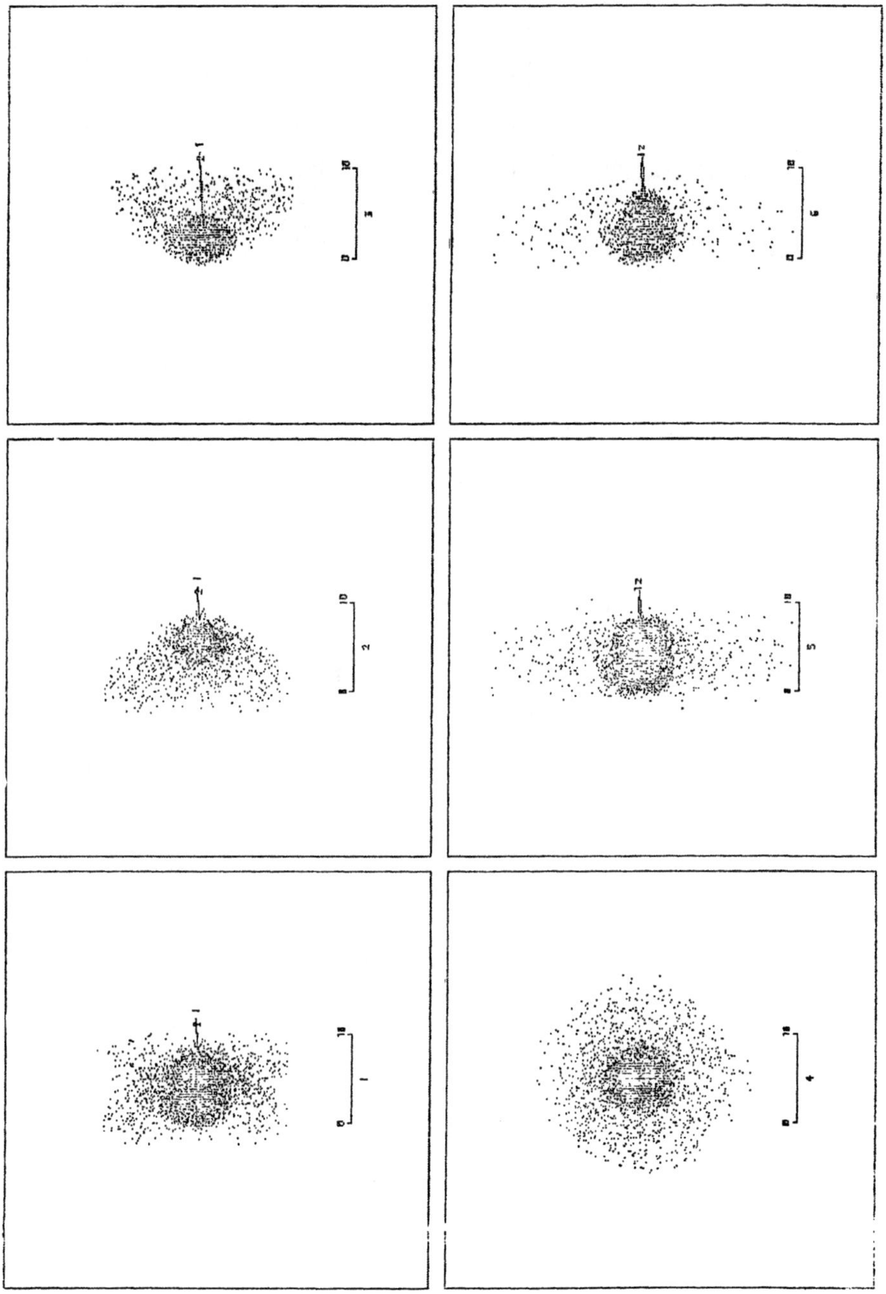

Figure 4. Slow Direct Collision & Merging (cont.)

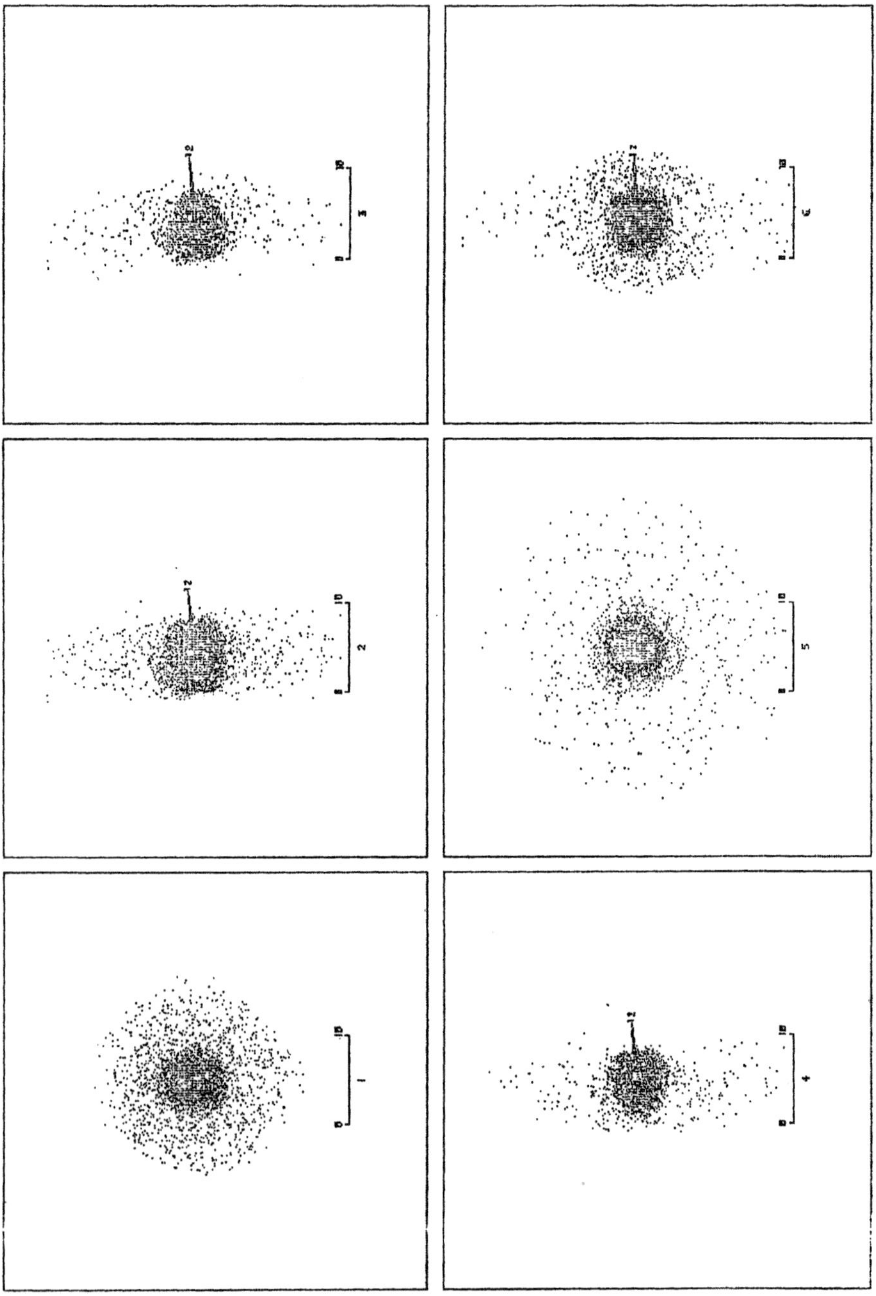

Figure 5. Slow Direct Collision & Merging (cont.)

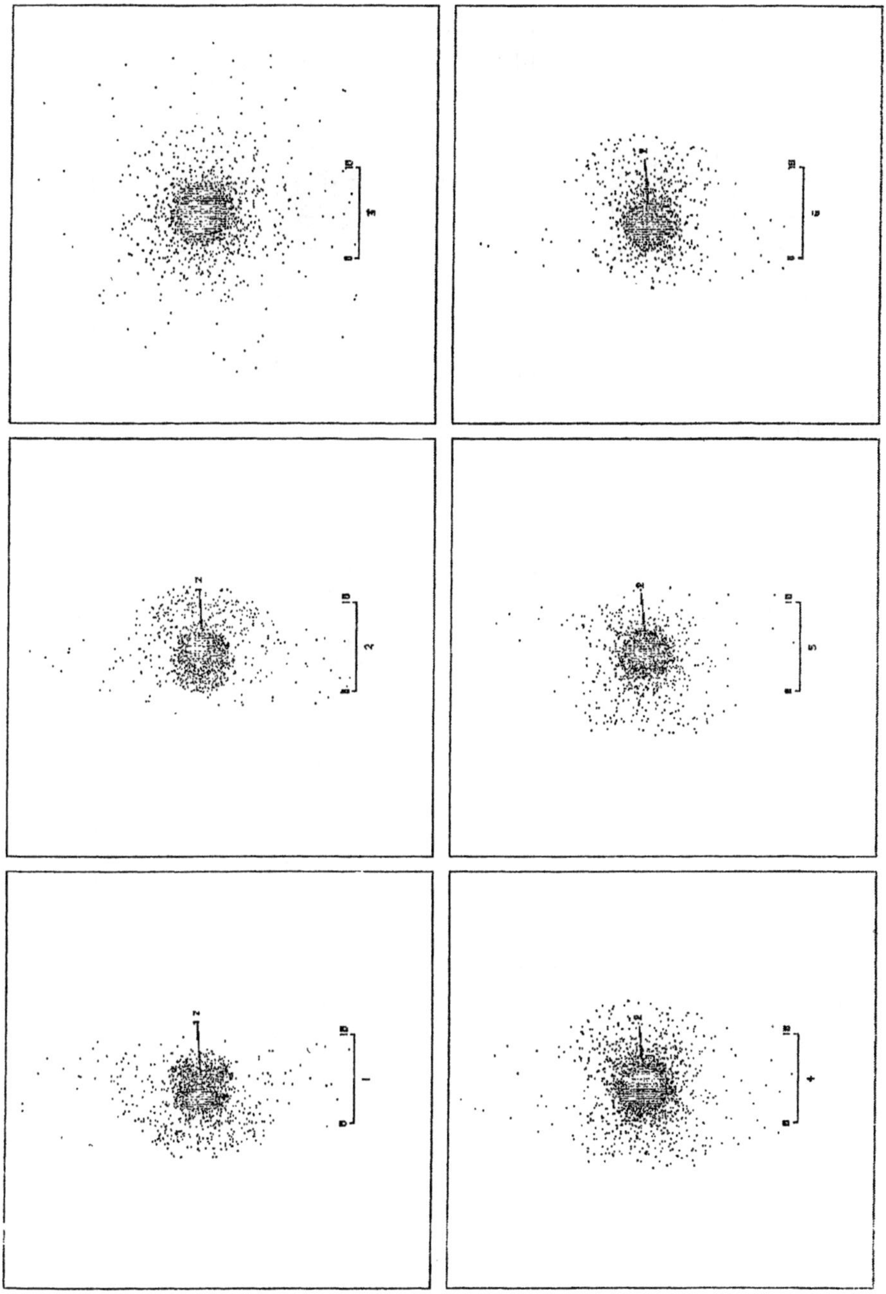

Figure 6. Slow Direct Collision & Merging (cont.)

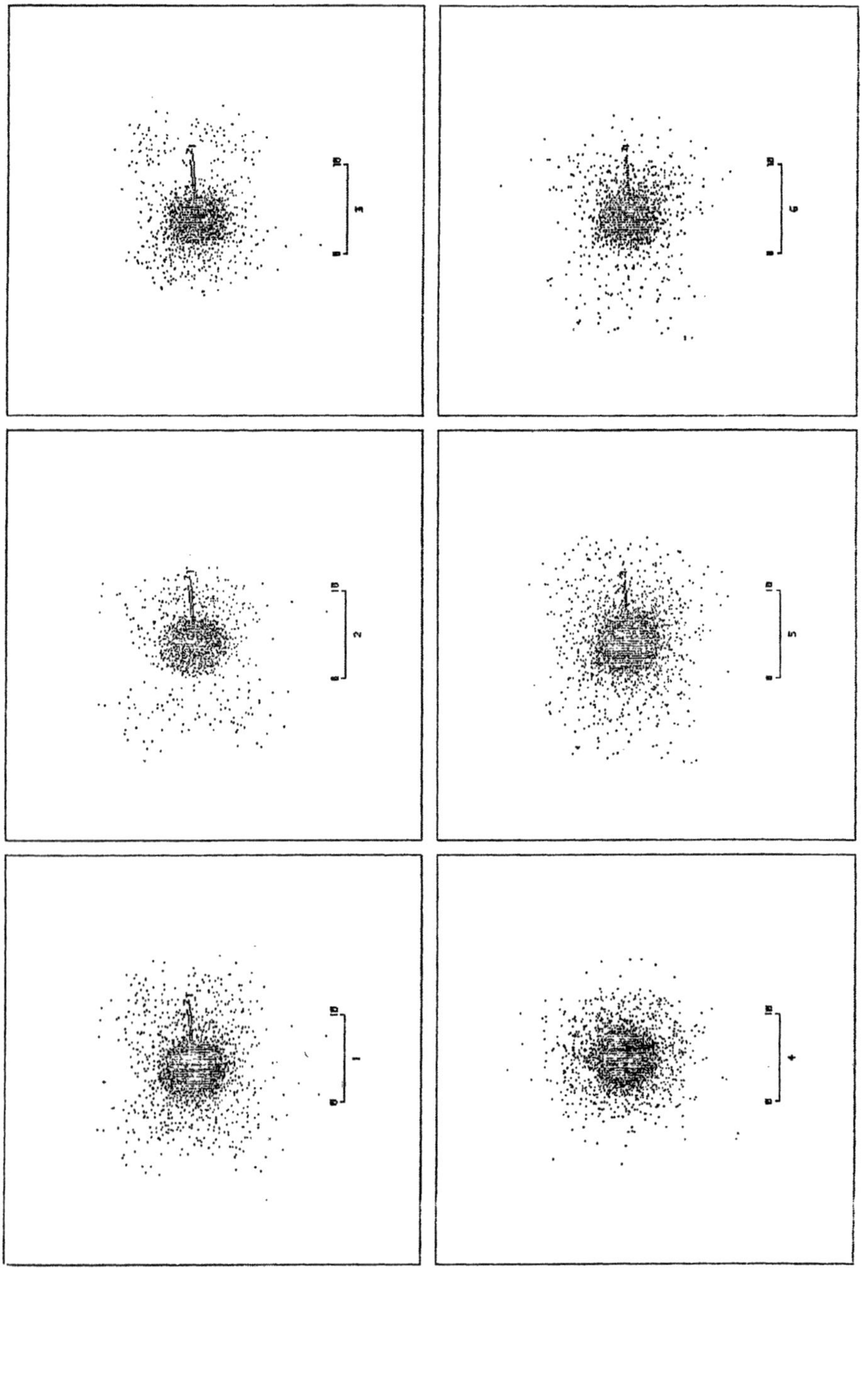

Figure 7. Slow Direct Collision & Merging (cont.)

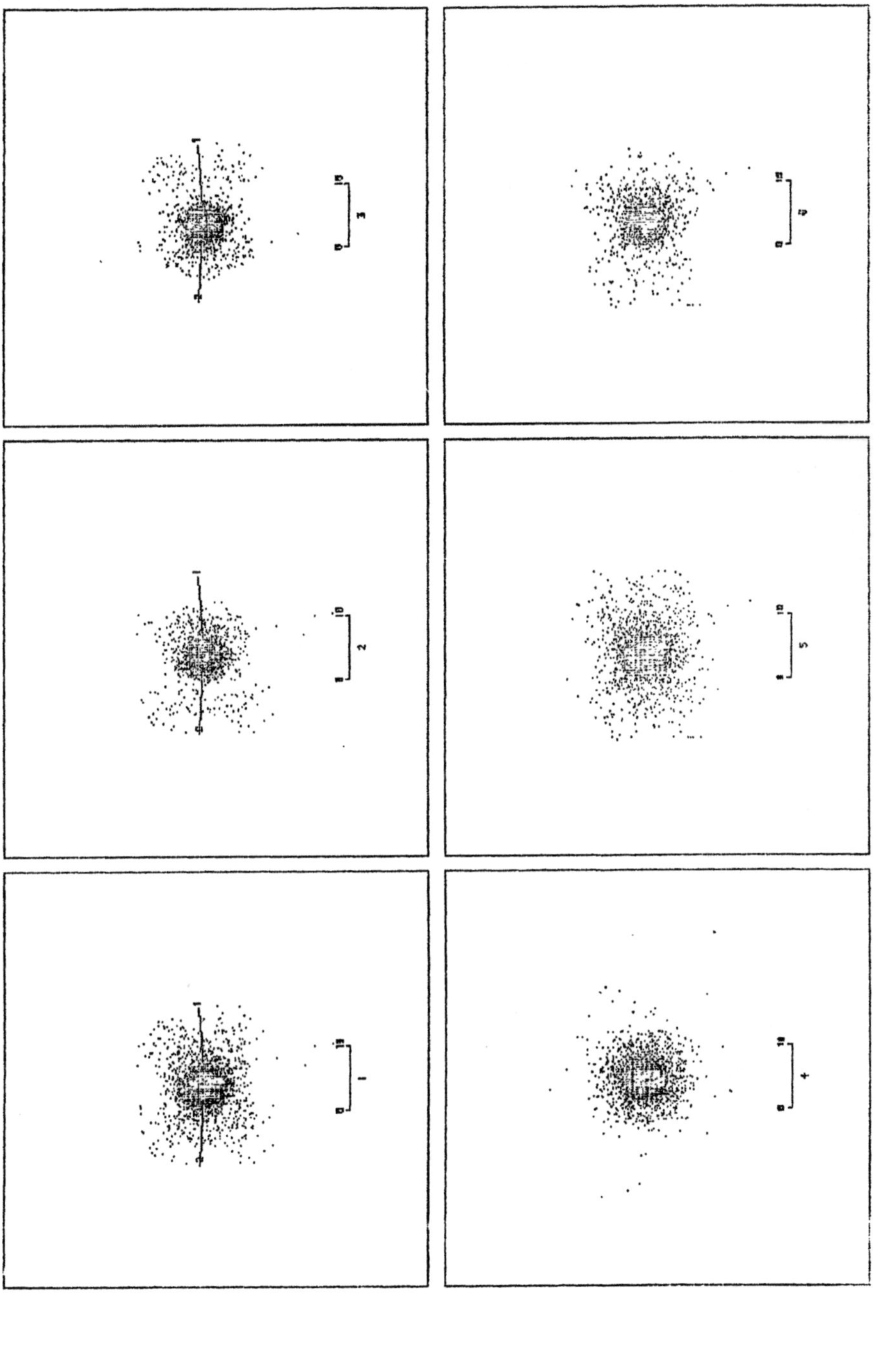

Figure 7a. Antiparallel case

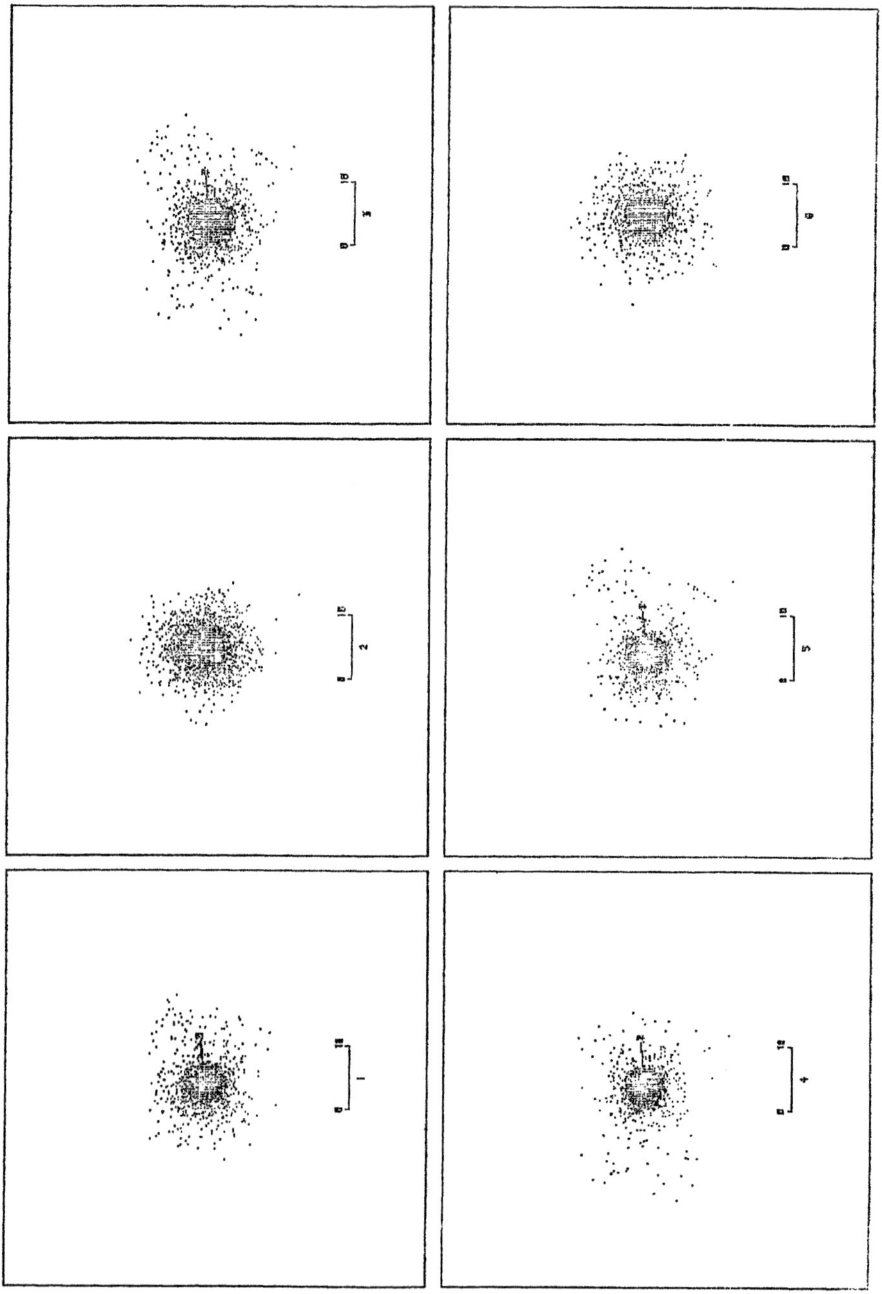

Figure 8. Slow Direct Collision & Merging (cont.)

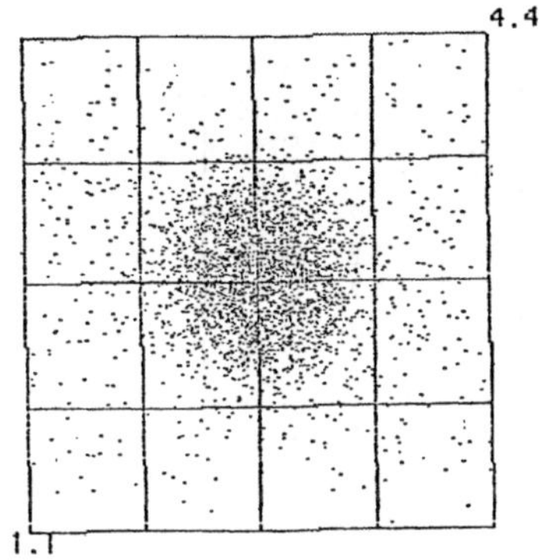

GAL 1,VELOCITY ANALYSIS

BOX	NSTS	NMAX	VMEAN	+-	SIGMA
1,1	10	4	-3.40	1.57	5.08
2,1	7	6	-3.64	0.81	2.60
3,1	21	11	-3.37	0.87	4.04
4,1	14	9	-4.85	0.75	3.12
1,2	11	4	-4.89	1.58	5.46
2,2	293	79	-3.68	0.44	7.61
3,2	453	116	-3.90	0.36	7.57
4,2	53	21	-3.26	0.76	5.55
1,3	20	7	0.57	1.32	5.92
2,3	349	90	2.75	0.45	8.46
3,3	534	135	2.54	0.36	8.41
4,3	41	15	-0.62	1.08	6.92
1,4	9	6	4.72	0.94	3.28
2,4	15	6	6.60	1.31	5.39
3,4	21	14	3.37	0.87	4.04
4,4	14	7	4.04	1.35	5.18

VMAX-+-28.30 BOX- 3.77 X 4.08 MESH UNITS
DEL= 500 VCMASS=(-0.45, -0.07, -0.53)
TOTAL VMEAN- -0.37

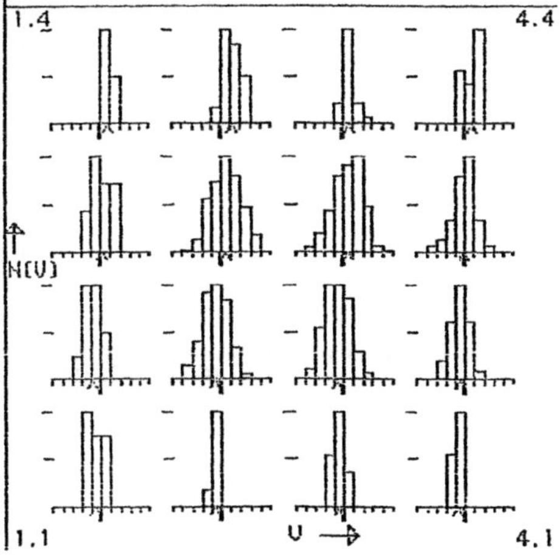

Figure 9. Slow Direct Collision & Merging
Spectra at $\tau = 6.56$, Galaxy 1

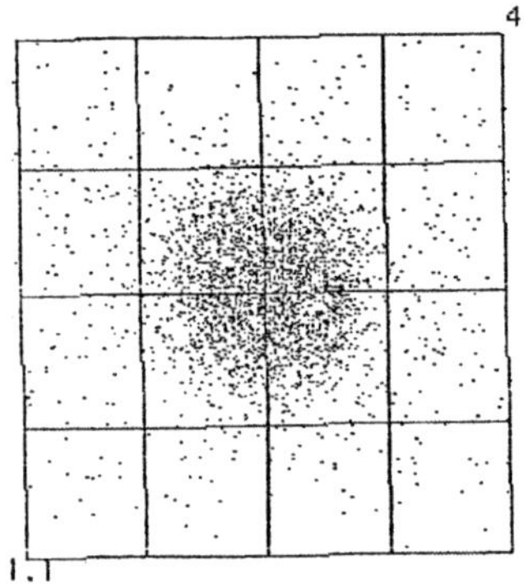

GAL 2,VELOCITY ANALYSIS

BOX	NSTS	NMAX	VMEAN	+-	SIGMA
1,1	10	7	-4.53	0.86	3.12
2,1	15	9	-5.09	0.74	3.18
3,1	10	5	-4.53	1.21	4.11
4,1	10	7	-4.53	0.86	3.12
1.2	32	9	-3.01	1.15	6.53
2.2	353	101	-3.57	0.40	7.49
3.2	371	108	-3.20	0.41	7.86
4,2	13	6	-3.27	1.63	5.95
1,3	57	21	4.12	0.69	5.27
2,3	476	125	2.57	0.37	8.02
3,3	418	113	1.90	0.41	8.31
4,3	11	6	2.32	1.20	4.03
1,4	19	11	4.62	0.76	3.47
2,4	29	14	3.61	0.83	4.52
3,4	16	8	2.12	1.02	4.11
4.4	15	8	5.85	0.75	3.31

VMAX-+-28.30 BOX- 3.77 X 4.08 MESH UNITS
DEL= 500 VCMASS=(0.45, 0.51, 0.47)
TOTAL VMEAN- -0.11

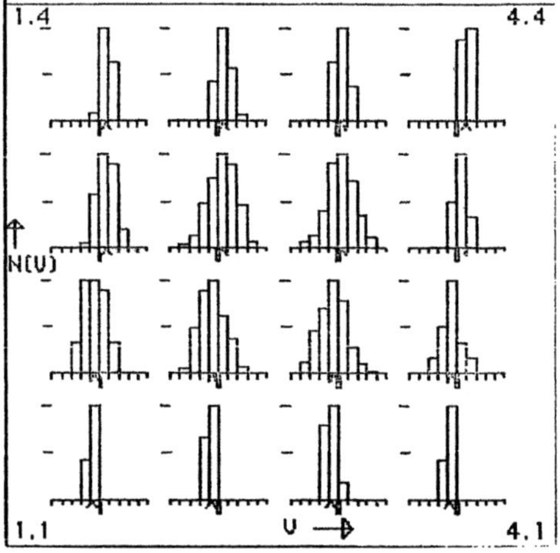

Figure 10. Slow Direct Collision & Merging
Spectra at τ = 6.56, Galaxy 2

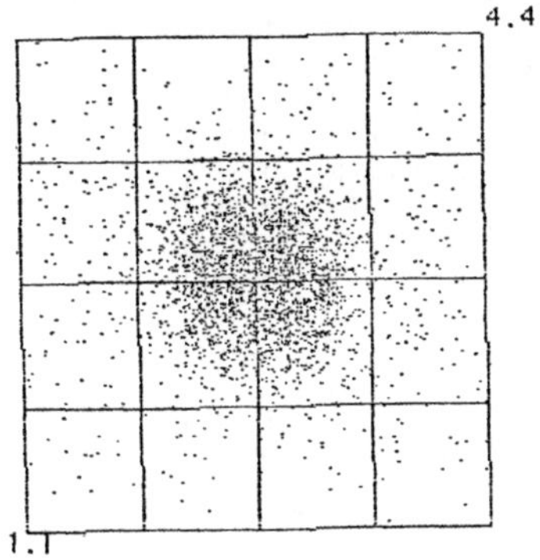

GAL 3.VELOCITY ANALYSIS					
BOX	NSTS	NMAX	VMEAN	+-	SIGMA
1,1	20	10	-3.96	0.68	4.04
2,1	22	15	-4.63	0.58	2.88
3,1	31	16	-3.74	0.70	3.95
4,1	24	16	-4.72	0.56	2.98
1,2	43	13	-3.49	0.95	6.22
2,2	646	180	-3.62	0.30	7.54
3,2	824	224	-3.59	0.27	7.70
4,2	66	27	-3.26	0.68	5.57
1,3	77	26	3.20	0.64	5.62
2,3	825	215	2.64	0.29	8.20
3,3	952	237	2.26	0.27	8.37
4,3	52	21	0.00	0.90	6.49
1,4	28	17	4.65	0.59	3.23
2,4	44	20	4.63	0.73	4.90
3,4	37	22	2.83	0.66	4.03
4,4	29	15	4.98	0.76	4.23

VMAX-+-28.30 BOX- 3.77 X 4.08 MESH UNITS
DEL= 500 VCMASS=(0.00, 0.22, -0.03)
TOTAL VMEAN- -0.24

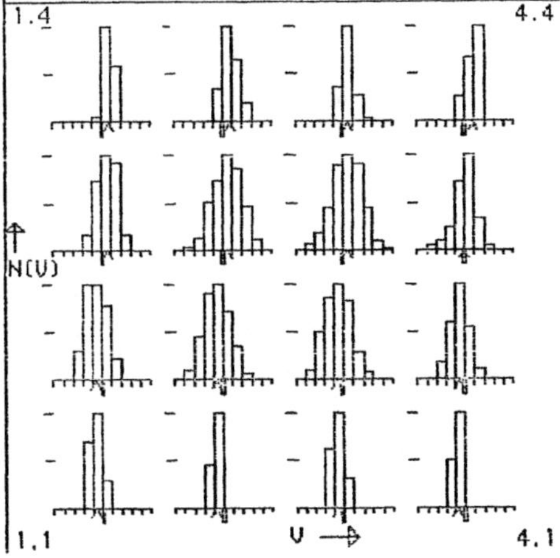

Figure 11. Slow Direct Collision & Merging
Spectra at $\tau = 6.56$, both Galaxies

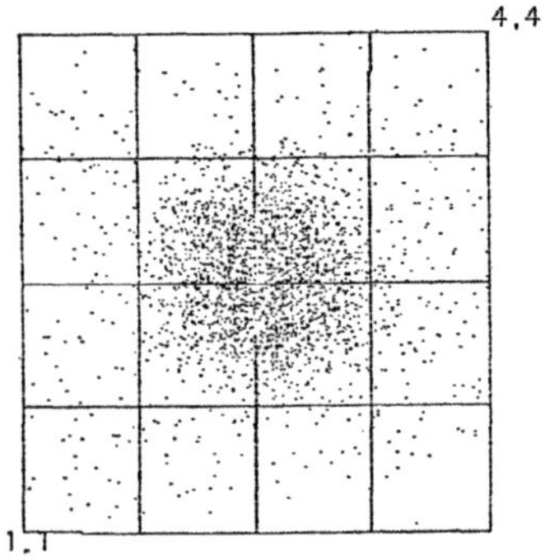

GAL 1.VELOCITY ANALYSIS					
BOX	NSTS	NMAX	VMEAN	+-	SIGMA
1,1	12	5	-3.61	1.31	4.65
2,1	12	6	-3.61	1.12	4.04
3,1	16	8	-4.40	1.07	4.45
4,1	20	14	-2.71	0.68	3.11
1,2	12	5	-3.61	1.47	5.20
2,2	293	69	-3.71	0.48	8.15
3,2	493	127	-3.63	0.34	7.59
4,2	58	21	-3.64	0.74	5.68
1,3	16	8	3.05	1.75	7.03
2,3	331	89	3.20	0.44	8.01
3,3	483	114	3.25	0.38	8.29
4,3	48	16	0.45	0.97	6.69
1,4	6	4	2.71	1.40	3.64
2,4	15	7	6.32	1.14	4.74
3,4	25	17	2.71	0.63	3.18
4,4	21	8	2.97	1.15	5.32

VMAX-+-27.09 BOX- 3.76 X 4.07 MESH UNITS
DEL= 504 VCMASS=(-0.51, -0.33, -0.63)
TOTAL VMEAN= -0.22

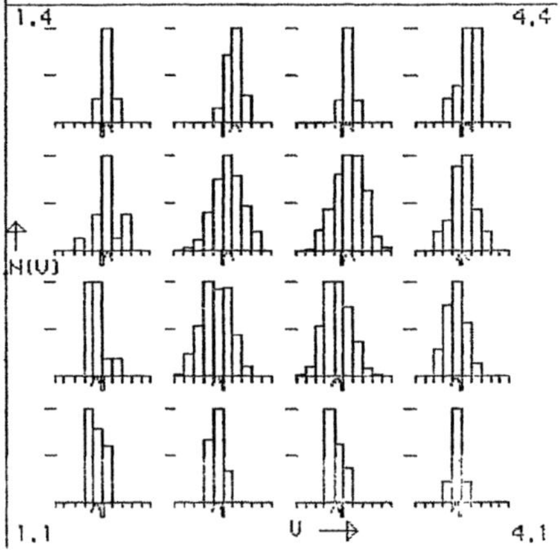

Figure 12. Slow Direct Collision & Merging Spectra at τ = 6.56, Galaxy 1 Antiparallel case

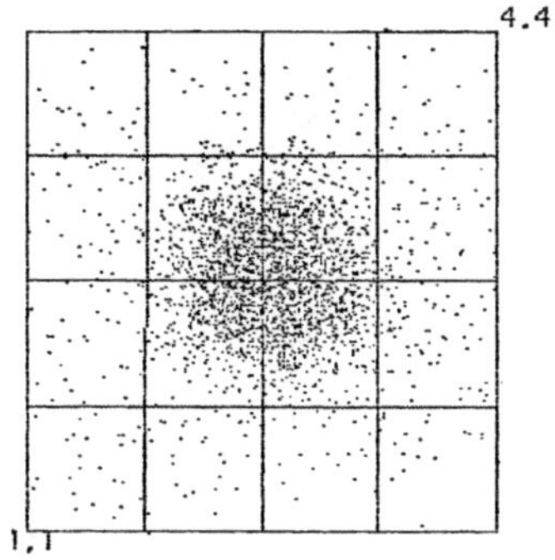

GAL 2.VELOCTTY ANALYSIS					
BOX	NSTS	NMAX	VMEAN	+-	SIGMA
1,1	23	13	3.65	0.73	3.61
2,1	17	8	4.30	0.90	3.87
3,1	6	4	4.52	1.14	3.45
4,1	4	1	5.42	3.50	7.66
1,2	36	14	3.01	0.71	4.32
2,2	321	88	3.92	0.45	7.88
3,2	430	105	2.99	0.38	7.98
4,2	18	6	3.91	1.28	5.52
1,3	37	21	-3.15	0.64	3.95
2,3	429	118	-3.05	0.39	8.02
3,3	449	109	-2.42	0.39	8.29
4,3	20	6	-0.27	1.55	6.52
1,4	20	12	-4.34	0.63	3.25
2,4	15	7	-5.24	1.04	4.26
3,4	15	11	-6.68	0.64	3.06
4,4	8	5	-4.74	1.76	5.28

VMAX-+-27.09 BOX- 3.76 X 4.07 MESH UNITS
DEL= 504 VCMASS=(0.54, 0.42, 0.34)
TOTAL VMEAN= 0.06

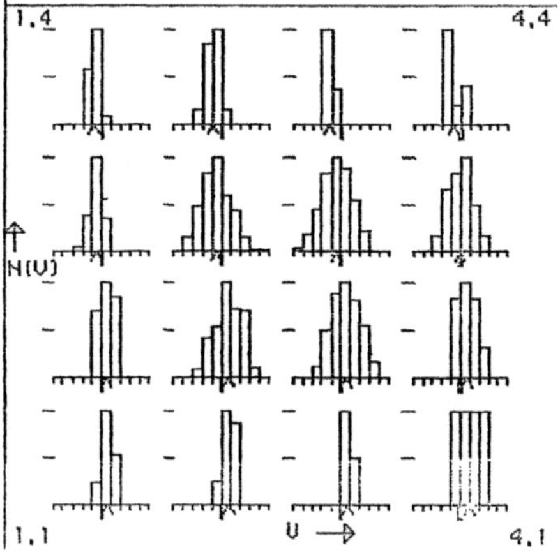

Figure 13. Slpw Direct Collision & Merging
Spectra at τ = 6.56, Galaxy 2
Antiparallel case

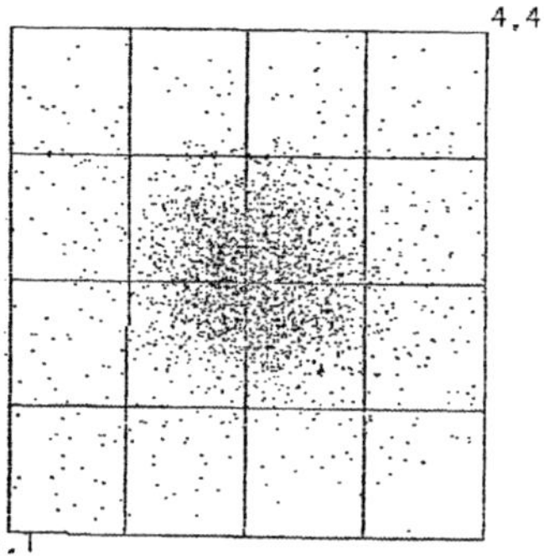

GAL 3,VELOCITY ANALYSIS

BOX	NSTS	NMAX	VMEAN	+-	SIGMA
1,1	35	16	1.16	0.88	5.19
2,1	29	10	1.03	1.01	5.44
3,1	21	8	-2.19	1.23	5.68
4,1	24	15	-1.35	0.99	4.87
1,2	48	15	1.35	0.77	5.31
2,2	609	153	0.30	0.36	8.92
3,2	920	220	-0.54	0.28	8.43
4,2	76	26	-1.85	0.74	6.45
1,3	53	24	-1.28	0.79	5.74
2,3	765	187	-0.34	0.31	8.59
3,3	936	210	0.51	0.29	8.76
4,3	68	22	0.24	0.81	6.71
1,4	25	13	-2.93	0.86	4.32
2,4	30	8	0.54	1.31	7.20
3,4	40	17	-0.81	0.86	5.43
4,4	29	10	0.84	1.15	6.19

VMAX=+-27.09 BOX= 3.76 X 4.07 MESH UNITS
DEL= 504 VCMASS=(0.02, 0.05, -0.15)
TOTAL VMEAN= -0.08

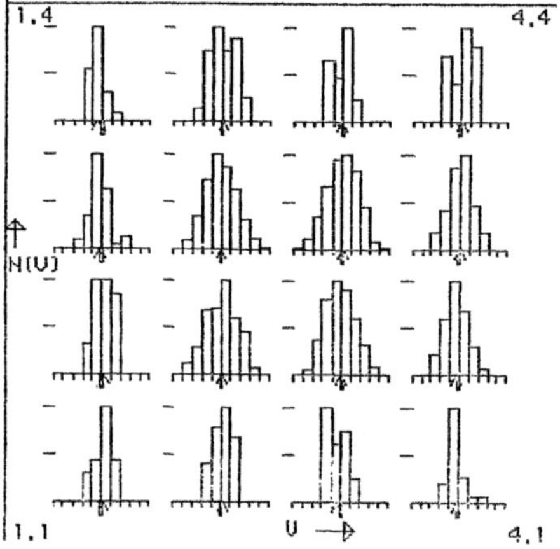

Figure 14. Slow Direct Collision & Merging Spectra at $\tau = 6.56$, both Galaxies Antiparallel case

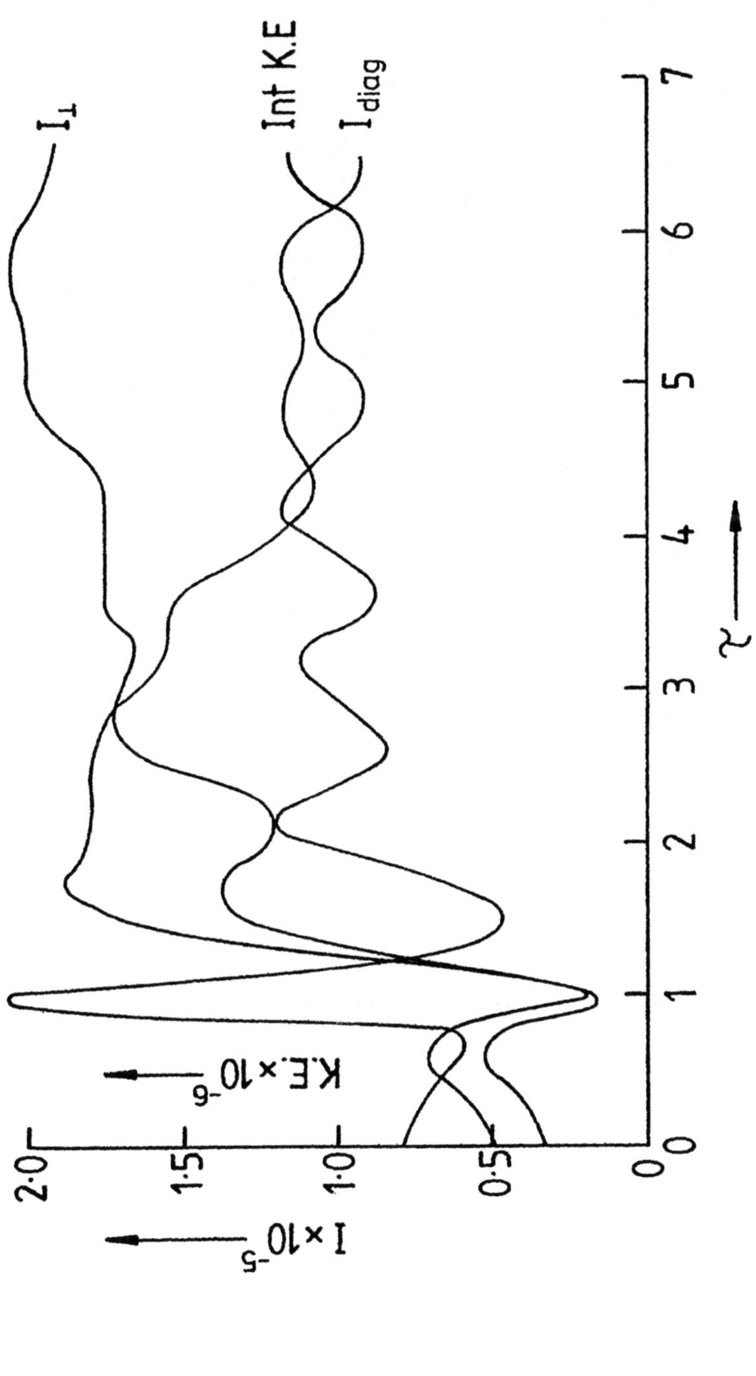

Figure 15. Slow Direct Collision & Merging

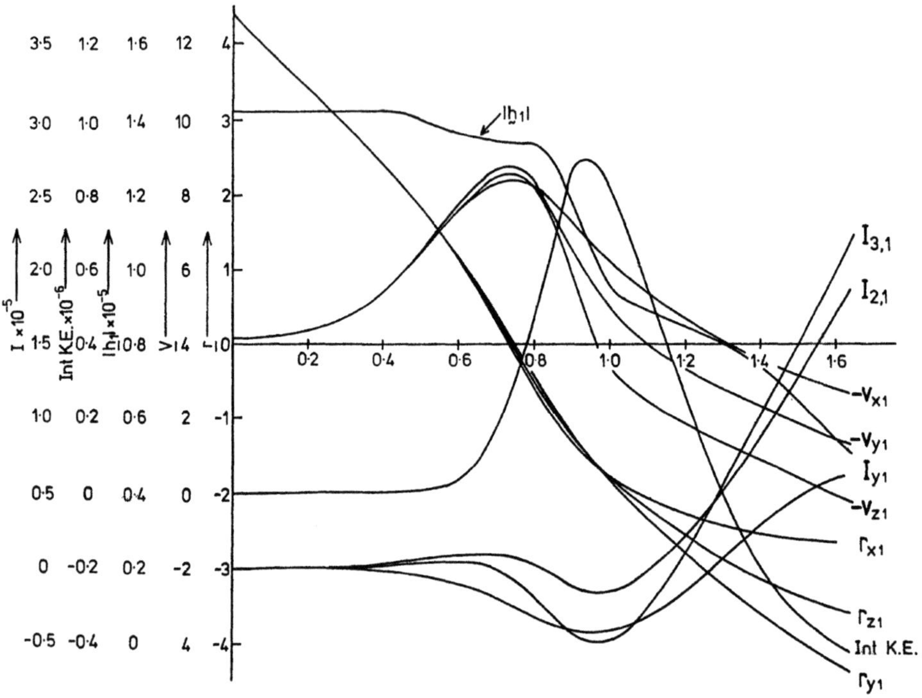

Figure 16. Direct Collision spins $\perp$ $\Delta \underline{r}$

Figure 17. Direct Collision spin $\perp$ Δr
Paths of Centroids

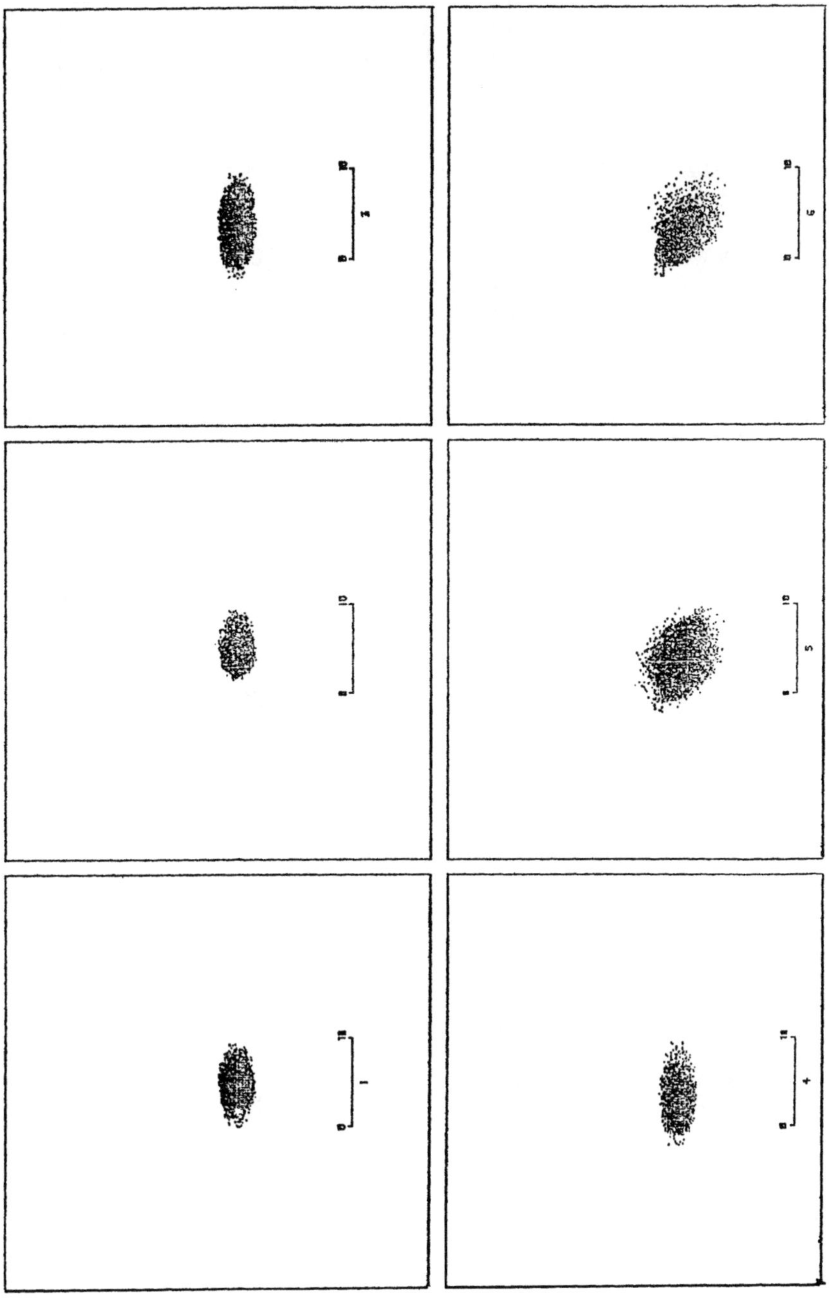

Figure 18. Direct Collision spin ⊥ $\Delta\underline{r}$

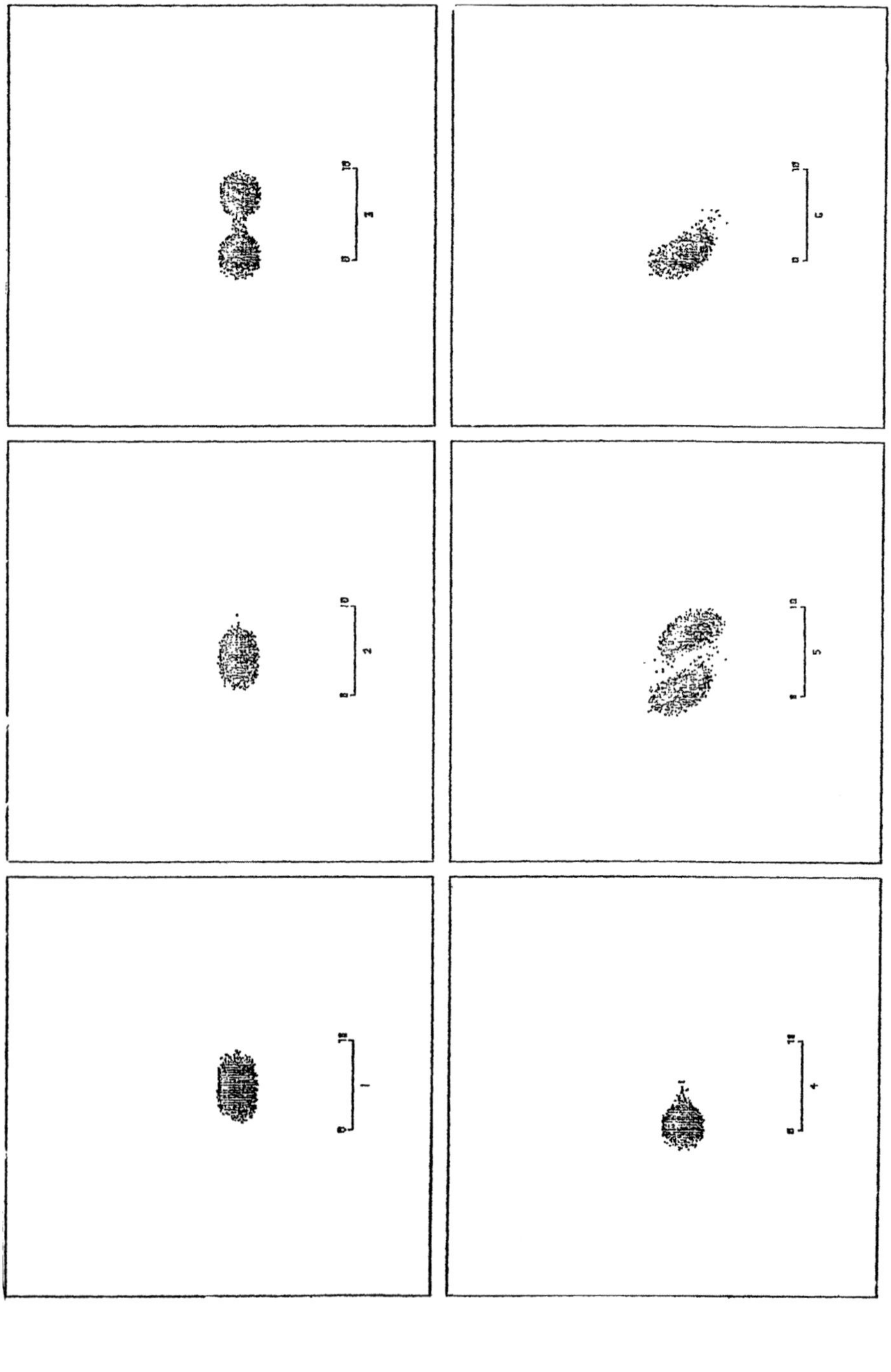

Figure 19. Direct Collision spin $\perp \Delta \underline{r}$ (cont.)

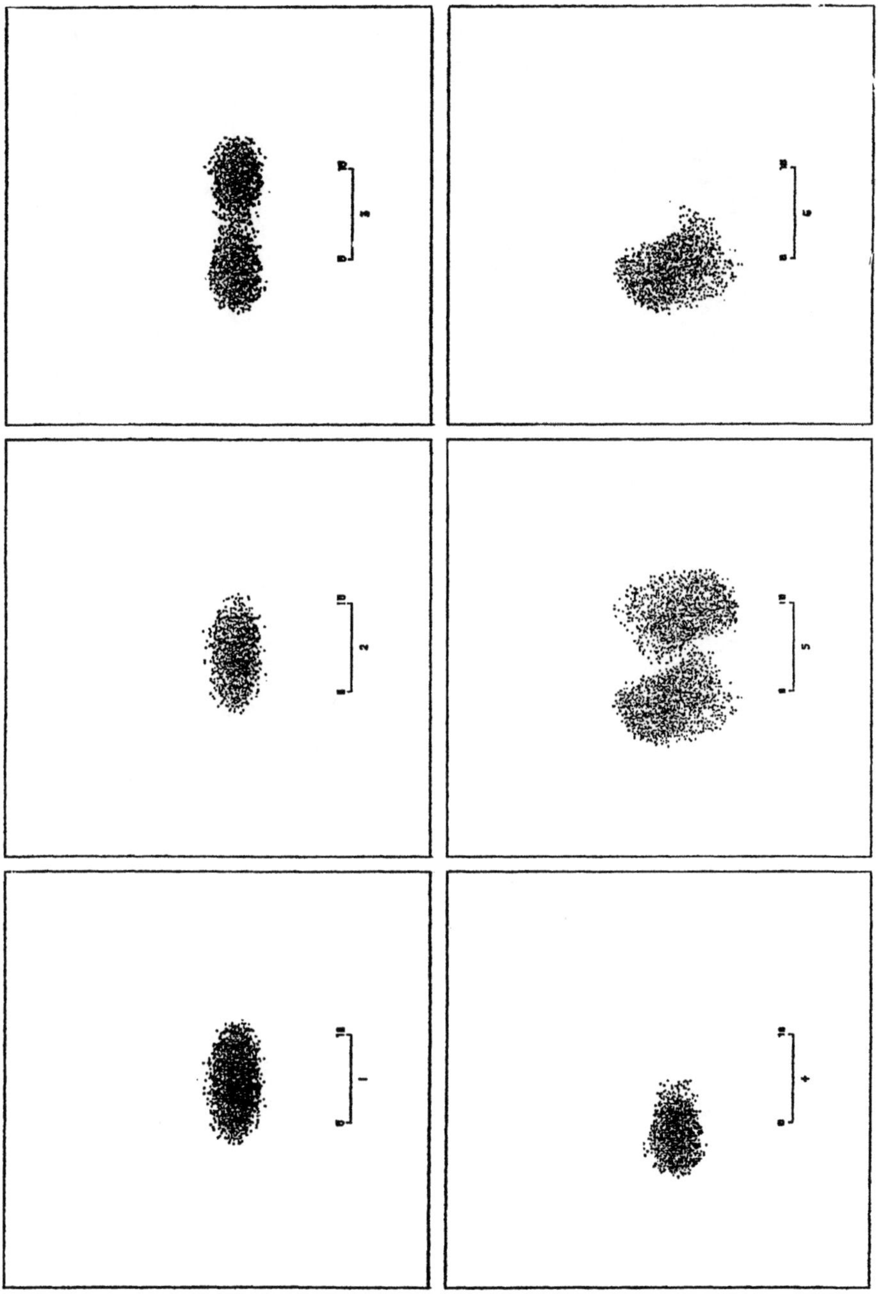

Figure 20. Direct Collision spin ⊥ $\Delta \underline{r}$ (cont.)

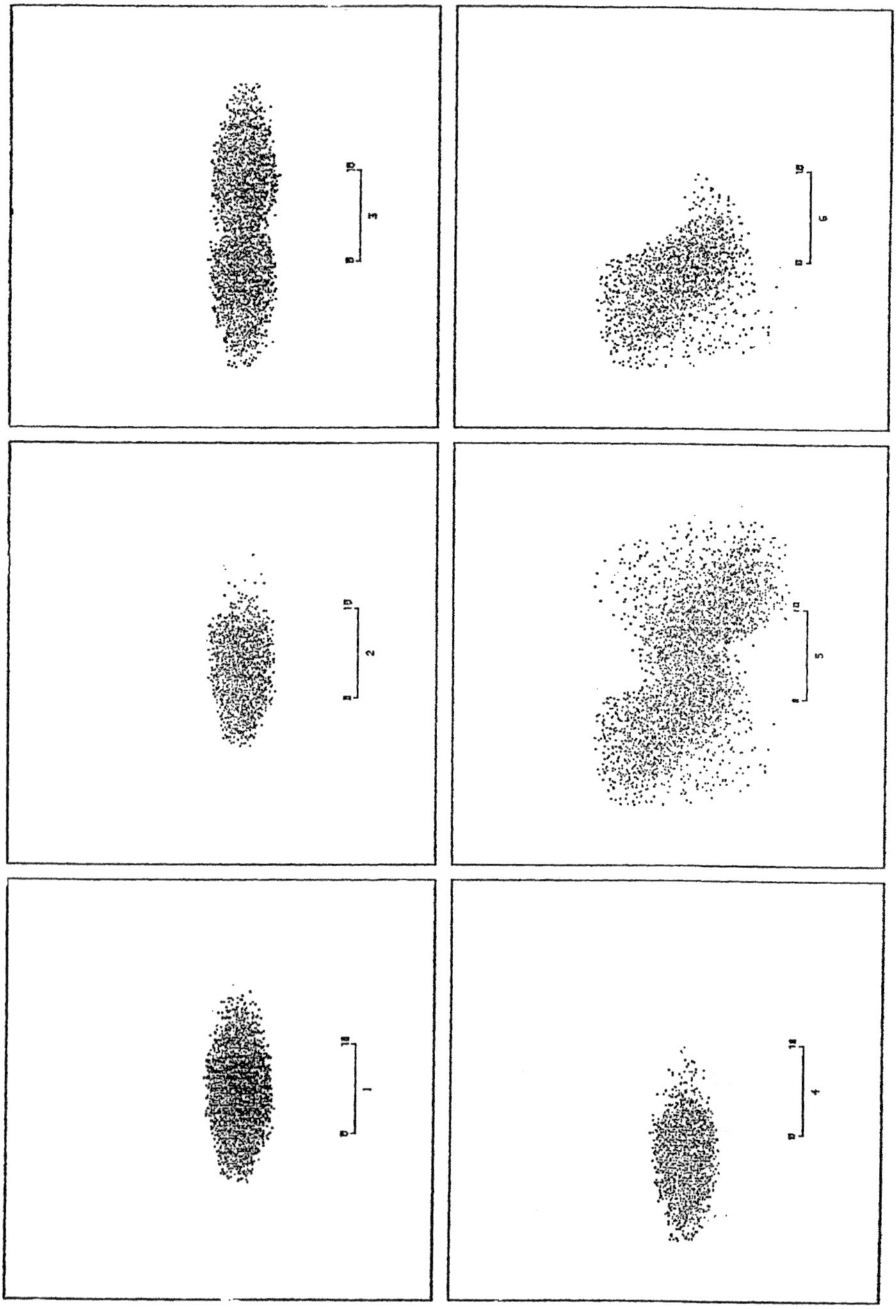

Figure 21. Direct Collision spin $\perp \Delta \underline{r}$ (cont.)

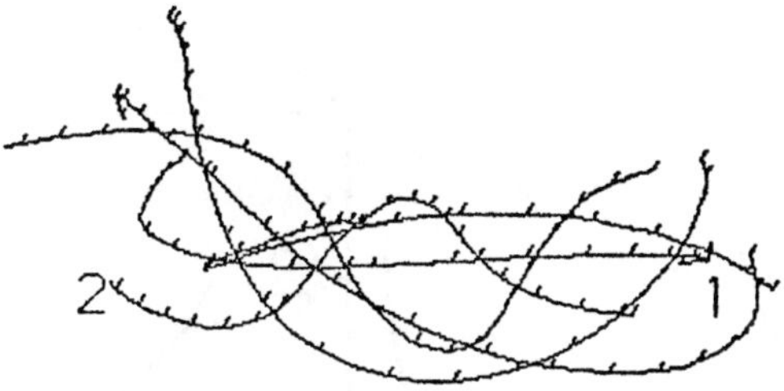

Figure 22a. Stellar Orbits

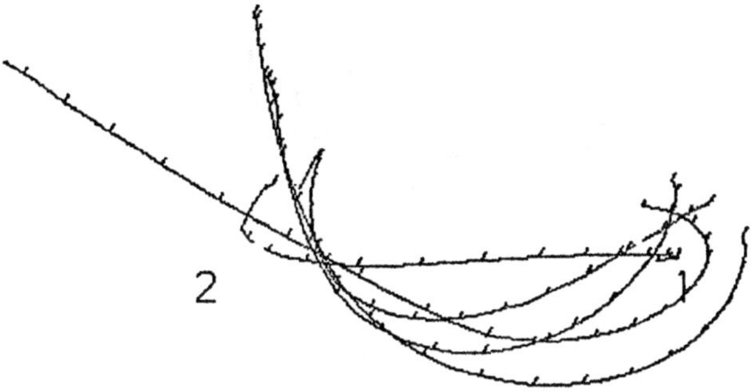

Figure 22b. Stellar Orbits

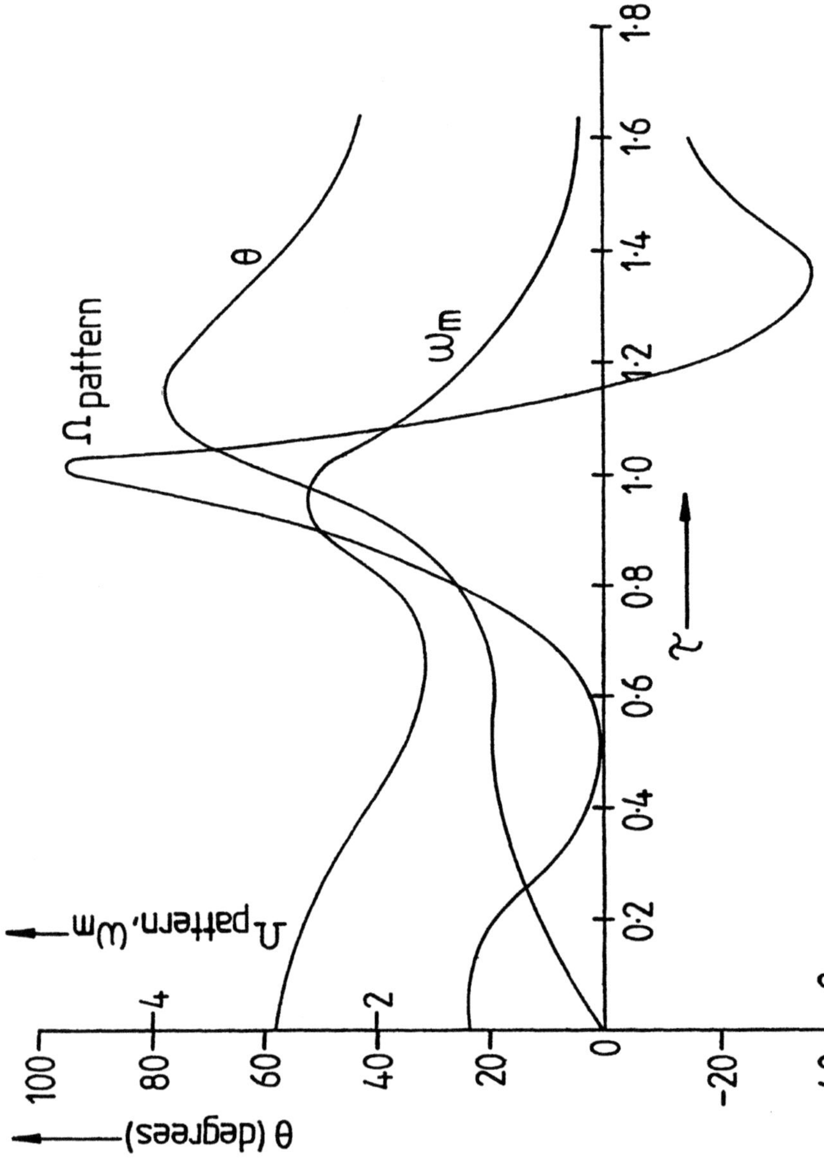

Figure 23. Direct Collision spin ⊥ $\Delta \underline{r}$

TECHNIQUES FOR SIMULATING GALACTIC COLLISIONS

R. A. James,
Department of Astronomy,
University of Manchester,
England.

ABSTRACT

Numerical simulation of the evolution of two or more interacting galaxies requires a fine spatial resolution to resolve detail in individual systems. The much larger regions separating the galaxies must be treated in the same detail with current fast Fourier transform based potential solvers. This paper describes a mesh segmentation technique which permits us to solve very large potential problems and to concentrate our effort on the regions where detailed potential maps are required. We discuss the need for this method in simulating single disc galaxies, and in the study of galactic collisions.

1. INTRODUCTION

Current N-body simulation techniques are not altogether satisfactory for the study of galactic collisions or of the dynamics of small groups of galaxies. It seems possible however to avoid many difficulties by segmenting the region occupied by a pair or a group of galaxies and obtaining the potential by an extension of the corrective charge procedure of James (1977). This extension enables us to restrict detailed potential determinations to regions where potentials are actually required, without falsifying interaction potentials between the various components of a system. Thus we may use a mesh which is fine enough to resolve detail in individual galaxies without incurring prohibitive costs for calculating the potential elsewhere.

The purpose of this paper is to describe the proposed technique and to demonstrate its advantages over the conventional application of the fast Fourier transform (or FFT). In Section 2 we summarise reasons for our interest in galactic interactions, and in Section 3 explain why existing methods are inadequate in this context. Section 4 outlines the use of FFT based methods in potential problems, drawing heavily on the review of Hockney (1970). We discuss the physical basis of the corrective charge technique in Section 5 and its application to a simple mesh segmen-

F. D. Kahn (ed.), Investigating the Universe, 423–441.

tation in Section 6. This application is developed further in Section 7. Sections 8 and 9 discuss two areas where the new method is likely to find application, and Section 10 summarises our conclusions.

Simulations have been performed with particle numbers as low as 30 and as high as 10^5 or more. The large calculations are expensive, and we seek to explain why it is desirable to proceed to the much larger calculations that the new method permits.

2. INTERACTING GALAXIES

At one time galactic collisions were believed to be rare. It has however recently become clear that galactic masses may exceed the mass of the visible material by factors of 5 or 10 (Ostriker, Peebles and Yahil, 1974, Yahil, 1977). Much of the additional dark matter appears to be associated with individual systems. The increase in the typical mass of a galaxy greatly enhances the collision rate.

Many multiple systems are known in which members are apparently linked by long narrow bridges or show extensive tails. The sharpness of these features is at first sight inconsistent with a tidal origin. However, Toomre and Toomre (1972 - to be referred to as TT) showed that this first reaction is unjustified. Their calculation is not self-consistent, as they represent galactic discs by massless particles. They note that a full N-body simulation is needed to model the disc response correctly. Such simulations would be a powerful tool in interpreting detailed observations such as those of Appleton, Davies and Stephenson (1981) for the M81/M82 group.

Mergers between bound pairs of galaxies, and the capture of satellite must depend on the response of individual galaxies to the external field. There seems no possibility of modelling such processes without performing a full N-body simulation for the combined system.

3. THE FAILURE OF SIMULATION TECHNIQUES FOR SMALL GROUPS OF GALAXIES

There are two procedures available for the self-consistent modelling of stellar systems. One is the direct method of Aarseth and his co-workers, in which each particle is followed with a time step appropriate to its circumstances. The field due to the rest of the system is calculated directly when required. The second, developed by Hockney, Hohl, Miller, Prendergast and others, uses a fixed time step for all particles and determines the gravitational potential at each time step by an algorithm based on the FFT. This technique is less flexible than Aarseth's scheme, but it allows us to use many more particles.

In applications to small groups of galaxies, both methods are unsatisfactory. The direct method becomes extremely expensive for more than about 1000 particles, and adequate resolution is hard to achieve. The FFT based method requires that potentials be calculated over a rectangular three-dimensional mesh, the spacing of which determines the spatial resolution. The dynamically important regions are typically an order of magnitude smaller than the cluster. Experience in modelling single galaxies shows that the mesh spacing should be less than ten per cent of the galactic radius. This implies a minimal mesh of 129 x 129 x 129 points, or slightly over 2×10^6 points, for a simulation. This would be on the borderline of possibility on the CRAY-1 or the ICL Distributed Array Processor (DAP). A more adequate mesh, say of 257 x 257 x 257 points or about 1.7×10^7 points, would be beyond possibility. The calculation is clearly most inefficient as the potential is only required in detail for the small regions of the mesh occupied by the galaxies. A coarser mesh would suffice in the regions occupied by collision debris, and potentials in the empty regions are not used at all.

4. DIRECT METHODS FOR POISSON'S EQUATION

We may diagonalise the Poisson equation in a rectangular region by Fourier transforming the density and potential along each of the co-ordinate axes. To solve a discrete form of the equation we use a discrete Fourier transform. If for example we take a three dimensional mesh with points at (rh, sh, th) for integral r, s, t, the simplest discrete representation of the Poisson equation

$$\nabla^2 \phi = -4 \pi G \rho \tag{4.1}$$

is

$$\phi_{rst} - \frac{1}{6} (\phi_{r+1,st} + \phi_{r-1,st} + \phi_{r,s+1,t} + \phi_{r,s-1,t} + \phi_{rs,t+1} + \phi_{rs,t-1}) = \frac{2}{3} \pi G h^2 \rho_{rst} \tag{4.2}$$

where ϕ_{rst}, ρ_{rst} are the potential and density at the appropriate mesh point. The reader can easily verify that the substitutions

$$\phi_{rst} = \sum_{\alpha,\beta,\gamma=1}^{N-1} \phi^{\alpha\beta\gamma} \sin \frac{\pi\alpha r}{N} \sin \frac{\pi\beta s}{N} \sin \frac{\pi\gamma t}{N} \tag{4.3}$$

and

$$\rho_{rst} = \sum_{\alpha,\beta,\gamma=1}^{N-1} \rho^{\alpha\beta\gamma} \sin \frac{\pi\alpha r}{N} \sin \frac{\pi\beta s}{N} \sin \frac{\pi\gamma t}{N} \tag{4.4}$$

reduce Equation (4.2) to the form

$$\left\{\frac{1}{3}(1 - \cos\frac{\pi\alpha}{N}) + \frac{1}{3}(1 - \cos\frac{\pi\beta}{N}) + \frac{1}{3}(1 - \cos\frac{\pi\gamma}{N})\right\}\phi^{\alpha\beta\gamma} = \frac{2}{3}\pi G h^2 \rho^{\alpha\beta\gamma} \tag{4.5}$$

If N is a power of 2 the expansions (4.3), (4.4) may be calculated and inverted efficiently by the FFT (Cooley and Tukey, 1965, Ziegler, 1972). Equation (4.2) may be solved by transforming ρ_{rst} to obtain $\rho^{\alpha\beta\gamma}$, solving Equations (4.5) for $\phi^{\alpha\beta\gamma}$ and then inverting to obtain ϕ_{rst}. As stated above the method gives the potential over a simple rectangular region with a zero boundary condition. We now discuss some simple extensions and refinements.

i) Non-zero Boundary Potential

We may calculate a set of equivalent charges which mimic the effect of changing the boundary potential. We apply Poisson's equation to a potential ϕ_1 taking prescribed values on the boundary and equal to zero elsewhere. This yields a set of charges which are non-zero at points adjacent to the boundary. We subtract equivalent charges from given charges and calculate a potential ϕ_0 (zero on the boundary) using the FFT procedure as above. We may restore the correct charge distribution by adding back the equivalent charges. The sum $(\phi_0 + \phi_1)$ thus arises from the correct charge distribution and takes the correct values on the boundary.

ii) Prescribed Boundary Field

For a zero boundary field we may replace sines by cosines in Equations (4.3), (4.4) for expansion in the appropriate direction. For a prescribed field it is easy to extend the procedure of the last paragraph.

iii) Irregular Boundaries

We may use Hockney's capacity-charge procedure (Hockney, 1970). This is a two-stage process in which we first solve for the potential in a rectangular region surrounding the region of interest. We also construct and invert the matrix relating potentials to charges at mesh points on the actual boundary. We use the inverted matrix to calculate the charges on this boundary that reduce the potentials there to those required. We then solve the Poisson equation using the new charge distribution. This procedure also enables us to deal with internal electrodes. It is efficient when we require many solutions of the Poisson equation for a given geometry. The corrective charge procedure of the next section uses a very similar approach.

iv) Boundary Condition at Infinity

There is no one-stage FFT based Poisson solver for this problem, but it is possible to obtain the same results by convolving the density with an arbitrary Greens function (Hockney, 1970). We double the mesh in each direction, and set the densities to zero at the additional points. We set the Greens function to zero at every point for which one component of the distance from the source exceeds the corresponding dimension of the original mesh. We then perform a periodic convolution on the extended mesh, using a procedure based on the FFT. The potentials on the original mesh are clearly those required, as spurious charges introduced by the periodicity are too far away to affect them. Potentials on the extensions of the mesh have no physical significance and must be discarded. Hockney shows how to apply the complex FFT to periodic convolution, and James (1977) gives an alternative description in terms of real transforms.

The introduction of additional mesh points increases the work involved in this process as compared with the case of a prescribed boundary potential. In three dimensions we perform slightly over five times the arithmetic for the simpler problem. Careful organisation allows us to manage with only twice the storage.

v) Improved Representations of the Poisson Equation

We derive Equation (4.2) by truncating the finite difference expansion for a second derivative with respect to any spatial direction:

$$h^2 y_r'' = (\delta^2 - \frac{1}{12}\delta^4 + \frac{1}{90}\delta^6 - \ldots) y_r \tag{4.6}$$

at the first term. This leads to errors of order h^2 in the derivative. The second term may be included, but it proves advantageous to replace the fourth difference terms by expressions involving $\delta_x^2 \delta_z^2 \phi_{rst}$ and so on, where δ_x, δ_y, δ_z denote central differences along the x, y, z axes respectively (Kantorovich and Krylov, 1958). It is even possible to include the sixth difference terms for a mesh with equal intervals along all three axes without involving potentials more than one mesh plane away from the source in any direction (James, 1977). These more accurate representations of Equation (4.1)diagonalise under Fourier transformation, and the only practical difference is that we must change the multipliers in Equation (4.5). For the sake of clarity we will continue to use Equation (4.2) in this paper.

vi) Cyclic Reduction

Hockney (1970) describes a variety of techniques for reducing the cost of determining a set of potentials. The cyclic reduction technique is particularly relevant to Section 6, so we outline it here for a three-dimensional problem.

Let us consider the case where the potential is known at the boundary. We introduce equivalent charges to account for the non-zero potentials on the 4 boundary planes parallel to the r-axis, but not on the other 2 boundaries. We transform ϕ_{rst}, ρ_{rst} along the s- and t-axes:

$$\phi_{rst} = \sum_{\beta,\gamma=1}^{N-1} \phi_r^{\beta\gamma} \sin\frac{\pi s\beta}{N} \sin\frac{\pi t\gamma}{N} \tag{4.7}$$

$$\rho_{rst} = \sum_{\beta,\gamma=1}^{N-1} \rho_r^{\beta\gamma} \sin\frac{\pi s\beta}{N} \sin\frac{\pi t\gamma}{N} \tag{4.8}$$

This reduces Equation (4.2) to the form

$$A^{\beta\gamma}\,\phi_r^{\beta\gamma} - B^{\beta\gamma}\,(\phi_{r+1}^{\beta\gamma} + \phi_{r-1}^{\beta\gamma}) = \frac{2}{3}\pi G h^2\,\rho_r^{\beta\gamma} \tag{4.9}$$

where

$$A^{\beta\gamma} = 1 - \frac{1}{3}\left(\cos\frac{\pi\beta}{N} + \cos\frac{\pi\gamma}{N}\right) \tag{4.10}$$

and

$$B^{\beta\gamma} = 1/6 \tag{4.11}$$

We now drop the superscripts in the discussion and consider the equation

$$\phi_r - C(\phi_{r+1} + \phi_{r-1}) = D_r \tag{4.12}$$

where

$$C = B^{\beta\gamma}/A^{\beta\gamma} \tag{4.13}$$

and

$$D_r = \frac{2}{3}\pi G h^2\,\rho_r^{\beta\gamma}/A^{\beta\gamma} \tag{4.14}$$

We note that $0<C<\frac{1}{2}$ for all β, γ.. We must solve Equations (4.12) for r in the range (1, N - 1) where ϕ_0, ϕ_N are known and $N = 2^n$.

We may rearrange Equation (4.12) to eliminate odd values of γ. We may use the equation in the form

$$\phi_r = D_r + C(\phi_{r+1} + \phi_{r-1}) \tag{4.15}$$

to eliminate ϕ_{r+1}, ϕ_{r-1} and obtain

$$D_r = \phi_r - C\left\{D_{r+1} + C(\phi_{r+2} + \phi_r) + D_{r-1} + C(\phi_r + \phi_{r-2})\right\} \tag{4.16}$$

Collecting terms we find

$$\phi_r - C_1(\phi_{r+2} + \phi_{r-2}) = D_{r,1} \tag{4.17}$$

with

$$C_1 = C^2/(1 - 2C^2) \tag{4.18}$$

$$D_{r,1} = \left\{D_r + C(D_{r+1} + D_{r-1})\right\} /(1 - 2C^2) \tag{4.19}$$

As $0< C<\frac{1}{2}$, the denominator is always non-zero and $0< C_1 < \frac{1}{2}$. We could solve Equations (4.17) to find ϕ_r for even r, and then use Equation (4.12) to fill in the missing potentials. Alternatively we can repeat the procedure on Equations (4.17) and obtain equations of the form

$$\phi_r - C_2(\phi_{r+4} + \phi_{r-4}) = D_{r,2} \tag{4.20}$$

Repeated applications of this procedure ultimately lead to the equation

$$\phi_{2^{n-1}} - C_{n-1}\ (\phi_0 + \phi_{2^n}) = D_{2^n,n-1} \tag{4.21}$$

which contains only one unknown. The divisors in Equations (4.18), (4.19) and their descendants are always non-zero. We may obtain the potentials on the centre plane of the mesh, then on the centre planes of the two half-meshes on each side, and so on until we have all the potentials. The technique is similar in its demands to the usual tri-diagonal elimination for Equation (4.12). Both methods are faster than Fourier transforming in the r-direction. We see later that the cyclic reduction method can be particularly efficient when the charge distribution in the region is restricted to a few planes - a situation which occurs in Section 6.

vii) Large meshes

Burridge and Temperton (1979) present an attractive algorithm for problems which are too large to fit in computer memory. They partially diagonalise the Poisson equation as for the cyclic reduction process and solve Equations (4.9) by tri-diagonal elimination. For consistency with their notation we write Equation (4.12) in the form

$$\phi_{r-1} + \lambda\phi_r + \phi_{r+1} = b_r \tag{4.22}$$

We assume that ϕ_0 and ϕ_N are known and apply Gaussian elimination to obtain the equations

$$\omega_0 = 0;\ \omega_r = (\lambda - \omega_{r-1})^{-1} \tag{4.23}$$

$$g_0 = \phi_0;\ g_r = \omega_r(b_r - g_{r-1}) \tag{4.24}$$

for $1 \leqslant r \leqslant N$, and

$$\phi_r = g_r - \omega_r \phi_{r+1} \tag{4.25}$$

for $1 \leqslant r \leqslant N - 1$. We solve Equations (4.23), (4.24) in the forward (r - increasing) direction, and Equations (4.25) backwards. The ω_r's are all in the range (-1, 0), so the process is numerically stable. When calculating ω_r, g_r we require only values relating to planes $(r - 1)$, r. The mesh values are stored in a file. They are read one plane at a time, and Fourier transformed in this plane prior to calculation of the g_r's. These are then sent to a file which is read in reverse order on the backward sweep. The ω_r's are pre-calculated and stored in another file. The process requires memory for two mesh planes, but its efficiency is improved if space is available for two more to permit buffering the input and output streams. The space required for a mesh of 129 x 129 x 129 points is about three per cent of the size of the mesh.

FFT based methods have considerable merits. In contrast to iterative procedures they do not rely on good initial guesses to the solution. The work per mesh point is roughly proportional to $\log_2 \mathcal{N}$, where $\mathcal{N}$ is the number of mesh points, and the methods are economical even for large meshes. However, they are not flexible. The restriction to $N = 2^n$ mesh intervals (or $2^n + 1$ mesh points) in each direction is not particularly inconvenient in practice. It can in any case be relaxed as we can construct fairly good FFT procedures for any composite N. The restriction to certain types of boundary condition is however a considerable disadvantage.

5. THE CORRECTIVE CHARGE ALGORITHM

This algorithm was applied to two-dimensional problems by von Hagenow and Lackner (1975) and to three-dimensional problems by James (1977). It permits the application of a boundary condition at infinity without the storage and computational costs of the convolution method.

We may describe the procedure in electrostatic terms as follows. Our system of charges occupies a finite section of the mesh, and may be enclosed by a rectangular boundary. We denote the potential it produces by ϕ. We now suppose that we introduce an earthed conducting box on the boundary. Charges appear on this box and screen the internal system of charges. We denote by ϕ_O the potential due to the prescribed charges together with the screening charges. ϕ_O is zero on and outside the boundary, and satisfies the Poisson equation with the prescribed charges inside. In a discrete problem we may generate it by the FFT based procedure of Section 4. We may then calculate the screening charges from ϕ_O by applying Equation (4.2) on the boundary.

If ϕ_s is the potential due to the screening charges alone, then clearly

$$\phi_O = \phi + \phi_s \tag{5.1}$$

ϕ_s satisfies the Laplace equation inside the boundary. We may calculate it efficiently by the procedure of Section 4(i) when we know its values on the boundary. These must be calculated by convolution, but we need only consider the surface points so the calculation is much cheaper than the full convolution of Section 4(iv).

To summarise we determine ϕ_O and thence obtain the screening charges. A subtraction gives the corrective charges generating $(\phi - \phi_O)$. A surface convolution gives the surface values of $(\phi - \phi_O)$, and thus the equivalent charges to supplement the original density distribution. The procedure of Section 4(i) generates the required potential from this distribution.

Formally the process has two stages. It is however possible to perform the boundary calculation and supplementation for the equivalent charges before transforming $\phi^{\alpha\beta\gamma}$ back to obtain ϕ_{rst}. The equivalent charge layer is confined to planes adjacent to the boundaries, and its Fourier transform is cheap to calculate. We require only one conversion of the densities to transform space, and one conversion of the potentials back. Some care is required in the boundary convolution to reduce overheads. It is not our purpose here to explore these details, which are fully discussed by James (1977). We note only that for a small mesh (17 x 33 x 33 points) on the CDC 7600, the boundary calculation and associated work cost nearly as much as the mesh transformations. The process is almost three times as fast as the direct convolution procedure, and the gain increases for larger meshes. The process is a little less attractive on the CRAY-1, as the complicated boundary calculation is not fully effective in exploiting the vector registers. However, it is still much better than direct convolution.

6. POTENTIAL DETERMINATION ON A MESH DIVIDED INTO TWO SEGMENTS

Let us consider an alternative approach to the problem of finding the potential on a finite discrete mesh with a zero boundary condition. We may split the mesh into two equal segments by means of an earthed conducting plane P parallel to a chosen pair of boundary planes. This forces the potential on P to be zero, and we may determine the potential separately for each segment by the procedure of Section 4(i). If the number of mesh intervals in a direction perpendicular to the conducting plane is 2^n for the original mesh, it is 2^{n-1} for each segment. The FFT based procedure is therefore efficient for the individual segments.

We denote by ϕ_O the potential on the mesh in the absence of the conducting plane P. This is of course the potential we require. In the presence of P we obtain a potential ϕ_p by juxtaposing the solutions in the two segments. In the interior of each segment, ϕ_O and ϕ_p both satisfy the form (4.2) for the Poisson equation. If we apply Equation (4.2) to ϕ_p on P we obtain a set of induced charges which differ from the prescribed charges. We may regard ϕ_p as arising from the induced charges on P, together with the prescribed charges elsewhere. ϕ_o of course arises from the prescribed charges everywhere on the mesh. The difference $\phi_c = \phi_o - \phi_p$ arises from the difference between the two charge systems. These difference (or corrective) charges are non-zero only on P.

We may thus calculate ϕ_o by adding a correction ϕ_c to ϕ_p. ϕ_c arises from the corrective charges on P only, with a zero condition at the mesh boundary. We may determine the potential on P from Equation (4.21). We note that in Equation (4.12) and its descendants the right hand side is non-zero only for $r = 2^{n-1}$.

Repeated use of Equation (4.19) shows that

$$D_{2^{n-1},\,n-1} = D_{2^{n-1}} / \left\{ (1 - 2C_{n-2}^{\;2}) \ldots (1 - 2C_1^2)(1 - 2C^2) \right\} \qquad (6.1)$$

The multipliers to obtain ϕ_p on P from the corrective charges may be calculated in advance, so the calculation of this part of ϕ_p is very cheap. The potentials in the two halves of the mesh may then be supplemented by the equivalent charge method.

7. APPLICATION TO MESH SEGMENTATION

If we use cyclic reduction on the two half meshes the procedure of Section 6 does not differ in any important way from normal cyclic reduction. There are differences of detail, which have little effect on the operation count or on any other aspect that concerns a user. The interest of the method is that it provides a basis for organising the potential calculation around a three-dimensional segmentation of the mesh. The procedure is to divide the mesh into a large number of rectangular blocks, in each of which we determine the (Fourier transformed) potential using a zero condition on the block boundary. We then determine the potentials on the partitions which bound the blocks and supplement the transformed potentials within each block, as in the original corrective charge algorithm, before transforming back to obtain the potentials we require.

The technique offers a number of benefits:

i) It provides a natural way to organise in serial form the data for a very large mesh, and so to reduce the computer memory required. The densities or potentials for each segment of the mesh can be stored as a single record of a serial file. Only one segment needs to reside in memory at any one time. Once the densities in a segment have been Fourier transformed the contributions on the boundary can be extracted and the segment returned to the file while subsequent segments are dealt with. When all segments have been processed the partition potentials may be calculated from the accumulated corrective charges, and the potentials calculated on a second scan through the segments. Careful overlapping of this stage with the particle handling would allow us to combine it with the first scan for the next timestep. The procedure is consistent with a sensible serial organisation for the particle data in the calculation. Computer memory is required for the partitions and for a single segment. An illustrative example is that of a mesh of 257 x 257 x 257 points, divided into segments with 32 mesh intervals in each direction. The total number of data points is about 1.7×10^7. Only 1.8×10^6 of these lie in the partitions, and careful organisation can ensure that at most a third of these need to be in memory at any time. The space for a segment is only 31^3 or about 30000 points, and is negligible in comparison with the partition requirement. The facility to handle such large meshes on a machine with (at first sight) less than four per cent of the necessary memory is of some value. Although computer memories are much larger nowadays than was the case a few years ago, memory is still a valuable resource. Machines with virtual store facilities appear to offer the user access to an extremely large memory. However, this memory is provided by automatic transfers to and from mass storage, and this would become a major bottleneck on a large potential problem. The segmentation process loses none of its attraction on such systems.

ii) Some processors, such as the DAP, impose limits on the size of mesh that can be handled conveniently. For example the system at Queen Mary College has a 64 x 64 array of processors and works most efficiently with arrays near this size in two of their dimensions. Our procedure offers a natural way of segmenting the mesh to this size, and extending the problem size in a reasonable way.

iii) Small groups of galaxies require computational meshes with large empty regions. Segments inside such regions do not need to be processed explicitly. Many segments contain only collision debris making a negligible contribution to the overall gravitational field. This debris moves in a field with linear scale much larger than that in individual galaxies. We can reduce the cost of our calculation by ignoring the density contributed by such regions, and by interpolating the potential there from a coarser mesh. Thus we may concentrate our effort on regions where the resolution is needed without falsifying the interactions between galaxies or the response of any one galaxy to perturbing fields from the others.

iv) For large meshes on which we require a full potential determination, the procedure reduces the length of the Fourier transforms required. The leading term in the operation count per mesh point for a fast Fourier sine transform, together with its inversion, is 5n for a set of $(2^n - 1)$ data points. However economies are possible in the first few folding stages as some of the multipliers are repeated and others are equal to unity. This, together with the elimination of a folding stage, implies that it is cheaper (for example) to carry out independent transformations on two adjacent data sets of length 15 than on a combined data set of length 31 (including the common end point, which does not appear in the lengths of the components). Actual operation counts for transformation plus inversion are given in Table 1 for sine transforms.

Table 1.

n	Transform Length	Operations Per Point	5n
2	3	3.33	10
3	7	7.14	15
4	15	11.93	20

v) For large problems requiring a full potential determination the method involves less input and output than the algorithm of Section 4(vii) The auxiliary array of coefficients ω_r used by that process is no longer required. The potential determinations in individual segments do not impose important storage overheads. It would be somewhat easier to overlap the second stage of the solution with the first stage of the solution for the next time step, as the mesh region occupied by the segment is more compact so fewer particles wander in or out. Burridge and Temperton do not discuss this particular matter, but it is clear that a similar procedure could be used with their algorithm if an alternative tri-diagonal elimination procedure were provided to perform the elimination in the reverse order to that of Section 4(vii). Such a reversal is not required with our technique.

Our algorithm has the effect of reducing the transformation process to several parallel processes with lengths of up to 16 points. An organisation of the mesh into prisms with a cross section of 8 x 8 or 16 x 16 mesh intervals, together with cyclic reduction along prism axes, leads to an overall operation count very similar to the best of Hockney's (1970) FACR algorithms.

8. APPLICATIONS TO DISC GALAXIES

Lynden-Bell (1979) presents a mechanism for the formation of bars in disc galaxies which agrees well with a subsequent simulation by James and Wilkinson (1981). We will show that a large increase in the resolution of this simulation is desirable for its application to observed galaxies.

Lynden-Bell's mechanism operates in interior regions of systems possessing a double inner Lindblad resonance. It is easy to see that such situations arise naturally when the rotation curve is well-behaved near the axis.

We require that the angular velocity Ω has a maximum Ω_0 on the rotation axis and is twice differentiable there. For large values of the axial distance ϖ we expect a nearly constant circular velocity V so that

$$\Omega = V/\varpi \tag{8.1}$$

The physically important frequency is the inner Lindblad frequency

$$\Omega_i = \Omega - \kappa/2 \qquad (8.2)$$

where κ is the epicyclic frequency for a star in a nearly circular orbit. The condition that a star is in inner Lindblad resonance with a pattern rotating at angular velocity Ω_p is just

$$\Omega_i = \Omega_p \qquad (8.3)$$

We may easily show that Equations (8.1), (8.2) imply

$$\Omega_i \sim -\tfrac{3}{4}\,\Omega_0''\varpi^2,\ \Omega_0'' < 0,\ \varpi \text{ small} \qquad (8.4)$$

$$\sim \left(1 - \frac{1}{\sqrt{2}}\right)\Omega\,\varpi \text{ large} \qquad (8.5)$$

We expect the behaviour shown in Figure 1. Ω_i has a single maximum at a radius somewhat inside that of the flat portion of the velocity curve.

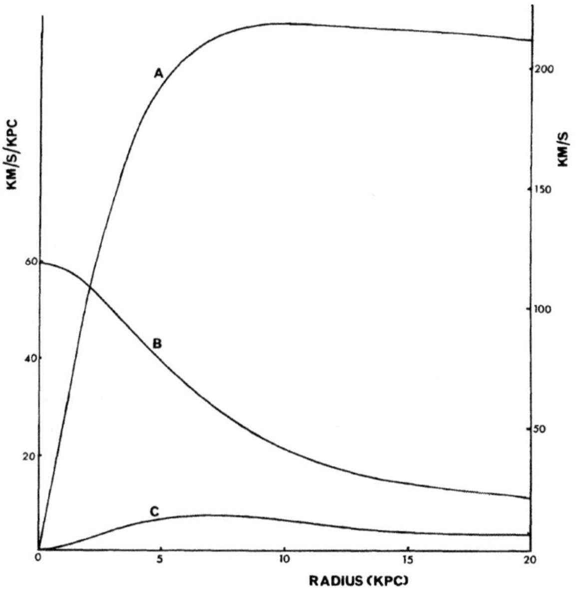

Figure 1: Curves of (A) circular velocity, (B) angular velocity Ω, (C) inner Lindblad frequency Ω_i for the longitudinally averaged potential in the simulation of James and Wilkinson (1981).

The response of an orbit to an asymmetric bar potential depends on the relation between Ω_i and the specific angular momentum h. Many authors have found that eccentric orbits align themselves at right angles to the bar in the external regions. Lynden-Bell considers the inner regions where Ω_i is an increasing function of h. An orbit such that $|\Omega_i - \Omega_p| \ll \Omega_p$ experiences a torque which increases h when the line of apocentres lags behind the pattern, and decreases h when it leads. The bar potential also changes the eccentricity of the orbit and we must account for this effect in the relationship between Ω_i and h. We introduce the adiabatic invariant

$$J_F = \tfrac{1}{2}h + \frac{1}{2\pi} \oint \dot{R} \, dR \qquad (8.6)$$

where the integral is taken about an (almost) closed loop in the $(R, \dot{R})$ plane. For any rotation law we may write

$$\Omega_i = \Omega_i(J_F, h) \qquad (8.7)$$

The bar potential has little effect on J_F for any orbit. We require that $\partial\Omega/\partial h > 0$. Lynden-Bell uses a simple analytic form for the rotation law and verifies this property in a substantial inner region.

The effect is that the bar potential increases h and Ω_i for orbits which lag behind and decreases them for orbits which lead. It is possible for an orbit to be trapped into oscillation about the bar and to reinforce the bar potential. We call the region where this occurs the aligning region.

As Lynden-Bell remarks, the story is incomplete because (i) some trapped orbits have large oscillation amplitudes and dilute the bar potential, and (ii) the shape and resonant frequency of the orbit may be affected by the bar. The simulation of James and Wilkinson offers a reasonable test of the mechanism, albeit with a slightly different rotation law. The Population I disc, represented by 25000 particles, is embedded in a rigid spherical halo containing eighty per cent of the mass interior to a radius of 20 kpc. The density distribution of this halo is similar to that of an isothermal sphere. The velocity curve is almost flat outside 7 kpc and from this point of view the model is more realistic than Lynden-Bell's. The radial distance of the turn-over point is larger than we would like, owing to the limited resolution of the experiment.

The software gives complete orbital information for a sample of 429 stars. Pole-on views of the galaxy (Figure 2) show alternately leading and trailing spiral structure setting in after about 7×10^9 years and persisting to the end of the run. A minority of the sample stars have well established values of Ω_i and a histogram of these values shows an extremely sharp peak at an angular velocity of 8.8 km s^{-1} kpc^{-1}. This agrees well with that of a visible bar in which the leading and trailing structure are embedded. A study of apocentre phases for the sample orbits,

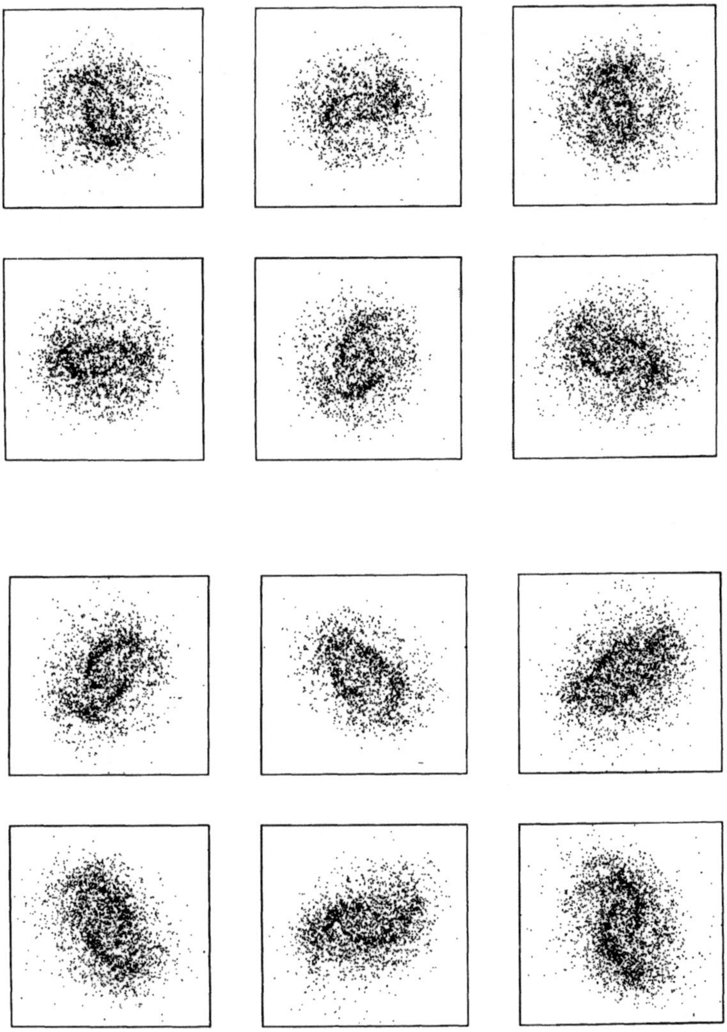

Figure 2: Pole-on views of the model galaxy at time steps 1800(50)2350. The time step is equal to 4.03×10^6 years. The galaxy rotates in a clockwise direction.

relative to a pattern rotating at this speed, shows that a large fraction of the orbits are trapped for appreciable times. Many orbits oscillate with amplitudes of up to 30°, and do not appear to have stable values of Ω_i. Their relationship to the pattern can be seen only when its rotation speed is known. The oscillations are related to the alternate episodes of leading and trailing structure. The number of orbits trapped for long periods increases throughout the run, although individual orbits can escape at any time. The bar potential has a considerable effect on the resonant frequencies. If we calculate Ω_i from the longitudinally averaged potential field, the maximum value is more than ten per cent below the values observed for the pattern and the sample orbits.

The simulation shows that well trapped orbits in the aligning region can dominate over orbits with large amplitudes of oscillation and over orbits in the outer anti-aligning region. Its deficiencies point to the necessity of running much larger experiments. The density concentrations which form the bar have inner and outer radii of 5 kpc and 11 kpc. The entire bar region has a linear extent of about 12 mesh spacings, and the resolution is minimal. The co-rotation and outer Lindblad resonances for the model have radii which place them well outside the regions of appreciable disc density. These resonances have no effect on the model, but are likely to be important in actual spirals. A spiral wake can be detected by examining interactions of untrapped orbits with the bar, but the mass involved is too small to be visible or to affect the bar dynamically.

The interest of the model lies in the support it gives to Lynden-Bell' picture of bar formation, and in the appearance of variable leading and trailing structure. A realistic application to genuine galaxies requires that the flat part of the rotation curve extend to smaller radii. The resolution of this curve should be maintained or improved in the important turn-over region. The Population I disc should be extended to provide reasonable density at the outer Lindblad resonance. A factor of (at least) 4 is needed in the mesh size. This can be provided on the CRAY-1 at the Daresbury Laboratory by using an extremely flattened mesh, at the cost of precluding studies of warping and of encounters with other systems. Any further extensions would require additional memory, or re-organisation of the calculation.

9. COLLISIONS AND MERGERS

Various techniques have been applied to modelling interacting galaxies. The simplest of them use massless particles to represent galactic discs. This can be very successful. For example, Toomre and Toomre (1972) were able to match several exotic objects with a tidal model, and to show that the sharpness of the bridges and tails in their chosen systems is not fatal. Lynds and Toomre (1976) achieved a remarkable match to a system with a double ring, though in this paper they also used a representation in terms of massive rings to show that ring formation arises naturally in a collision.

Analytic treatments usually depend on the impulse approximation

(Alladin, 1965, Sastry and Alladin, 1970, 1977). In principle this applies to extremely hyperbolic collisions but numerical simulations (Toomre, 1977, van Albada and van Gorkom, 1977) show that the predictions are remarkably good for slower encounters.

Fully self-consistent calculations with small numbers of particles have been carried out by White (1978, 1979), Roos and Norman (1979) and Dekel, Lecar and Shaham (1980). These calculations can start while the participating galaxies are well separated and so can treat the approach phase in a consistent way. The main technical problem is that a Newtonian interaction between the model particles exaggerates the importance of binary relaxation. It is necessary to soften the potential, replacing r in the Newtonian expression by $(r^2 + a^2)$ for some suitable a. Unfortunately the value of a needed to extend the relaxation time beyond the calculation time is comparable with the core radii of the initial galaxies. The limited resolution restricts the range of problems to which such simulations can be applied.

Partially self-consistent simulations (Theys and Spiegel, 1977) allow us to use a reasonable number of particles to represent the target galaxy. The intruder is represented by a single very massive particle with a variable softening parameter for the potential. This approximation permits detailed examination of the target's response and need not limit the initial separation or orbit of the intruder. Theys and Spiegel applied it to the formation of ring galaxies. Its main limitation is that it does not include the intruder's dynamical response in a self-consistent way.

To see why the deformations of the participating galaxies matter, we examine the case of a head-on collision. The mean free path of a star against binary encounters greatly exceeds the dimensions of the system, and we need only consider collective effects. For a short period the two galaxies are superimposed and each feels the gravitational field due to the combined system. The velocity dispersions (or rotational velocities for disc galaxies) are appropriate to the smaller individual fields, so both galaxies are forced to contract. If they subsequently separate, the contraction must be followed by oscillation and ultimately relaxation by the mechanism of Lynden-Bell (1967). The final result must be that both galaxies become more distended and less tightly bound.

The energy for these motions comes from the orbital energy of the participants. This part of the mechanism is also easy to visualise. At a given separation in the recession phase the systems are more compact than they were at the same separation on approach, provided that the transit time is small compared with individual oscillation times. The two potential wells are accordingly deeper during separation.

A fully self-consistent treatment of both participants in a collision requires that the mesh be scaled to hold them at the beginning of the run. This limits the initial separation and also the spatial resolution. Miller and Smith (1980) use just under 10^5 particles on a mesh of about

2.5×10^5 points. By current standards this is a large calculation and it yields a wealth of information. Despite the size of the experiment the authors have been forced to compromise. Their initial configurations are spherical and follow the density law for a polytrope of index 3. This understates the central condensation for elliptical galaxies, but with a half-mass radius of only 6.8 mesh units there is little room for improvement. The bounding radius of the models is 24 mesh units, and the bounding spheres overlap at the start of a run. The impact parameter is limited to about 3.3 times the half-mass radius. The model is well adapted to studying deeply penetrating collisions.

Miller and Smith find that the objects are strongly non-spherical after the collision, and damage extends throughout both systems. The cores of the initial galaxies do not survive and there are no cases of individual particles remaining bound to a core region. This result raises difficulties in interpreting complex bridges and tails in colliding systems If we regard visible galaxies as luminous tracers embedded in massive dark halos, then for dynamical purposes the halo is the only important component. We may apply Miller and Smith's results and conclude that collisions completely disrupt the halo cores and thus the luminous tracers. It is no longer obvious that bridge and tail structures can be formed by collisions.

The question could possibly be resolved by experiments involving larger impact parameters and greater density contrasts in the original galaxies. However the work of Miller and Smith is already uncomfortably close to the practical limit with current techniques and computers. The method of Section 7 offers the possibility of improving the calculation. It should no longer be necessary to restrict the central condensation of the participants so severely or to constrain the initial separations and impact parameters by the need to fit the calculation onto a single mesh.

10. CONCLUSION

The mesh segmentation procedure described in this paper is likely to find its first application in problems involving collisions and interaction in pairs or small multiple systems of galaxies. Such systems require a fine mesh to resolve local detail, but do not require that the potential be calculated everywhere. Full calculations on very large meshes would be feasible with the storage economies that the method permits. However the author's estimate of the CPU time for a complete potential determination on a mesh of 257 x 257 x 257 points on the CRAY-1 is about 50 seconds. File transfer would be a significant additional overhead. It would not be easy to justify such a calculation at present.

Applications to smaller (but still large) problems seem promising. The method allows us to arrange large potential problems to fit neatly on specialised processors such as the DAP. Its arithmetic economy makes it a reasonable choice for such problems, but would not alone justify the work of development in comparison with, for example, the FACR procedure.

REFERENCES

Alladin, S. M., 1965, Astrophys. J., 141, p. 768.
Appleton, P. N., Davies, R. D. and Stephenson, R. J., 1981, Mon. Not. Roy. Astron. Soc., 195, p. 327.
Burridge, D. M. and Temperton, C., 1979, J. Computat. Phys., 30, p. 145.
Cooley, J. W. and Tukey, J. W., 1965, Math. Comp., 19, p. 297.
Hockney, R. W., 1970, Methods in Computational Physics, 9, p. 135.
James, R. A., 1977, J. Computat. Phys., 25, p. 71.
James, R. A. and Wilkinson, A., 1981, in preparation.
Kantorovich, L. V. and Krylov, V. I., 1958, Approximate Methods of Higher Analysis, p. 185, Groningen, Noordhoff.
Lynden-Bell, D., 1967, Mon. Not. Roy. Astron. Soc., 136, p. 101.
Lynden-Bell, D., 1979, Mon. Not. Roy. Astron. Soc., 187, p. 101.
Ostriker, J. P., Peebles, R. J. E. and Yahil, A., 1974, Astrophys. J., 193, L1.
Sastry, K. S. and Alladin, S. M., 1970, ASS 7, p. 261.
Sastry, K. S. and Alladin, S. M., 1977, ASS 46, p. 285.
Theys, J. C. and Spiegel, E. A., 1977, Astrophys. J., 212, p. 616.
Toomre, A. and Toomre, J., 1972, Astrophys. J., 178, p. 623.
van Albada, T. S. and van Gorkom, J. H., 1977, Astron. and Astrophys., 54, p. 121.
von Hagenow, K. and Lackner, K., 1975, Proc. 7th Conference on Numerical Simulation of Plasmas, New York.
White, S. D. M., 1978, Mon. Not. Roy. Astron. Soc., 184, p. 185.
White, S. D. M., 1979, Mon. Not. Roy. Astron. Soc., 189, p. 831.
Yahil, A., 1977, Astrophys. J., 217, p. 27.
Ziegler, H., 1972, IEEE Trans AU-20, p. 353.

NAME INDEX

SUBJECT INDEX

ASTROPHYSICS AND SPACE SCIENCE LIBRARY

Edited by

J. E. Blamont, R. L. F. Boyd, L. Goldberg, C. de Jager, Z. Kopal, G. H. Ludwig, R. Lüst, B. M. McCormac, H. E. Newell, L. I. Sedov, Z. Švestka, and W. de Graaff

1. C. de Jager (ed.), *The Solar Spectrum, Proceedings of the Symposium held at the University of Utrecht, 26–31 August, 1963.* 1965, XIV + 417 pp.
2. J. Orthner and H. Maseland (eds.), *Introduction to Solar Terrestrial Relations, Proceedings of the Summer School in Space Physics held in Alpbach, Austria, July 15–August 10, 1963 and Organized by the European Preparatory Commission for Space Research.* 1965, IX + 506 pp.
3. C. C. Chang and S. S. Huang (eds.), *Proceedings of the Plasma Space Science Symposium, held at the Catholic University of America, Washington, D.C., June 11–14, 1963.* 1965, IX + 377 pp.
4. Zdeněk Kopal, *An Introduction to the Study of the Moon.* 1966, XII + 464 pp.
5. B. M. McCormac (ed.), *Radiation Trapped in the Earth's Magnetic Field. Proceedings of the Advanced Study Institute, held at the Chr. Michelsen Institute, Bergen, Norway, August 16–September 3, 1965.* 1966, XII + 901 pp.
6. A. B. Underhill, *The Early Type Stars.* 1966, XII + 282 pp.
7. Jean Kovalevsky, *Introduction to Celestial Mechanics.* 1967, VIII + 427 pp.
8. Zdeněk Kopal and Constantine L. Goudas (eds.), *Measure of the Moon. Proceedings of the 2nd International Conference on Selenodesy and Lunar Topography, held in the University of Manchester, England, May 30–June 4, 1966.* 1967, XVIII + 479 pp.
9. J. G. Emming (ed.), *Electromagnetic Radiation in Space. Proceedings of the 3rd ESRO Summer School in Space Physics, held in Alpbach, Austria, from 19 July to 13 August, 1965.* 1968, VIII + 307 pp.
10. R. L. Carovillano, John F. McClay, and Henry R. Radoski (eds.), *Physics of the Magnetosphere, Based upon the Proceedings of the Conference held at Boston College, June 19–28, 1967.* 1968, X + 686 pp.
11. Syun-Ichi Akasofu, *Polar and Magnetospheric Substorms.* 1968, XVIII + 280 pp.
12. Peter M. Millman (ed.), *Meteorite Research. Proceedings of a Symposium on Meteorite Research, held in Vienna, Austria, 7–13 August, 1968.* 1969, XV + 941 pp.
13. Margherita Hack (ed.), *Mass Loss from Stars. Proceedings of the 2nd Trieste Colloquium on Astrophysics, 12–17 September, 1968.* 1969, XII + 345 pp.
14. N. D'Angelo (ed.), *Low-Frequency Waves and Irregularities in the Ionosphere. Proceedings of the 2nd ESRIN-ESLAB Symposium, held in Frascati, Italy, 23–27 September, 1968.* 1969, VII + 218 pp.
15. G. A. Partel (ed.), *Space Engineering. Proceedings of the 2nd International Conference on Space Engineering, held at the Fondazione Giorgio Cini, Isola di San Giorgio, Venice, Italy, May 7–10, 1969.* 1970, XI + 728 pp.
16. S. Fred Singer (ed.), *Manned Laboratories in Space. Second International Orbital Laboratory Symposium.* 1969, XIII + 133 pp.
17. B. M. McCormac (ed.), *Particles and Fields in the Magnetosphere. Symposium Organized by the Summer Advanced Study Institute, held at the University of California, Santa Barbara, Calif., August 4–15, 1969.* 1970, XI + 450 pp.
18. Jean-Claude Pecker, *Experimental Astronomy.* 1970, X + 105 pp.
19. V. Manno and D. E. Page (eds.), *Intercorrelated Satellite Observations related to Solar Events. Proceedings of the 3rd ESLAB/ESRIN Symposium held in Noordwijk, The Netherlands, September 16–19, 1969.* 1970, XVI + 627 pp.
20. L. Mansinha, D. E. Smylie, and A. E. Beck, *Earthquake Displacement Fields and the Rotation of the Earth, A NATO Advanced Study Institute Conference Organized by the Department of Geophysics, University of Western Ontario, London, Canada, June 22–28, 1969.* 1970, XI + 308 pp.
21. Jean-Claude Pecker, *Space Observatories.* 1970, XI + 120 pp.
22. L. N. Mavridis (ed.), *Structure and Evolution of the Galaxy. Proceedings of the NATO Advanced Study Institute, held in Athens, September 8–19, 1969.* 1971, VII + 312 pp.

23. A. Muller (ed.), *The Magellanic Clouds. A European Southern Observatory Presentation: Principal Prospects, Current Observational and Theoretical Approaches, and Prospects for Future Research, Based on the Symposium on the Magellanic Clouds, held in Santiago de Chile, March 1969, on the Occasion of the Dedication of the European Southern Observatory.* 1971, XII + 189 pp.
24. B. M. McCormac (ed.), *The Radiating Atmosphere. Proceedings of a Symposium Organized by the Summer Advanced Study Institute, held at Queen's University, Kingston, Ontario, August 3–14, 1970.* 1971, XI + 455 pp.
25. G. Fiocco (ed.), *Mesospheric Models and Related Experiments. Proceedings of the 4th ESRIN-ESLAB Symposium, held at Frascati, Italy, July 6–10, 1970.* 1971, VIII + 298 pp.
26. I. Atanasijević, *Selected Exercises in Galactic Astronomy.* 1971, XII + 144 pp.
27. C. J. Macris (ed.), *Physics of the Solar Corona. Proceedings of the NATO Advanced Study Institute on Physics of the Solar Corona, held at Cavouri-Vouliagmeni, Athens, Greece, 6–17 September 1970.* 1971, XII + 345 pp.
28. F. Delobeau, *The Environment of the Earth.* 1971, IX + 113 pp.
29. E. R. Dyer (general ed.), *Solar-Terrestrial Physics/1970. Proceedings of the International Symposium on Solar-Terrestrial Physics, held in Leningrad, U.S.S.R., 12–19 May 1970.* 1972, VIII + 938 pp.
30. V. Manno and J. Ring (eds.), *Infrared Detection Techniques for Space Research. Proceedings of the 5th ESLAB-ESRIN Symposium, held in Noordwijk, The Netherlands, June 8–11, 1971.* 1972, XII + 344 pp.
31. M. Lecar (ed.), *Gravitational N-Body Problem. Proceedings of IAU Colloquium No. 10, held in Cambridge, England, August 12–15, 1970.* 1972, XI + 441 pp.
32. B. M. McCormac (ed.), *Earth's Magnetospheric Processes. Proceedings of a Symposium Organized by the Summer Advanced Study Institute and Ninth ESRO Summer School, held in Cortina, Italy, August 30–September 10, 1971.* 1972, VIII + 417 pp.
33. Antonin Rükl, *Maps of Lunar Hemispheres.* 1972, V + 24 pp.
34. V. Kourganoff, *Introduction to the Physics of Stellar Interiors.* 1973, XI + 115 pp.
35. B. M. McCormac (ed.), *Physics and Chemistry of Upper Atmospheres. Proceedings of a Symposium Organized by the Summer Advanced Study Institute, held at the University of Orléans, France, July 31–August 11, 1972.* 1973, VIII + 389 pp.
36. J. D. Fernie (ed.), *Variable Stars in Globular Clusters and in Related Systems. Proceedings of the IAU Colloquium No. 21, held at the University of Toronto, Toronto, Canada, August 29–31, 1972.* 1973, IX + 234 pp.
37. R. J. L. Grard (ed.), *Photon and Particle Interaction with Surfaces in Space. Proceedings of the 6th ESLAB Symposium, held at Noordwijk, The Netherlands, 26–29 September, 1972.* 1973, XV + 577 pp.
38. Werner Israel (ed.), *Relativity, Astrophysics and Cosmology. Proceedings of the Summer School, held 14–26 August, 1972, at the BANFF Centre, BANFF, Alberta, Canada.* 1973, IX + 323 pp.
39. B. D. Tapley and V. Szebehely (eds.), *Recent Advances in Dynamical Astronomy. Proceedings of the NATO Advanced Study Institute in Dynamical Astronomy, held in Cortina d'Ampezzo, Italy, August 9–12, 1972.* 1973, XIII + 468 pp.
40. A. G. W. Cameron (ed.), *Cosmochemistry. Proceedings of the Symposium on Cosmochemistry, held at the Smithsonian Astrophysical Observatory, Cambridge, Mass., August 14–16, 1972.* 1973, X + 173 pp.
41. M. Golay, *Introduction to Astronomical Photometry.* 1974, IX + 364 pp.
42. D. E. Page (ed.), *Correlated Interplanetary and Magnetospheric Observations. Proceedings of the 7th ESLAB Symposium, held at Saulgau, W. Germany, 22–25 May, 1973.* 1974, XIV + 662 pp.
43. Riccardo Giacconi and Herbert Gursky (eds.), *X-Ray Astronomy.* 1974, X + 450 pp.
44. B. M. McCormac (ed.), *Magnetospheric Physics. Proceedings of the Advanced Summer Institute, held in Sheffield, U.K., August 1973.* 1974, VII + 399 pp.
45. C. B. Cosmovici (ed.), *Supernovae and Supernova Remnants. Proceedings of the International Conference on Supernovae, held in Lecce, Italy, May 7–11, 1973.* 1974, XVII + 387 pp.
46. A. P. Mitra, *Ionospheric Effects of Solar Flares.* 1974, XI + 294 pp.
47. S.-I. Akasofu, *Physics of Magnetospheric Substorms.* 1977, XVIII + 599 pp.

48. H. Gursky and R. Ruffini (eds.), *Neutron Stars, Black Holes and Binary X-Ray Sources.* 1975, XII + 441 pp.
49. Z. Švestka and P. Simon (eds.), *Catalog of Solar Particle Events 1955–1969. Prepared under the Auspices of Working Group 2 of the Inter-Union Commission on Solar-Terrestrial Physics.* 1975, IX + 428 pp.
50. Zdeněk Kopal and Robert W. Carder, *Mapping of the Moon.* 1974, VIII + 237 pp.
51. B. M. McCormac (ed.), *Atmospheres of Earth and the Planets. Proceedings of the Summer Advanced Study Institute, held at the University of Liège, Belgium, July 29–August 8, 1974.* 1975, VII + 454 pp.
52. V. Formisano (ed.), *The Magnetospheres of the Earth and Jupiter. Proceedings of the Neil Brice Memorial Symposium, held in Frascati, May 28–June 1, 1974.* 1975, XI + 485 pp.
53. R. Grant Athay, *The Solar Chromosphere and Corona: Quiet Sun.* 1976, XI + 504 pp.
54. C. de Jager and H. Nieuwenhuijzen (eds.), *Image Processing Techniques in Astronomy. Proceedings of a Conference, held in Utrecht on March 25–27, 1975.* XI + 418 pp.
55. N. C. Wickramasinghe and D. J. Morgan (eds.), *Solid State Astrophysics. Proceedings of a Symposium, held at the University College, Cardiff, Wales, 9–12 July 1974.* 1976, XII + 314 pp.
56. John Meaburn, *Detection and Spectrometry of Faint Light.* 1976, IX + 270 pp.
57. K. Knott and B. Battrick (eds.), *The Scientific Satellite Programme during the International Magnetospheric Study. Proceedings of the 10th ESLAB Symposium, held at Vienna, Austria, 10–13 June 1975.* 1976, XV + 464 pp.
58. B. M. McCormac (ed.), *Magnetospheric Particles and Fields. Proceedings of the Summer Advanced Study School, held in Graz, Austria, August 4–15, 1975.* 1976, VII + 331 pp.
59. B. S. P. Shen and M. Merker (eds.), *Spallation Nuclear Reactions and Their Applications.* 1976, VIII + 235 pp.
60. Walter S. Fitch (ed.), *Multiple Periodic Variable Stars. Proceedings of the International Astronomical Union Colloquium No. 29, held at Budapest, Hungary, 1–5 September 1976.* 1976, XIV + 348 pp.
61. J. J. Burger, A. Pedersen, and B. Battrick (eds.), *Atmospheric Physics from Spacelab. Proceedings of the 11th ESLAB Symposium, Organized by the Space Science Department of the European Space Agency, held at Frascati, Italy, 11–14 May 1976.* 1976, XX + 409 pp.
62. J. Derral Mulholland (ed.), *Scientific Applications of Lunar Laser Ranging. Proceedings of a Symposium held in Austin, Tex., U.S.A., 8–10 June, 1976.* 1977, XVII + 302 pp.
63. Giovanni G. Fazio (ed.), *Infrared and Submillimeter Astronomy. Proceedings of a Symposium held in Philadelphia, Penn., U.S.A., 8–10 June, 1976.* 1977, X + 226 pp.
64. C. Jaschek and G. A. Wilkins (eds.), *Compilation, Critical Evaluation and Distribution of Stellar Data. Proceedings of the International Astronomical Union Colloquium No. 35, held at Strasbourg, France, 19–21 August, 1976.* 1977, XIV + 316 pp.
65. M. Friedjung (ed.), *Novae and Related Stars. Proceedings of an International Conference held by the Institut d'Astrophysique, Paris, France, 7–9 September, 1976.* 1977, XIV + 228 pp.
66. David N. Schramm (ed.), *Supernovae. Proceedings of a Special IAU-Session on Supernovae held in Grenoble, France, 1 September, 1976.* 1977, X + 192 pp.
67. Jean Audouze (ed.), *CNO Isotopes in Astrophysics. Proceedings of a Special IAU Session held in Grenoble, France, 30 August, 1976.* 1977, XIII + 195 pp.
68. Z. Kopal, *Dynamics of Close Binary Systems,* XIII + 510 pp.
69. A. Bruzek and C. J. Durrant (eds.), *Illustrated Glossary for Solar and Solar-Terrestrial Physics.* 1977, XVIII + 204 pp.
70. H. van Woerden (ed.), *Topics in Interstellar Matter.* 1977, VIII + 295 pp.
71. M. A. Shea, D. F. Smart, and T. S. Wu (eds.), *Study of Travelling Interplanetary Phenomena.* 1977, XII + 439 pp.
72. V. Szebehely (ed.), *Dynamics of Planets and Satellites and Theories of Their Motion. Proceedings of IAU Colloquium No. 41, held in Cambridge, England, 17–19 August 1976.* 1978, XII + 375 pp.
73. James R. Wertz (ed.), *Spacecraft Attitude Determination and Control.* 1978, XVI + 858 pp.

74. Peter J. Palmadesso and K. Papadopoulos (eds.), *Wave Instabilities in Space Plasmas. Proceedings of a Symposium Organized Within the XIX URSI General Assembly held in Helsinki, Finland, July 31–August 8, 1978.* 1979, VII + 309 pp.
75. Bengt E. Westerlund (ed.), *Stars and Star Systems. Proceedings of the Fourth European Regional Meeting in Astronomy held in Uppsala, Sweden, 7–12 August, 1978.* 1979, XVIII + 264 pp.
76. Cornelis van Schooneveld (ed.), *Image Formation from Coherence Functions in Astronomy. Proceedings of IAU Colloquium No. 49 on the Formation of Images from Spatial Coherence Functions in Astronomy, held at Groningen, The Netherlands, 10–12 August 1978.* 1979, XII + 338 pp.
77. Zdeněk Kopal, *Language of the Stars. A Discourse on the Theory of the Light Changes of Eclipsing Variables.* 1979, VIII + 280 pp.
78. S.-I. Akasofu (ed,), *Dynamics of the Magnetosphere. Proceedings of the A.G.U. Chapman Conference 'Magnetospheric Substorms and Related Plasma Processes' held at Los Alamos Scientific Laboratory, N.M., U.S.A., October 9-13, 1978.* 1980, XII + 658 pp.
79. Paul S. Wesson, *Gravity, Particles, and Astrophysics. A Review of Modern Theories of Gravity and G-variability, and their Relation to Elementary Particle Physics and Astrophysics.* 1980, VIII + 188 pp.
80. Peter A. Shaver (ed.), *Radio Recombination Lines. Proceedings of a Workshop held in Ottawa, Ontario, Canada, August 24-25, 1979.* 1980, X + 284 pp.
81. Pier Luigi Bernacca and Remo Ruffini (eds.), *Astrophysics from Spacelab*, 1980, XI + 664 pp.
82. Hannes Alfvén, *Cosmic Plasma*, 1981, X + 160 pp.
83. Michael D. Papagiannis (ed.), *Strategies for the Search for Life in the Universe*, 1980, XVI + 254 pp.
84. H. Kikuchi (ed.), *Relation between Laboratory and Space Plasmas*, 1981, XII + 386 pp.
85. Peter van der Kamp, *Stellar Paths*, 1981, forthcoming.
86. E. M. Gaposchkin and B. Kołaczek (eds.), *Reference Coordinate Systems for Earth Dynamics*, 1981, XIV + 396 pp.
87. R. Giacconi (ed.), *X-Ray Astronomy with the Einstein Satellite. Proceedings of the High Energy Astrophysics Division of the American Astronomical Society Meeting on X-Ray Astronomy held at the Harvard-Smithsonian Center for Astrophysics, Cambridge, Mass., U.S.A., January 28–30, 1980.* 1981, VII + 330 pp., forthcoming.
88. Icko Iben Jr. and Alvio Renzini (eds.), *Physical Processes in Red Giants. Proceedings of the Second Workshop, held at the Ettore Majorana Centre for Scientific Culture, Advanced School of Agronomy, in Erice, Siciliy, Italy, September 3–13, 1980.* 1981, XV + 488 pp.
89. C. Chiosi and R. Stalio (eds.), *Effect of Mass Loss on Stellar Evolution. IAU Colloquium No. 59 held in Miramare, Trieste, Italy, September 15–19, 1980.* 1981, XXII + 532 pp., forthcoming.

GPSR Compliance

The European Union's (EU) General Product Safety Regulation (GPSR) is a set of rules that requires consumer products to be safe and our obligations to ensure this.

If you have any concerns about our products, you can contact us on ProductSafety@springernature.com

In case Publisher is established outside the EU, the EU authorized representative is:

Springer Nature Customer Service Center GmbH
Europaplatz 3
69115 Heidelberg, Germany

Batch number: 10815974

Printed by Printforce, the Netherlands